Birkhäuser

Abhijit Dasgupta

Set Theory

With an Introduction to Real Point Sets

Birkhäuser

Abhijit Dasgupta
Department of Mathematics
University of Detroit Mercy
Detroit, MI, USA
dasgupab@udmercy.edu

ISBN 978-1-4614-8853-8 ISBN 978-1-4614-8854-5 (eBook)
DOI 10.1007/978-1-4614-8854-5
Springer New York Heidelberg Dordrecht London

Library of Congress Control Number: 2013949160

Mathematics Subject Classification (2000): 03E10, 03E20, 03E04, 03E25, 03E15, 28A05, 54H05, 03E30, 03E75, 03E50, 03E55, 03E60, 03E02

Printed on acid-free paper

Springer is part of Springer Science+Business Media (www.birkhauser-science.com)

To my mother
and
to the memory of my father

Preface

Most modern set theory texts, even at the undergraduate level, introduce specific formal axiom systems such as ZFC relatively early, perhaps because of the (understandably real) fear of paradoxes. At the same time, most mathematicians and students of mathematics seem to care little about special formal systems, yet may still be interested in the part of set theory belonging to "mathematics proper," i.e., cardinals, order, ordinals, and the theory of the real continuum. There appears to be a gulf between texts of mainstream mathematics and those of set theory and logic.

This undergraduate set theory textbook regards the core material on cardinals, ordinals, and the continuum as a subject area of classical mathematics interesting in its own right. It separates and postpones all foundational issues (such as paradoxes and special axioms) into an optional part at the end. The main material is thus developed informally—not within any particular axiom system—to avoid getting bogged down in the details of formal development and its associated metamathematical baggage. I hope this will make this text suitable for a wide range of students interested in any field of mathematics and not just for those specializing in logic or foundations. At the same time, students with metamathematical interests will find an introduction to axiomatic ZF set theory in the last part, and some glimpses into key foundational topics in the postscript chapters at the end of each part.

Another feature of this book is that its coverage of the real continuum is confined exclusively to the real line **R**. All abstract or general concepts such as topological spaces, metric spaces, and even the Euclidean spaces of dimension 2 or higher are completely avoided. This may seem like a severe handicap, but even this highly restricted framework allows the introduction of many interesting topics in the theory of real point sets. In fact, not much substance in the theory is lost and a few deeper intuitions are gained. As evidenced by the teaching of undergraduate real analysis, the student who is first firmly grounded in the hard and concrete details of **R** will better enjoy and handle the abstraction found in later, more advanced studies.

The book grew out of an undergraduate course in introductory set theory that I taught at the University of Detroit Mercy. The prerequisite for the core material of the book is a post-calculus undergraduate US course in discrete mathematics or linear algebra, although precalculus and some exposure to proofs should technically suffice for Parts I and II.

The book starts with a "prerequisites" chapter on sets, relations, and functions, including equivalence relations and partitions, and the definition of linear order. The rest is divided into four relatively independent parts with quite distinct mathematical flavors. Certain basic techniques are emphasized across multiple parts, such as Cantor's back-and-forth method, construction of perfect sets, Cantor–Bendixson analysis, and ordinal ranks.

Part I is a problem-based short course which, starting from Peano arithmetic, constructs the real numbers as Dedekind cuts of rationals in a routine way with two possible uses. A student of mathematics not going into formal ZF set theory will work out, once and for all, a detailed existence proof for a complete ordered field. And for a student who might later get into axiomatic ZF set theory, the redevelopment of Peano arithmetic and the theory of real numbers formally within ZF will become largely superfluous. One may also decide to skip Part I altogether and go directly to Part II.

Part II contains the core material of the book: The Cantor–Dedekind theory of the transfinite, especially order, the continuum, cardinals, ordinals, and the Axiom of Choice. The development is informal and naive (non-axiomatic), but mathematically rigorous. While the core material is intended to be interesting in its own right, it also forms the folklore set-theoretic prerequisite needed for graduate level topology, analysis, algebra, and logic. Useful forms of the Axiom of Choice, such as Zorn's Lemma, are covered.

Part III of the book is about point sets of real numbers. It shows how the theory of sets and orders connects intimately to the continuum and its topology. In addition to the basic theory of **R** including measure and category, it presents more advanced topics such as Brouwer's theorem, Cantor–Bendixson analysis, Sierpinski's theorem, and an introduction to Borel and analytic sets—all in the context of the real line. Thus the reader gets access to significant higher results in a concrete manner via powerful techniques such as Cantor's back-and-forth method. As mentioned earlier, all development is limited to the reals, but the apparent loss of generality is mostly illusory and the special case for real numbers captures much of the essential ideas and the central intuitions behind these theorems.

Parts II and III of the book focus on gaining intuition rather than on formal development. I have tried to start with specific and concrete cases of examples and theorems before proceeding to their more general and abstract versions. As a result, some important topics (e.g., the Cantor set) appear multiple times in the book, generally with increasing levels of sophistication. Thus, I have sacrificed compactness and conciseness in favor of intuition building and maintaining some independence between the four parts.

Part IV deals with foundational issues. The paradoxes are first introduced here, leading to formal set theory and the Zermelo–Fraenkel axiom system. Von Neumann ordinals are also first presented in this part.

Each part ends with a postscript chapter discussing topics beyond the scope of the main text, ranging from philosophical remarks to glimpses into modern landmark results of set theory such as the resolution of Lusin's problems on projective sets using determinacy of infinite games and large cardinals.

Problems form an integral and essential part of the book. While some of them are routine, they are generally meant to form an extension of the text. A harder problem will contain hints and sometimes an outline for a solution. Starred sections and problems may be regarded as optional.

The book has enough material for a one-year course for advanced undergraduates. The relative independence of the four parts allows various possibilities for covering topics. In a typical one-semester course, I usually briefly cover Part I, spend most of the time in Part II, and finish with a brief overview of Part IV. For students with more foundational interests, more time can be spent on the material of Part IV and the postscripts. On the other hand, for less foundationally inclined but more mathematically advanced students with prior exposure to advanced calculus or real analysis, only Parts II and III may be covered with Parts I and IV skipped altogether.

Acknowledgments. I want to thank the University of Detroit Mercy for supporting me with a sabbatical leave during 2011–2012 which made the writing of this book possible, and Professor László Kérchy, Editor of Acta Sci. Math. (Szeged), for kindly giving me permission to translate a section of von Neumann's original German paper [80]. I also wish to thank my students, my teacher Pinaki Mitra for introducing me to set theory and logic, and the late set theorist R. Michael Canjar, who attended all my lectures in 2010 and provided daily feedback. Sadly, Mike left us before this book could be finished. Professor Tarun Mukherjee, to whom my lifelong debt is beyond measure, painstakingly read the entire manuscript, correcting many errors and giving his invaluable suggestions for improvement. Professor Andreas Blass was unbelievably kind and quick to give his expert comments on a part of the book even though I (shamelessly) asked him at the last moment to check it in an unreasonably short time. Immensely valuable were the extensive and deeply engaging feedback on the whole manuscript that came from the anonymous referees, which led me to drop and rewrite some sections and add the later postscripts. I am also indebted to Professor Prasanta Bandyopadhyay and Professor Kallol Paul for their help, and to Professor Ioannis Souldatos for his thoughtful comments on a chapter. At Birkhäuser, I got patient help from Tom Grasso in initial planning, from Kate Ghezzi and Mitch Moulton during writing, and above all from Dr. Allen Mann, a logician himself, in resolving many crucial difficulties. My friend and colleague Dr. Shuvra Das provided guidance, advice, and wisdom. Very special concern, care, encouragement, and inspiration came from my friend Sreela Datta. My brother Anirban and his family were truly supportive. My daughter Malini frequently helped me with my writing. Finally, nothing would be possible without my wife Soma who is always by my side supporting me in every possible way.

Ann Arbor and Kolkata — Abhijit Dasgupta
2011–2013

Contents

Chapter 1
Preliminaries: Sets, Relations, and Functions

Abstract This preliminary chapter informally reviews the prerequisite material for the rest of the book. Here we set up our notational conventions, introduce basic set-theoretic notions including the power set, ordered pairs, Cartesian product, relations, functions, and their properties, sequences, strings and words, indexed and unindexed families, partitions and equivalence relations, and the basic definition of linear order. Much of the material of this chapter can be found in introductory discrete mathematics texts.

1.1 Introduction

Note. *In this preliminary chapter, we informally use the familiar number systems* **N**, **Z**, **R**, *and their properties to provide illustrative examples for sets, relations, and functions. In the next three chapters all of these notions will be formally defined. Thus all our assumptions about these number systems are temporary and will be dropped at the end of this chapter.*

We assume basic familiarity with sets and functions, e.g., as found in elementary calculus. Some examples of sets are the *real intervals*: The *open interval* (a, b) consists of real numbers lying strictly between a and b, and the *closed interval* $[a, b]$ consists of real numbers x satisfying $a \leq x \leq b$. The interval $(-\infty, \infty)$ is the entire real line and is denoted by the special symbol **R**:

$$\mathbf{R} = (-\infty, \infty).$$

In addition we will be using the special symbols **N** and **Z**, where

- **N** consists of the *natural numbers starting from* 1 (positive integers).[1]
- **Z** consists of *all integers*—positive, negative, or zero.

---

[1]Usage varies for the interpretation of the term "natural number" and the symbol **N**. Many texts include 0 as a natural number, but we will not follow that convention.

A. Dasgupta, *Set Theory: With an Introduction to Real Point Sets*,
DOI 10.1007/978-1-4614-8854-5_1,

The Principle of Induction

We will also assume some familiarity with *the principle of induction for the positive integers* **N**. Let P be a property of natural numbers. We will use the notation "$P(n)$" to stand for the assertion "n has the property P." For example, $P(n)$ may stand for "$n(n^2+2)$ is divisible by 3."

The Principle of Induction. Let P be a property of natural numbers such that

- $P(1)$ is true.
- For any natural number n, if $P(n)$ is true then $P(n+1)$ is true.

Then $P(n)$ is true for all natural numbers n.

Problem 1. *Show that the principle of induction is equivalent to the* principle of strong induction for **N** *which is as follows:*

> *Let P be a property of natural numbers such that*
>
> - *For any natural number n, if $P(m)$ is true for all natural numbers $m < n$ then $P(n)$ is true.*
>
> *Then $P(n)$ is true for all natural numbers n.*

The natural numbers and the principle of induction will be studied in detail in Chap. 2.

1.2 Membership, Subsets, and Naive Axioms

Naively speaking, a set A is a collection or group of objects such that membership in A is definitely determined in the sense that given any x, exactly one of "$x \in A$" or "$x \notin A$" is true, where the notation

$$x \in A$$

is used to denote that x *is a member of the set* A, and the notation

$$x \notin A$$

stands for x *is not a member of* A. For example, we have $3 \in (2, \infty)$, $\frac{1}{2} \notin \mathbf{Z}$, $1 \in \mathbf{N}$, $0 \notin \mathbf{N}$, etc.

We say that A *is a subset of* B, denoted by $A \subseteq B$, if every member of A is a member of B. We write $A \nsubseteq B$ to denote that A is not a subset of B. $A \subseteq B$ is also expressed by saying that A *is contained in* B or B *contains* A. Thus we have

$$A \subseteq B \;\Leftrightarrow\; \text{for all } x, \text{if } x \in A \text{ then } x \in B, \quad \text{and}$$

$$A \not\subseteq B \;\Leftrightarrow\; \text{there is some } x \in A \text{ such that } x \notin B.$$

We are using the symbol "$\Leftrightarrow$" as a short-hand for the phrase "if and only if" (or *equivalence* of statements). Similarly, the symbol "$\Rightarrow$" will stand for *implication,* that is "$P \Rightarrow Q$" means "if P then Q" or "P implies Q."

We will also often use the abbreviations "$\forall x(\dots)$" for "for all x, ..." (the *universal quantifier*) and "$\exists x(\dots)$" for "there is some x such that ..." (the *existential quantifier*). With such abbreviations, the lines displayed above can be shortened to:

$$A \subseteq B \;\Leftrightarrow \forall x\,(x \in A \Rightarrow x \in B), \qquad \text{and}$$

$$A \not\subseteq B \;\Leftrightarrow \exists x\,(x \in A \text{ and } x \notin B).$$

The *principle of extensionality* says that *two sets having the same members must be identical,* that is:

$$A = B \;\Leftrightarrow\; \forall x\,(x \in A \Leftrightarrow x \in B),$$

which can also be stated in terms of subsets as:

$$A = B \;\Leftrightarrow\; A \subseteq B \text{ and } B \subseteq A \qquad (\textbf{Extensionality}).$$

The *naive principle of comprehension* is used to form new sets. Given any property P, we write $P(x)$ for the assertion "x has property P." Then the naive principle of comprehension says that, given any property P, *there is a set A consisting precisely of those x for which $P(x)$ is true.* In symbols:

$$\exists A\; \forall x\; (x \in A \Leftrightarrow P(x)) \qquad (\textbf{Comprehension}).$$

We use the qualifier "naive" to indicate that the principle of comprehension uses the vague notion of "property," and unrestricted use of the comprehension principle can cause problems that will be discussed later.[2] For now, we follow the naive approach of Cantor's classical set theory, and the axioms of extensionality and comprehension (together with a couple more axioms such as the Axiom of Choice to be introduced in Chap. 5) will form the basis of development for our central topics of study.[3]

---

[2]Such difficulties lead to consideration of *metamathematical* issues. We will be confining ourselves to purely *mathematical* aspects of set theory in the first three parts of the book.

[3]This is a satisfactory approach for most areas of mathematics (and for most mathematicians) since the natural ways of forming new sets out of old ones such as taking subsets, forming the power set, and taking unions, do not seem to lead to difficulties.

Set Builder Notation

By extensionality, the set A whose existence is given by comprehension from a property P is unique, so we can introduce the *set builder notation*

$$\{x \mid P(x)\} \quad \text{or} \quad \{x\colon P(x)\}$$

to denote the unique set A consisting precisely of those x for which $P(x)$ is true, i.e., the set A defined by the condition: $x \in A \Leftrightarrow P(x)$ (for all x). So,

$$A = \{x \mid P(x)\} \quad \text{if and only if:} \quad \text{for all } x,\, x \in A \Leftrightarrow P(x).$$

For example, we have:

$$[a, \infty) = \{x \mid x \in \mathbf{R} \text{ and } a \leq x\}.$$

In this example the resulting set $[a, \infty)$ is a subset of $\mathbf{R}$. In general, when a new set B is defined as a subset of an old set A as those members of A which have the property P, that is when

$$B = \{x \mid x \in A \text{ and } P(x)\},$$

we will often use the alternative notation

$$B = \{x \in A \mid P(x)\}.$$

The Empty Set and Singletons

Perhaps the simplest set is the *empty set* $\varnothing$ which has no members. (Its existence can be proved using the naive comprehension principle by taking $P(x)$ to be "$x \neq x$.") The empty set is a subset of every set:

$$\varnothing \subseteq A \text{ for all sets } A.$$

For any a, the *singleton set* $\{a\}$ is the set whose only member is a:

$$\{a\} := \{x \mid x = a\}.$$

We will often use the notation ":=" when definitions are introduced.

The singleton set $\{a\}$ should be distinguished from the element a. For example $\{\varnothing\}$ and $\varnothing$ cannot be the same since the first one has a member while the second has no members, and a set which has a member cannot be identical with a set with no members (by extensionality).

Problem 2 (Royden). *Prove that if* $x \in \emptyset$, *then* x *is a green-eyed lion.*

Problem 3. *Prove that* $\{a\} = \{b\}$ *if and only if* $a = b$.

Problem 4. *Prove that the sets* $\{\{\{\{\emptyset\}\}\}\}$ *and* $\{\{\emptyset\}\}$ *are distinct. For each of these two sets, determine if it is a singleton.*

The Brace-List Notation

More generally, we can denote sets consisting of multiple members using the *brace-list notation,* as in:

$$\{a, b\} := \{x \mid x = a \text{ or } x = b\},$$
$$\{a, b, c\} := \{x \mid x = a \text{ or } x = b \text{ or } x = c\}, \quad \text{etc.}$$

The set $\{a, b\}$ is sometimes called an *unordered pair.*

Problem 5. *Prove that* $\{a, b\} = \{b, a\} = \{a, a, b, a\}$. *Can* $\{a, b\}$ *be a singleton?*

More informally, we often use the brace-list notation together with dots "..." (ellipsis) where not all the elements are listed, but the missing elements can readily be understood from the notation. For example,

$$\{1, 2, \ldots, 100\} \quad \text{stands for} \quad \{x \in \mathbf{N} \mid x \leq 100\}.$$

This is used for infinite sets as well, as in

$$\mathbf{N} = \{1, 2, 3, \ldots, n, \ldots\}.$$

If A is a set and $\alpha(x)$ is an expression involving x which is uniquely determined for each $x \in A$, then we use the notation

$$\{\alpha(x) \mid x \in A\}$$

as a convenient abbreviation for the set $\{y \mid y = \alpha(x) \text{ for some } x \in A\}$. For example,

$$\{2n - 1 \mid n \in \mathbf{N}\}$$

denotes the set of all odd positive integers. This notation is also extended for expressions with multiple variables. For example,

$$\{\alpha(u, v) \mid u \in A,\ v \in B\}$$

stands for $\{x \mid x = \alpha(u, v) \text{ for some } u \in A \text{ and } v \in B\}$. Thus

$$\{m^2 + n^2 \mid m, n \in \mathbf{N}\}$$

denotes the set of integers which can be expressed as a sum of two perfect squares.

1.3 The Power Set and Set Operations

The Power Set

The *power set* $\mathrm{P}(A)$ of a set A is defined to be the set of all subsets of A:

$$\mathrm{P}(A) := \{x \mid x \subseteq A\}.$$

Problem 6. *What is* $\mathrm{P}(\emptyset)$*? Find* $\mathrm{P}(\mathrm{P}(\emptyset))$ *and* $\mathrm{P}(\mathrm{P}(\mathrm{P}(\emptyset)))$*. If* a, b, c *are distinct elements, find* $\mathrm{P}(\{a, b, c\})$.

Problem 7. *Show that if a set* A *has* n *elements then* $\mathrm{P}(A)$ *has* 2^n *elements.*

Problem 8. *A set* A *is called* transitive *if every element of* A *is also a subset of* A.

1. *Among* $\emptyset$, $\{\emptyset\}$, *and* $\{\{\emptyset\}\}$, *which ones are transitive?*
2. *Find a transitive set with five elements.*
3. *Prove that a set is transitive if and only if its power set is transitive.*
4. *Can you find an example of an infinite transitive set?*

Set Operations

Given sets A and B we define their

Union:	$A \cup B := \{x \mid x \in A \text{ or } x \in B\}$
Intersection:	$A \cap B := \{x \mid x \in A \text{ and } x \in B\}$
Difference:	$A \smallsetminus B := \{x \mid x \in A \text{ and } x \notin B\}$.

We will assume basic familiarity with these operations which are often illustrated by means of Venn diagrams.

Problem 9. *Show that* $(A \smallsetminus B) \cup (B \smallsetminus A) = (A \cup B) \smallsetminus (A \cap B)$.

The set $(A \smallsetminus B) \cup (B \smallsetminus A)$ is called the *symmetric difference of* A *and* B, and denoted by $A \triangle B$.

Universal Sets and Complementation: In many situations there will be a fixed set U such that all sets under consideration will be subsets of U. The set U then becomes the largest set among those being considered, and is sometimes called a *universal set,* with the power set $\mathbf{P}(U)$ being the universe. For example, in the context of the real line, all sets being discussed may be subsets of $\mathbf{R}$. In such cases, where the fixed largest set U does not change and can be understood from context, it is convenient to write $U \smallsetminus A$ simply as A', called the *complement of A,* giving rise to the set operation of *complementation* relative to U.

The Algebra of Subsets of a Fixed Set

In the last situation described, where all sets being considered are subsets of a universal set U, the collection $\mathbf{P}(U)$, together with the operations of union, intersection, and complementation, is known as *the Boolean Algebra of Subsets of U,* or simply *the Algebra of Subsets of U*. For the algebra of subsets of U, the set operations satisfy many properties, most of which are readily derived. We list them in the following problems.

Problem 10. *Show that, for all sets A, B, C:*

1. $A \cup A = A = A \cap A$.
2. $A \cup B = B \cup A$ *and* $A \cap B = B \cap A$.
3. $A \cup (B \cup C) = (A \cup B) \cup C$ *and* $A \cap (B \cap C) = (A \cap B) \cap C$.
4. $A \cap (B \cup C) = (A \cap B) \cup (A \cap C)$ *and* $A \cup (B \cap C) = (A \cup B) \cap (A \cup C)$.
5. $A \smallsetminus (B \cup C) = (A \smallsetminus B) \cap (A \smallsetminus C)$ *and* $A \smallsetminus (B \cap C) = (A \smallsetminus B) \cup (A \smallsetminus C)$.
6. $A \cap B \subseteq A \subseteq A \cup B$.
7. $A \subseteq B \Leftrightarrow A \cup B = B \Leftrightarrow A \cap B = A \Leftrightarrow A \smallsetminus B = \emptyset$.

Problem 11. *Suppose that U is a set such that all sets $A, B, C, \ldots$ under consideration are subsets of U, and write A' for $U \smallsetminus A$. Show that*

1. $A \cup \emptyset = A = A \cap U$.
2. $\emptyset' = U$, $U' = \emptyset$, *and* $(A')' = A$.
3. $A = B' \Leftrightarrow B = A' \Leftrightarrow A \cap B = \emptyset$ *and* $A \cup B = U$.
4. $A \cap A' = \emptyset$ *and* $A \cup A' = U$.
5. $A \cap \emptyset = \emptyset$ *and* $A \cup U = U$.
6. $A \cap B = \emptyset \Leftrightarrow A' \cup B' = U \Leftrightarrow A \subseteq B' \Leftrightarrow B \subseteq A'$
7. $(A \cup B)' = A' \cap B'$ *and* $(A \cap B)' = A' \cup B'$.

The last equalities in the list are called the *DeMorgan laws.*

Problem 12. *1.* $A \smallsetminus B = A \smallsetminus (A \cap B)$ *and* $A \smallsetminus (A \smallsetminus B) = A \cap B$.
2. $A \cup B = \emptyset \Leftrightarrow A = B = \emptyset$, *and* $A = B \Leftrightarrow A \triangle B = \emptyset$.

1.4 Ordered Pairs and Relations

We will let $\langle a,b \rangle$ denote the *ordered pair* consisting of a and b in the order of appearance. Its exact definition will not matter to us; any definition will be satisfactory so long as $\langle a,b \rangle$ is uniquely defined for all a,b, and satisfies the following property:

$$\langle a,b \rangle = \langle c,d \rangle \;\Rightarrow\; a = c \text{ and } b = d \qquad \text{(for all } a,b,c,d\text{).}$$

We will leave the ordered pair as an undefined primitive notion satisfying this characterizing condition.[4]

It is important to distinguish the *unordered* pair $\{a,b\}$, for which the "commutative property" $\{a,b\} = \{b,a\}$ always holds, from the *ordered* pair $\langle a,b \rangle$, for which $\langle a,b \rangle = \langle b,a \rangle$ will hold only if $a = b$.

Cartesian Product

The *Cartesian product* $A \times B$ of two sets A and B is defined as the set of all ordered pairs $\langle a,b \rangle$ with $a \in A$ and $b \in B$:

$$A \times B \;:=\; \{\langle a,b \rangle \mid a \in A,\; b \in B\}.$$

We abbreviate $A \times A$ as A^2. For example, the familiar Cartesian plane $\mathbf{R}^2 = \mathbf{R} \times \mathbf{R}$ is the set of all ordered pairs $\langle a,b \rangle$ where a and b are real numbers.

Relations

A *relation* is defined to be any set consisting only of ordered pairs. Thus:

$$R \text{ is a relation} \;\Leftrightarrow\; \text{for all } x \in R,\, x = \langle a,b \rangle \text{ for some } a,b.$$

We will use the notation xRy to denote $\langle x,y \rangle \in R$, and $\neg xRy$ to denote $\langle x,y \rangle \notin R$.

The *domain* and *range* of a relation R, denoted by $\operatorname{dom}(R)$ and $\operatorname{ran}(R)$ respectively, are defined as:

$$\operatorname{dom}(R) \;:=\; \{x \mid xRy \text{ for some } y\}, \quad \text{and} \quad \operatorname{ran}(R) \;:=\; \{y \mid xRy \text{ for some } x\}.$$

---

[4]This was the standard practice until Wiener showed in 1914 that the notion of ordered pair can be reduced to a definition in terms of sets by taking $\langle a,b \rangle := \{\{a\},\{b,\varnothing\}\}$. This was later improved by Kuratowski in 1921 to the current standard definition $\langle a,b \rangle := \{\{a\},\{a,b\}\}$. The interested reader may want to verify as an exercise that both of these are satisfactory definitions for the ordered pair.

Problem 13. *Find the domain and range for each of the following relations.*

$$R_1 := \{\langle x, y\rangle \in \mathbf{R}^2 \mid xy = 1\},$$
$$R_2 := \{\langle x, y\rangle \in \mathbf{R}^2 \mid x^2 + y^2 = 1\},$$
$$R_3 := \{\langle x, y\rangle \in \mathbf{R}^2 \mid x = \sin y\},$$
$$R_4 := \{\langle x, y\rangle \in \mathbf{R}^2 \mid x^2 < y\}.$$

If R is a relation then $R \subseteq \mathrm{dom}(R) \times \mathrm{ran}(R)$ and so a relation could also have been defined using the condition in the following problem.

Problem 14. *R is a relation $\Leftrightarrow$ $R \subseteq A \times B$ for some sets A and B.*

We say that R *is a relation on* A if $R \subseteq A \times A$.

Problem 15. *If R is a relation, then R is a relation on some set A.*

If R is a relation, then its *inverse relation* R^{-1} is defined as

$$R^{-1} := \{\langle u, v\rangle \mid \langle v, u\rangle \in R\}.$$

For example, if $R = \{\langle x, y\rangle \in \mathbf{R}^2 \mid x < y\}$ and $S = \{\langle x, y\rangle \in \mathbf{R}^2 \mid x > y\}$, then $R^{-1} = S$ and $S^{-1} = R$. It is easily verified that $(R^{-1})^{-1} = R$.

If R and S are relations, their *relative product* (or *composition*), denoted by $R \cdot S$ or by RS, is defined as $R \cdot S := \{\langle x, y\rangle \mid \text{For some } u, xRu \text{ and } uRy\}$.

Problem 16. *Let S be the relation on $\mathbf{R}$ defined as*

$$S := \{\langle x, y\rangle \in \mathbf{R} \mid -1 < x - y < 1\}.$$

What is $S \cdot S$? Draw figures showing S and $S \cdot S$ on the Cartesian plane.

Problem 17. *Show that $(R \cdot S)^{-1} = S^{-1} \cdot R^{-1}$.*

Properties of Relations

Let R be a relation on A. We say that

1. *R is reflexive on A* if xRx, for all $x \in A$;
2. *R is irreflexive on A* if $\neg xRx$, for all $x \in A$;
3. *R is symmetric on A* if $xRy \Rightarrow yRx$, for all $x, y \in A$;
4. *R is asymmetric on A* if $xRy \Rightarrow \neg yRx$, for all $x, y \in A$;
5. *R is antisymmetric on A* if xRy and $yRx \Rightarrow x = y$, for all $x, y \in A$;
6. *R is transitive on A* if xRy and $yRz \Rightarrow xRz$, for all $x, y, z \in A$;
7. *R is connected on A* if $x \neq y \Rightarrow xRy$ or yRx, for all $x, y \in A$;

Problem 18. *Given a set A, put $\Delta_A := \{\langle x, x\rangle \mid x \in A$. ($\Delta_A$ is called the* diagonal *or* identity *on A.) If R is a relation on A, show that:*

1. *R is reflexive on $A \Leftrightarrow \Delta_A \subseteq R$;*
2. *R is irreflexive on $A \Leftrightarrow R \cap \Delta_A = \emptyset$;*
3. *R is symmetric on $A \Leftrightarrow R \subseteq R^{-1} \Leftrightarrow R = R^{-1} \Leftrightarrow R^{-1} \subseteq R$;*
4. *R is asymmetric on $A \Leftrightarrow R \cap R^{-1} = \emptyset$;*
5. *R is antisymmetric on $A \Leftrightarrow R \cap R^{-1} \subseteq \Delta_A$;*
6. *R is transitive on $A \Leftrightarrow R \cdot R \subseteq R$;*
7. *R is connected on $A \Leftrightarrow R \cup R^{-1} \cup \Delta_A = A \times A$.*

Problem 19. *Let S be the relation of non-equality on the set A, that is, $xSy \Leftrightarrow x, y \in A$ and $x \neq y$. Then S is irreflexive, symmetric, and connected on A. Moreover, if R is any relation on A which is irreflexive, symmetric, and connected on A, then $R = S$.*

Problem 20. *Given a set A with n elements, how many relations are there on A which are both symmetric and connected? How many are reflexive? How many are irreflexive? How many are neither?*

Transitivity is an important property of relations. For transitive relations, the properties of irreflexivity and asymmetry coincide.

Problem 21. *If R is a transitive relation on A, show that R is irreflexive on A if and only if R is asymmetric on A.*

1.5 Functions

A relation F is said to be a *function* if xFy and $xFz \Rightarrow y = z$, for all x, y, z. If F is a function and $x \in \text{dom}(F)$, then there is a unique y such that xFy, and we denote this y by $F(x)$, the usual functional notation.

We also say that F *is a function from A to B*, written using the standard notation

$$F\colon A \to B,$$

to mean that F is function with $\text{dom}(F) = A$ and $\text{ran}(F) \subseteq B$. *In this case, it is common to abuse terminology and refer to the triplet $\langle F, A, B\rangle$ as "the function $F\colon A \to B$."* The set B is then sometimes referred to as the *converse domain* or *co-domain* of the function $F\colon A \to B$ (more precisely, B is the co-domain of the triplet $\langle F, A, B\rangle$).

Functions are also called *mappings*. If A is a set and $\alpha(x)$ is an expression involving the variable x which is uniquely determined for each $x \in A$, then the relation

$$F := \{\langle x, y\rangle \mid x \in A \text{ and } y = \alpha(x)\}$$

is a function with domain A. This function F is sometimes referred to as "*the mapping* $x \mapsto \alpha(x)$ *where* $x \in A$," and can be written more simply as

$$F = \{\langle x, \alpha(x)\rangle \mid x \in A\}.$$

Function Builder Notation

It is very convenient to further simplify the notation and to denote the function F using the *function builder notation*:

$$F = \langle \alpha(x) \mid x \in A\rangle .$$

Notice the use of angle-brackets in place of the curly braces. The function builder notation is highly useful in defining new functions. For example, the relation

$$R = \{\langle x, y\rangle \in \mathbf{R}^2 \mid x \in [0,1],\ y = x^2 + 1\}$$

is a function with domain $[0, 1]$ which is denoted by $\langle x^2 + 1 \mid x \in [0,1]\rangle$.

Two functions F and G are said to *agree on a set* A if $A \subseteq \text{dom}(F)$, $A \subseteq \text{dom}(G)$, and $F(x) = G(x)$ for all $x \in A$.

If $F\colon A \to B$ and $C \subseteq A$, then the *restriction of* F *to* C, denoted by $F|_C$ or $F \restriction C$, is the function with domain C defined as

$$F|_C = \{\langle x, y\rangle \mid x \in C \text{ and } \langle x, y\rangle \in F\}.$$

In this case we also say that F is an *extension of* G. Note that $F|_C$ could also be defined as $\langle F(x) \mid x \in C\rangle$ and is the unique function with domain C which agrees with F on A.

Let $F\colon A \to B$. For each $C \subseteq A$ we define the *(forward) image of* C *under* F, denoted by $F[C]$, as the set

$$F[C] := \{F(x) \mid x \in C\}.$$

Thus $F[C] = \text{ran}(F|_C)$. Similarly, for each $D \subseteq B$, we define the *inverse image of* D *under* F, denoted by $F^{-1}[D]$, as the set

$$F^{-1}[D] := \{x \in A \mid F(x) \in D\}.$$

Problem 22. *Let* $F\colon X \to Y$, $A, B \subseteq X$, *and* $C, D \subseteq Y$. *Show that*

1. $F[A \cup B] = F[A] \cup F[B]$ *and* $F[A \cap B] \subseteq F[A] \cap F[B]$.
2. *The equality* $F[A \cap B] = F[A] \cap F[B]$ *may fail.*

3. $F^{-1}[C \cup D] = F^{-1}[C] \cup F^{-1}[D]$ *and* $F^{-1}[C \cap D] = F^{-1}[C] \cap F^{-1}[D]$.
4. $F^{-1}[Y \smallsetminus C] = X \smallsetminus F^{-1}[C]$.

If $F: A \to B$ and $G: B \to C$ are functions, then their *composition* $G \circ F$ is the function $G \circ F: A \to C$ defined by

$$G \circ F := \langle G(F(x)) \mid x \in A \rangle ,$$

which is well defined since $\operatorname{ran}(F) \subseteq \operatorname{dom}(G)$.

Problem 23. *Show that function composition is associative.*

Problem 24. *If* $F: A \to B$, $G: B \to C$, $X \subseteq A$, *and* $Y \subseteq C$, *then we have* $(G \circ F)[X] = G[F[X]]$ *and* $(G \circ F)^{-1}[Y] = F^{-1}[G^{-1}[Y]]$.

A function $F: A \to B$ is said to be *one-to-one* or *injective* if

$$F(u) = F(v) \Rightarrow u = v \qquad (\text{for all } u, v \in A).$$

Note that a function F is one-to-one if and only if the inverse relation F^{-1} is a function (in this case we will have $\operatorname{dom}(F^{-1}) = \operatorname{ran}(F)$).

Problem 25. *Show that* $F: A \to B$ *is injective if and only if for all* $X, Y \subseteq A$ *we have* $F[X \cap Y] = F[X] \cap F[Y]$.

A function $F: A \to B$ is *onto* or *surjective* if $\operatorname{ran}(F) = B$, i.e., if

$$\text{for each } y \in B \text{ there is } x \in A \text{ such that } y = F(x).$$

Note the terminological abuse mentioned earlier. The term "onto" or "surjective" really applies to the triplet $\langle F, A, B \rangle$.

A function $F: A \to B$ which is both one-to-one and onto is called a *one-to-one correspondence* or a *bijection from* A *onto* B (or a bijection between A and B). When $A = B$, that is if $F: A \to A$ is a bijection, we say that F is a bijection *on* A.

For example, if $\mathbf{Z}$ is the set of all integers, positive, negative, or zero, A is the set of all odd integers in $\mathbf{Z}$, and B is the set of all even integers in $\mathbf{Z}$, then the function $F := \langle n + 1 \mid n \in \mathbf{Z} \rangle$ is a bijection on $\mathbf{Z}$, while $F\restriction_A$ is a bijection from A onto B and $F\restriction_B$ is a bijection from B onto A.

Problem 26. *For any set* A, *define a bijection between the set of all reflexive relations on* A *and the set of all irreflexive relations on* A.

Problem 27. *Show that*

1. *For any set* A, *the identity mapping on* A *given by* $\langle x \mid x \in A \rangle$ *is a bijection on* A.
2. *If* $F: A \to B$ *is an injection, then* $F: A \to F[A]$ *is a bijection (where* $F[A] = \operatorname{ran}(F)$*).*
3. *If* F *is a bijection from* A *onto* B, *then* F^{-1} *is a bijection from* B *onto* A.

Problem 28. *Let $F: A \to B$, $G: B \to C$, and let $G \circ F: A \to C$ be their composition. If F and G are injective, so is $G \circ F$, and if F and G are surjective, so is $G \circ F$. Conclude that a composition of bijections is a bijection.*

Problem 29. *Let $F: X \to Y$. Show that*

1. *F is injective if and only if there is a $G: Y \to X$ such that the composition $G \circ F$ equals the identity function on X.*
2. *If there is a $G: Y \to X$ such that the composition $F \circ G$ equals the identity function on Y, then $F: X \to Y$ is surjective.*

The converse of the second result in the last problem holds under the *axiom of choice* which will be introduced and studied in a later chapter.

Problem 30. *Prove that for any infinite subset A of the set $\mathbf{N}$ of positive integers, there is bijection between $\mathbf{N}$ and A.*

If A, B are sets, then B^A denotes the collection of all functions from A to B:

$$B^A := \{f \mid f: A \to B\}.$$

Thus $f \in B^A \Leftrightarrow f: A \to B$.

1.6 Families and Partitions

Indexed Families

A function E with domain $I = \mathrm{dom}(E)$ will also be called an *indexed family with index set I*. In this case it is customary to denote $E(i)$ by E_i for each $i \in I$, and denote the entire indexed family, that is the function E, as:

$$E = \langle E_i \mid i \in I \rangle .$$

If E_i is a set for each $i \in I$, then we say that $\langle E_i \mid i \in I \rangle$ is an *indexed family of sets* (with index set I). Thus $\mathbf{P}(X)^I$ is the collection of all indexed families of subsets of X with index set I.

Given an indexed family of sets $\langle E_i \mid i \in I \rangle$, we define its union

$$\bigcup_{i \in I} E_i := \{x \mid x \in E_i \text{ for some } i \in I\},$$

and intersection

$$\bigcap_{i \in I} E_i := \{x \mid x \in E_i \text{ for all } i \in I\}.$$

An indexed family of sets $\langle E_i \mid i \in I \rangle$ is said to be *pairwise disjoint* if

$$i \neq j \Rightarrow E_i \cap E_j = \emptyset \qquad \text{(for all } i, j \in I\text{)}.$$

When the index set is $\mathbf{N}$, the indexed family $\langle E_n \mid n \in \mathbf{N} \rangle$ is called a *sequence of sets,* and we use the following notations:

$$\bigcup_{n=1}^{\infty} E_n := \bigcup_{n \in \mathbf{N}} E_n, \quad \text{and} \quad \bigcap_{n=1}^{\infty} E_n := \bigcap_{n \in \mathbf{N}} E_n.$$

Of course, this notation can naturally be extended to the case where the starting index is an integer other than 1.

Problem 31 (De Morgan's Laws). *Let U be a fixed "universal" set, and for $E \subseteq U$, let $E' := U \smallsetminus E$ denote the complement of E. If $\langle E_i \mid i \in I \rangle$ is any indexed family of subsets of U, show that*

$$\Big(\bigcup_{i \in I} E_i\Big)' = \bigcap_{i \in I} E_i' \quad \textit{and} \quad \Big(\bigcap_{i \in I} E_i\Big)' = \bigcup_{i \in I} E_i'.$$

Problem 32. *If $f : X \to Y$ and $\langle E_i \mid i \in I \rangle$ is an indexed family of subsets of X, then*

$$f\Big[\bigcup_{i \in I} E_i\Big] = \bigcup_{i \in I} f[E_i].$$

Problem 33. *If $f : X \to Y$ and $\langle F_j \mid j \in J \rangle$ is an indexed family of subsets of Y, then*

$$f^{-1}\Big[\bigcup_{j \in J} F_j\Big] = \bigcup_{j \in J} f^{-1}[E_j] \quad \textit{and} \quad f^{-1}\Big[\bigcap_{j \in J} F_j\Big] = \bigcap_{j \in J} f^{-1}[E_j].$$

Unindexed Families (or Collections) of Sets

If C is a set whose every member is itself a set, we say that C is a *family of sets* (unindexed) or a *collection of sets.* For example, $\mathbf{P}(A)$ is a family of sets. If each member of C is a subset of a fixed set X, or equivalently, if $\mathsf{C} \subseteq \mathbf{P}(X)$, we say that C is a *collection of subsets of X* or a *family of subsets of X*. Thus $\mathbf{P}(X)$ is the largest family of subsets of X.

If C is any (unindexed) family of sets, we can convert C into the indexed family $\langle A \mid A \in \mathrm{C}\rangle$, which is the identity function on C. Hence the notation

$$\bigcup_{A\in\mathrm{C}} A$$

can be used to denote the union of members of C, but in this case we can use the simpler notation

$$\bigcup\mathrm{C} := \bigcup_{A\in\mathrm{C}} A.$$

Thus $\bigcup\mathrm{C}$ is the set of *members of members of* C, that is:

$$x \in \bigcup\mathrm{C} \;\Leftrightarrow\; x \in A \text{ for some } A \in \mathrm{C}.$$

Similarly, we define:

$$\bigcap\mathrm{C} := \bigcap_{A\in\mathrm{C}} A, \qquad \text{and so} \qquad x \in \bigcap\mathrm{C} \;\Leftrightarrow\; x \in A \text{ for every } A \in \mathrm{C}.$$

Problem 34. *Show that* $\bigcup\mathrm{C}$ *is the "smallest set containing every set in* C*" in the sense that it contains every set in* C *and is contained in any set which contains every set in* C.

Similarly show that $\bigcap\mathrm{C}$ *is the largest set contained in every set in* C*, that is, it is contained in every set in* C *and contains any set which is contained in every set in* C.

Problem 35*. *What is* $\bigcap\mathrm{C}$ *if* C *is empty?*

Partitions

A family C of sets is called (pairwise) *disjoint* if any pair of sets in C are either identical or disjoint, i.e., if for all $A, B \in \mathrm{C}$, $A = B$ or $A \cap B = \emptyset$.

We say that a family C *covers* or *exhausts* a set X if every element of X belongs to some set in C (i.e., if for all $x \in X$ there is $A \in \mathrm{C}$ such that $x \in A$), or equivalently if $X \subseteq \bigcup\mathrm{C}$.

We say that C *is a partition of* X if C is a disjoint family of nonempty subsets of X which covers X. More precisely, C is a partition of X if

- C is a family of subsets of X ($\mathrm{C} \subseteq \mathbf{P}(X)$);
- No member of C is empty ($B \in \mathrm{C} \Rightarrow B \neq \emptyset$);
- Distinct sets in C are disjoint ($A, B \in \mathrm{C}$ and $A \neq B \Rightarrow A \cap B = \emptyset$);
- C covers X (for all $x \in X$, $x \in A$ for some $A \in \mathrm{C}$, or $X = \bigcup\mathrm{C}$).

Problem 36. *List all possible partitions of* $\{a, b, c\}$ *(with* a, b, c *distinct).*

Problem 37. *Let* $E_1 := \{2k - 1 \mid k \in \mathbf{N}\}$ *be the set of odd positive integers, and inductively define, for each* $n = 1, 2, \ldots,$

$$E_{n+1} := \{2k \mid k \in E_n\}.$$

Show that $\{E_n \mid n \in \mathbf{N}\}$ *is a partition of* $\mathbf{N}$.

Problem 38. *For each of the following, determine if* C *is a partition or not.*

(a) $\mathsf{C} = \emptyset$; (b) $\mathsf{C} = \{\emptyset\}$; (c) $\mathsf{C} = \{\{\emptyset\}\}$.

1.7 Finite and Infinite Sequences and Strings

The notion of ordered pair can be generalized to that of a *finite sequence,* which can have any finite number of entries instead of just two. For example, $\langle a, b, c\rangle$ denotes the *ordered triple* consisting of the entries a, b, c in the order of appearance. In general, we will use the notation $\langle a_1, a_2, \ldots, a_n\rangle$ to denote the *ordered n-tuple* or the *finite sequence of length n* consisting of the entries $a_1, a_2, \ldots, a_n$ in the displayed order. Its defining property is:

$$\langle a_1, a_2, \ldots, a_n\rangle = \langle b_1, b_2, \ldots, b_n\rangle \Rightarrow a_1 = b_1, a_2 = b_2, \ldots, a_n = b_n.$$

A finite sequence of length n can be officially defined as a *function whose domain is the set* $\{1, 2, \ldots, n\} = \{k \in \mathbf{N} \mid 1 \le k \le n\}$ *of the first n natural numbers.* So an n-tuple a is a function $a\colon \{1, 2, \ldots, n\} \to \operatorname{ran}(a)$, with k-th entry being $a(k)$. It is then customary to abbreviate $a(k)$ as a_k, and we have

$$a = \langle a(1), a(2), \ldots, a(n)\rangle = \langle a_1, a_2, \ldots, a_n\rangle = \langle a_k \mid 1 \le k \le n\rangle,$$

where the last expression on the right uses the function-builder notation.[5]

Cartesian products are also generalized by defining

$$A_1 \times A_2 \times \cdots \times A_n := \{\langle a_1, a_2, \ldots, a_n\rangle \mid a_1 \in A_1,\ a_2 \in A_2,\ \ldots\ a_n \in A_n\}.$$

---

[5]The case $n = 2$ causes a notational conflict between the *ordered pair* and the *finite sequence of length* 2, as both are denoted by $\langle a, b\rangle$. To be pedantic, we could use a separate notation for n-tuples (such as $[a_1, a_2, \ldots, a_n]$), but our ambiguous notation will hardly cause any real trouble in informal set theory. A similar remark applies to the case $n = 1$ where we identify A^1 with A by confusing $\langle a\rangle$ as a.

If all the "factors" of a Cartesian product are equal, we use the abbreviations:

$$A^2 := A \times A, \quad A^3 := A \times A \times A, \quad \ldots, \quad A^n := \underbrace{A \times A \times \cdots \times A}_{n \text{ factors}}.$$

Thus A^n is the set $A^{\{1,2,\ldots,n\}}$ of all finite sequences of length n *with entries from* A, which is consistent with the familiar notations $\mathbf{R}^2$ for the Cartesian plane and $\mathbf{R}^3$ for the usual 3-space. We also identify A^1 with A. Finally, when $n = 0$ the set $\{1, 2, \ldots, n\}$ is empty and the only function with empty domain is the empty set $\emptyset$ itself, so we have $A^0 = A^{\emptyset} = \{\emptyset\}$.

The collection of all finite sequences from A of all possible lengths will be denoted by A^*:

$$A^* := \bigcup_{n=0}^{\infty} A^n.$$

In particular, A^* includes the empty sequence $\emptyset$ which has length 0.

We will also consider non-terminating *infinite sequences* of the form

$$\langle a_1, a_2, \ldots, a_k, \ldots \rangle = \langle a_k \mid k \in \mathbf{N} \rangle .$$

As suggested by the function-builder notation on the right above, an *infinite sequence from a set* A is officially defined to be a function $a: \mathbf{N} \to A$. Thus $A^{\mathbf{N}}$ is the set of all infinite sequences from A. Once again, it is customary to abbreviate $a(k)$ as a_k, so that $a = \langle a_k \mid k \in \mathbf{N} \rangle$ for $a \in A^{\mathbf{N}}$. Also, we will sometimes abbreviate the set $\{0, 1\}^{\mathbf{N}}$ of all infinite binary sequences as $\mathbf{2}^{\mathbf{N}}$.

To summarize, a member of $a \in A^*$ (a finite sequence) can be written as

$$a = \langle a_1, a_2, \ldots, a_n \rangle = \langle a_k \mid 1 \leq k \leq n \rangle ,$$

for some $n \geq 0$ (we get the empty sequence by taking $n = 0$), while a member of $a \in A^{\mathbf{N}}$ (an infinite sequence) can be written as

$$a = \langle a_1, a_2, \ldots, a_k, \ldots \rangle = \langle a_k \mid k \in \mathbf{N} \rangle .$$

Alphabets and Strings

Sometimes it is more convenient to regard the set A as an *alphabet* whose elements are *symbols* or *letters,* and write the sequence $a = \langle a_1, a_2, \ldots, a_n \rangle$ more simply as a *word* or *string of symbols,* as in:

$$a = a_1 a_2 \cdots a_n.$$

In such contexts, the empty sequence is called the *empty string* or the *empty word*, and is denoted by ε (instead of $\emptyset$); it is the unique string of length zero. The length of a string u is denoted by $\text{len}(u)$.

For example, when $A = \{0, 1\}$, we say that A is the *binary alphabet* consisting of the two binary digits (or bits) 0 and 1. A string from A will now be a word composed of the symbols 0 and 1, such as "10001110" or "00101," and we have the set of *binary strings*:

$$\{0,1\}^* := \{\varepsilon, 0, 1, 00, 01, 10, 11, 000, 001, 010, 011, 100, 101, 110, \dots\},$$

where for each n there are 2^n binary words of length n.

If $a = a_1a_2\cdots a_m$ and $b = b_1b_2\cdots b_n$ are finite strings of length m and n respectively, we say that a is an *initial segment* or *prefix* of b if $m \le n$ and $a_k = b_k$ for all $k \le m$. In this case, we also say that b is an *extension* of a (or b *extends* a). When b extends a and $\text{len}(b) = \text{len}(a) + 1$, we say that b is an *immediate extension* of a.

If $u = u_1u_2\cdots u_m$ is a string of length m and $v = v_1v_2\cdots v_n$ is a string of length n, we can form their *concatenation*, denoted by $u * v$, to be the string of length $m+n$ obtained by "writing u followed by v," as in:

$$u * v := u_1u_2\cdots u_mv_1v_2\cdots v_n.$$

Thus u is a prefix of w if and only if $w = u*v$ for some string v. Note that $\text{len}(u*v) = \text{len}(u) + \text{len}(v)$.

Let A be an alphabet. If $u \in A^*$ is a finite string from A and $s \in A$ is a letter in A, we use the notation $u^\frown s$ to denote the immediate extension of u obtained by suffixing it with the letter s, that is, if $u = u_1u_2\cdots u_m$ with $u_1, u_2, \dots, u_m \in A$, then

$$u^\frown s := u * \langle s\rangle = u_1u_2\cdots u_ms.$$

Thus $\text{len}(u^\frown s) = \text{len}(u) + 1$.

It is often useful to regard an infinite sequence $a \in A^{\mathbf{N}}$ as an infinite string $a = a_1a_2\cdots a_n\cdots$ of letters from the alphabet A. If $u = u_1u_2\cdots u_m$ is a finite string and $v = v_1v_2\cdots v_k\cdots$ is an infinite string, then we say that u is a (finite) *initial prefix* of v, or that v *extends* u, if $u_k = v_k$ for $k = 1, 2, \dots, m$.

If $a = a_1a_2\cdots a_k\cdots$ is an infinite string, we use the notation

$$a|n := a_1a_2\cdots a_n,$$

to denote the finite initial prefix of a obtained by truncating a to its first n letters. Thus $a|0 = \varepsilon$, $a|1 = a_1$, $a|2 = a_1a_2$, $a|3 = a_1a_2a_3$, etc, and a extends $a|n$ for every n.

Finally, given a finite string $a = a_1a_2\cdots a_m$ and an infinite string $b = b_1b_2\cdots b_k\cdots$, we can concatenate them to form the infinite string $a * b$ as

$$a * b := a_1a_2\cdots a_m b_1b_2\cdots b_k\cdots,$$

or more formally as the infinite string $c = c_1c_2\ldots c_k\cdots$, where

$$c_k := \begin{cases} a_k & \text{if } k \leq m, \\ b_{k-m} & \text{if } k > m. \end{cases}$$

1.8 Partitions and Equivalence Relations

A relation R on a set A is said to be an *equivalence relation* on A if R is reflexive, symmetric, and transitive on A. The symbols $\sim$ and $\equiv$ are often used to denote equivalence relations, and we say x *is equivalent to* y to express $x \sim y$. Thus $\sim$ is an equivalence relation on A if and only if:

1. Reflexivity: $x \sim x$ (for all $x \in A$);
2. Symmetry: $x \sim y \Rightarrow y \sim x$ (for all $x, y \in A$); and
3. Transitivity: $x \sim y$ and $y \sim z \Rightarrow x \sim z$ (for all $x, y, z \in A$).

The identity relation $=$ is an equivalence relation. In fact, the notion of equivalence relation can be viewed as a generalization of the notion of identity.

Problem 39. *Let F be a function with* $\text{dom}(F) = A$, *and for $x, y \in A$, define*

$$x \sim y \ \Leftrightarrow\ F(x) = F(y).$$

Show that $\sim$ is an equivalence relation on A.

Given any equivalence relation $\sim$ on a set A, a function F with domain A is called a *complete invariant for the relation* $\sim$ if "F reduces $\sim$ to the identity $=$" in the following sense:

$$x \sim y \ \Leftrightarrow\ F(x) = F(y) \qquad (\text{for all } x, y \in A).$$

If $\sim$ is an equivalence relation on A and $x \in A$, the $\sim$*-equivalence class of* x, denoted by $[x]_\sim$, or simply by $[x]$ if there is no risk of confusion, is defined as:

$$[x] := [x]_\sim := \{y \in A \mid x \sim y\}.$$

Thus $[x]$ is the set of elements equivalent to x, and so $y \in [x] \Leftrightarrow y \sim x$.

Problem 40. *Let $\sim$ be an equivalence relation on A. Prove that $x \in [x]$ for all $x \in A$. Prove also that for any $x, y \in A$, either $[x] = [y]$ or $[x] \cap [y] = \emptyset$.*

Problem 41. *Let $\sim$ be an equivalence relation on A. Prove that if $x, y \in A$, then $[x] = [y]$ if and only if $x \sim y$.*

Therefore *every equivalence relation has a natural complete invariant*:

Theorem 42 (Principle of Abstraction). *Given any equivalence relation $\sim$ on a set A, the mapping $x \mapsto [x]_\sim$, which assigns to every element its own equivalence class, is a natural complete invariant for $\sim$, i.e.,*

$$x \sim y \Leftrightarrow [x]_\sim = [y]_\sim \qquad \text{(for all } x, y).$$

The mapping $x \mapsto [x]_\sim$ is called the quotient map *given by $\sim$.*

The following theorem exhibits a natural one-to-one correspondence between equivalence relations and partitions over a given set A, thus bringing out the fact that the two notions "equivalence relation" and "partition" in a sense represent the same concept (i.e., each can be regarded as form of the other).

Theorem 43 (Identifying Equivalence Relations with Partitions). *Given an equivalence relation $\sim$ on A, the family $\boldsymbol{\Pi}(\sim)$ of all distinct $\sim$-equivalence classes forms a partition of A such that*

$$x \sim y \Leftrightarrow x \text{ and } y \text{ belong to some common set in the partition } \boldsymbol{\Pi}(\sim).$$

Conversely, given any partition C of A, the relation $\mathsf{E}(\mathsf{C})$ on A defined by

$$x\,\mathsf{E}(\mathsf{C})\,y \Leftrightarrow \text{there is some } B \in \mathsf{C} \text{ such that } x, y \in B$$

is an equivalence relation on A such that C equals the family of all $\mathsf{E}(\mathsf{C})$-equivalence classes.

Moreover, we have

$$\mathsf{E}(\boldsymbol{\Pi}(\sim)) = \sim, \quad \text{for any equivalence relation } \sim, \quad \text{and}$$
$$\boldsymbol{\Pi}(\mathsf{E}(\mathsf{C})) = \mathsf{C}, \quad \text{for any partition } \mathsf{C}.$$

Problem 44. *Prove Theorem 43.*

Given an equivalence relation $\sim$ on A, the partition $\boldsymbol{\Pi}(\sim)$ of A consisting of all the $\sim$-equivalence classes is often denoted by $A/\sim$, and is called the *quotient of A modulo $\sim$*. The quotient map $x \mapsto [x]$ of Theorem 42 (the natural complete invariant for $\sim$) is then a surjection of A onto $A/\sim$.

Problem 45. *Let $\mathbf{Z}$ be the set of all integers, positive, negative, and zero. We write $x \mid y$ to express "x divides y," i.e., $y = xz$ for some $z \in \mathbf{Z}$. Define two relations $\equiv$ and $\sim$ on $\mathbf{Z}$ by the conditions*

$$x \equiv y \Leftrightarrow 4 \mid (x - y), \quad \textit{and} \quad x \sim y \Leftrightarrow x \mid y \text{ and } y \mid x,$$

Show that both $\equiv$ and $\sim$ are equivalence relations on $\mathbf{Z}$. In each case, what are the equivalence classes and what is the partition?

Problem 46. *Let $\mathbf{R}^2 := \{(x, y) \mid x, y \in \mathbf{R}\}$ be the usual plane, and define a relation $\sim$ on the plane by*

$$(x_1, y_1) \sim (x_2, y_2) \Leftrightarrow x_1 + y_2 = x_2 + y_1.$$

1. *Show that $\sim$ is an equivalence relation on $\mathbf{R}^2$.*
2. *Describe the equivalence classes and the partition.*
3. *Find a complete invariant for $\sim$.*

Problem 47. *Let $\mathbf{N} := \{1, 2, 3, \dots\}$ be the set of natural numbers. Define an equivalence relation $\equiv$ on $\mathbf{N}$ by:*

$$m \equiv n \quad \Leftrightarrow \quad \text{for all } k \in \mathbf{N}\text{:} \quad 2^k \mid m \Leftrightarrow 2^k \mid n.$$

Describe the equivalence classes and the partition given by $\equiv$. Can you find a complete invariant for this equivalence relation?

Problem 48. *Let $\mathbf{N} := \{1, 2, 3, \dots\}$ be the set of natural numbers. Define an equivalence relation $\sim$ on $\mathbf{N}$ by:*

$$m \sim n \quad \Leftrightarrow \quad \text{for every prime } p\text{:} \quad p \mid m \Leftrightarrow p \mid n.$$

Describe the equivalence classes and the partition given by $\sim$. Can you find a complete invariant for this equivalence relation?

Problem 49. *Define an equivalence relation $\sim$ on the set $\mathbf{R}$ of reals by*

$$x \sim y \quad \Leftrightarrow \quad \cos x = \cos y.$$

Precisely describe the equivalence classes and the corresponding partition.

1.9 Orders (Linear Orders)

We will study orders in detail starting from Chap. 7, but a few basic notions needed before Chap. 7 will be introduced here.

We say that a relation R *is a (linear) order on a set* X, or that $\langle X, R\rangle$ *is a (linear) order,* or simply that R *orders* X, if R is a transitive relation on X satisfying the *trichotomy property*: For all $a, b \in X$, exactly one of

$$aRb, \qquad a = b, \qquad bRa,$$

holds. (Thus if R orders X and $a, b \in X$ are distinct ($a \neq b$), then either aRb or bRa, but not both.)

Problem 50. *If R orders X, then R is irreflexive and asymmetric on X. Moreover, we have: R orders X if and only if R is a relation on X which is transitive, asymmetric, and connected on X.*

Notation and Terminology. If R orders X, we write $x <_R y$ to denote xRy. When there is no chance of confusion, we even drop the subscript R and simply write $x < y$ for xRy, and so $x \leq y$ means xRy or $x = y$. We will also say that "X is order" in place of "$\langle X, <\rangle$ is an order."

In a general order X, *intervals* are defined using the familiar notations for the real number line. Subsets such as $(a, b) := \{x \in X \mid a < x < b\}$ and $(a, \infty) := \{x \in X \mid a < x\}$ are *open intervals*, while examples of *closed intervals* are $[a, b] := \{x \in X \mid a \leq x \leq b\}$ and $(-\infty, a] := \{x \in X \mid x \leq a\}$. Similarly, we could also define *half-open intervals* such as $[a, b)$. In a general order, the interval $(-\infty, a)$ will usually be denoted by $\mathrm{Pred}_<(a)$ or $\mathrm{Pred}(a)$.

Let $<$ be an order on a set X.

If $A \subseteq X$, then an element $a \in X$ is a *first* or *least* element of A if $a \in A$ and for all $x \in A$, $x \neq a \Rightarrow a < x$. Similarly we define *last* and *greatest* elements. An element $a \in X$ is called an *endpoint* of the order X if a is either a first or a last element of the entire set X.

If $x, y \in X$, we say that x *is an immediate predecessor of* y, or equivalently that y *is an immediate successor of* x, if $x < y$ and there is no $z \in X$ such that $x < z < y$. It is easily seen that each element has at most one immediate successor or immediate predecessor. We also say that two elements are *consecutive elements* if one of them is an immediate successor of the other.

The order $<$ on X is said to be a *dense order* if for all $x, y \in X$, if $x < y$ then there is some $z \in X$ with $x < z$ and $z < y$. Thus an order is a dense order if and only if it does not have any pair of consecutive elements.

The order $<$ on X is said to be a *well-order* if every nonempty subset $A \subseteq X$ has a least element $a \in A$.

For example, let X be the set $\mathbf{N}$ of natural numbers with their usual order of magnitude. Then the element 1 is the first element of X, but X does not have a last element. If $m, n \in X$, then m is an immediate predecessor of n (or equivalently n is an immediate successor of m) if and only if $n = m + 1$, and pair elements are consecutive if their difference equals 1.

Many interesting examples of orders are obtained by fixing X to be a subset of $\mathbf{R}$ and taking $<$ to be the usual order of magnitude among the elements of X. For example, for any $a < b$ in **R**, the real interval $[a, b]$ with the usual order is a dense order with first element a and last element b.

Problem 51. *For each of the following, give an example of an order satisfying the stated condition.*

1. *A dense order (i.e., without consecutive elements) which has a last element but no first element.*
2. *An order on an infinite set which has a first and a last element and such that each element except the last has an immediate successor and each element except the first has an immediate predecessor.*
3. *An order having a unique element which has neither an immediate successor nor an immediate predecessor while every other element has both an immediate successor and an immediate predecessor.*

Part I
Dedekind: Numbers

Introduction to Part I

The primary goal of Part I is to construct the real numbers starting with the natural numbers as the only foundation.

Chapter 2 derives standard properties of naturals numbers from the Dedekind–Peano axioms and then develops the ratios (positive rationals). The chapter ends with an optional section on Dedekind's general method of recursive definition (primitive recursion).

Chapter 3 covers the definition of continuity in the context of linear orders, leading to the notion of a linear continuum and the satisfaction of the intermediate value theorem. It gives a construction of the real numbers using the method of Dedekind cuts.

The philosophical postscript to this part (Chap. 4) discusses two different approaches, namely Frege–Russell absolutism and Dedekindian structuralism, which are applicable not only in the conception of the natural numbers, but also more generally in the wider context of mathematics.

A great deal of the material of this part is due to Dedekind. This includes the Dedekind–Peano axioms, definition by primitive recursion, categoricity of Dedekind–Peano systems, definition of a linear continuum, the construction of irrational numbers via cuts in rationals, and the structuralist approach to the natural numbers. Most of the material of Chap. 2 correspond to Dedekind's 1888 work [11] (*Was sind und was sollen die Zahlen?*), and that of Chap. 3 to his earlier 1872 work [10] (*Stetigkeit und irrationale Zahlen*).

Note. *In the informal preliminary Chap. 1, we temporarily assumed the existence and properties of integers and real numbers to provide examples for sets, relation, and functions. In this part we will drop all such assumptions and derive everything from the Dedekind–Peano axioms. Familiar notions, like addition, are not assumed to be known until formally introduced.*

Chapter 2
The Dedekind–Peano Axioms

Abstract This chapter develops the theory of natural numbers based on Dedekind–Peano Axioms, also known as *Peano Arithmetic*. The basic theory of *ratios* (positive rational numbers) is also developed. It concludes with a section on formal definition by primitive recursion.

2.1 Introduction

With the real numbers and their properties as a starting point, a large part of classical mathematics known as *analysis* can be developed deductively. This includes analytic geometry, calculus, the theory of sequences and series of real and complex numbers and functions, differential equations, and so on.

Mathematicians in the nineteenth century such as Weierstrass, Dedekind, and Cantor produced further analysis and construction of the real numbers which reduced everything down to the notion of natural numbers **N**.

It thus became clear that (with the aid of a certain amount of set theoretic and logical apparatus) the entire body of traditional pure mathematics can be constructed rigorously starting from the theory of natural numbers.[1]

Dedekind, in his profound work [11], and Peano, in his clear and highly modern axiomatic development [59], showed how, in turn, the entire theory of natural numbers could be derived from a few basic axioms and primitive notions. The resulting deductive theory is known today as *Peano Arithmetic*. This chapter develops parts of Peano Arithmetic dealing with properties of natural numbers, fractions, and ratios.

Throughout Part I, we assume that only the primitive Dedekind–Peano notions and axioms are given and that nothing else about any kinds of numbers or their

[1] "God created the natural numbers, all else is work of man," said Kronecker.

A. Dasgupta, *Set Theory: With an Introduction to Real Point Sets*,
DOI 10.1007/978-1-4614-8854-5_2,

properties are known. All familiar notions, like addition, will be formally introduced, and their properties will be derived from the axioms.

2.2 The Dedekind–Peano Axioms

The three primitive Dedekind–Peano notions are: "natural number," "1," and "successor," where the successor of a natural number n is denoted by $S(n)$.

The five axioms involving these primitive notions are the following.

The Dedekind-Peano Axioms. The natural numbers satisfy the axioms:

1. 1 is a natural number.
2. Every natural number n has a unique successor $S(n)$ which also is a natural number.
3. 1 is not the successor of any natural number.
4. No two distinct natural numbers have the same successor (i.e., for all natural numbers m, n, $S(m) = S(n)$ implies $m = n$).
5. Induction: If P is a property of natural numbers such that

   a. 1 has property P, and
   b. whenever a natural number has property P so does its successor,

   then all natural numbers have property P.

Mathematicians found it remarkable that all known properties of natural numbers can be derived from the Dedekind–Peano Axioms.

We define:

$$2 := S(1), \quad 3 := S(2), \quad 4 := S(3), \quad 5 := S(4), \quad 6 := S(5)$$
$$7 := S(6), \quad 8 := S(7), \quad 9 := S(8), \quad 10 := S(9), \quad \text{etc.},$$

adopting the usual decimal notation as a shorthand to replace long formal expressions of the form "$S(\cdots S(S(1))\cdots)$."

Notational convention. Natural numbers will be denoted by lowercase Roman letters such as $a, b, c, m, n, p, x, y, z$, without or with subscripts and/or superscripts. Quantifiers involving these variables will be assumed to range over natural numbers. Thus "for every m there exists n" stands for "for every natural number m there exists a natural number n."

Problem 52. $3 \neq 5$.

Problem 53. *No natural number is its own successor: $S(n) \neq n$ for any n.*

Problem 54 (Converse of Axiom 3). *Every natural number other than* 1 *is the successor of some natural number, i.e., if* $n \neq 1$ *then* $n = S(m)$ *for some* m.

At this point, expression such as $1 + 3$ or $(5 + 6) \cdot 7$ or statements like $3 < 5$ cannot be used; such expressions do not even make sense yet, since the operations $+$ and $\cdot$ and the relation $<$ have not been defined.

2.3 Addition, Order, and Multiplication

Addition

Definition 55. The *sum* $m + n$ of two natural numbers m and n is defined "by induction on n" as follows (for any m):

1. $m + 1 := S(m)$, and
2. $m + S(n) := S(m + n)$.

In other words, define $m + 1$ to be $S(m)$ (this is the case $n = 1$), and once $m + n$ is defined, define $m + S(n)$ to be $S(m + n)$. This defines the sum of any two numbers.[2]

Problem 56. $2 + 2 = 4$.

Problem 57. $n + 1 = 1 + n$ *for all* n.

Problem 58. *Addition (as defined above) is associative:*

$$m + (n + p) = (m + n) + p.$$

[Hint: Use induction on p*.]*

Problem 59. *Addition is commutative:* $m + n = n + m$, *for all* m *and* n.

Problem 60. *Cancellation law for addition: If* $m + p = n + p$, *then* $m = n$.

Order

Definition 61. Define $m < n$ if and only if $n = m + p$ for some p. Also, write $m > n$ for $n < m$.

---

[2]This can be done more rigorously using *the method of definition by primitive recursion* due to Dedekind, covered in the last section of this chapter.

The next few results can be proved without induction.

Problem 62. $n < S(n)$ *for all* n.

Problem 63. $n \not< n$ *for all* n*; that is, there are no* n, p *such that* $n = n + p$.

Problem 64. *For all* n*, either* $1 < n$ *or* $1 = n$*. Also, there is no* n *with* $n < 1$.

Thus 1 is "the least natural number" (less than all other natural numbers).

Problem 65. $m < n$ *if and only if either* $S(m) < n$ *or* $S(m) = n$.

Problem 66. $m + k < n + k$ *if and only if* $m < n$.

Recall from the previous chapter that a relation on a set is called a *linear order* if it is transitive, irreflexive, and connected on the set.

Theorem 67. $<$*, as defined above, is a linear order on the natural numbers.*

Proof. Transitivity of $<$ is an easy consequence of the associative property of addition: If $m < n$ and $n < p$, then $n = m + r$ and $p = n + s$ for some r, s. Hence $p = n + s = (m + r) + s = m + (r + s) = m + t$, where $t := r + s$, so $m < p$.

Irreflexivity is a direct consequence of Problem 63.

Finally, to show that $<$ is connected, define a property P as follows, writing "$P(k)$" as a shorthand for "k has property P":

$$P(k) \text{ is true if and only if for every } n, \text{ either } k < n, \text{ or } k = n, \text{ or } k > n.$$

We establish $<$ is connected on the set of natural numbers by showing that $P(k)$ is true for all k, which is proved by induction:

First, $P(1)$ is true, as 1 is less than all other natural numbers (Problem 64).

Next, suppose that $P(k)$ is true. Then for every n, either $k < n$ in which case $S(k) < n$ or $S(k) = n$, and so $P(S(k))$ is true; or $k = n$ in which case $S(k) > n$ so $P(S(k))$ is true; or $k > n$ in which case $S(k) > k > n$ by transitivity so again $P(S(k))$ is true. Thus $P(S(k))$ is true if $P(k)$ is true.

Therefore, by induction $P(k)$ is true for every natural number k. □

Theorem 68 (The Well-Ordering Property). *Every nonempty set* A *of natural numbers has a "least" element* $m \in A$ *such that for all* $k \in A$ *either* $m < k$ *or* $m = k$.

Proof. We prove the equivalent statement that if A has no least element then A must be empty. So suppose that A does not have a least element.

Let P be the property of being less than every member of A, that is, a natural number n has property P if and only if $n < k$ for all $k \in A$.

First, since 1 is less than all other natural numbers and A has no least element, so $1 \notin A$. Hence 1 has property P, again since 1 is less than all other natural numbers.

Next, suppose that n has P. Then $S(n) \notin A$, since otherwise $S(n)$ would be the least element of A: for any $k \in A$ we have $n < k$, so by Problem 65 $S(n) < k$ or

$S(n) = k$. Hence $S(n)$ has P: for any $k \in A$, $n < k$, so $S(n) < k$ or $S(n) = k$ (by Problem 65), but we cannot have $S(n) = k$ since $S(n) \notin A$, so $S(n) < k$. Thus we have shown that if n has P then $S(n)$ has P.

By induction, every natural number has property P, so A is empty. □

Remark. This theorem is actually equivalent to the general induction axiom. It is said to be phrased in the language of *second order arithmetic,* since, unlike most other results of this chapter, it talks about *all* sets of natural numbers.

Problem 69. *If $m > n$ there is a unique k such that $m = n + k$.*

Definition 70 (Subtraction). If $m > n$, define $m - n$ to be the unique k with $m = n + k$.

Multiplication

Definition 71. The *product* $m \cdot n$ of two natural numbers m and n is defined by induction on n as follows (for any m):

1. $m \cdot 1 := m$, and
2. $m \cdot S(n) := (m \cdot n) + m$.

We write mn for $m \cdot n$.

Problem 72. $2 \cdot 3 = 6$.

Problem 73. *Multiplication (as defined above) is distributive over addition:*

$$m(n + p) - mn + mp.$$

[Hint: Use induction on p.]

Problem 74. *Multiplication is associative:*

$$m(np) = (mn)p.$$

[Hint: Use induction on p.]

Problem 75. *Multiplication is commutative: $mn = nm$, for all m and n.*

Problem 76. *Cancellation law for multiplication: If $mp = np$ then $m = n$.*

Problem 77. *$mp < np$ if and only if $m < n$.*

Problem 78. $m < 2m$.

Notation. We write n^2 for nn.

Problem 79. *$m^2 < n^2$ if and only if $m < n$.*

Definition 80. Define n to be *even* if and only if $n = 2m$ for some m. Define n to be *odd* if and only if $n = 1$ or $n = S(2m)$ for some m.

Problem 81. *For every n, either n is odd or n is even but not both. Moreover, n is even if and only if $S(n)$ is odd, and n is odd if and only if $S(n)$ is even.*

Problem 82. *n is even if and only if n^2 is even, and n is odd if and only if n^2 is odd.*

Theorem 83. *There do not exist m, n such that $m^2 = 2n^2$.*

Proof. Let $A := \{m \mid m^2 = 2n^2 \text{ for some } n\}$. The result will follow if we show that A is empty, so we assume A is nonempty and derive a contradiction. By the Well-Ordering Property, fix a least member $m \in A$. Then we can fix p such that $m^2 = 2p^2$. Then $p^2 < m^2$ (Problem 78), hence by Problem 79, $p < m$. Also, since m^2 is even, so m is even by the last result. Hence $m = 2q$ for some q. So $2q \cdot 2q = 2p^2$, or $p^2 = 2q^2$. So $p \in A$. But this is impossible since $p < m$ and m is the least member of A. □

Remark. In this proof, we had to avoid number theoretic properties such as reduced fractions, gcds, relatively prime numbers, etc., which are not available to us at this point.

2.4 Fractions and Ratios

Definition 84. A *fraction* is an ordered pair of natural numbers $\langle m, n \rangle$.

Thus $\mathbf{N} \times \mathbf{N}$ is the set of all fractions. For a fraction $\langle m, n \rangle$, m and n are called the *numerator* and *denominator*, respectively.

Definition 85 (Equivalent Fractions). We say that the fractions $\langle m, n \rangle$ and $\langle p, q \rangle$ are *equivalent,* and write $\langle m, n \rangle \sim \langle p, q \rangle$ if and only if $mq = np$.

Problem 86. $\langle mk, nk \rangle \sim \langle m, n \rangle$. *for all m, n, k.*

Problem 87. *$\sim$ is an equivalence relation on the set $\mathbf{N} \times \mathbf{N}$ of all fractions, and so $\mathbf{N} \times \mathbf{N}$ is partitioned into $\sim$-equivalence classes.*

Problem 88. *Find the equivalence classes* $[\langle 1, 1 \rangle]$, $[\langle 3, 1 \rangle]$, *and* $[\langle 2, 4 \rangle]$.

Definition 89. $\dfrac{m}{n}$ denotes the $\sim$-equivalence class of the fraction $\langle m, n \rangle$:

$$\frac{m}{n} := [\langle m, n \rangle] = \{\, \langle p, q \rangle \mid \langle p, q \rangle \sim \langle m, n \rangle \,\}.$$

Such an equivalence class of fractions is called a *ratio* (or *positive rational*):

$$\rho \text{ is a } \textit{ratio} \text{ if and only if } \rho = \frac{m}{n} \text{ for some } m, n.$$

Thus the collection of all ratios is identical to the partition determined by the equivalence relation $\sim$ (equivalence of fractions).

Ratios will be denoted by lowercase Greek letters such as $\rho, \sigma, \tau, \alpha, \beta, \gamma, \xi, \eta$, and ζ, and quantifiers involving these variables will be assumed to range over ratios. Thus "for every ρ there exists σ" really means "for every ratio ρ there exists a ratio σ."

Note. The fraction $\langle m, n\rangle$ should be distinguished from the ratio $\frac{m}{n}$. The fraction $\langle m, n\rangle$ is simply an ordered pair, and therefore is a *member* of $\mathbf{N} \times \mathbf{N}$. The ratio $\frac{m}{n}$ is the set of all fractions equivalent to the fraction $\langle m, n\rangle$, so $\frac{m}{n}$ is an entire set of fractions, and thus is a *subset* of $\mathbf{N} \times \mathbf{N}$.

Problem 90. *Explain what is wrong with the claim:*

$$\frac{n}{1} = n.$$

Problem 91. $\rho = \dfrac{m}{n}$ *if and only if* $\langle m, n\rangle \in \rho$.

Problem 92. $\dfrac{m}{n} = \dfrac{p}{q}$ *if and only if* $\langle m, n\rangle \sim \langle p, q\rangle$.

2.5 Order, Addition, and Multiplication of Fractions and Ratios

Order for Fractions and Ratios

To "compare" two fractions $\langle m, n\rangle$ and $\langle p, q\rangle$, we can (by Problem 86) find corresponding equivalent fractions $\langle mq, nq\rangle \sim \langle m, n\rangle$ and $\langle np, nq\rangle \sim \langle p, q\rangle$ with a "common denominator" nq, and compare just the numerators.

Definition 93. Define $\langle m, n\rangle < \langle p, q\rangle$ if and only if $mq < np$.

Problem 94. *If* $\langle m, n\rangle < \langle p, q\rangle$ *and* $\langle p, q\rangle < \langle r, s\rangle$, *then* $\langle m, n\rangle < \langle r, s\rangle$.

Problem 95. *Given fractions* $\langle m, n\rangle$ *and* $\langle p, q\rangle$, *exactly one of the conditions*

$$\langle m, n\rangle < \langle p, q\rangle, \qquad \langle m, n\rangle \sim \langle p, q\rangle, \qquad \langle m, n\rangle > \langle p, q\rangle,$$

is true.

Problem 96. *If* $\langle m, n\rangle \sim \langle m', n'\rangle$, $\langle p, q\rangle \sim \langle p', q'\rangle$, *and* $\langle m, n\rangle < \langle p, q\rangle$, *then* $\langle m', n'\rangle < \langle p', q'\rangle$.

Thus if a fraction in one class is less than a fraction in another class, then the same is true for all pairs of representatives from the two classes. Hence the following is well defined:

Definition 97. Define $\rho < \sigma$ if and only if there are m, n, p, q with $\rho = \dfrac{m}{n}$, $\sigma = \dfrac{p}{q}$, and $\langle m, n\rangle < \langle p, q\rangle$.

Problem 98. *$<$, as defined in the last definition for ratios, is a linear order on the set of ratios (i.e., transitive, irreflexive, and connected).*

Addition and Multiplication of Fractions and Ratios

To add two fractions $\langle m, n\rangle$ and $\langle p, q\rangle$, we can as before take the corresponding equivalent fractions $\langle mq, nq\rangle \sim \langle m, n\rangle$ and $\langle np, nq\rangle \sim \langle p, q\rangle$ with the common denominator nq, and then add the numerators. For multiplication, the numerators, and separately the denominators, are simply multiplied together.

Definition 99 (Addition of Fractions). The sum of two fractions is defined as

$$\langle m, n\rangle + \langle p, q\rangle \ := \ \langle mq + np, nq\rangle\,.$$

Problem 100. *If $\langle m, n\rangle \sim \langle m', n'\rangle$ and $\langle p, q\rangle \sim \langle p', q'\rangle$, then*

$$\langle m, n\rangle + \langle p, q\rangle \ \sim \ \langle m', n'\rangle + \langle p', q'\rangle\,.$$

Thus the class of the sum depends only on the classes to which the summands belong, making the following definition for addition of ratios *well defined*:

Definition 101 (Addition of Ratios). The sum of two ratios $\rho = \frac{m}{n}$ and $\sigma = \frac{p}{q}$ is defined as

$$\rho + \sigma = \frac{m}{n} + \frac{p}{q} \ := \ \frac{mq + np}{nq}.$$

Definition 102 (Multiplication of Fractions). The product of two fractions is defined as

$$\langle m, n\rangle \cdot \langle p, q\rangle \ := \ \langle mp, nq\rangle\,.$$

Problem 103. *If $\langle m, n\rangle \sim \langle m', n'\rangle$ and $\langle p, q\rangle \sim \langle p', q'\rangle$, then*

$$\langle m, n\rangle \cdot \langle p, q\rangle \ \sim \ \langle m', n'\rangle \cdot \langle p', q'\rangle\,.$$

Thus the class of the product depends only on the classes to which the factors belong. Hence the following is well defined.

Definition 104 (Multiplication of Ratios). The product of two ratios $\rho = \frac{m}{n}$ and $\sigma = \frac{p}{q}$, denoted by $\rho \cdot \sigma$ or $\rho\sigma$, is defined as

$$\rho \cdot \sigma = \frac{m}{n} \cdot \frac{p}{q} := \frac{mp}{nq}.$$

2.6 Properties of Addition and Multiplication of Ratios

Problem 105. $\frac{m}{p} + \frac{n}{p} = \frac{m+n}{p}$, *and* $\frac{m}{p} < \frac{n}{p}$ *if and only if* $m < n$.

Problem 106 (Commutative Laws). $\rho + \sigma = \sigma + \rho$, *and* $\rho\sigma = \sigma\rho$.

Problem 107 (Associative Laws). $(\rho + \sigma) + \tau = \rho + (\sigma + \tau)$, *and* $(\rho\sigma)\tau = \rho(\sigma\tau)$.

Problem 108 (Cancellation Laws). *If* $\rho + \tau = \sigma + \tau$ *or if* $\rho\tau = \sigma\tau$, *then* $\rho = \sigma$.

Problem 109 (Distributive Law). $\rho(\sigma + \tau) = \rho\sigma + \rho\tau$.

Problem 110. $\rho < \rho + \xi$.

Problem 111. *If* $\rho < \sigma$, *then there is a unique* ξ *such that* $\rho + \xi = \sigma$.

Corollary 112. $\rho < \sigma$ *if and only if* $\sigma = \rho + \xi$ *for some (unique)* ξ.

Definition 113 (Subtraction). If $\rho < \sigma$, define $\sigma - \rho$ to be the unique ξ with $\rho + \xi = \sigma$.

Problem 114. $\rho < \sigma$ *if and only if* $\rho + \tau < \sigma + \tau$. *if and only if* $\rho\tau < \sigma\tau$.

Problem 115 (Identity and Reciprocal). $\rho \cdot \frac{1}{1} = \rho$ *and* $\frac{m}{n} \cdot \frac{n}{m} = \frac{1}{1}$.

Problem 116. *For any* ρ, σ *there is a unique* ξ *such that* $\xi \cdot \rho = \sigma$.

Definition 117 (Division). σ/ρ denotes the unique ξ such that $\xi \cdot \rho = \sigma$.

Corollary 118. $(\sigma/\rho)\rho = \sigma$.

Problem 119. *If* $\rho_1 < \sigma_1$ *and* $\rho_2 < \sigma_2$, *then* $\rho_1 + \rho_2 < \sigma_1 + \sigma_2$ *and* $\rho_1\rho_2 < \sigma_1\sigma_2$.

Problem 120 (Difference of Squares). *If* $\alpha < \beta$ *so that* $\beta - \alpha$ *is defined, then* $\beta^2 = \alpha^2 + (\beta - \alpha)(\beta + \alpha)$, *where* σ^2 *stands for* $\sigma \cdot \sigma$.

2.7 Integral Ratios and the Embedding of the Natural Numbers

Definition 121. A ratio ρ is said to be *integral* if $\rho = \frac{m}{1}$ for some m.

Problem 122. ρ *is integral if and only if* $\langle m, 1\rangle \in \rho$ *for some* m.

Problem 123. $\dfrac{m}{n}$ *is integral if and only if* $m = nk$ *for some* k.

We will now see that the integral ratios form a subset of the ratios which is structurally identical, or "isomorphic," to the natural numbers in the following sense: There is a one-to-one correspondence between the natural numbers and the integral ratios which preserves the operations of addition and multiplication as well as the order relation (Problem 125). Such a bijection is called an *isomorphism.*

Problem 124. $\dfrac{m}{1} = \dfrac{n}{1}$ *if and only if* $m = n$. *Thus the mapping* $n \mapsto \dfrac{n}{1}$ *is a bijection from the set of natural numbers onto the set of integral ratios.*

Problem 125 (Isomorphism of Natural Numbers with Integral Ratios). *For any* m, n*:*

$$\frac{m}{1} + \frac{n}{1} = \frac{m+n}{1}, \quad \frac{m}{1} \cdot \frac{n}{1} = \frac{m \cdot n}{1}, \quad \text{and} \quad \frac{m}{1} < \frac{n}{1} \text{ if and only if } m < n.$$

Problem 126. *The integral ratios satisfy the five Dedekind–Peano axioms when*

- 1 *is interpreted as* $\dfrac{1}{1}$, *and*
- $S\left(\dfrac{n}{1}\right)$ *is interpreted as* $\dfrac{S(n)}{1}$.

At this point, the natural numbers and the integral ratios become interchangeable since all the properties of the natural numbers listed in the initial sections are possessed by the integral ratios.

Therefore, we throw away the natural numbers[3] and use the corresponding integral ratios in their place. The old natural numbers are not used directly anymore, and so we now deal with only one type of numbers, namely the ratios, which include the "new natural numbers" (really the integral ratios) as a subset.

This process is known as *embedding the natural numbers into the ratios.*

Definition 127 (New Meaning for the Natural Number Symbols). With the old natural numbers thrown out, the integral ratio $\dfrac{n}{1}$ will now be denoted simply by the letter n and called *the natural number* n (similarly for other lowercase Roman letters). Not only lowercase Roman letters now denote the new natural numbers (integral ratios) by default, but also any other symbol previously used for a natural number will now denote the corresponding new natural number.

For example, the symbol 1 now stands for the integral ratio $\frac{1}{1}$, the symbol 2 for the integral ratio $\frac{2}{1}$, etc. The resulting notational ambiguity is not a real problem, as the intended interpretation can be determined from context.

[3]Phrase of Edmund Landau [47].

This allows us to mix symbols that were previously assigned to different types, and "$n + \rho$" and "$n \cdot \rho$" now become valid terms. But we remind the reader again that *lowercase Roman letters will denote the new natural numbers (really the integral ratios), and lowercase Greek letters will continue to denote arbitrary ratios.* Fractions will no longer be used.

Problem 128. $\dfrac{m}{n} = m/n$. *Also, since* $\sigma \cdot 1 = \sigma = 1 \cdot \sigma$, *so* $\sigma/1 = \sigma$ *and* $\sigma/\sigma = 1$. *Finally,* $\sigma(1/\sigma) = 1$.

2.8 The Archimedean and Fineness Properties

Problem 129 (The Archimedean Property for Ratios). *For any* ρ, σ *there is* n *such that* $n\rho > \sigma$.

Problem 130. *For any* ρ, *there exists* $\sigma > \rho$, *and also there exists* $\tau < \rho$.

Problem 131 (Density). *If* $\rho < \sigma$, *then there is* τ *such that:*

$$\rho < \tau < \sigma.$$

The last two results express the fact that *the ratios form a dense linear order without end points.*

Definition 132. We say that the pair L, U is a *Dedekind partition* of the ratios if L and U are nonempty sets forming a partition of the ratios such that every ratio in L is smaller than every ratio in U, that is, $\rho < \sigma$ for all $\rho \in L$ and $\sigma \in U$.

For example, if $L := \{\rho \mid \rho < 1\}$ and $U := \{\rho \mid \rho \geq 1\}$, then L, U forms a Dedekind partition of the ratios.

Problem 133. *If* L, U *is a Dedekind partition of the ratios, then* L *is "downward closed under* $<$*" meaning that if* $\rho \in L$ *and* $\rho' < \rho$ *then* $\rho' \in L$, *and similarly* U *is upward closed under* $>$.

The following property, which we call the *Fineness Property* for ratios, is closely related to the Archimedean property.[4] It will be used in the next section and in the next chapter when we study Dedekind partitions in detail.

Theorem 134 (Fineness Property for Ratios). *If* L, U *is a Dedekind partition of the ratios, then for any* ϵ *there are* $\rho \in L$ *and* $\sigma \in U$ *such that* $\sigma - \rho < \epsilon$, *that is,* $\sigma < \rho + \epsilon$.

---

[4]The notion can be defined for (the positive elements of) any ordered field, where it will hold if and only if the field is Archimedean. A Dedekind partition L, U satisfying the condition of Theorem 134 is sometimes called a *Scott cut.*

Remark. Like Theorem 68, this is a result of *second order arithmetic*, since, unlike most other results of this chapter, it quantifies over *sets* of ratios.

Proof. Let ϵ be given. Fix $\alpha \in L$ and $\beta \in U$. By the Archimedean property fix a natural number n with $n > 1/\alpha$ and also $n > 1/\epsilon$. Then $1/n < \alpha$, and so $1/n \in L$. Also $1/n < \epsilon$. There is k such that $k/n > \beta$ (by the Archimedean property again), and so there is k with $k/n \in U$, hence by the Well-Ordering property we can fix the least natural number m such that $m/n \in U$. Then $m \neq 1$ since $1/n \notin U$. Hence $m = p + 1$ for some p. Put $\rho = p/n$ and $\sigma = m/n$. Then $\sigma \in U$ and since m is the least natural number for which $m/n \in U$, so $\rho = p/n \in L$. Finally, $\sigma = \rho + 1/n < \rho + \epsilon$. □

2.9 Irrationality of $\sqrt{2}$ and Density of Square Ratios

Definition 135. We write ρ^2 for $\rho\rho$. A ratio σ is said to be a *square ratio* if $\sigma = \rho^2$ for some ρ. A ratio σ is said to be a *nonsquare ratio* if it is not a square ratio, i.e., if there is no ρ such that $\sigma = \rho^2$.

For example, 1 is a square ratio since $1 = 1^2$, but 2 is a nonsquare ratio by Theorem 83:

Problem 136. *There is no ρ such that $\rho^2 = 2$.*

Problem 137. *$\rho < \sigma$ if and only if $\rho^2 < \sigma^2$.*

The following says that the square ratios are "dense" in the set of all ratios:

Theorem 138 (Density of Square Ratios). *Given $\rho < \sigma$, there is β such that $\rho < \beta^2 < \sigma$.*

Proof. Let $\rho < \sigma$, and put $L := \{\gamma \mid \gamma^2 \leq \rho\}$ and $U := \{\gamma \mid \gamma^2 > \rho\}$. Then L, U is a Dedekind partition by Problem 137, so by the fineness property we can fix $\alpha \in L$ and $\beta \in U$ with $\beta - \alpha < (\sigma - \rho)/(2(\sigma + 1))$. We can assume $\beta < \sigma + 1$ (since otherwise we could have replaced β by $\sigma + 1/2$), and so $\beta + \alpha < 2\beta < 2(\sigma + 1)$. Hence by Problem 120:

$$\rho < \beta^2 = \alpha^2 + (\beta - \alpha)(\beta + \alpha) < \rho + (\beta - \alpha)(2(\sigma + 1)) < \rho + (\sigma - \rho) = \sigma.$$

□

Corollary 139. *If $\rho^2 < 2$, then there is $\sigma > \rho$ with $\rho^2 < \sigma^2 < 2$. Similarly, if $\rho^2 > 2$, then there is $\sigma < \rho$ with $\rho^2 > \sigma^2 > 2$.*

Corollary 140. *$L := \{\rho \mid \rho^2 < 2\}$ and $U := \{\rho \mid \rho^2 > 2\}$ form a Dedekind partition of the ratios with L having no largest element and U having no smallest element.*

In the last corollaries, we could obviously replace 2 by any nonsquare ratio.

Like the square ratios, the nonsquare ratios are also dense in the set of all ratios:

Corollary 141. *Given $\rho < \sigma$, there is some nonsquare ratio τ such that $\rho < \tau < \sigma$.*

Proof. Let $\rho < \sigma$. Then $\rho/2 < \sigma/2$, so $\rho/2 < \beta^2 < \sigma/2$, or $\rho < 2\beta^2 < \sigma$ for some β. But $2\beta^2$ is a nonsquare ratio, as otherwise $(2\beta^2)/\beta^2 = 2$ would be a square ratio. □

Remark. Although $\sqrt{2}$ does not exist (as a ratio), we do have arbitrarily close approximations to it both from below and from above: Given any ϵ, we can apply the fineness property to the Dedekind partition $L := \{\rho \mid \rho^2 < 2\}$ and $U := \{\rho \mid \rho^2 > 2\}$ to get $\rho \in L$ and $\sigma \in U$ with $\sigma - \rho < \epsilon$. Since we expect $\sqrt{2}$ (whatever it may be) to lie between ρ and σ, we can regard ρ and σ as approximations differing from the target $\sqrt{2}$ by an amount less than ϵ.

Problems Using Concepts from Abstract Algebra

The following problems are meant for students with prior exposure to abstract algebra.

Problem 142. *The ratios form an abelian group under multiplication.*

Problem 143. *Generalize the fineness property for the positive elements of an ordered field. Then show that the positive elements of an ordered field has the fineness property if and only if the field is Archimedean.*

Our method of going from the natural numbers to the ratios is a basic method in algebra in which one embeds a given commutative cancellative semigroup A into a group constructed from a pairs of elements of A and forming a quotient. The semigroup we started with was $\mathbf{N}$ with the operation of multiplication, but addition could have been incorporated as well.

Problem 144. *Construct the integral domain $\mathbf{Z}$ of signed integers from $\mathbf{N}\times\mathbf{N}$, where $\langle m, n\rangle$ is identified with $\langle p, q\rangle$ if and only if $m + q = n + p$, by defining addition and multiplication appropriately.*

The following result of Dedekind shows that the Dedekind–Peano axioms characterize the natural numbers with the successor function up to isomorphism:

Problem 145 (Dedekind). *If $\overline{S}\colon \overline{N} \to \overline{N}$ with $\overline{1} \in \overline{N}$, $\underline{S}\colon \underline{N} \to \underline{N}$ with $\underline{1} \in \underline{N}$, and if both structures satisfy the Dedekind–Peano axioms, then there is a unique bijection h from $\overline{N}$ onto $\underline{N}$ which preserves 1 and the successor functions, that is such that $h(\overline{1}) = \underline{1}$ and $h(\overline{S}(n)) = \underline{S}(h(n))$ for all $n \in \overline{N}$.*

2.10 Recursive Definitions*

Recall that we had "defined" addition of natural numbers by the following *recursion equations*:

$$m + 1 := S(m), \qquad \text{and} \qquad m + S(n) := S(m + n).$$

But this is not an explicit definition! We took it for granted (as was done in the work of Peano) that a two-place function $+$ (the mapping $(m, n) \mapsto m + n$) satisfying the above equations exists, without giving any rigorous justification for its existence. Similarly, multiplication of natural numbers was "defined" by recursion equations without proper justification.

Dedekind introduced a general method, known as *primitive recursion*, which provides such justification. It assures the *existence and uniqueness* of functions which are defined implicitly using recursion equations having forms similar to the ones for addition and multiplication.

We will formulate and prove a general version of Dedekind's principle of recursive definition, from which the existence and uniqueness for the addition and multiplication functions can be immediately derived.

Principles of Recursive Definition

The following *Basic Principle of Recursive Definition* is perhaps the simplest yet very useful result for defining functions recursively.

Theorem 146 (Basic Principle of Recursive Definition). *If Y is a set, $a \in Y$, and $h: Y \to Y$, then there is a unique $f: \mathbf{N} \to Y$ such that*

$$f(1) = a, \qquad \textit{and} \qquad f(n + 1) = h(f(n)) \textit{ for all } n \in \mathbf{N}.$$

Informally, this says that given $a \in Y$ and $h: Y \to Y$, we can form the infinite sequence $\langle a, h(a), h(h(a)), \ldots \rangle$.

Proof. First note that the uniqueness of the function f can be established by an easy and routine induction, so let us prove existence.

Let $I_n := \{1, 2, \ldots, n\} = \{k \in \mathbf{N} \mid 1 \leq k \leq n\}$ denote the set of first n natural numbers. The proof uses *functions $u: I_n \to Y$ having domain I_n*, i.e., finite sequences from Y of length n (with varying n).

Let us say that a function u *is partially h-recursive with domain I_n* if $u: I_n \to Y$, $u(1) = a$, and $u(k + 1) = h(u(k))$ for all k with $1 \leq k < n$.

We first prove by induction that for every $n \in \mathbf{N}$ there is a unique partially h-recursive u with domain I_n.

Basis step ($n = 1$): Let $v: \{1\} \to Y$ be defined by setting $v(1) = a$. Then v is partially h-recursive with domain I_1. Moreover, if $u, u': I_1 \to Y$ are partially h-recursive functions with domain I_1, then $u(1) = a = u'(1)$, so $u = u'$ since 1 is the only element in their domain $I_1 = \{1\}$. So there is a unique partially h-recursive v with domain I_1, establishing the basis step.

Induction step: Suppose that $n \in \mathbf{N}$ is such that there is a unique partially h-recursive v with domain I_n (induction hypothesis). We fix this v for the rest of this step, and define $w: I_{n+1} \to Y$ by setting $w(k) := v(k)$ for $k \leq n$ and $w(k) := h(v(n))$ if $k = n+1$. Then w is easily seen to be partially h-recursive with domain I_{n+1}. Moreover, if $u, u': I_{n+1} \to Y$ are partially h-recursive with domain I_{n+1}, then the restrictions $u \restriction_{I_n}$ and $u' \restriction_{I_n}$ are partially h-recursive with domain I_n, so they must be identical by induction hypothesis, i.e., $u(k) = u'(k)$ for $1 \leq k \leq n$. In particular, $u(n) = u'(n)$, so $u(n+1) = h(u(n)) = h(u'(n)) = u'(n+1)$, which gives $u = u'$. Thus there is a unique partially h-recursive w with domain I_{n+1}, which finishes the induction step.

Thus for each n there is a unique partially h-recursive function with domain I_n; let us denote this function by u_n.

Now define $f: \mathbf{N} \to Y$ by setting:

$$f(n) := u_n(n).$$

First, $f(1) = a$ since $u_1(1) = a$. Next, the restriction of u_{n+1} to I_n equals u_n (by uniqueness, since the restriction is partially h-recursive), so $u_{n+1}(n) = u_n(n)$. Hence $f(n+1) = u_{n+1}(n+1) = h(u_{n+1}(n)) = h(u_n(n)) = h(f(n))$. Thus f satisfies the recursion equations of the theorem. □

To handle functions of multiple variables, the following theorem is used.

Theorem 147 (General Principle of Recursive Definition). *For any $g: X \to Y$ and $h: X \times \mathbf{N} \times Y \to Y$, there is a unique function $f: X \times \mathbf{N} \to Y$ such that for all $x \in X$ and $n \in \mathbf{N}$:*

$$f(x, 1) = g(x) \quad \textit{and} \quad f(x, n+1) = h(x, n, f(x, n)).$$

Here f is being defined by recursion on the second variable n, that is, n is the *variable of recursion* ranging over $\mathbf{N}$, while x is a *parameter* ranging over the set X. This is the most general form of recursive definition, where both the parameters (in X) and the values (in Y) come from arbitrary sets.

Proof. The proof is essentially the same as that of Theorem 146, since the additional parameter does not play any significant role in the recursion. The details are left as an exercise for the reader. □

Theorem 148 (Course of Values Recursion). *Let Y be a nonempty set and Y^* denote the set of all finite sequences (strings) of elements from Y. Given any $G: Y^* \to Y$ there is a unique $f: \mathbf{N} \to Y$ such that*

$$f(n) = G(\langle f(k) \mid k < n \rangle) \text{ for all } n \in \mathbf{N}.$$

Denoting the empty string by ε, this means that $f(1) = G(\varepsilon)$, $f(2) = G(\langle f(1) \rangle)$, $f(3) = G(\langle f(1), f(2) \rangle)$, etc.

Proof. Let $h_G: Y^* \to Y^*$ be the function defined by

$$h_G(u) := u^\frown G(u), \quad \text{i.e.} \quad h_G(\langle u_1, \ldots, u_n \rangle) := \langle u_1, \ldots, u_n, G(u) \rangle .$$

Here $u^\frown y = u * \langle y \rangle$ denotes the string obtained from the string u by appending the element $y \in Y$, so that $\text{len}(u^\frown y) = \text{len}(u) + 1$. The Basic Principle of Recursive Definition (Theorem 146) gives a unique function $\phi: \mathbf{N} \to Y^*$ with

$$\phi(1) = h_G(\varepsilon), \qquad \text{and} \qquad \phi(n+1) = h_G(\phi(n)) \quad \text{for all } n \in \mathbf{N}.$$

Now note that $\phi(n)$ is a finite sequence of length n for every n, and put $f(n) := \phi(n)(n) =$ the last coordinate of the finite sequence $\phi(n)$. □

The form of recursion in the above theorem generalizes to transfinite ordinals, where it is called *transfinite recursion* (see Theorem 622 and Theorem 650).

Primitive Recursion

We start with a special case, which is an immediate corollary of Theorem 147.

Theorem 149 (Primitive Recursion for Two-Place Functions). *Given a one-variable function $g: \mathbf{N} \to \mathbf{N}$ and a three-variable function $h: \mathbf{N}^3 \to \mathbf{N}$, there is a unique two variable function $f: \mathbf{N} \to \mathbf{N}$ such that for all $m, n \in \mathbf{N}$:*

$$f(m, 1) = g(m), \quad \textit{and} \quad f(m, S(n)) = h(m, n, f(m, n)).$$

The result of this theorem is often expressed by saying that *the function f is obtained from the function g and h by primitive recursion.*

Proof. This is simply Theorem 147 with $X = Y = \mathbf{N}$. □

We can now give a full justification for our original recursive definition of addition, by showing that the two-place function $+$ can be obtained from the successor function by primitive recursion as follows:

Let $g = S$ be the successor function, and let h be the function defined by $h(m, n, p) = S(p)$. Applying the last theorem with these g and h gives a two-place function f satisfying

$$f(m, 1) = S(m), \qquad \text{and} \qquad f(m, S(n)) = S(f(m, n)).$$

But these are the same as our original recursion equations for defining addition, as is easily verified by writing $m + n$ for $f(m,n)$:

$$m + 1 = S(m), \qquad \text{and} \qquad m + S(n) = S(m + n).$$

Once we have justified the addition function, we can use it to obtain the multiplication function ($\langle m, n\rangle \mapsto mn$) by primitive recursion.

Problem 150. *Prove that the multiplication function can be obtained from the identity function and the addition function using primitive recursion, verifying that it gives our original recursion equations for defining multiplication.*

The most general version of the primitive recursion principle, which is again an immediate corollary of Theorem 147, is formulated as follows:

Theorem 151 (The General Principle of Primitive Recursion). *Given a $(k-1)$-place function g and a $(k+1)$-place function h on* **N**, *there is a unique k-place function f on* **N** *such that for all $x_1, x_2, \ldots, x_k \in \mathbf{N}$:*

$$f(x_1, \ldots, x_{k-1}, 1) = g(x_1, \ldots, x_{k-1}), \quad \textit{and}$$
$$f(x_1, \ldots, x_{k-1}, S(x_k)) = h(x_1, \ldots, x_k, f(x_1, \ldots, x_k)).$$

Proof. This is Theorem 147 with $X = \mathbf{N}^{k-1}$ and $Y = \mathbf{N}$. □

As before, the function f in the above theorem is said to be *defined by primitive recursion from g and h.*

After obtaining the addition and multiplication functions, one can keep applying primitive recursion repeatedly to define more and more functions on **N**. Essentially all commonly used functions, such as exponentiation, the factorial function, the gcd function, and so on, can be obtained via primitive recursion.

Problem 152. *Define the factorial function (one-place) as well as the exponentiation function (two-place) from the multiplication function using primitive recursion.*

Problem 153. *What familiar single-variable function is defined using the following primitive recursion equations?*

$$f(1) = 1 \quad \textit{and} \quad f(S(m)) = h(m, f(m)), \qquad \textit{where } h(m,n) := nS(m).$$

Remark. Principles of primitive recursion, such as Theorem 151, are results of *second order arithmetic* which involve quantification over *functions* of natural numbers: Functions are defined *implicitly* by assertions of the form "there is a unique function satisfying such and such recursion equations." This is unavoidable in the Dedekind–Peano system $\langle \mathbf{N}, 1, S\rangle$. However, if $+$ and $\cdot$ are also added as primitives to obtain the extended system $\langle \mathbf{N}, 1, S, +, \cdot\rangle$, then primitive recursion is no longer necessary and functions such as exponentiation can be explicitly defined.

The reason for this is that $+$ and $\cdot$ have sufficient power to express the notion of *finite sequences* (represented in a coded form as natural numbers), and so one can essentially replicate the process given in the proof of Theorem 146 to produce explicit definitions.

Chapter 3
Dedekind's Theory of the Continuum

Abstract This chapter constructs the real numbers from the rational numbers using the method of Dedekind cuts and discusses properties of general linear continuums, such as the Intermediate Value Theorem, in the process.

3.1 Introduction

Modern Set Theory was born in late nineteenth century primarily due to the work of Richard Dedekind and Georg Cantor. Among many remarkable things, they independently found two distinct methods for rigorously constructing the real numbers from the ratios or rational numbers. Here we will follow Dedekind's method,[1] whose central idea is the geometric intuition of a *linearly ordered continuum*.[2] The size of a continuum proved to be a difficult problem, and it dominated a large part of twentieth century set theory.

3.2 Linear Continuum in Geometry

The idea of a linear continuum is embodied in geometric notions such as a *line*, *segment*, or a *ray*. The points of a ray are ordered naturally if we declare that for points P and Q on a ray that P *precedes* Q (symbolically $P < Q$) if and only if P is between the initial point of the ray and Q (using "betweenness" as a primitive notion), as in:

[1]See Stoll [76] or Suppes [77] for Cantor's method based on *Cauchy sequences of rationals*. See also the remarks on Cantor's method at the end of this chapter.

[2]Also known as a *linear continuum*, or an *ordered continuum*, or simply a *continuum*.

A. Dasgupta, *Set Theory: With an Introduction to Real Point Sets*,

DOI 10.1007/978-1-4614-8854-5_3,

Classical axioms of geometry ensure that the relation expressed by "P precedes Q" on the points of the ray satisfies two properties as follows.

1. **Axiom of Order.** *The relation "$P < Q$" on the points of a ray is a transitive relation satisfying the law of trichotomy, i.e. $<$ is an ordering of the points of the ray.*
2. **Axiom of Order-Density.** *If $P < Q$ then there is R such that $P < R < Q$.*

These axioms are two *necessary* conditions for a linear continuum, but mathematicians since Pythagoras knew (see Sect. 3.3) that they are not sufficient to ensure a linear continuum. Until Dedekind, however, it was not clear what exactly is needed to capture the intuitive notion of a linear continuum.

Analytic Geometry: Modeling the Ray by Ratios

Analytic Geometry uses a correspondence between points and numbers (or n-tuples of numbers for n-dimensions, called coordinates) to transform geometric problems into problems of algebra and analysis (and back). If the points of a line or a ray correspond to a system of numbers, then the system of numbers in question must satisfy the two axioms above.

Note that the ratios ordered by magnitude satisfy the two axioms above. Moreover, ratios are used to measure lengths of line segments and for all practical purposes suffice in this role. Thus one can think of a correspondence between the points of an open ray and the ratios, i.e., use the ratios as the system of numbers to assign "coordinates" to the points of the ray. This is done in such a way that if two points P and Q on the ray correspond to the ratios ρ and σ, respectively, then (a) P precedes Q if and only if $\rho < \sigma$, and moreover (b) if $\sigma = \rho + \lambda$, then the length[3] of the segment $\overline{PQ}$ equals the ratio λ.

3.3 Problems with the Ratios

Even though the ratios are sufficient for all direct measurements of lengths in practice and even though the ratios appear to provide a system of numbers adequate

[3]This implies the important additivity property for lengths of segments: If P, Q, R are points on a line with Q between P and R then $\text{Len}(\overline{PR}) = \text{Len}(\overline{PQ}) + \text{Len}(\overline{QR})$. However, this does not imply that geometric line segments are *a priori* associated to lengths in an invariant fashion. The fact that physical line segments have lengths (rigidity) is essentially empirical. See Carnap [7], *An Introduction to the Philosophy of Science,* especially Chaps. 6–9, for more details.

for analytic geometry, problems arise in their theoretical use in geometry and algebra—problems which indicate that the system of ratios is inadequate for its purpose.

Ratios Are Inadequate for Measuring Lengths

A famous Pythagorean result states that the hypotenuse of a right-angled isosceles triangle is *incommensurable* relative to its legs:

Problem 154. *Use the Pythagorean theorem to show that if the length of each of the legs of a right-angled isosceles triangle is measured by the ratio ρ, then there is no ratio σ measuring the length of the hypotenuse.*

In particular if each of the legs has a length of 1, then there is no ratio which gives the exact length of the hypotenuse, even though ratios can approximate the length of the hypotenuse with arbitrarily small errors.

Ratios Are Inadequate for Analytic Geometry

A consequence of the above problem of measuring lengths is that if we use ratios as the system of coordinates for analytic geometry, it can sometimes fail to represent points of intersection. For example, it is a theorem of geometry that a ray originating at the center of any circle must intersect it at a unique common point. However, in the picture shown below,

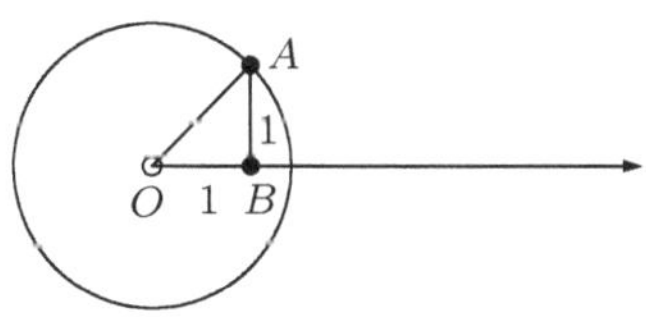

if both legs AB and OB of the right triangle OAB have length 1, then the point of the ray $\overrightarrow{OB}$ intersected by the circle is not represented by any ratio.

Ratios Are Inadequate for Solving Algebraic Equations: Failure of the Intermediate Value Theorem

We have earlier seen that the equation $x^2 = 2$ has no rational solution, even though approximate solutions can be found in the ratios with arbitrarily small errors. More general algebraic equations face similar problems, due to the lack of a standard tool called the *Intermediate Value Theorem* (IVT), which guarantees existence of roots in the case of real numbers.

To discuss the Intermediate Value Theorem, we need the notion of a *continuous function* defined on an order. The reader may already be familiar with continuous functions as encountered in elementary calculus. Roughly speaking, F is continuous, or equivalently the value $F(x)$ depends continuously on x, if "small changes in x produce small changes in $F(x)$." This vague description will be made precise in the context of general orders, but for simplicity, we will restrict our attention to orders without endpoints.

If a is in the domain of the function so that $F(a)$ is the value of the function at the input a, then continuity of the function at a amounts to the following: If we desire that the values of the function should not differ from $F(a)$ by more than a certain small amount, say by requiring that the values of $F(x)$ remain above a value p and below a value q where $p < F(a) < q$, then we can always find an interval in the domain containing the point a, say (r, s) with $r < a < s$, such that throughout this interval (r, s) the values of the function remain within the prescribed limits—i.e., we have $p < F(x) < q$ for $r < x < s$. This leads to the following precise definition:

Definition 155 (Continuous Function). Let X be an order without first or last elements. $F: X \to X$ is called *continuous* if for any $\omega \in X$ and any $\eta, \zeta \in X$ with $\eta < F(\omega) < \zeta$, there are $\alpha, \beta \in X$ with $\alpha < \omega < \beta$ such that for all $\xi \in X$, $\alpha < \xi < \beta \Rightarrow \eta < F(\xi) < \zeta$.

Problem 156. *Show that the squaring function $\rho \mapsto \rho^2$ defined on the set of ratios is continuous.*

[Hint: Use Problem 137 and Theorem 138.]

Definition 157 (Intermediate Value Theorem, or IVT). An order X (without first or last elements) is said to satisfy the *IVT* (Intermediate Value Theorem) if whenever $F: X \to X$ is continuous, $\alpha < \beta$, and γ lies strictly between $F(\alpha)$ and $F(\beta)$ (i.e., either $F(\alpha) < \gamma < F(\beta)$ or $F(\beta) < \gamma < F(\alpha)$), then $F(\xi) = \gamma$ for some ξ with $\alpha < \xi < \beta$.

Problem 158. *Show that the set of ratios with their usual ordering does not satisfy the IVT.*

The IVT is the main tool for formalizing and establishing results which claim, roughly, that if two continuous curves "cross," then they must have at least one common point of intersection. It is a workhorse for guaranteeing existence of roots in many algebraic equations.

3.4 Irrationals: Dedekind's Definition of the Continuum

While the geometric axioms of order and order-density (and practical use of ratios in measurement) appear to provide an adequate correspondence between the points of a ray and the ratios, Dedekind realized that we assume a more fundamental underlying continuity property when we believe that "two continuous crossing curves must intersect," or even that a ray originating at the center of a circle must intersect it at a unique common point.

Dedekind Cuts

Dedekind's method of isolating this continuity property is to partition or "cut" a given ordering into two nonempty pieces with one piece completely preceding the other. Formally:

Definition 159. A *Dedekind cut* in an ordering X is a partition of X consisting two nonempty disjoint sets L and U such that $x \in L, y \in U \Rightarrow x < y$, i.e., every member of L precedes every member of U, as pictured below.

In other words, all elements of U are upper bounds for L, all elements of L are lower bounds for U, as well as $L \neq \emptyset \neq U$, $L \cap U = \emptyset$, and $L \cup U = X$.

If L, U is a Dedekind cut, exactly one of the following four possibilities hold:

1. Both L has a largest element and U has a smallest element. In this case we call the cut a *Dedekind jump*, or simply a *jump*.
2. L does not have a largest element but U has a smallest element. In this case the smallest element of U is viewed as a "limit" of the elements of L, and is called the (unique) boundary of thc cut.
3. L has a largest element but U does not have a smallest element. In this case the largest element of L is viewed as a "limit" of the elements of U and is called the (unique) boundary of the cut.
4. Neither L has a largest element nor U has a smallest element. In this case we call the cut a *Dedekind gap*, or simply a *gap*.

A cut as in case (2) or case (3) is called a cut with a *unique limiting boundary*, or simply a *boundary cut*.

We now briefly discuss each type of cut.

Jump

Case (1). This case, the possibility of a Dedekind jump, is ruled out in the presence of order-density. In fact, order-density is equivalent to being "jumpless," i.e., an ordering is order-dense if and only if no Dedekind cut for it is a jump. The ratios, e.g., are order-dense and no Dedekind cut over them will be a jump. We will therefore not consider this case anymore, and assume that all orderings in this chapter will be order-dense and hence will have no jumps.

Boundary Cut

Cases (2) or (3). If σ is any ratio, put

$$L_\sigma = \{\rho \,|\, \rho < \sigma\}, \;\; U_\sigma = \{\rho \,|\, \rho \geq \sigma\}; \quad L'_\sigma = \{\rho \,|\, \rho \leq \sigma\}, \;\; U'_\sigma = \{\rho \,|\, \rho > \sigma\}.$$

Then L_σ, U_σ form a Dedekind cut over the ratios, and so does L'_σ, U'_σ. The cuts L_σ, U_σ and L'_σ, U'_σ are essentially equivalent, since both correspond to the ratio σ: For both these cuts the ratio σ is the *boundary* of the cut.

Gap

Case (4). For the ratios ordered by magnitude, put

$$L = \{\rho \,|\, \rho^2 < 2\}, \qquad \text{and} \qquad U = \{\rho \,|\, \rho^2 > 2\}.$$

This is a Dedekind cut over the ratios which is a gap. This follows from the results of the previous chapter; recall the density of square ratios.

Another example of a gap is provided if we remove a single fixed point on an open ray: The remaining set of points on the ray breaks apart into two pieces L and U, forming a gap.

Thus *for an order-dense ordering, any Dedekind cut is either a boundary cut with a unique boundary (cases (2) or (3)), or is a gap (case (4)).*

Problem 160 (Density of Gaps). *Prove that for the ratios ordered by magnitude, if $\rho < \sigma$ then there is a gap between ρ and σ, i.e., there is a Dedekind cut L, U for the ratios such that (a) $\rho \in L$, (b) $\sigma \in U$, and (c) L, U is a gap (L has no maximum and U has no minimum).*

Dedekind's Definition of a Linear Continuum

As we saw earlier, Dedekind found that the root cause behind the inadequacy of the ratios lies in the fact that the ratios have lots of gaps.

The hypotenuse of an isosceles right triangle is incommensurable relative to its legs because there is no ratio whose square is two, causing the gap $L = \{\rho \mid \rho^2 < 2\}$ and $U = \{\rho \mid \rho^2 > 2\}$.

The point of intersection between two geometric curves which appear to cross may lack coordinate representation by ratios because the "location of the crossing" may correspond to a gap in the ratios.

The IVT for ratios fails for the same reason.

For geometry, we thus postulate that in addition to the axioms of order and order-density, the points of a ray or a line must satisfy:

> **Axiom of Continuity.** *The ordered set of points of a ray has no gaps. In other words, if all the points of the ray is partitioned into two disjoint nonempty sets L and U with all points of L preceding all the points of U, then the Dedekind cut L, U is a boundary cut, i.e., there is a point of the ray which is the boundary of the cut.*

Finally, we have the definition for a linear continuum.

Definition 161 (Dedekind). An ordering is *order-complete* (or simply *complete*) if it has no Dedekind gaps. A *linear continuum* is an ordering with at least two points which is order-dense and order-complete.[4]

Thus an ordering is a linear continuum if and only if (a) it has at least two points, and (b) every Dedekind cut is a boundary cut.

Anybody familiar with "limits and continuity of real functions" as studied in elementary calculus will recall examples of *removable discontinuity*, as in the function $x \mapsto (x^2 - 1)/(x - 1)$ at $x = 1$.

One of Dedekind's simple but fundamental intuition was this: For an order to be a continuum, an order must be "continuous," and so must not have such discontinuities. By not allowing gaps, Dedekind's definition of continuum precisely avoids discontinuities of this type and achieves continuity.[5]

Problem 162. *Prove that an ordering without first or last elements forms a linear continuum if and only if it satisfies the IVT.*

[Hint: Given $a < b$ in a linear continuum and $f(a) < c < f(b)$ with f continuous, let L be the set of all x such that $x < y$ for some $y \in [a, b]$ with $f(y) < c$, and let U be the complement of L. Then L, U is a Dedekind cut, $a \in L$, $b \in U$, and if $z = \max L$ or $z = \min U$ then $f(z) = c$.]

Once we postulate the axiom of continuity for the geometric ray, we see that the ratios are unable to label all the points of the ray, and infinitely many points on the ray (the "irrational points" on it) do not get labeled by any ratio at all. (This is

---

[4]Order completeness will be studied in a more general setting in Sect. 8.2, where we will see that completeness is equivalent to the *least upper bound property* (see Problem 518).

[5]Following Dedekind, all rigorous axiomatizations of geometry, first by Hilbert and later by Tarski and Birkhoff, postulate this property as the *Axiom of Continuity*.

because the ratios, ordered by magnitude, have an abundance of Dedekind gaps.) To see this, suppose that $\sigma \leftrightarrow P_\sigma$ is a correspondence between the ratios and certain points of the ray preserving order, so that $\sigma < \rho$ if and only if P_σ precedes P_ρ. We then say that the point P_σ is labeled by or represented by the ratio σ. Now if P_{σ_0} precedes P_{ρ_0} then $\sigma_0 < \rho_0$, so there is a Dedekind gap L, U of the ratios with $\sigma_0 \in L$ and $\rho_0 \in U$. Let A be the set of points on the ray which precede P_σ for some $\sigma \in L$, and let B be the set of points of the ray which are preceded by P_ρ for some $\rho \in U$. Since L, U form a Dedekind gap of the ratios and since $\sigma \leftrightarrow P_\sigma$ is an order-preserving correspondence, A has no last point and B has no first point. Hence by the Axiom of Continuity there must be a point R on the ray which is neither in A nor in B, and so R is not represented by any ratio.

We can thus divide the points on the ray into two disjoint sets: (a) Each point on the ray that does not correspond to any ratio is called an *irrational point*, while (b) the points that correspond to ratios are called the *rational points*. Furthermore, the rational and irrational points on the ray are intermixed in a dense fashion: Between any two points on the ray, there are an infinite number of rational points and also an infinite number of irrational points.

If a system of numbers has to serve as an adequate system of coordinates for analytic geometry, then they will need to be in one-to-one correspondence with all the points of the ray in an order-preserving way, and so they must satisfy the axiom of continuity as well. While each rational point on the ray is represented by a ratio, for each irrational point on the ray we are missing the "irrational number" to represent it, as the ratios as a number system is full of gaps. We thus look for "irrational numbers" to fill all these gaps—the removable discontinuities—to extend the system of ratios to a number system satisfying the axiom of continuity and adequate for analytic geometry.

It is now crucial to notice that the points of the ray correspond to *Dedekind cuts over the ratios*: The rational points on the ray correspond to *Dedekind cuts with boundary,* while each irrational point on the ray corresponds to a *Dedekind gap* over the ratios. This natural one-to-one correspondence between the irrational points on the ray and the gaps over the ratios led Dedekind to define an irrational number simply as a Dedekind cut over the ratios which is a gap.

Our construction will be a slight variant of Dedekind's original one. First, notice that a Dedekind cut L, U over the ratios get determined by the lower set L alone, since U can be found from L by taking its complement. Thus instead of a pair L, U, we will only use L. Second, the two "rational" cuts with the same boundary σ mentioned above are essentially equivalent; we will use the cut where the lower set L has no maximum. Thus our "numbers" will be just the lower parts L of Dedekind cuts L, U where L has no maximum (U may or may not have a minimum).

3.5 Lengths (Magnitudes)

We now define "length" or "magnitude" in a way so that the length of a line segment will consist of all ratios representing lengths shorter than the given line segment.

Definition 163 (Length). We say that Γ is a *length* if and only if

1. Γ is a set of ratios.
2. Γ contains at least one ratio but does not contain all ratios.
3. $\rho \in \Gamma$ and $\sigma < \rho \Rightarrow \sigma \in \Gamma$. (So far, the conditions say that Γ forms the lower piece of a Dedekind cut over the ratios.)
4. Γ does not contain a largest ratio, i.e., $\rho \in \Gamma \Rightarrow \exists \sigma \in \Gamma\ (\sigma > \rho)$.

Lengths will in general be denoted by uppercase non-Roman Greek letters such as $\Gamma, \Delta, \Phi, \Psi, \Lambda, \Xi, \Theta, \Upsilon$, and Ω.

Definition 164. For any length Γ, we define $\sim\Gamma := \{\rho \mid \rho \notin \Gamma\}$.

Problem 165. *The pair* $\Gamma, \sim\Gamma$ *forms a Dedekind cut over the ratios.*

Definition 166. Given a ratio ρ, we define ρ^* as

$$\rho^* := \{\sigma \mid \sigma < \rho\}.$$

Problem 167. *For any* ρ, ρ^* *is a length.*

Definition 168. A length Γ is *rational* if and only if $\Gamma = \rho^*$ for some ρ. Otherwise, Γ is *irrational*.

Problem 169. *A length* Γ *is rational if and only if* $\Gamma, \sim\Gamma$ *is a boundary cut; and* Γ *is irrational if and only if* $\Gamma, \sim\Gamma$ *is a gap.*

Problem 170. *There are rational and irrational lengths.*

Definition 171. For lengths Γ, Δ, we write

$$\Gamma < \Delta \quad \text{if and only if} \quad \Gamma \subseteq \Delta \text{ and } \Gamma \neq \Delta.$$

Problem 172. *The relation* $<$ *is an ordering of the set of all lengths.*

Problem 173 (Density of Rational and Irrational Lengths). *If* $\Gamma < \Delta$, *then (a) there exists a rational* Φ *such that* $\Gamma < \Phi < \Delta$, *and (b) there exists an irrational* Ξ *such that* $\Gamma < \Xi < \Delta$.

Definition 174 (Addition). $\Gamma + \Delta := \{\rho + \sigma \mid \rho \in \Gamma, \sigma \in \Delta\}$.

Theorem 175. $\Gamma + \Delta$ *is a length.*

Proof. First, we show that neither $\Gamma + \Delta$ nor its complement is empty: Fix $\gamma \in \Gamma$, $\gamma' \in\sim \Gamma$, $\delta \in \Delta$, and $\delta' \in\sim \Delta$. Then $\gamma + \delta \in \Gamma + \Delta$, but $\gamma' + \delta' \notin \Gamma + \Delta$, since $\rho \in \Gamma$ and $\sigma \in \Delta$ implies $\rho < \gamma'$ and $\sigma < \delta'$ and so $\rho + \sigma < \gamma' + \delta'$, so $\rho + \sigma \neq \gamma' + \delta'$ for all $\rho \in \Gamma$ and $\sigma \in \Delta$ so $\gamma' + \delta' \notin \Gamma + \Delta$.

Next, $\Gamma + \Delta$ is "closed under taking lower members": Let $\zeta \in \Gamma + \Delta$, so that $\zeta = \gamma + \delta$ with $\gamma \in \Gamma$ and $\delta \in \Delta$. Given $\alpha < \zeta$, put $\alpha/\zeta = \tau$. Then $\tau < 1$ and so $\gamma\tau \in \Gamma$ and $\delta\tau \in \Delta$, so $\gamma\tau + \delta\tau \in \Gamma + \Delta$, but $\gamma\tau + \delta\tau = \tau(\gamma + \delta) = \tau\zeta = \alpha$, so $\alpha \in \Gamma + \Delta$.

Finally, $\Gamma + \Delta$ has no maximum ratio: Let $\zeta \in \Gamma + \Delta$, so $\zeta = \gamma + \delta$ for some $\gamma \in \Gamma$ and $\delta \in \Delta$. There is $\gamma' \in \Gamma$ such that $\gamma' > \gamma$. Then $\gamma' + \delta \in \Gamma + \Delta$ and $\gamma' + \delta > \gamma + \delta = \zeta$. □

Problem 176. *Addition of lengths is associative and commutative.*

Theorem 177. *For any ρ and Γ, there is σ such that $\sigma \in \Gamma$ but $\sigma + \rho \notin \Gamma$.*

Proof. Given ρ and Γ, as $\Gamma, \sim \Gamma$ form a Dedekind partition, we can use the Fineness property (previous chapter) to find $\sigma \in \Gamma$ and $\tau \in \sim \Gamma$ such that $\tau < \sigma + \rho$. Hence $\sigma + \rho \in \sim \Gamma$, as $\sim \Gamma$ is "upward closed under >." □

Theorem 178. $\Gamma < \Gamma + \Delta$.

Proof. Let $\gamma \in \Gamma$. Fix $\delta \in \Delta$. Fix $\tau < \min(\gamma, \delta)$. Then $\gamma = \gamma' + \tau$ for some $\gamma' < \gamma$ and $\tau < \delta$. Then $\gamma' \in \Gamma$ and $\tau \in \Delta$, so $\gamma \in \Gamma + \Delta$. So $\Gamma \subseteq \Gamma + \Delta$. Next fix $\rho \in \Delta$, and by Theorem 177 find $\sigma \in \Gamma$ such that $\sigma + \rho \in \sim \Gamma$. Then $\sigma + \rho \in \Gamma + \Delta$ but $\sigma + \rho \notin \Gamma$. □

Problem 179. *If* $\quad \Gamma < \Delta, \quad$ *or* $\quad \Gamma = \Delta, \quad$ *or* $\quad \Gamma > \Delta, \quad$ *, then*

$$\Gamma + \Xi < \Delta + \Xi, \quad \text{or} \quad \Gamma + \Xi = \Delta + \Xi, \quad \text{or} \quad \Gamma + \Xi > \Delta + \Xi,$$

respectively, and conversely.

Theorem 180. *If $\Gamma < \Delta$, then $\Gamma + \Xi = \Delta$ for a unique Ξ.*

Proof. Let $\Xi := \{\alpha \mid (\exists \beta \in \sim \Gamma)(\beta + \alpha \in \Delta)\}$. We show that $\Gamma + \Xi = \Delta$. It is easy to see that $\Gamma + \Xi \subseteq \Delta$. Now let $\zeta \in \Delta$. If $\zeta \in \Gamma$, then $\zeta \in \Gamma + \Xi$ by Theorem 178. So assume $\zeta \in \Delta \smallsetminus \Gamma$. Pick $\tau \in \Delta$ such that $\zeta < \tau$. Let $\tau = \zeta + \rho$. By Theorem 177, find $\sigma \in \Gamma$ such that $\sigma + \rho \in \sim \Gamma$. Since $\sigma < \zeta$, so $\zeta = \sigma + \alpha$ for some α. Now put $\beta = \sigma + \rho$. Then $\beta \in \sim \Gamma$ and $\beta + \alpha = \sigma + \rho + \alpha = \sigma + \alpha + \rho = \zeta + \rho = \tau \in \Delta$, so $\alpha \in \Xi$. So $\zeta = \sigma + \alpha \in \Gamma + \Xi$. □

Definition 181 (Proper Subtraction). If $\Gamma < \Delta$, we define $\Delta \dot{-} \Gamma$ to be the unique Ξ such that $\Delta = \Gamma + \Xi$.

Definition 182 (Multiplication). $\Gamma\Delta := \{\rho\sigma \mid \rho \in \Gamma, \sigma \in \Delta\}$.

Definition 183 (Reciprocal). $\Gamma^{-1} := \{\rho \mid (\exists \sigma \in \sim \Gamma)(\rho\sigma < 1)\}$.

Problem 184. *$\Gamma\Delta$ is a length and Γ^{-1} is a length.*

Problem 185. *Multiplication (of lengths) as defined above is associative and commutative, and for any length Γ we have $\Gamma 1^* = \Gamma$ and $\Gamma\Gamma^{-1} = 1^*$. (The lengths form a multiplicative "commutative group" with unity 1^*.)*

Also, multiplication is distributive over addition.

Problem 186. *If* $\quad \Gamma < \Delta, \quad$ *or* $\quad \Gamma = \Delta, \quad$ *or* $\quad \Gamma > \Delta, \quad$ *, then*

$$\Gamma\Xi < \Delta\Xi, \quad \text{or} \quad \Gamma\Xi = \Delta\Xi, \quad \text{or} \quad \Gamma\Xi > \Delta\Xi,$$

respectively, and conversely.

Definition 187. $\sqrt{2} := \{\rho \mid \rho^2 < 2\}$.

Problem 188. *$\sqrt{2}$ is an irrational length, and $\sqrt{2}\sqrt{2} = 2^*$.*

Problem 189 (Existence of Square Roots). *Given any length Γ there is a unique length Δ such that $\Delta\Delta = \Gamma$.*

Problem 190 (Isomorphic Embedding of Ratios). *The mapping $\rho \to \rho^*$ is a bijection from the set of ratios onto the set of rational lengths which preserves order, addition, and multiplication, i.e.,*

$$\rho < \sigma \Leftrightarrow \rho^* < \sigma^*; \qquad (\rho + \sigma)^* = \rho^* + \sigma^*; \qquad (\rho\sigma)^* = \rho^*\sigma^*.$$

At this point, the ratios and the rational lengths become interchangeable since all the properties of the ratios listed in earlier sections are possessed by the rational lengths.

Therefore, we throw away the ratios[6] and use the corresponding rational lengths in their place. So only one type of numbers remain, namely the lengths, which include the "ratios" (really the rational lengths), and therefore in turn also the "natural numbers" (integral lengths), as subsets.

Definition 191 (New Meaning for Symbols for Ratios). With the old ratios thrown out, the rational length ρ^* will now be denoted simply by the letter ρ (and similarly for other Greek letters). Not only do Greek letters now exclusively stand for rational lengths, but also other symbols that were previously used to denote a ratio will now denote the corresponding rational length (e.g., 2 now stands for what was being called 2^*). Similarly, lowercase Roman letters will denote integral lengths.

This allows us to mix symbols that were previously assigned to different types, and "$n + \rho + \Gamma$" and "$n \cdot \rho \cdot \Gamma$" now become valid terms.

Problem 192 (Dedekind's Theorem for the Real Continuum). *The collection of lengths ordered by the relation $<$ forms a linear continuum containing the ratios as a subset. Thus, it is an ordering which is order-dense and order-complete (has no Dedekind gaps), and so every Dedekind cut for the lengths is a boundary cut.*

[Hint: In a Dedekind cut of the ordered collection of all lengths into two pieces, the set-theoretic union of the lengths in the left piece is itself a length.]

We now have a system of numbers (the *lengths*) which can uniquely represent every point of the geometric open ray and serve as the basis for analytic geometry. The operations of addition, multiplication, and division are possible between an arbitrary pair of these numbers. However, we are still missing "negative magnitudes," and so "the subtraction $\Gamma - \Delta$" is defined only when $\Gamma > \Delta$. In the next

---

[6] Phrase of Edmund Landau [47].

section, we extend the system of lengths to a system of "signed lengths," or the *field of real numbers*, in which the subtraction of two arbitrary real numbers produces a well-defined real number.

3.6 The Ordered Field R of Real Numbers

To get signed real numbers, we regard a pair of lengths $\langle \Gamma, \Delta \rangle$ as the "signed magnitude" $\Gamma - \Delta$. This means that the length-pairs $\langle \Gamma, \Delta \rangle$ and $\langle \Xi, \Theta \rangle$ will define the same signed real if $\Gamma + \Theta = \Xi + \Delta$, which of course results in a lot of duplication. More precisely, this condition defines an equivalence relation on the set of pairs of lengths and we could use the approach of forming the "quotient structure" by defining signed real numbers as equivalence classes. It is easy, however, to choose canonical representatives by considering those pairs in which the smaller member equals 1. Then 1 acts as a reference length and the magnitude of the signed real is determined by how much the other length of the pair exceeds 1. Thus positive reals are precisely the pairs of the form $\langle \Gamma, 1 \rangle$ with $\Gamma > 1$, and negative reals are the pairs $\langle 1, \Gamma \rangle$ with $\Gamma > 1$. Zero is defined as the pair $\langle 1, 1 \rangle$.

Definition 193 (Real Numbers). A *real number* is a pair of lengths $\langle \Gamma, \Delta \rangle$ such that $\min(\Gamma, \Delta) = 1$. The set of all real numbers is denoted by **R**.

Thus a real number is an ordered pair of lengths none of which is less than 1 and at least one of which equals 1.

Definition 194 (Zero, Negative, and Positive Reals). A real number $\langle \Gamma, \Delta \rangle$ is called *positive* if $\Gamma > 1$ (and so $\Delta = 1$), and $\langle \Gamma, \Delta \rangle$ is *negative* if $\Delta > 1$ (and so $\Gamma = 1$). Define $0 := \langle 1, 1 \rangle$.

The set of positive real numbers will be denoted by $\mathbf{R}^+$, and the set of negative real numbers will be denoted by $\mathbf{R}^-$.

Thus $\langle \Gamma, \Delta \rangle$ is positive if $\Gamma > \Delta$, is negative if $\Gamma < \Delta$, and is zero if $\Gamma = \Delta$.

Definition 195. For any pair of lengths Γ, Δ, define:

1. $\langle \Gamma, \Delta \rangle \sim \langle \Xi, \Theta \rangle$, or $\langle \Gamma, \Delta \rangle$ *is equivalent to* $\langle \Xi, \Theta \rangle$, if $\Gamma + \Theta = \Xi + \Delta$.

2. $$*(\Gamma, \Delta) = \begin{cases} \langle 1 + \Gamma \dot{-} \Delta,\ 1 \rangle & \text{if } \Gamma > \Delta, \\ \langle 1,\ 1 + \Delta \dot{-} \Gamma \rangle & \text{if } \Gamma < \Delta, \\ \langle 1, 1 \rangle = 0 & \text{if } \Gamma = \Delta. \end{cases}$$

Note that $*(\Gamma, \Delta)$ *is always a real number.* We now have:

Problem 196. *For all lengths* Γ, Δ,

1. $*(\Gamma, \Delta)$ *is the unique real number satisfying* $*(\Gamma, \Delta) \sim \langle \Gamma, \Delta \rangle$.
2. $\langle \Gamma, \Delta \rangle$ *is a real number if and only if* $*(\Gamma, \Delta) = \langle \Gamma, \Delta \rangle$.
3. $*(\Gamma, \Delta) = *(\Gamma + \Xi, \Delta + \Xi)$ *for any length* Ξ.
4. $*(\Gamma, \Gamma) = \langle 1, 1 \rangle = 0$.

Problem 197. *$\langle \Gamma, \Delta \rangle \sim \langle \Xi, \Theta \rangle$ if and only if $*(\Gamma, \Delta) = *(\Xi, \Theta)$, and so equivalence between pairs of lengths is an equivalence relation for which the mapping $\langle \Gamma, \Delta \rangle \mapsto *(\Gamma, \Delta)$ is a complete invariant.*

Definition 198 (Order, Addition, Multiplication). Given real numbers $\langle \Gamma, \Delta \rangle$ and $\langle \Xi, \Theta \rangle$, define

1. Order: $\langle \Gamma, \Delta \rangle < \langle \Xi, \Theta \rangle \Leftrightarrow \Gamma + \Theta < \Xi + \Delta$.
2. Sum: $\langle \Gamma, \Delta \rangle + \langle \Xi, \Theta \rangle := *(\Gamma + \Xi, \Delta + \Theta)$.
3. Product: $\langle \Gamma, \Delta \rangle \cdot \langle \Xi, \Theta \rangle := *(\Gamma\Xi + \Delta\Theta, \Gamma\Theta + \Delta\Xi)$.

Uppercase Roman letters A, B, C, X, Y, Z, etc. will denote real numbers.

Problem 199. *The relation $<$ defined above is an order on* **R**. *Also, $A \in$* **R** *is positive if and only if $0 < A$, and A is negative if and only if $A < 0$.*

Problem 200. *If $A, B \in$* **R** *are positive, then so are $A + B$ and $A \cdot B$.*

Problem 201 (Additive Inverse). *For each real number $A = \langle \Gamma, \Delta \rangle$, define $-A := \langle \Delta, \Gamma \rangle$. Show that for any real number A,*

1. *$-A$ is a real number.*
2. *$A + (-A) = 0$, and $-(-A) = A$.*
3. *A is positive if and only if $-A$ is negative.*
4. *$A = -A$ if and only if $A = 0$.*

Problem 202 (Isomorphic Embedding of Lengths). *For each length Γ, let $\overline{\Gamma} := \langle \Gamma + 1, 1 \rangle$. The mapping $\Gamma \mapsto \overline{\Gamma}$ is a bijection from the set of lengths onto the set* $\mathbf{R}^+$ *of positive real numbers which preserves order, addition, and multiplication, i.e.,*

$$\Gamma < \Delta \Leftrightarrow \Gamma < \Delta; \qquad \overline{\Gamma + \Delta} = \overline{\Gamma} + \overline{\Delta}; \qquad \overline{\Gamma \cdot \Delta} = \overline{\Gamma} \cdot \overline{\Delta}.$$

At this point, the lengths and the positive reals $\mathbf{R}^+$ become interchangeable since all the properties of the lengths listed earlier are possessed by the positive real numbers.

Therefore, we throw away the lengths[7] and use the corresponding positive real numbers in their place. In other words, we identify the lengths with the positive reals $\mathbf{R}^+$, and a length now means a positive real, i.e., a member of $\mathbf{R}^+$. So from now on we deal with only one type of numbers, namely the real numbers, which include the "lengths" (really the positive reals) as a subset, as well as all previously defined types such as the ratios and the natural numbers.

Definition 203. As subsets of **R**, the natural numbers will be denoted by **N**, and the ratios (positive rationals) by $\mathbf{Q}^+$. Thus we have:

$$\mathbf{N} \subset \mathbf{Q}^+ \subset \mathbf{R}^+ \subset \mathbf{R}.$$

[7] Phrase of Edmund Landau [47].

Also put $\mathbf{Z} := \mathbf{N} \cup \{0\}\{-A \mid A \in \mathbf{N}\}$ and $\mathbf{Q} := \mathbf{Q}^+ \cup \{0\}\{-A \mid A \in \mathbf{Q}^+\}$, so:

$$\mathbf{N} \subset \mathbf{Z} \subset \mathbf{Q} \subset \mathbf{R}, \qquad \mathbf{N} = \mathbf{Z} \cap \mathbf{R}^+, \qquad \mathbf{Q}^+ = \mathbf{Q} \cap \mathbf{R}^+.$$

Problem 204. $\mathbf{Q}$ *is order-dense in* $\mathbf{R}$, *i.e., if* $A < B$ *are in* $\mathbf{R}$, *then there is a* $C \in \mathbf{Q}$ *with* $A < C < B$.

[Hint: Use Problem 173.]

Since positive reals are identified with the lengths and since reciprocals have already been defined for lengths, reciprocals of arbitrary nonzero reals are defined in the following way.

Problem 205 (Multiplicative Inverse). *For a positive real* $A = \langle \Gamma, 1 \rangle$ *with* $\Gamma > 1$, *define* $A^{-1} := \overline{(\Gamma \dot{-} 1)^{-1}}$. *For* $A < 0$, *define* $A^{-1} := -((-A)^{-1})$. *If* $A = 0$, *we leave* A^{-1} *undefined. Then, for any real number* $A \neq 0$,

1. A^{-1} *is a nonzero real number, and* $A > 0 \Leftrightarrow A^{-1} > 0$.
2. $A \cdot A^{-1} = 1$ *and* $(A^{-1})^{-1} = A$.

We now have all the operations to develop the theory of real numbers, but the algebraic theory of real numbers can be derived from the properties listed in the following definition.

Definition 206 (Ordered Fields). A set with an order $<$ containing two distinct elements 0 and 1 and with two operations addition $(+)$ and multiplication $(\cdot)$ is an *ordered field* if, to each A there corresponds an element $-A$, and for $A \neq 0$ an element A^{-1}, such that for all elements A, B, C we have:

1. $A + B = B + A$ and $AB = BA$.
2. $A + (B + C) = (A + B) + C$ and $A(BC) = (AB)C$.
3. $A(B + C) = AB + AC$.
4. $A + 0 = A = A \cdot 1$.
5. $A + (-A) = 0$ and if $A \neq 0$ then $AA^{-1} = 1$.
6. $A > 0$ if and only if $-A < 0$, and $A, B > 0 \Rightarrow A + B > 0$ and $AB > 0$.

The ordered field is called *complete* if the ordering forms a linear continuum.

The following theorem is the main and central result of this chapter.

Theorem 207 (R is a Complete Ordered Field). *The set* $\mathbf{R}$, *with the ordering* $<$ *and the operations* $+$ *and* $\cdot$, *forms an ordered field in which the ordering relation* $<$ *and the operations* $+$ *and* $\cdot$ *extend the corresponding relation and operations originally defined for* $\mathbf{R}^+$, $\mathbf{Q}^+$, *and* $\mathbf{N}$.

Moreover, $\mathbf{R}$ *is a linear continuum (no Dedekind cut is a gap), and hence* $\mathbf{R}$ *satisfies the Intermediate Value Theorem (IVT).*

Problem 208. *Prove Theorem 207.*

[Hints: The algebraic properties listed in the definition of ordered field are all proved by expressing real numbers in terms of lengths and exploiting the corresponding property for lengths, making frequent use of the construct $*(\Gamma, \Delta)$. For example, the commutative property of addition is proved as:

$$\langle \Gamma, \Delta \rangle + \langle \Xi, \Theta \rangle = *(\Gamma + \Xi, \Delta + \Theta) = *(\Xi + \Gamma, \Theta + \Delta) = \langle \Xi, \Theta \rangle + \langle \Gamma, \Delta \rangle,$$

and the property $A + 0 = A$ is proved, by taking $A = \langle \Gamma, \Delta \rangle = *(\Gamma, \Delta)$, as:

$$\langle \Gamma, \Delta \rangle + 0 = \langle \Gamma, \Delta \rangle + \langle 1, 1 \rangle = *(\Gamma + 1, \Delta + 1) = *(\Gamma, \Delta) = \langle \Gamma, \Delta \rangle.$$

To show that **R** is a linear continuum, use the corresponding fact for $\mathbf{R}^+$.]

Problem 209. **R** *satisfies the* Archimedean property, *that is, for all* $x, y \in \mathbf{R}$ *if* $x > 0$ *then there is a positive integer* n *such that* $nx > y$.

A *bounded closed interval* is a set of the form $I = [a, b] = \{x \in \mathbf{R} \mid a \leq x \leq b\}$ $(a \leq b)$, whose *length* is defined as $\text{len}(I) := b - a$. A sequence $\langle I_n \mid n \in \mathbf{N} \rangle$ of intervals is said to be a *nested sequence* if $I_n \supseteq I_{n+1}$ for all $n \in \mathbf{N}$.

Theorem 210 (The Nested Interval Property). *For a nested sequence*

$$I_1 \supseteq I_2 \supseteq \cdots \supseteq I_n \supseteq \cdots$$

of nonempty bounded closed intervals in **R**, *we have* $\bigcap_{n \in \mathbf{N}} I_n \neq \emptyset$.

Proof. If $I_n = [a_n, b_n]$ is a nested sequence of nonempty closed intervals, then $a_n \leq a_{n+1} \leq b_{n+1} \leq b_n$ for all n. Let $L := \{x \mid x < a_n \text{ for some } n\}$ and $U := \{y \mid y > b_n \text{ for some } n\}$. Note that we have $x < y$ for all $x \in L$ and $y \in U$. Also L has no maximum and U has no minimum. Now if $\cap_n I_n$ were empty, then we would have $L \cup U = \mathbf{R}$, and so L, U would be a Dedekind gap in **R**, which is a contradiction. Hence $\cap_n I_n$ must be nonempty. □

This theorem is true in any linear continuum, not just **R** (Theorems 579 and 578). However, the intersection $\cap_n I_n$ here may contain multiple points. The following version ensures that the intersection contains a unique point under the *additional restriction that the lengths of the intervals approach zero in the limit,* i.e., for any $\epsilon > 0$ there is n with $\text{len}(I_n) < \epsilon$.

Theorem 211 (The Cauchy Nested Interval Property). *If for a nested sequence* $\langle I_n \mid n \in \mathbf{N} \rangle$ *of nonempty closed intervals in* **R** *we have* $\text{len}(I_n) \to 0$ *as* $n \to \infty$, *then the intersection* $\cap_n I_n$ *contains a unique point.*

Proof. By Theorem 210, $\cap_n I_n \neq \emptyset$. Having $p, q \in \cap_n I_n$ with $p < q$ would imply $\text{len}(I_n) \geq q - p$ for all n, contrary to the assumption $\text{len}(I_n) \to 0$. □

Theorem 207 and the above results are the foundations for developing "real variable theories" such as calculus. But such a development is beyond the scope of this text and belongs to the subject of mathematical analysis.

3.7 Additional Facts on Ordered Fields*

We state without proof some important results about **R** and ordered fields.

Examples of basic algebraic properties that can be derived from the ordered-field axioms of Definition 206 are:

1. $A + B = A + C \Rightarrow B = C$, $A \cdot 0 = 0$, *and* $(-A)(-B) = AB$.
2. $0 < 1$. *Also,* $A \neq 0 \Rightarrow A^2 > 0$, *and so there is no* A *such that* $A^2 + 1 = 0$ *(i.e., the polynomial* $x^2 + 1$ *has no zero in an ordered-field).*
3. $A < B \Rightarrow A < (A + B) \cdot 2^{-1} < B$, *so every ordered field is order-dense.*
4. *Every ordered field contains a subfield isomorphic to* **Q**.

Deriving such algebraic results from the ordered-field axioms is done in elementary abstract algebra. The reader may wish to try to derive these results as exercises, or find them in standard abstract algebra texts.

Unlike the purely algebraic properties which do not depend on order-completeness, the following properties need the fact that **R** is a linear continuum. Proofs use the IVT and can be found in standard real analysis texts.

1. *Every positive real number has an* n*-th root* ($n \in$ **N**).
2. *Any odd degree polynomial over* **R** *has a root in* **R**.

Theorem 212. *An ordered field is order complete (a linear continuum) if and only if it satisfies both the Archimedean Property and the Cauchy Nested Interval Property.*

The qualifier "Cauchy" in the theorem may be dropped, since in Archimedean fields the NIP (Nested Interval Property) is equivalent to the weaker property of having the Cauchy NIP.[8]

Also, the conditions of being Archimedean and satisfying the NIP are independent: There are ordered fields (such as **Q**) which are Archimedean but satisfies neither of the NIPs, and there are non-Archimedean fields which satisfy both the NIPs.[9]

Theorem 213 (Categoricity of R). *If* F *is any order-complete ordered field, then* F *must be "isomorphic to"* **R**, *i.e., there is a (unique) one-to-one correspondence* $x \leftrightarrow x'$ *between the elements* $x \in F$ *and the elements* $x' \in$ **R** *such that for all* $x, y \in F$:

$$x < y \Leftrightarrow x' < y', \qquad (x + y)' = x' + y', \quad \textit{and} \quad (xy)' = x'y'.$$

This result says that **R** is essentially ("up to isomorphism") the unique order-complete ordered field: Two such fields will have identical structural properties and so will be structurally indistinguishable.

---

[8]However, there are non-Archimedean fields satisfying the Cauchy NIP in which the (unrestricted) NIP fails, such as the formal Laurent series field over **R**.

[9]Examples for the second kind are given by hyperreal fields of type η_1.

3.8 Alternative Development Routes*

To build the field **R** of real numbers from the natural numbers **N**, three different paths may be followed depending on which intermediate class of numbers are built on the way, as shown in the following diagram.

$$\begin{array}{ccccc}
\mathbf{N} & \longrightarrow & \mathbf{Q}^+ & \longrightarrow & \mathbf{R}^+ \\
\downarrow & & \downarrow & & \downarrow \\
\mathbf{Z} & \longrightarrow & \mathbf{Q} & \longrightarrow & \mathbf{R}
\end{array}$$

In our development, we followed the topmost path $\mathbf{N} \to \mathbf{Q}^+ \to \mathbf{R}^+ \to \mathbf{R}$, which avoids negative numbers and zero until the last step.

Suppes [77] follows the middle route $\mathbf{N} \to \mathbf{Q}^+ \to \mathbf{Q} \to \mathbf{R}$.

Stoll [76] uses the "algebraic" bottom route $\mathbf{N} \to \mathbf{Z} \to \mathbf{Q} \to \mathbf{R}$, where the *ordered integral domain* **Z** is first built from **N**. One then builds **Q** as the *field of fractions* of **Z**, a process applicable to any integral domain.

In the last step of building **R** from **Q**, both Suppes and Stoll use an alternative way of building **R** due to Cantor, which is quite distinct from the method of Dedekind cuts (due to Dedekind) that we used in this text.

Cantor's method represents real numbers as "Cauchy sequences of rationals." A sequence $\langle \rho_n \rangle$ of rational numbers is a *Cauchy sequence* if for any $\epsilon > 0$, there is a k such that $|\rho_m - \rho_n| < \epsilon$ for all $m, n \geq k$. Two Cauchy sequences of rationals $\langle \rho_n \rangle$ and $\langle \sigma_n \rangle$ are called *equivalent* if for any $\epsilon > 0$, there is a k such that $|\rho_n - \sigma_n| < \epsilon$ for all $n \geq k$. Real numbers are then defined as equivalence classes of Cauchy sequences of rationals, with operations on them defined by performing the operation term-wise on the sequences.

Cantor's method leads to a far-reaching generalization known as the *metric completion*, applicable to a class of spaces called *metric spaces*. It captures the intuitive idea of "filling in" the "missing spatial points"—points to which a sequence "tries but fails" to converge. Furthermore, it is applicable to an arbitrary ordered field, giving what is known as the *Cauchy-completion* of the field. To see how Cantor's method is carried out in the general context of ordered fields, see Hewitt and Stromberg [30].

On the other hand, Dedekind's method of cuts is based on order and captures the geometric notion of a linear continuum in a highly intuitive manner. Dedekind's idea of continuity amounts to the condition that if the line is partitioned into two pieces then at least one of the pieces must contain a limit point of the other. This idea itself leads to a direct but far-reaching generalization—a concept known as *connectedness,* which is a form of continuity applicable to the most general types of spaces called *topological spaces.*[10]

3.9 Complex Numbers*

We define a *complex number* to be an ordered pair of real numbers. Complex numbers form a field (unordered) with sum and product defined as follows.

Definition 214 (Addition and Multiplication of Complex Numbers).

$$\langle A, B\rangle + \langle C, D\rangle := \langle A + C, B + D\rangle\,,$$
$$\langle A, B\rangle \cdot \langle C, D\rangle := \langle AC - BD, BC + AD\rangle\,.$$

A complex number of the form $\langle A, 0\rangle$ is called a *real complex number.*

Problem 215. *The mapping $A \to \langle A, 0\rangle$ is a bijection from the set of real numbers onto the set of real complex numbers and it preserves both the operations $+$ and $\cdot$:*

$$\langle A + B, 0\rangle = \langle A, 0\rangle + \langle B, 0\rangle\,, \quad \text{and} \quad \langle AB, 0\rangle = \langle A, 0\rangle\,\langle B, 0\rangle\,.$$

At this point, the real numbers and the real complex numbers become interchangeable since all the properties of the real numbers are possessed by the real complex numbers.

Therefore, we throw away the real numbers[11] and use the corresponding real complex numbers in their place. In particular, the real complex number $\langle A, 0\rangle$ will be denoted simply by A.

Definition 216. $i := \langle 0, 1\rangle$.

Problem 217. *With our convention of using A as an abbreviation for $\langle A, 0\rangle$, prove that*

$$i^2 = -1 \quad \text{and} \quad \langle A, B\rangle = A + Bi.$$

One problem with the real numbers is that one cannot solve equations like $x^2 + 1 = 0$. Complex numbers guarantee the existence of roots for not only such equations but also any arbitrary polynomial equation. We have the following basic theorem.

---

[10]Dedekind's condition can be used word for word to define the notion of connectedness: A topological space is connected if and only if whenever it is partitioned into two pieces then at least one of the pieces contains a limit point of the other.

[11]Phrase of Edmund Landau [47].

Theorem 218 (Fundamental Theorem of Algebra). *Any non-constant polynomial with complex coefficients has a complex root.*

The proof of this theorem is beyond the scope of this book. A proof can be found in Birkhoff and Mac Lane [4], *A Survey of Modern Algebra.*

Chapter 4
Postscript I: What Exactly Are the Natural Numbers?

Abstract This postscript to Part I consists of philosophical and historical remarks concerning the nature of the natural numbers. It contrasts the *absolutist* approach requiring absolute constructions of individual natural numbers such as those given by Frege, Russell, Zermelo, and von Neumann, with Dedekind's *structuralist* approach in which the natural numbers can be taken as members of *any Dedekind–Peano system.*

Note: *In this postscript we will often use the variant convention that the natural numbers include* 0 *and so start from* 0 *instead of* 1.

4.1 Russell's Absolutism?

In the last couple of chapters we outlined how the field of real and complex numbers, and thus essentially the entire body of traditional pure mathematics, can be deductively developed starting from only the natural numbers based on the Dedekind–Peano Axioms. However, the Dedekind–Peano Axioms do not specify what the natural numbers themselves really are, and thus leave the *interpretation* of the notion of natural numbers open.

The following passage is quoted from Russell (1920) [68] to illustrate the problem of finding an absolute interpretation for the natural numbers.

> ... Peano's three primitive ideas—namely, "0," "number," and "successor"—are capable of an infinite number of different interpretations, all of which will satisfy the five primitive propositions. We will give some examples.
>
> (1) Let "0" be taken to mean 100, and let "number" be taken to mean the numbers from 100 onward in the series of natural numbers. Then all our primitive propositions are satisfied, even the fourth, for, though 100 is the successor of 99, 99 is not a "number" in the sense which we are now giving to the word "number." It is obvious that any number may be substituted for 100 in this example.

A. Dasgupta, *Set Theory: With an Introduction to Real Point Sets*,
DOI 10.1007/978-1-4614-8854-5_4,

(2) Let "0" have its usual meaning, and let "number" mean what we usually call "even numbers," and let the "successor" of a number be what results from adding two to it. Then "1" will stand for the number two, "2" will stand for the number four, and so on; the series of "numbers" now will be

0, two, four, six, eight, ...

All Peano's five premisses are satisfied still.

(3) Let "0" mean the number one, let "number" mean the set

$$1, \tfrac{1}{2}, \tfrac{1}{4}, \tfrac{1}{8}, \tfrac{1}{16}, \ldots$$

and let "successor" mean "half." Then all Peano's five axioms will be true of this set.

It is clear that such examples might be multiplied indefinitely. In fact, given any series

$$x_0, x_1, x_2, x_3, \ldots, x_n, \ldots$$

which is endless, contains no repetitions, has a beginning, and has no terms that cannot be reached from the beginning in a finite number of steps, we have a set of terms verifying Peano's axioms.

...

In Peano's system there is nothing to enable us to distinguish between these different interpretations of his primitive ideas. It is assumed that we know what is meant by "0," and that we shall not suppose that this symbol means 100 or Cleopatra's Needle or any of the other things that it might mean.

This point, that "0" and "number" and "successor" cannot be defined by means of Peano's five axioms, but must be independently understood, is important. We want our numbers not merely to verify mathematical formulas, but to apply in the right way to common objects. We want to have ten fingers and two eyes and one nose. A system in which "1" meant 100, and "2" meant 101, and so on, might be all right for pure mathematics, but would not suit daily life. We want "0" and "number" and "successor" to have meanings which will give us the right allowance of fingers and eyes and noses. We have already some knowledge (though not sufficiently articulate or analytic) of what we mean by "1" and "2" and so on, and our use of numbers in arithmetic must conform to this knowledge. We cannot secure that this shall be the case by Peano's method; all that we can do, if we adopt his method, is to say "we know what we mean by '0' and 'number' and 'successor,' though we cannot explain what we mean in terms of other simpler concepts." ...

It might be suggested that, instead of setting up "0" and "number" and "successor" as terms of which we know the meaning although we cannot define them, we might let them stand for any three terms that verify Peano's five axioms. They will then no longer be terms which have a meaning that is definite though undefined: they will be "variables," terms concerning which we make certain hypotheses, namely, those stated in the five axioms, but which are otherwise undetermined. If we adopt this plan, our theorems will not be proved concerning an ascertained set of terms called "the natural numbers," but concerning all sets of terms having certain properties. Such a procedure is not fallacious; indeed for certain purposes it represents a valuable generalization. But from two points of view it fails to give an adequate basis for arithmetic. In the first place, it does not enable us to know whether there are any sets of terms verifying Peano's axioms; it does not even give the faintest suggestion of any way of discovering whether there are such sets. In the second place, as already observed, we want our numbers to be such as can be used for counting common objects, and this requires that our numbers should have a definite meaning, not merely that they should have certain formal properties. [68, pages 7–10]

4.2 Interpretations for the Natural Numbers

The Frege–Russell Natural Numbers. In 1884 Frege had already built an interpretation for the natural numbers satisfying Russell's requirements above (which was later re-invented by Russell himself). It is based on the natural principle of abstraction which defines a complete invariant for a given equivalence relation by assigning to each object its own equivalence class. The "Frege–Russell invariant" is obtained by applying this principle to the relation of one-to-one correspondence between sets. Two sets are called *equinumerous* (i.e., they have the same "number" of elements) if there is a one-to-one correspondence between them.[1] Equinumerosity is easily seen to be an equivalence relation, and *the number of elements a set* A is then defined as *the equivalence class* $[A]$ *of* A, i.e., the collection of *all* sets equinumerous to A.[2] For example, the first few Frege–Russell numbers are

$$0 := [\emptyset] = \{\emptyset\}$$
$$1 := [\{a\}] = \text{the collection of all singletons}$$
$$2 := [\{a, b\}]\ (a, b \text{ distinct}) = \text{the collection of all doubletons, etc.}$$

The Zermelo Natural Numbers. In 1908, Zermelo [85] gave a definition of the natural numbers in his framework of axiomatic set theory as follows:

$$0 := \emptyset,\quad 1 := \{\emptyset\},\quad 2 := \{\{\emptyset\}\},\quad 3 := \{\{\{\emptyset\}\}\},\ \ldots,\ n+1 := \{n\},\ \ldots$$

The Von Neumann Natural Numbers. In 1923, von Neumann built another interpretation for the natural numbers which has become standard in modern axiomatic set theories. His interpretation is as follows:

$$\begin{aligned} 0 &:= \emptyset \\ 1 &:= \{0\} = \{\emptyset\} \\ 2 &:= \{0, 1\} = \{\emptyset, \{\emptyset\}\} \\ 3 &:= \{0, 1, 2\} = \{\emptyset, \{\emptyset\}, \{\emptyset, \{\emptyset\}\}\} \\ &\ldots \\ n+1 &:= \{0, 1, 2, \ldots n\} = n \cup \{n\} \\ &\ldots \end{aligned}$$

[1] Equinumerosity and cardinal numbers will be studied in Chap. 5.

[2] Problems arise with the naive Frege–Russell invariant (Chap. 20) which can only be addressed by using less natural approaches such as Quine's New Foundations or Scott's modified invariant (Definition 1297) in the context of ZF set theory.

Here every natural number n is defined as a simple and canonical n-element set consisting precisely of the smaller natural numbers, and the successor function is defined as $S(x) := x \cup \{x\}$. Von Neumann's method also extends to the transfinite, giving a canonical interpretation for the *ordinal numbers* (which was the original purpose of von Neumann, see Chap. 21).

Each of the above definitions of natural numbers (Frege–Russell, Zermelo, and Von Neumann) provides a valid interpretation for the three primitive notions of Dedekind–Peano so that in each framework the five Dedekind–Peano axioms can be derived as theorems.

4.3 Dedekind's Structuralism

The view expressed by Russell's comments above on finding an absolute interpretation for the natural numbers as the "real one" is sometimes called *Frege–Russell absolutism*. This is in sharp contrast to Dedekind's *structuralism*, expressed by Dedekind in 1888, which we now discuss.

As illustrated by Russell's comments above and by Zermelo and von Neumann's definition of the natural numbers, there are many possible interpretations for the Dedekind–Peano axioms. Dedekind proved that all interpretations for the natural numbers which satisfy the Dedekind–Peano axioms have *the same structure*, or are "isomorphic." This is known as the *categoricity of the Dedekind–Peano axioms* and is given in the theorem below. It suggests that there is no reason to prefer any interpretation over any other. Thus, according to Dedekind's structuralism, one cannot take any specific interpretation of the natural numbers as the real one; rather, the true concept of natural number is given by the abstract *common structure* present in all interpretations which satisfy the Dedekind–Peano axioms.

Definition 219. A *Dedekind–Peano system* $N, 1_N, \phi$ consists of a set N, an element 1_N, and a function ϕ which satisfy:

1. $1_N \in N$
2. $\phi: N \to N$
3. $1_N \notin \phi[N] = \text{ran}(\phi)$
4. ϕ is injective
5. If P is a subset of N such that

   a. $1_N \in P$, and
   b. For all $x \in N, x \in P \Rightarrow \phi(x) \in P$,

   then $P = N$.

Clearly, the five conditions above correspond precisely to the five Dedekind–Peano axioms, with N playing the role of **N**, 1_N that of 1, and ϕ that of the successor function. The following theorem shows that any two Dedekind–Peano systems are "isomorphic," that is, they have the same structure:

Theorem 220 (Dedekind). *If $N, 1_N, \phi$ and $\Omega, 1_\Omega, \theta$ are two Dedekind–Peano systems then there is a unique bijection $\psi: N \to \Omega$ such that $\psi(1_N) = 1_\Omega$ and $\psi(\phi(x)) = \theta(\psi(x))$ for all $x \in N$.*

The function ψ in the theorem is the *isomorphism* between the two systems. The reader is invited to construct a proof of this categoricity theorem using the method of recursive definition given in Sect. 2.10.

What we are calling Dedekind–Peano systems were called *simply infinite systems* by Dedekind himself. For a Dedekind–Peano system $N, 1_N, \phi$, Dedekind uses the terminology that "the simply infinite system N is set in order by this function ϕ" with 1_N being the "base-element of N." Dedekind wrote in 1888 [11]:

> 73. Definition. If in the consideration of a simply infinite system N set in order by a function ϕ we entirely neglect the special character of the elements, merely retaining their distinguishability and taking into account only the relations to one another in which they are placed by the order-setting function ϕ, then are these elements called *natural numbers* or *ordinal numbers* or simply *numbers,* and the base-element 1 is called the *base-number* of the *number-series* N. With reference to this freeing the elements from every other content (abstraction) we are justified in calling numbers a free creation of the human mind. The relations or laws which are derived entirely from the conditions [...], and which are therefore always the same in all ordered simply infinite systems, whatever names may happen to be given to the individual elements (compare 134), form the first object of the *science of numbers* or *arithmetic.* [12, p. 68]

He continues later:

> 132. Theorem. All simply infinite systems are similar to the number-series N and consequently [...] also to one another.
>
> ...
>
> 133. Theorem. Every system that is similar to a simply infinite system and therefore [...] to the number-series N is simply infinite.
>
> ...
>
> 134. Remark. By the two preceding theorems (132), (133) all simply infinite systems form a class in the sense of [an equivalence class for the isomorphism relation]. At the same time, [...] it is clear that every theorem regarding numbers, i.e., regarding the elements n of the simply infinite system N set in order by the transformation ϕ, and indeed every theorem in which we leave entirely out of consideration the special character of the elements n and discuss only such notions as arise from the arrangement ϕ, possesses perfectly general validity for every other simply infinite system Ω set in order by a transformation θ and its elements ν, and that the passage from N to Ω (e.g., also the translation of an arithmetic theorem from one language into another) is effected by the transformation ψ considered in (132), (133), which changes every element n of N into an element ν of Ω, i.e., into $\psi(n)$. This element ν can be called the nth element of Ω and accordingly the number n is itself the nth number of the number-series N. The same significance which the transformation ϕ possesses for the laws in the domain N, in so far as every element n is followed by a determinate element $\phi(n) = n'$, is found, after the change effected by ψ, to belong to the transformation θ for the same laws in the domain Ω, in so far as the element $\nu = \psi(n)$ arising from the change of n is followed by the element $\theta(\nu) = \psi(n')$ arising from the

> change of n'; we are therefore justified in saying that by ψ, ϕ is changed into θ, which is symbolically expressed by $\theta = \psi\phi\bar{\psi}$, $\phi = \bar{\psi}\theta\psi$. By these remarks, as I believe, the definition of the notion of numbers given in (73) is fully justified. [12, p. 92–96]

In Dedekind's structuralist approach, the existence of natural numbers is tantamount to the existence of at least one Dedekind–Peano system. Dedekind observed that this follows from the existence of an infinite set in his sense, also called a *reflexive* or *Dedekind infinite* set, that is, a set for which there is an injective function mapping the set into a proper subset of itself.[3] Such an existence proof again has a much more structuralist flavor than the absolutist presentations of Frege–Russell, Zermelo, or von Neumann, where natural numbers are constructed in a unique canonical way with every natural number having a specific absolute definition.

Most regular mathematicians (as opposed to set theorists or logicians) do not think of natural numbers to be absolute constructs as presented in the Frege–Russell, Zermelo, or von Neumann definitions. It is fair to say that Dedekind's viewpoint above had a tremendous impact on later mathematicians such as Hilbert, and has overwhelmingly dominated the approach found in modern mathematics.[4]

Other Mathematical Notions. The distinction between the absolutist and the structuralist approaches applies not only to natural numbers but also to many other mathematical concepts. For example, the notion of ordered pair was reduced to an absolutist definition in terms of sets first by Wiener in 1914 as $\langle a,b\rangle := \{\{a\},\{b,\emptyset\}\}$, and then again by Kuratowski in 1921 as $\langle a,b\rangle := \{\{a\},\{a,b\}\}$. Both definitions satisfy the characterizing criterion for the ordered pair, namely: $\langle a,b\rangle = \langle c,d\rangle \Rightarrow a = c$ and $b = d$. For a structuralist, it is the characterizing criterion that matters the most.

The real numbers also can be considered either in absolutist or in structuralist terms. The construction of the real numbers that we presented using "Dedekind cuts of ratios" is an example of the absolutist approach. On the other hand, it is common for modern analysis texts to take a structuralist approach to the real numbers, where the system of real numbers is simply taken to be any complete ordered field.[5] Such a structuralist definition is sound because of the corresponding categoricity theorem: *Any two complete ordered fields are isomorphic as ordered fields* (Theorem 213). However, there is no simple "structuralist" existence proof in this case, and all known constructions of the real numbers, either using Dedekind's method or using Cantor's method, require some "hard work."

---

[3]This existence result is related to the Axiom of Infinity and its equivalent forms. See the part of Sect. 21.5 dealing with the Axiom of Infinity where this topic is further discussed, particularly Theorem 1263, as well as Problem 1225.

[4]The literature is large on the topics of this postscript. See, e.g., [63, 71], and the chapters by Hellman in [72] and [49], where further references can be found.

[5]Even in our absolute constructions of the previous chapters for extending the system of natural numbers to larger and larger systems of numbers such as the ratios, the lengths, the real numbers, and the complex numbers, we had already used the structuralist approach by throwing away old entities and replacing them with "isomorphic copies" found within the new extensions.

Part II
Cantor: Cardinals, Order, and Ordinals

Introduction to Part II

This part contains the core material of the book.

Chapters 5–10 cover cardinals, finitude, countability and uncountability, cardinal arithmetic, the theory of order types, dense and complete orders, well-orders, transfinite induction, ordinals, and alephs—almost all of which are due to Cantor.

In addition, Chaps. 10 and 11 cover the basic facts about the Axiom of Choice and equivalent maximal principles such as Zorn's Lemma, as well as well-founded relations and trees.

The postscript to this part (Chap. 12) briefly presents a selection of some of the most elementary topics of Infinitary Combinatorics.

Chapter 5
Cardinals: Finite, Countable, and Uncountable

Abstract This chapter introduces the basic idea of cardinal numbers, comparability, and operations, and next covers the theory of finite sets and natural numbers, from which the Dedekind–Peano axioms are derived as theorems. Dedekind infinite sets and reflexive cardinals are also defined. It then presents the Axiom of Choice and contrasts it with effective choice, using the notion of effectiveness informally. The rest of the chapter is about countability and uncountability: It focuses on the two specific cardinals $\aleph_0 = |\mathbf{N}|$ and $\mathfrak{c} = |\mathbf{R}|$, and gives the first proof of $\aleph_0 < \mathfrak{c}$ (uncountability of $\mathbf{R}$). In the process, the principles of countable and dependent choice are encountered.

5.1 Cardinal Numbers

Recall that f is said to be a *one-to-one correspondence between A and B* if $f: A \to B$ is a bijection (i.e., . f is a one-to-one function mapping A onto B).

Definition 221 (Similar or Equinumerous Sets). Two sets A and B are called *similar*, or *equinumerous*, written $A \sim_c B$ (or simply $A \sim B$) if there is a one-to-one correspondence between A and B.

Problem 222. *We have: (a) $A \sim A$, (b) $A \sim B \Rightarrow B \sim A$, and (c) $A \sim B$ and $B \sim C \Rightarrow A \sim C$. Thus equinumerosity, $\sim$, is an equivalence relation.*

We now permanently fix—once and for all—a specific complete invariant $A \mapsto |A|$ for the equivalence relation of similarity (equinumerosity). For any set A, we call $|A|$ the *cardinal number of A*.

Definition 223 (Cardinal Number, Cantor). For each set A, $|A|$ denotes its *cardinal number* and satisfies the condition:

$$|A| = |B| \quad \text{if and only if} \quad A \sim B, \qquad \text{for all sets } A \text{ and } B.$$

A. Dasgupta, *Set Theory: With an Introduction to Real Point Sets*,
DOI 10.1007/978-1-4614-8854-5_5,

We say that α is a *cardinal number* if α is the cardinal number of some set.

Discussion. At this point, "the cardinal number of a set" is simply a *primitive notion* serving as a complete invariant for the relation of similarity of sets. The main reason for introducing it is that *the cardinal numbers form a generalization of the natural numbers which extends into the transfinite.*

Historically, cardinal numbers—first introduced by Cantor in their full generality—were defined in two main ways, one known as the *Frege–Russell definition* and the other we call the *Cantor–Von Neumann definition.*

The Frege–Russell definition uses the natural complete invariant associated with the equivalence relation of similarity of sets—the quotient map given by the Principle of Abstraction (Theorem 42)—to define cardinals: $|A|$ is defined as the equivalence class $[A]$ of A under the similarity relation, i.e., $|A|$ equals the collection of all sets similar to A. Although a natural definition, this becomes problematic as the "collection" of *all* sets similar to A is so large that it is questionable whether it is a legitimate collection at all (see Chap. 20). In certain formal set theories such as the Zermelo–Fraenkel system (ZF), such a collection does not even exist as a set (Problem 1296), although the definition works in some other systems such as Quine's New Foundations. In ZF, a modified definition by Dana Scott, called the *Frege–Russell–Scott definition*, handles the problem by significantly reducing the collection which serves as the cardinal number of a set. We will discuss the Frege–Russell–Scott definition in Sect. 21.8 (Definition 1297).

The Cantor–Von Neumann definition is technically more complicated and needs the notions of well-orders and ordinal numbers which will be defined later. Still, the definition goes as follows: $|A|$ is defined as the least (von Neumann) ordinal α such that A can be well-ordered with type α. So in order for $|A|$ to exist, the set A needs to be well-orderable, which in turn requires a special axiom called the *Axiom of Choice.* Thus the Cantor–Von Neumann method cannot be used to effectively define cardinal numbers of arbitrary sets (such as the set of real numbers $\mathbf{R}$) without the use of Axiom of Choice. It is however the one found in Cantor's original conception of the transfinite and still is the more common definition of cardinal number used in formal set theory with the Axiom of Choice (such as ZFC). We will present it as Definition 1266 in Sect. 21.4 on von Neumann ordinals.

Definition 224 (0 **and** 1)**.** We define 0 to be the cardinal number of the empty set and 1 to be the cardinal number of the singleton set $\{0\}$:

$$0 := |\emptyset|, \qquad \text{and} \qquad 1 := |\{0\}|.$$

We remind the reader that a singleton was defined as a set of the form $\{a\}$, so A is a singleton if A contains an element a and no other element, i.e., if there is some a such that for all x, $x \in A \Leftrightarrow x = a$.

Problem 225. $|A| = 0$ *if and only if* $A = \emptyset$ *and* $|A| = 1$ *if and only if* A *is a singleton.*

Comparison of Cardinals and Sets

Definition 226 (Comparison of Cardinals and Sets). For sets A and B, we write $A \preccurlyeq B$ if there is a one-to-one function $f: A \to B$ (which may or may not be onto). We also write $A \not\preccurlyeq B$ for "not $A \preccurlyeq B$."

For cardinal numbers α and β, we write $\alpha \leqslant \beta$ if $A \preccurlyeq B$ for some sets A and B with $|A| = \alpha$ and $|B| = \beta$.

General cardinal and set comparison, where both finite and infinite sets are allowed, behaves quite differently from the familiar situation of finite sets.

Problem 227. *If $|A| = \alpha$, $|B| = \beta$, and $f: A \to B$ is one-to-one but not onto, then which of the following statements must necessarily be true?*

(a) $A \preccurlyeq B$ and $\alpha \leqslant \beta$; (b) $\alpha \leqslant \beta$ but $\alpha \neq \beta$; (c) $A \preccurlyeq B$ but not $A \sim B$.

So it is quite possible that for sets A and B, A is similar to some proper subset of B, while at the same time B is similar to some proper subset of A.

Problem 228. *Give examples showing that the last statement is correct.*

This is very different from the case of finite sets—in fact, we will see that this is impossible if A or B is finite, when we formally define finite sets below.

To analyze the general situation for two sets A and B with cardinal numbers $\alpha = |A|$ and $\beta = |B|$, the following four possibilities are mutually exclusive and exhaustive (meaning that exactly one of these holds):

Definition 229. Let A and B be sets with $\alpha = |A|$ and $\beta = |B|$. Then exactly one of the following cases holds, and in each case we give a definition:

1. Both $A \preccurlyeq B$ and $B \preccurlyeq A$. We then say A is *weakly equivalent* to B, and write this as $A \sim^* B$, and also write $\alpha =^* \beta$.
2. $A \preccurlyeq B$ but not $B \preccurlyeq A$. We then write $A \prec B$ and $\alpha < \beta$, and say α is *less than* β.
3. $B \preccurlyeq A$ but not $A \preccurlyeq B$. Here we write $A \succ B$ and $\alpha > \beta$, and say α is *more than* β.
4. Neither $A \preccurlyeq B$ nor $B \preccurlyeq A$. In this case we say that A is *not comparable* to B and α and β are *incomparable cardinals*, writing this as $\alpha || \beta$.

Each of these relations is invariant over equinumerosity $\sim$, i.e., if $A \sim A'$ and $B \sim B'$, then any of the above four relations will hold between A' and B' if and only if it holds between A and B. Hence the four cardinal relations

(1) $\alpha =^* \beta$, (2) $\alpha < \beta$, (3) $\alpha > \beta$, (4) $\alpha || \beta$,

are well defined, *exactly one* of which always holds between any pair of cardinals α and β. It is also easy to see that $\beta > \alpha$ if and only if $\alpha < \beta$.

Problem 230. $0 < 1$, *and so* $0 \neq 1$.

By entirely routine arguments we have:

Problem 231. *The relation* $A \preccurlyeq B$ *for sets and the corresponding relation* $\alpha \leqslant \beta$ *for cardinals are both reflexive and transitive. The relations* $A \prec B$ *for sets and* $\alpha < \beta$ *for cardinals are asymmetric, and therefore irreflexive.*

The following is only slightly more interesting.

Problem 232. *If* $A \prec B$ *and* $B \preccurlyeq C$, *or if* $A \preccurlyeq B$ *and* $B \prec C$, *then* $A \prec C$. *So for cardinals* α, β, γ, *if* $\alpha < \beta$ *and* $\beta \leqslant \gamma$, *or if* $\alpha \leqslant \beta$ *and* $\beta < \gamma$, *then* $\alpha < \gamma$. *The relations* $A \prec B$ *for sets and* $\alpha < \beta$ *for cardinals are transitive.*

Problem 233. *Suppose that* A *and* B *are sets,* $a \notin A$, *and* $b \notin B$. *Then*

1. $A \sim B$ *if and only if* $A \cup \{a\} \sim B \cup \{b\}$.
2. $A \preccurlyeq B$ *if and only if* $A \cup \{a\} \preccurlyeq B \cup \{b\}$.
3. $A \prec B$ *if and only if* $A \cup \{a\} \prec B \cup \{b\}$.

Can We Get Trichotomy?

We would like to establish that the relation $<$ is an ordering of the cardinals, and so we need the law of trichotomy for $<$. But all we have at this point is that exactly one of

$$(1)\,\alpha =^* \beta, \qquad (2)\,\alpha < \beta, \qquad (3)\,\alpha > \beta, \qquad (4)\,\alpha || \beta,$$

holds, which is a long way from trichotomy.

Our goal to obtain trichotomy can be realized if, for the four conditions above, we can

- Replace condition (1) $\alpha =^* \beta$ by the condition $\alpha = \beta$; and
- Prove that condition (4), incomparability, cannot hold.

Later, using the Axiom of Choice, we will see that condition (4) is in fact impossible. For now, we discuss how we can replace "$\alpha =^* \beta$" by "$\alpha = \beta$," i.e., how to prove the equivalence

$$\alpha =^* \beta \Leftrightarrow \alpha = \beta.$$

The implication $\alpha = \beta \Rightarrow \alpha =^* \beta$ holds trivially. The converse implication ($\alpha =^* \beta \Rightarrow \alpha = \beta$) is also true, but the proof is nontrivial. We can restate it as "weakly equivalent sets are equinumerous" (i.e., $A \sim^* B \Rightarrow A \sim B$), a result called the *Cantor–Bernstein Theorem* or *Schröder–Bernstein Theorem.*

Theorem (Cantor–Bernstein). If $A \preccurlyeq B$ and $B \preccurlyeq A$, then $A \sim B$. Therefore, the relation $\leqslant$ defined on the cardinals is antisymmetric.

This theorem will be established in the next chapter. However, it is instructive for the reader to attempt a proof at this point.

Proving that $<$ is connected requires (in fact is equivalent to) the Axiom of Choice, and will be given much later in Theorem 719.

5.2 Sum and Product of Cardinal Numbers

Problem 234 (Disjoint Copies of sets). *Given any sets A, B there exist disjoint sets A', B' with $A \sim A'$ and $B \sim B'$.*

[Hint: Take $A' = \{0\} \times A$ and $B' = \{1\} \times B$.]

Problem 235 (Uniqueness of Sum). *If $A \sim A'$, $B \sim B'$, and $A \cap B = \emptyset = A' \cap B'$, then $(A \cup B) \sim (A' \cup B')$.*

Problem 236. *Given cardinals α and β there is a unique cardinal γ such that there are disjoint sets A and B with $|A| = \alpha$, $|B| = \beta$, and $|A \cup B| = \gamma$.*

Definition 237 (Sum of two Cardinal Numbers). Given cardinal numbers α and β, the unique cardinal number γ whose existence is guaranteed by Problem 236 is called the *sum of α and β* and is denoted by $\alpha + \beta$.

Problem 238. *The sum of cardinal numbers is an associative and commutative operation with 0 as the identity. In other words, for any cardinals α, β, γ:*

$$\alpha + (\beta + \gamma) = (\alpha + \beta) + \gamma, \quad \alpha + \beta = \beta + \alpha, \quad \alpha + 0 = \alpha.$$

Hence it follows that: $\alpha + (\beta + 1) = (\alpha + \beta) + 1$.

Problem 239. *For cardinals α, β, we have $\beta = \alpha + 1$ if and only if there is a set A with $|A| = \alpha$ and $x \notin A$ such that $\beta = |A \cup \{x\}|$.*

Problem 240. *If α, β are cardinals, then $\alpha \leqslant \beta$ if and only if $\beta = \alpha + \gamma$ for some cardinal γ.*

Problem 241. *If α, β are cardinals, then:*

1. *$\alpha + 1 = \beta + 1$ if and only if $\alpha = \beta$.*
2. *$\alpha + 1 \leqslant \beta + 1$ if and only if $\alpha \leqslant \beta$.*
3. *$\alpha + 1 < \beta + 1$ if and only if $\alpha < \beta$.*

[Hint: Use Problem 233.]

Problem 242 (Uniqueness of Product). *$A \sim A'$ and $B \sim B' \Rightarrow A \times B \sim A' \times B'$.*

By the last Problem, given $\alpha = |A|$ and $\beta = |B|$, the product $\alpha\beta := |A \times B|$ is well defined.

Definition 243 (Product of two Cardinal Numbers). Given cardinals α and β, the *product* $\alpha\beta$ is the unique cardinal number γ such that $\gamma = |A \times B|$ for some A, B with $|A| = \alpha$ and $|B| = \beta$.

Problem 244. *Cardinal product is associative and commutative, with* 1 *as the identity. In other words, for any cardinals α, β, γ:*

$$\alpha(\beta\gamma) = (\alpha\beta)\gamma, \quad \alpha\beta = \beta\alpha, \quad 1\alpha = \alpha.$$

Note that $A \times B$ can be naturally partitioned into the pairwise disjoint family $\langle A \times \{b\} \mid b \in B \rangle$ indexed by B:

$$A \times B = \bigcup_{b \in B} A \times \{b\},$$

where $(A \times \{b\}) \sim A$ for all $b \in B$. In other words, with $\alpha = |A|$ and $\beta = |B|$, $A \times B$ is the union of β-many pairwise disjoint sets, each having cardinality α. Hence $\alpha\beta$ may be regarded as the result of "repeatedly summing α, repeated β times" (see Definition 355 and Problem 357).

Problem 245. *The distributive law for cardinals holds: $\alpha(\beta + \gamma) = \alpha\beta + \alpha\gamma$. Hence it follows that $\alpha(\beta + 1) = \alpha\beta + \alpha$.*

5.3 Finite Sets and Dedekind Infinite Sets

The concepts of finitude and infinity have been used in mathematics since antiquity—recall Euclid's proof that there are an infinity of primes—but precise definitions were given only in relatively modern times.

First Definition: Finite Sets as Inductive Sets

Our first definition of finite sets closely matches the intuition of "being similar to the set $\{1, 2, \ldots n\}$ for some natural number n," but does not presuppose the notion of a natural number. The key idea here is the principle of induction: The definition will roughly be that the finite sets are precisely those sets which satisfy induction, in a sense to be seen below.

Definition 246 (Inductive Families). Let A be a set. We say that a collection C of subsets of A is an *inductive family over* A if it satisfies

1. $\emptyset \in \mathsf{C}$;
2. $E \in \mathsf{C}$ and $a \in A \Rightarrow E \cup \{a\} \in \mathsf{C}$;

Informal discussion. Here are some informal examples of inductive families:

1. For any set A, the power set of A, $\mathbf{P}(A)$ is an inductive family over A.
2. The family of all bounded subsets of $\mathbf{R}$ is inductive over $\mathbf{R}$ (a subset E of $\mathbf{R}$ is called bounded if we have $-a < x < a$ for some real number a). This inductive family does not include $\mathbf{R}$ itself as a member.
3. The family C of subsets of $\mathbf{N}$ not containing any arithmetic progression is inductive over $\mathbf{N}$ (E contains an arithmetic progression if there are $a, b \in \mathbf{N}$ with $a + bn \in E$ for all $n \in \mathbf{N}$). Here $\mathbf{N} \notin \mathsf{C}$ but many infinite sets, e.g., the set $\{1, 4, 9, \dots\}$ of perfect squares, are members of C.

Informally using the word "finite," note the following "test for finitude":

1. *If A is finite, then any inductive family* C *over A contains A as a member.* Reason: $\emptyset \in \mathsf{C}$ by the first clause of the definition, so by the second clause C contains all singleton subsets of A, then all the doubletons, and so on, picking up every finite subset of A, and so A itself, in the process.
2. *If A is not finite, then there is an inductive family over A which does not contain A as a member,* namely the family F_A of all finite subsets of A. (F_A is an inductive family over A since $\emptyset$ is finite, and for any finite set E the set $E \cup \{x\}$ has at most one more member and so is finite.)

We now invert this test to get our first formal definition of finite sets.

Definition 247 (Finite Sets). A set A is *finite*, or *inductive*, if A is a member of every inductive family over A. A is *infinite* if A is not finite, i.e., if there is an inductive family over A which does not contain A as a member.

An immediate corollary of the definition is the following principle which is very useful for establishing properties of finite sets.

Theorem 248 (The Principle of Induction over Finite Sets). *Let P be a property of sets such that (a) the empty set $\emptyset$ has property P and (b) if any set E has property P, then so does the set $E \cup \{x\}$ obtained from E by adjoining any single element x. Then every finite set has property P.*

Proof. Let A be any finite set. Define C to be the collection of all those subsets of A which has property P. By the given conditions C is an inductive family over A, and so $A \in \mathsf{C}$ since A is finite. Hence A has property P. □

Another useful fact is the following.

Theorem 249. *The empty set is finite. If A is finite then so is $A \cup \{b\}$ for any b. In particular, every singleton is finite.*

Proof. The empty set is a member of every inductive family, so is finite.

Suppose A is finite. Let C be an inductive family over $A \cup \{b\}$. We show that $A \cup \{b\} \in \mathrm{C}$.

Put $\mathrm{C}_A := \{E \in \mathrm{C} \mid E \subseteq A\}$. Then C_A is an inductive family over A, and so $A \in \mathrm{C}_A$ since A is finite. Hence $A \in \mathrm{C}$. But C is inductive over $A \cup \{b\}$ and $b \in A \cup \{b\}$, so $A \cup \{b\} \in \mathrm{C}$. □

The following results about finite sets are proved by induction over finite sets (Theorem 248) and using Theorem 249. Let us give a typical example.

Problem 250. *Any subset of a finite set is finite.*

Proof. We prove the result by induction on finite sets (Theorem 248).

First, any subset of $\emptyset$, being equal to $\emptyset$ itself, is finite.

Next, assume that every subset of E is finite (induction hypothesis). Then given any x, a subset S of $E \cup \{x\}$ is either a subset of E and so is finite by induction hypothesis, or S has the form $S = T \cup \{x\}$ for some $T \subseteq E$ and so is finite by Theorem 249 since T is finite by induction hypothesis. □

Problem 251. *The image of a finite set under any function is finite. If A is finite and $B \sim A$ then B is finite. If A is finite and $B \preccurlyeq A$ then B is finite.*

Problem 252. *A set A is infinite if and only $B \prec A$ for every finite set B.*

Problem 253. *If A and B are finite then so is $A \cup B$.*

[Hint: Use induction on B. The induction step consists of showing that if $A \cup B$ is finite (induction hypothesis) then so is $A \cup (B \cup \{x\})$ for any x.]

Problem 254. *If A is finite then so is its power set $\mathrm{P}(A)$. If A and B are finite then so is the Cartesian product $A \times B$.*

Problem 255. *A finite union of finite sets is finite. That is, if C is finite and every member of C is a finite set, then $\bigcup \mathrm{C}$ is finite.*

[Hint: In the induction step, use the fact that $\bigcup(\mathrm{C} \cup \{E\}) = (\bigcup \mathrm{C}) \cup E$.]

Problem 256 (Transitive Closure, Frege–Russell). *Let R be a relation. We say that y* is an R-successor of x *if xRy. A set A is called* R-hereditary *if for all x, y, $x \in A$ and $xRy \Rightarrow y \in A$. Define a relation R_* as follows: xR_*y if and only if y is a member of every R-hereditary set containing all R-successors of x. Then R_* is the least transitive relation containing R, i.e.,*

1. *$R \subseteq R_*$, i.e., $xRy \Rightarrow xR_*y$.*
2. *R_* is a transitive relation.*
3. *If T is any transitive relation with $R \subseteq T$, then $R_* \subseteq T$.*

Second Definition: Dedekind Finite Sets

Galileo found that, paradoxically, a "small" *part* of a collection may in fact be of the same size as the *whole* collection. He took the natural numbers as the whole collection and then formed a strictly "smaller" part of the whole by taking only the perfect squares, which are rather sparsely distributed in the natural numbers since they become rarer among larger numbers. But, strangely enough, a one-to-one correspondence between the whole and the strictly smaller part is established by $n \leftrightarrow n^2$, showing that the size of the part is equal to the size of the whole, not smaller!

Dedekind turned Galileo's paradox into a precise definition of infinity: The Dedekind infinite sets are precisely the ones showing this "paradoxical" behavior. This is our second definition of finite and infinite sets.

Definition 257 (Dedekind). A set A is said to be *Dedekind infinite* or *reflexive* if $A \sim B$ for some proper subset $B \subsetneq A$, i.e., if there is a function $f: A \to A$ which is one-to-one but not onto. A set will be called *Dedekind finite* or *non-reflexive* if it is not Dedekind infinite.

A *reflection* of a set A is a one-to-one map of A into a proper subset of A.

In our first version (Definition 247), we gave a direct natural definition of finite sets, and infinite sets were then defined indirectly—as sets which are not finite. In Dedekind's definition, the opposite is done: A simple direct definition of infinite sets is given, and finite sets are defined indirectly—as sets which are not Dedekind infinite.

Problem 258. *If $A \sim B$ then A is Dedekind finite if and only if B is.*

Problem 259. *Let $A \subseteq B$. If A is Dedekind infinite then so is B. Equivalently, if B is Dedekind finite then so is A.*

[Hint: A reflection $f: A \to A$ can be extended to $f^*: B \to B$ by setting $f^*(x) = x$ for all $x \in B \smallsetminus A$.]

Corollary 260. *Let $A \preccurlyeq B$. If A is Dedekind infinite then so is B.*

Proposition 261. *Suppose that $x \notin A$. Then A is Dedekind infinite if and only if $A \cup \{x\} \preccurlyeq A$.*

Proof. Suppose first that A is Dedekind infinite. Let $f: A \to A$ be one-to-one but not onto, fix $y \in A \smallsetminus f[A]$, and extend f to $f^*: A \cup \{x\} \to A$ by setting $f^*(x) = y$. Then f^* is injective, so $A \cup \{x\} \preccurlyeq A$.

For the converse, assume that $A \cup \{x\} \preccurlyeq A$, and let $f: A \cup \{x\} \to A$ be an injection. Then $f(x) \in A \smallsetminus f[A]$ since f is injective. Hence $f \restriction_A: A \to A$ is a reflection, and so A is Dedekind infinite. □

Corollary 262. *If A is Dedekind finite and $A \subsetneq B$ then $A \prec B$.*

[Hint: $B \preccurlyeq A$ would imply $A \cup \{b\} \preccurlyeq A$ for some $b \in B \smallsetminus A$.]

Proposition 263. *If A is Dedekind finite then so is $A \cup \{b\}$.*

Proof. Let $B := A \cup \{b\}$ be Dedekind infinite. We show that then so is A.

This follows from Problem 233, but let us give a direct proof.

Let $f: B \to B$ be a reflection and fix $c \in B \smallsetminus f[B]$. If $b \notin \operatorname{ran}(f)$ then $f[B] \subseteq A$ and we are done by Proposition 261, so let us fix $a \in B$ with $f(a) = b$. Now modify f to a function $g: B \to B$ by redefining the value of f at a to be c, i.e., let $g(a) = c$ and $g(x) = f(x)$ for all $x \neq a$. Then g is one-to-one with $g[B] \subseteq A$, so A is Dedekind infinite by Proposition 261. □

This gives the following basic result by induction over finite sets.

Theorem 264. *Any finite set is Dedekind finite.*

Finite Cardinals

Definition 265 (Finite Cardinals). A cardinal μ is called a *finite cardinal* if $\mu = |A|$ for some finite set A; otherwise, μ is called an *infinite cardinal.* The set of all finite cardinals will be denoted by $\mathbf{J}$, and for each cardinal κ, $\mathbf{J}_\kappa$ will denote the set of all finite cardinals less than κ:

$$\mathbf{J} := \{|A| : A \text{ is finite}\}, \qquad \text{and} \qquad \mathbf{J}_\kappa := \{\mu \in \mathbf{J} \mid \mu < \kappa\}.$$

Problem 266. $0 \in \mathbf{J}$ *and if* $\mu \in \mathbf{J}$ *then* $\mu + 1 \in \mathbf{J}$. *So* $1 \in \mathbf{J}$, $1 + 1 \in \mathbf{J}$, *etc. Moreover if* $\mu, \nu \in \mathbf{J}$ *then* $\mu + \nu \in \mathbf{J}$ *and* $\mu\nu \in \mathbf{J}$.

Theorem 267 (Principle of Induction for J). *Suppose that K is a set of cardinal numbers such that $0 \in K$ and $\mu \in K \Rightarrow \mu + 1 \in K$. Then $\mathbf{J} \subseteq K$.*

[Hint: To get $\mathbf{J} \subseteq K$, show that $|A| \in K$ for every finite set A by induction over finite sets. Use the fact that if $x \notin A$ then $|A \cup \{x\}| = |A| + 1$.]

We have the following series of corollaries to Corollary 262:

Corollary 268. *If ν, κ are cardinals with ν finite, then $\nu < \kappa \Leftrightarrow \kappa = \nu + \mu$ for some nonzero cardinal μ. In particular, $\nu < \nu + 1$ for all $\nu \in \mathbf{J}$.*

Corollary 269. *If ν, κ are cardinals with ν finite, then $\nu < \kappa \Leftrightarrow \kappa = \nu + 1$ or $\kappa > \nu + 1$.*

Corollary 270 (Strong Trichotomy for Finite Cardinals). *If ν, κ are cardinals with ν finite, then exactly one of $\nu < \kappa$, $\nu = \kappa$, or $\nu > \kappa$ holds.*

[Hint: Use induction on ν.]

Corollary 271. *If ν, κ are cardinals with ν finite, then $\kappa < \nu + 1$ if and only if $\kappa < \nu$ or $\kappa = \nu$.*

Corollary 272. *If ν is an finite cardinal, then the set of all cardinals smaller than ν is a finite set of cardinality ν. That is, $|\mathbf{J}_\nu| = \nu$ for all $\nu \in \mathbf{J}$.*

5.4 Natural Numbers and Reflexive Cardinals

In Part I, we defined real numbers in terms of ratios, and ratios in terms of natural numbers, but the natural numbers themselves were left undefined—we only assumed they are primitive entities satisfying the Dedekind–Peano axioms. We now officially *define* the natural numbers as the nonzero finite cardinals, and, from this definition, *derive* the Dedekind–Peano axioms, i.e., *prove* them as theorems. This gives an *interpretation* for the natural numbers, or a *model for the Dedekind–Peano axioms,* in terms of our current primitive of cardinal numbers.

Definition 273 (The Natural Numbers N). A *natural number* is a nonzero finite cardinal. The set of all natural numbers is denoted by $\mathbf{N}$:

$$\mathbf{N} := \mathbf{J} \smallsetminus \{0\} = \{|A| : A \text{ is finite}, A \neq \emptyset\},$$

We also define the successor map S by $S(\kappa) := \kappa + 1$ (for any cardinal κ).

Theorem 274. *The successor function S restricted to the set $\mathbf{J}$ of finite cardinals maps $\mathbf{J}$ bijectively onto the set $\mathbf{N}$ of natural numbers.*

[Hint: To show that S is injective, use Problem 241.]

Corollary 275. *We have $\mathbf{N} \sim \mathbf{J}$ with $\mathbf{N} \subsetneq \mathbf{J}$ and $\mathbf{J} \smallsetminus \mathbf{N} = \{0\}$. Hence $\mathbf{J}$ and $\mathbf{N}$ are Dedekind infinite, and so infinite. So $A \prec \mathbf{N}$ for every finite A.*

Proving the Dedekind–Peano Axioms

Problem 276. $1 \in \mathbf{N}$, *and if $\kappa \in \mathbf{N}$ then $S(\kappa) \in \mathbf{N}$. Moreover if $\alpha, \beta \in \mathbf{N}$ then $\alpha + \beta \in \mathbf{N}$ and $\alpha\beta \in \mathbf{N}$.*

It is now routine to verify that the above interpretation of the natural numbers satisfies the Dedekind–Peano Axioms:

Theorem 277 ($\langle \mathbf{N}, S, 1 \rangle$ Models the Dedekind–Peano Axioms). *With the natural numbers $\mathbf{N}$, the successor map S, and the cardinal number 1 as defined above, all five Dedekind–Peano axioms are satisfied.*

Proof. Since $1 \in \mathbf{N}$, the first axiom holds. The second axiom holds because if κ is a nonzero finite cardinal then so is $S(\kappa) = \kappa + 1$. To verify the third axiom, suppose, if possible, that $1 = S(\kappa)$ for some $\kappa \in \mathbf{N}$. Then $S(0) = S(\kappa)$, so $0 = \kappa$

(since S is injective), which is impossible since $0 \notin \mathbf{N}$. The fourth axiom again follows from injectivity of S. Finally, the induction axiom is essentially the same as Theorem 267. □

Problem 278. *Addition and multiplication of natural numbers as defined recursively in the Dedekind–Peano system coincide with the corresponding operations for cardinal numbers restricted to* **N**. *Similarly, the ordering relation defined for the natural numbers via the Dedekind–Peano system coincides with the cardinal "less than" relation restricted to* **N**.

[Hint: Let $+$ denote cardinal addition, and $+'$ denote addition as defined in the Dedekind–Peano system via the recursion equations $m +' 1 = S(m)$ and $m +' S(n) = S(m +' n)$. By the associative law, cardinal addition satisfies the same recursion equations, and a routine induction on the second variable shows that $+$ and $+'$ coincide. Multiplication is handled similarly.

The ordering relation $<'$ in the Dedekind–Peano system was defined as $m <' n \Leftrightarrow n = m+k$ for some $k \in \mathbf{N}$. But the same criterion has been already established for finite cardinals, so the two relations coincide.]

Now that we have proved the Dedekind–Peano axioms with the natural numbers defined as the nonzero finite cardinals, the entire theory developed in Part I becomes available to us as a corollary. In particular, the principles of recursive definition, as well as the complete ordered field **R** of real numbers, its subsets **Q** (rational numbers) and **Z** (integers), and their general properties can be officially used. For example, by Theorem 68 we have:

Corollary 279 (The Well-Ordering Property). *Every nonempty subset of* **N** *contains a smallest element.*

Dedekind Infinite Sets, Reflexive Cardinals, and $\aleph_0$

Theorem 280. *A set A is Dedekind infinite if and only if* $\mathbf{N} \preccurlyeq A$.

Proof. If $\mathbf{N} \preccurlyeq A$ then A is Dedekind infinite since **N** is.

For the other direction, suppose there is a one-to-one reflection $h: A \to A$, and fix $a \in A \smallsetminus h[A]$. The main idea behind the proof is this: Since h is a reflection, we have the strictly decreasing sequence of sets

$$A \supsetneq h[A] \supsetneq h[h[A]] \supsetneq h[h[h[A]]] \supsetneq \cdots .$$

Since h is injective, from $a \in A \smallsetminus h[A]$ we get $h(a) \in h[A] \smallsetminus h[h[A]]$, and so $h(h(a)) \in h[h[A]] \smallsetminus h[h[h[A]]]$, etc. This makes the elements $a, h(a), h(h(a)), \ldots$ all distinct, as shown in the figure below.

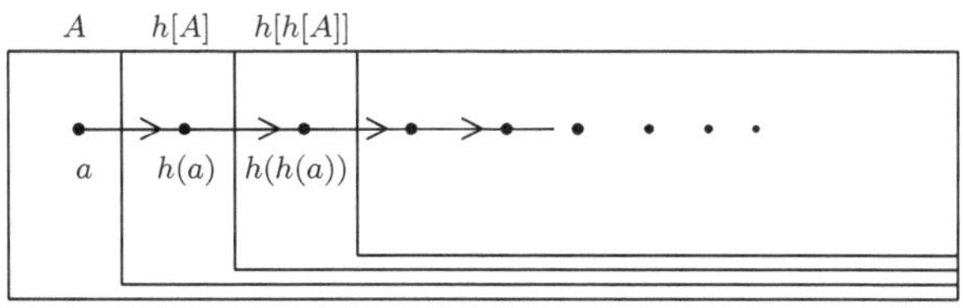

We thus get an injective mapping $f:\mathbf{N} \to A$ if we set $f(1) = a$, $f(2) = h(a)$, $f(3) = h(h(a))$, etc. We now formalize this idea into a rigorous proof.

By the basic principle of recursive definition, there is $f:\mathbf{N} \to A$ such that $f(1) = a$ and $f(n+1) = h(f(n))$ for all $n \in \mathbf{N}$. We claim that f is injective, i.e., $f(m) \neq f(n)$ for $m \neq n$. By trichotomy, it suffices to show that *for any n, $f(m) \neq f(n)$ for all $m < n$*. We prove this *by induction on n*.

For $n = 1$ this is vacuously true. Assume that $f(m) \neq f(n)$ for all $m < n$ (induction hypothesis). We show that $f(m) \neq f(n+1)$ for all $m < n+1$. Let $m < n+1$. If $m = 1$, then $f(m) = f(1) = a \notin \operatorname{ran}(h)$ while $f(n+1) = h(f(n)) \in \operatorname{ran}(h)$, so $f(m) \neq f(n+1)$. If $m > 1$, then $m = k+1$ for some $k \in \mathbf{N}$. Since $m < n+1$, we get $k+1 < n+1$, and so $k < n$. By induction hypothesis, $f(k) \neq f(n)$, so $h(f(k)) \neq h(f(n))$ (since h is one-to-one), i.e., $f(k+1) \neq f(n+1)$ or $f(m) \neq f(n+1)$ as desired. $\square$

Definition 281 (Reflexive Cardinals). A cardinal κ is called a *reflexive cardinal* if $\kappa = |A|$ for some Dedekind infinite set A.

Corollary 282. *Every reflexive cardinal is an infinite cardinal.*

We now define $\aleph_0$ to be the cardinal number of the set $\mathbf{N}$ of natural numbers. ($\aleph_0$ is called aleph-nought, or aleph-null, or aleph-zero.)

Definition 283. $\aleph_0 := |\mathbf{N}|$.

By Corollary 275 we have:

Corollary 284. $\aleph_0 = |\mathbf{J}|$, *so* $\aleph_0 + 1 = \aleph_0$. *Also $\aleph_0$ is a reflexive cardinal and therefore an infinite cardinal. Thus $n < \aleph_0$ for all finite cardinals $n \in \mathbf{J}$.*

We have the following characterizations of a reflexive cardinal.

Proposition 285. *For any cardinal κ the following conditions are equivalent:*

1. *κ is reflexive.*
2. *$\aleph_0 \leqslant \kappa$.*
3. *$\kappa + 1 = \kappa$.*

[Hint: $1 \Rightarrow 2$ by Theorem 280. For $2 \Rightarrow 3$, note that if $\aleph_0 \leqslant \kappa$ then $\kappa = \alpha + \aleph_0$ for some cardinal α, so $\kappa + 1 = (\alpha + \aleph_0) + 1 = \alpha + (\aleph_0 + 1) = \alpha + \aleph_0 = \kappa$. Finally, $3 \Rightarrow 1$ follows from definition (or use Proposition 261).]

Corollary 286. *$\aleph_0$ is the* smallest *reflexive cardinal.*

Can an Infinite Set be Dedekind Finite?

Since no Dedekind infinite set is finite, we have the following picture.

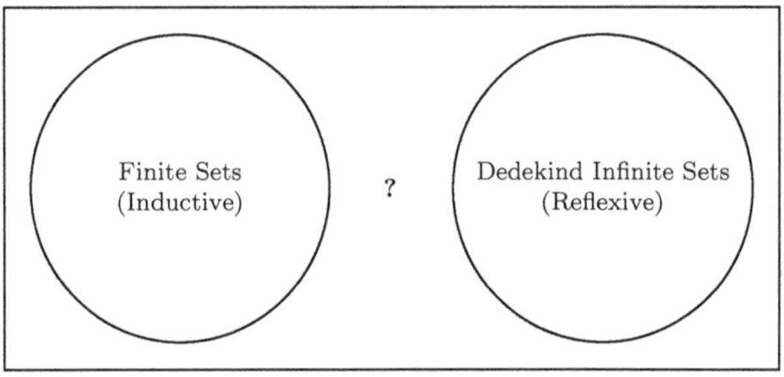

So the question is: Is every infinite set Dedekind infinite? That would imply that the region marked by "?" in the diagram above is empty and so the two notions of finitude would coincide: A set will be Dedekind finite if and only if it is finite. Or: Are there sets which are infinite but Dedekind finite?

If there were such a set A, then $\{1, 2, \ldots, n\} \preccurlyeq A$ for all n, i.e., A has finite subsets $\{a_1, a_2, \ldots, a_n\}$ with n distinct elements *for every n,* yet $\mathbf{N} \not\preccurlyeq A$, i.e., there is no infinite sequence $\langle a_1, a_2, \ldots, a_n, \ldots \rangle$ of distinct elements from A. As we do not have a clear intuition about such sets, we can perhaps show that this is impossible—resulting in the "clean solution" that the two notions of finitude coincide.

Let us recall the proof of this when A was Dedekind infinite (Theorem 280): The presence of a reflection $h: A \to A$ and an element $a \in A \smallsetminus h[A]$ allowed us to define a sequence of distinct elements as $a_1 = a$, $a_2 = h(a_1)$, $a_3 = h(a_2)$, etc. Notice that this infinite sequence is *specified uniquely in terms of the reflection h and the element a.*

When A is infinite but not known to be Dedekind infinite, no such reflection is available but we can try to argue as follows. Fix $a_1 \in A$, then pick $a_2 \in A \smallsetminus \{a_1\}$, $a_3 \in A \smallsetminus \{a_1, a_2\}$, etc, and in general choose $a_{n+1} \in A \smallsetminus \{a_1, a_2, \ldots, a_n\}$. Since A is infinite, a finite set $\{a_1, a_2, \ldots, a_n\}$ cannot exhaust A, so it will always be possible to choose a_{n+1}, and the induction seems to go through.

However, the problem here is that—unlike the Dedekind infinite case where a reflection was available—there is no mechanism to specify a_{n+1} uniquely in terms of $a_1, a_2, \ldots, a_n$. Thus the argument requires infinitely many arbitrary choices—a process that can only be formalized using the *Axiom of Choice.*

5.5 The Axiom of Choice vs Effectiveness

Recall from Sect. 1.6 that a *partition* is a family of pairwise-disjoint nonempty sets, and that we say P *is a partition of A* if P is a partition whose union equals A.

Definition 287 (Choice Set). A *choice set* for a partition P is a set containing exactly one element from each set of the partition P. More precisely, C *is a choice set for* P if $C \subseteq \bigcup \mathrm{P}$ and $C \cap E$ is a singleton for every $E \in \mathrm{P}$.

We now state the *Axiom of Choice,* henceforth referred to as "AC."

AC (The Axiom of Choice). Every partition has a choice set.

Whether AC is a "self-evident mathematical principle" or not was initially a matter of controversy, although many mathematicians do find it acceptable. However, the introduction of AC as a separate explicit axiom (by Zermelo [84, 86]) eventually helped to mitigate the debate, since now one could sharply distinguish between mathematical results which use the AC and the ones which do not, and so mathematicians could individually make (or postpone) the choice to accept or reject the Axiom of Choice.

In the case where the partition P is finite,[1] the validity of the principle AC can formally be derived by induction (on the size of P), but since nobody objects to making finitely many choices from finitely many sets, it is common to encounter informal proof-fragments such as

> ...Since the sets $A_1, A_2, \dots, A_n$ are nonempty, let us choose and fix elements $a_1 \in A_1,\ a_2 \in A_2,\ \dots,\ a_n \in A_n$.

The use of AC therefore is necessary only when the partition P is infinite. For many standard results of mathematics, the full general form needs AC, while special "finite" cases can be proved without AC. For example the proof that an arbitrary vector space has a basis requires AC, but one can prove that every finite dimensional vector space has a basis without using AC.

Effectiveness and Effectively Defined Choice Sets

In some cases, even when the partition P is infinite one can (by exploiting some additional structure and properties of the underlying set or partition) explicitly state a rule which determines a unique member of E for each E in P. This is expressed by saying that a member of E can be "effectively and uniquely determined" from E. In this case, a choice set C can be *effectively specified* or *effectively defined* from P, and the use of AC is not needed. Such a choice set C will be called an *effective choice set.*

Example 288. If P is any partition of the set **N** of natural numbers (which is naturally well-ordered), then one can effectively define a choice set C by choosing

[1]Note that if P is finite, the sets *in* P may be infinite. We should be careful to distinguish between "a partition being finite" and "the sets in the partition being finite." For example, the partition of the set of integers into even and odd integers is a finite partition consisting of infinite sets, while the partition $\{\{2n, 2n+1\} \mid n \in \mathbf{Z}\}$ of the integers is an infinite partition consisting of finite sets.

the least element of each set of the partition P, i.e., by setting $C = \{\min(E) \mid E \in \mathsf{P}\}$. Notice how C is effectively specified from P.

Thus *every partition of the set of natural numbers has an effective choice set.*

The concept of *effectiveness* (or *effective specification*) will be encountered occasionally throughout this book. We will use it as an informal intuitive notion, without attempting to give a formal definition.[2] Kuratowski [45, p. 254] explains that effectiveness concerns ways of proving existence theorems, i.e., theorems of the form "there exists x having property P." Such a theorem is said to be proved effectively if one can explicitly define a specific object a and prove that a has property P.

We have already used effective choice sets in proving that $\mathbf{N} \preccurlyeq A$ if one is given a one-to-one reflection $h: A \to A$ and $a \in A \smallsetminus h[A]$. The sets $A \supsetneq h[A] \supsetneq h[h[A]] \supsetneq \cdots$ then keep decreasing, producing pairwise disjoint sets $A_1 = A \smallsetminus h[A]$, $A_2 = h[A] \smallsetminus h[h[A]]$, etc. Since $h(a) \in A_2$, $h(h(a)) \in A_3$, etc, so the elements $a, h(a), h(h(a)), \ldots$ form an effective choice set for the family $\{A_n \mid n \in \mathbf{N}\}$, thereby *effectively proving* $\mathbf{N} \preccurlyeq A$.

Problem 289. *Let* $\mathbf{R}/\mathbf{Z}$ *denote the partition of* $\mathbf{R}$ *consisting of all sets of the form* $a + \mathbf{Z}$ *with* a *any real, where* $a + \mathbf{Z} := \{a + x \mid x \in \mathbf{Z}\}$. *In other words,* $\mathbf{R}/\mathbf{Z}$ *is the partition given by the equivalence relation* $\sim_{\mathbf{Z}}$ *on* $\mathbf{R}$, *where we define* $x \sim_{\mathbf{Z}} y \Leftrightarrow x - y \in \mathbf{Z}$, *for* $x, y \in \mathbf{R}$.[3] *Find an effective choice set for this partition* $\mathbf{R}/\mathbf{Z}$ *of* $\mathbf{R}$.

Problem 290. *Find effective choice sets for the partitions of the equivalence relations in Problems 46 and 49 of Chap. 1.*

Among the equivalence relations studied by the ancient Greek geometers (e.g., congruence and similarity mappings) was *commensurability of length.* Say that the positive reals $x, y \in \mathbf{R}^+$ are *commensurable* if x/y is rational. We can then ask: Can we define a choice set for the partition determined by the commensurability relation?

Commensurability also has an essentially equivalent "additive" version where two reals in $\mathbf{R}$ are defined to be equivalent if they differ by a rational number. The question of defining a choice set for commensurability then becomes equivalent to the following problem.

Problem 291. *Let* $\mathbf{R}/\mathbf{Q}$ *denote the partition of* $\mathbf{R}$ *consisting of all sets of the form* $a + \mathbf{Q}$ *with* a *any real, where* $a + \mathbf{Q} := \{a + x \mid x \in \mathbf{Q}\}$. *In other words,* $\mathbf{R}/\mathbf{Q}$ *is the partition given by the equivalence relation* $\sim_{\mathbf{Q}}$ *on* $\mathbf{R}$, *where we define* $x \sim_{\mathbf{Q}} y \Leftrightarrow x - y \in \mathbf{Q}$, *for* $x, y \in \mathbf{R}$.[4] *Can you define a choice set for this partition* $\mathbf{R}/\mathbf{Q}$ *of* $\mathbf{R}$?

---

[2]Effectiveness is a *metamathematical* notion, and degrees of effectiveness (which depends on the complexity of the specification or rule) are studied in areas of mathematical logic such as recursion theory.

[3]This is the coset decomposition of the additive group $\mathbf{R}$ modulo the subgroup $\mathbf{Z}$.

[4]This is the coset decomposition of the additive group $\mathbf{R}$ modulo the subgroup $\mathbf{Q}$, and is sometimes called the *Vitali partition.*

It is instructive for the reader to try to define a choice set for $\mathbf{R}/\mathbf{Q}$, but there is no reason to get discouraged if such a choice set seems too elusive to define. From the work of Feferman, it is now known that the existence of a choice set for the partition $\mathbf{R}/\mathbf{Q}$ cannot be proved without appealing to the Axiom of Choice, and even if the use of AC is allowed, no effectively defined set can be proved (without additional axioms) to be a choice set for $\mathbf{R}/\mathbf{Q}$.

Thus for some partitions effective choice sets can be defined without using AC, but there are partitions which have no effective choice sets, making the use of AC essential to obtain choice sets in such partitions.

This is illustrated by Russell's example of the millionaire who bought $\aleph_0$ pairs of socks and $\aleph_0$ pairs of boots. The question is: Can we make a selection of socks with exactly one sock from each pair, and similarly for boots? Russell, who called the Axiom of Choice the *multiplicative axiom,* wrote:

> [I]t can be done with the boots, but not with the socks ... The reason for the difference is this: Among boots we can distinguish right and left, and therefore we can make a selection of one out of each pair, namely, we can choose all the right boots or all the left boots; but with socks no such principle of selection suggests itself, and we cannot be sure, unless we assume the multiplicative axiom, that there is any class consisting of one sock out of each pair. ... [W]ith the socks we shall have to choose arbitrarily, with each pair, which to put first; and an infinite number of arbitrary choices is an impossibility. Unless we can find a *rule* for selecting, i.e., a relation which is a selector, we do not know that a selection is even theoretically possible. [68, p. 126]

Thus AC may be needed even when every set in the partition is finite (or even of cardinality 2), unless some additional structure can be exploited.

Problem 292. *Let* C *be a collection of pairwise disjoint nonempty finite sets of complex numbers. Show that* C *has an effective choice set.*

Problem 293. *For an equivalence relation* $\sim$ *on a set* A*, a function* $F\colon A \to A$ *is called a* selector *for* $\sim$ *if* F *is a complete invariant for* $\sim$ *(i.e.,* $x \sim y \Leftrightarrow F(x) = F(y)$*) and* $F(x) \sim x$ *for all* $x \in A$*. Prove that*

1. *AC holds if and only if every equivalence relation has a selector.*
2. *An equivalence relation has an effective selector if and only if the corresponding partition has an effective choice set.*

Problem 294. *Show that AC is equivalent to the following statement:* If $F\colon X \to Y$ is surjective then there is $G\colon Y \to X$ such that the function $F \circ G\colon Y \to Y$ is the identity mapping on Y.

If R is a relation we say that F is a *uniformization of* R if $F \subseteq R$, F is a function, and $\mathrm{dom}(F) = \mathrm{dom}(R)$.

Problem 295. *Show that AC is equivalent to the statement every relation has a uniformization.*

AC via Choice Functions

The form of AC we have been using so far is called the *partition version* of AC, where the sets from which elements are chosen are required to be pairwise disjoint. It is generally more useful to use a formulation of AC where the sets from which elements are to be chosen are not required to be disjoint.

Definition 296 (Choice Functions). If C is any family of nonempty sets (i.e., if $E \in \mathsf{C} \Rightarrow E \neq \emptyset$), then a *choice function* for C is any function $F\colon \mathsf{C} \to \bigcup\mathsf{C}$ such that $F(E) \in E$ for all $E \in \mathsf{C}$.

AC1 (Choice Function Version of AC). Every family of nonempty sets has a choice function.

Problem 297. *The principle AC1 is equivalent to the principle AC.*

[Hint: If Y is any collection of nonempty sets, then $\{\{E\} \times E \mid E \in \mathsf{Y}\}$ is a closely related collection of pairwise disjoint sets.]

A special case of AC1 is obtained by taking $\mathsf{C} = \mathbf{P}^*(A)$, where A is any set and $\mathbf{P}^*(A) := \mathbf{P}(A) \smallsetminus \{\emptyset\}$ denotes the collection of all nonempty subsets of A. In this case a choice function $F\colon \mathbf{P}^*(A) \to A$ is also called a *choice function for the set* A. Note that this special case is actually equivalent to AC1, as we can restrict any choice function for the set $A := \bigcup\mathsf{C}$ to the subfamily C. Thus AC1 is sometimes expressed by saying *every set has a choice function.*

AC1 can also be restated in terms of indexed families of sets as follows.

AC1 (Indexed Family Version). If $\langle A_i \mid i \in I\rangle$ is an indexed family of sets with $A_i \neq \emptyset$ for all $i \in I$, then there is "choice function" $\varphi\colon I \to \bigcup_i A_i$ such that $\varphi(i) \in A_i$ for all $i \in I$.

Problem 298. *The indexed family version above is also equivalent to AC.*

[Hint: If $\langle A_i \mid i \in I\rangle$ is an indexed family of nonempty sets then $\{\{i\} \times A_i \mid i \in I\}$ is partition, and any choice set for this partition is a function which satisfies the needed condition.]

5.6 $\aleph_0$ and Countable Sets

Theorem 299. *If $A \subseteq \mathbf{N}$ and A is infinite then $A \sim \mathbf{N}$.*

Proof. Using the well-ordering property of $\mathbf{N}$, define $f\colon \mathbf{N} \to \mathbf{N}$ by recursion as follows: Let $f(1)$ be the least element of A and $f(n+1)$ be the least element of $A \smallsetminus \{f(1), f(2), \dots, f(n)\}$. The recursion proceeds without halting since A is infinite. It is then easily verified that f is a bijection.

For a more formal proof, let a be the least element of A, let $h: \mathbf{N} \to \mathbf{N}$ be the function defined by $h(n) :=$ the least element of A greater than n, and apply the basic principle of recursive definition (Theorem 146). □

Corollary 300. *For a cardinal κ, $\kappa \leqslant \aleph_0$ if and only if κ is finite or $\kappa = \aleph_0$.*

Definition 301 (Denumerable and Countable Sets). A is *denumerable* if $A \sim \mathbf{N}$, i.e., if $|A| = \aleph_0$. A set is *countable* if it is denumerable or finite.

Corollary 302. *A is countable if and only if $A \preccurlyeq \mathbf{N}$ if and only if $|A| \leqslant \aleph_0$. Thus a subset of a countable set is countable.*

Corollary 303. *Any infinite subset of a denumerable set must have cardinality $\aleph_0$ and so must itself be denumerable.*

Effectiveness for equinumerosity and cardinal equality. The notion of effectiveness plays an important role in similarity (equinumerosity) of sets and equalities between cardinals. We say that a set A is *effectively equinumerous* or *effectively similar* to a set B if one can effectively define a bijection between A and B. Two cardinals κ and μ are said to be *effectively equal,* expressed by saying that "$\kappa = \mu$ effectively" if some set of cardinality κ is effectively equinumerous to some set of cardinality μ. In particular, a set A is said to be *effectively denumerable* if there is an effectively defined bijection between A and $\mathbf{N}$.

The map $n \mapsto n+1$ establishes a bijection between $\mathbf{J}$ and $\mathbf{N}$ so $\mathbf{J} \sim \mathbf{N}$ effectively. But $\mathbf{J}$ is the disjoint union of $\mathbf{N}$ and the singleton $\{0\}$, so $\aleph_0 + 1 = \aleph_0$ effectively. It follows by induction that

Problem 304. $\aleph_0 + n = \aleph_0$ *effectively, for all $n \in \mathbf{J}$.*

Problem 305. *Prove that $\aleph_0 + \aleph_0 = \aleph_0$ effectively, and so $2\aleph_0 = \aleph_0$ effectively. Prove that $n\aleph_0 = \aleph_0$, for all $n \in \mathbf{J}$.*

Problem 306. *Prove that the range of any function with countable domain is countable. Prove that a nonempty set is countable if and only if it is the range of a function with domain* $\mathbf{N}$.

Terminology overview. Recall that an *infinite sequence* is a function with domain $\mathbf{N}$. The *terms* of an infinite sequence are the elements of its range. The terms are said to be arranged *without repetition* if the sequence is one-to-one; otherwise we say that the sequence has *repeated terms.*

Definition 301 and Problem 306 can then be restated as follows.

> *A set is denumerable if and only if its elements can be arranged in an infinite sequence without repetition. A nonempty set is countable if and only if its members can be arranged in an infinite sequence, with repetitions allowed.*

Definition 307. An *enumeration* of a nonempty countable set A is a sequence whose range equals A.

If $a\colon \mathbf{N} \to A$ is an enumeration of A (i.e., $\operatorname{ran}(a) = A$), we informally express the fact by putting $a_n := a(n)$ and writing

$$A = \{a_1, a_2, \ldots, a_n, \ldots\},$$

or by saying that "A is enumerated as $a_1, a_2, \ldots, a_n, \ldots$."

Examples of denumerable sets are readily obtained. Any infinite subset of **N** is denumerable. Moreover the set **Z** of all integers—positive, negative, or zero—is also effectively denumerable (why?).

A more interesting example of a denumerable set is the following.

Theorem 308 (Cantor). *The set of ratios is effectively denumerable.*

Proof. Let the *rank* of ratio be defined as the sum of the numerator and denominator when it is expressed in lowest terms (reduced form). The smallest possible rank is 2, and it is easy to show that there are at most $n - 1$ ratios of rank n. Now arrange the ratios not by their order of magnitude, but so that the ratios with a smaller rank come before the ones with a larger rank, and if two ratios have the same rank, then put the one with a smaller numerator before the one with a larger numerator. This arranges the ratios in the sequence

$$\overbrace{\frac{1}{1}}^{2},\ \overbrace{\frac{1}{2}, \frac{2}{1}}^{3},\ \overbrace{\frac{1}{3}, \frac{3}{1}}^{4},\ \overbrace{\frac{1}{4}, \frac{2}{3}, \frac{3}{2}, \frac{4}{1}}^{5},\ \overbrace{\frac{1}{5}, \frac{5}{1}}^{6},\ \overbrace{\frac{1}{6}, \frac{2}{5}, \frac{3}{4}, \frac{4}{3}, \frac{5}{2}, \frac{6}{1}}^{7},\ \ldots$$

where the number above the brace in a group is the rank for that group. □

Remark. The function φ defined as $\varphi(n) :=$ the number of reduced ratios of rank n is important in number-theory and is known as *Euler's φ-function.*

Problem 309. *Show that the set* **Q** *of all rational numbers, positive, negative, or zero, is effectively denumerable. In particular,* $|\mathbf{Q}| = \aleph_0$.

Effective Enumeration of N x N

One can naturally arrange the members of $\mathbf{N} \times \mathbf{N}$, the set of all pairs of natural numbers, into the following "infinite matrix," where the ordered pair $\langle m, n\rangle$ occupies the entry in row m column n:

$$\begin{matrix}
\langle 1,1\rangle & \langle 1,2\rangle & \langle 1,3\rangle & \langle 1,4\rangle & \ldots & \langle 1,n\rangle & \ldots\\
\langle 2,1\rangle & \langle 2,2\rangle & \langle 2,3\rangle & \langle 2,4\rangle & \ldots & \langle 2,n\rangle & \ldots\\
\langle 3,1\rangle & \langle 3,2\rangle & \langle 3,3\rangle & \langle 3,4\rangle & \ldots & \langle 3,n\rangle & \ldots\\
\vdots & & & & \ddots & & \ldots
\end{matrix}$$

We can also arrange the natural numbers $\mathbf{N}$ into a similar infinite matrix as shown below, in which the top row consists of the odd natural numbers, and every other row is obtained by doubling the previous row:

$$\begin{array}{ccccccc} 1 & 3 & 5 & 7 & \dots & 2n-1 & \dots \\ 2 & 6 & 10 & 14 & \dots & 2(2n-1) & \dots \\ 4 & 12 & 20 & 28 & \dots & 4(2n-1) & \dots \\ \vdots & & & & \ddots & & \dots \end{array}$$

Notice that here the entry in row m column n is $2^{m-1}(2n-1)$.

Thus, by letting the ordered pair $\langle m, n\rangle$ in row m and column n of the first matrix correspond to the natural number $2^{m-1}(2n-1)$ occurring at the same position (row m and column n) of the second matrix, we get an effective enumeration of $\mathbf{N} \times \mathbf{N}$.

Problem 310. *The mapping $\langle m, n\rangle \mapsto 2^{m-1}(2n-1)$ is an effective bijection from $\mathbf{N} \times \mathbf{N}$ onto $\mathbf{N}$. Thus $\mathbf{N} \times \mathbf{N}$ is effectively denumerable and $\aleph_0\aleph_0 = \aleph_0$.*

Problem 311. *If A and B are denumerable then so is $A \times B$. If A and B are effectively denumerable then so is $A \times B$. (And similarly with "denumerable" replaced by "countable.")*

We write $\aleph_0^2$ is an abbreviation for $\aleph_0\aleph_0$, so that $\aleph_0^2 = \aleph_0$. We can inductively define $\aleph_0^n$ for $n \in \mathbf{N}$ by letting $\aleph_0^1 := \aleph_0$ and $\aleph_0^{n+1} := \aleph_0^n\aleph_0$.

Problem 312. *Show that $\aleph_0^n = \aleph_0$ for all $n \in \mathbf{N}$.*

We thus get many examples of countable sets. For example, the set of all points (x, y) in the Cartesian plane with rational coordinates (i.e., with both $x, y \in \mathbf{Q}$) is countable, and similarly for n-dimensional space for $n \in \mathbf{N}$. The set of all triples of natural numbers (or rational numbers) is countable.

The next problem gives another effective enumeration of $\mathbf{N} \times \mathbf{N}$.

Problem 313. *Consider the following arrangement of $\mathbf{N} \times \mathbf{N}$*

$$\langle 1,1\rangle, \langle 1,2\rangle, \langle 2,1\rangle, \langle 1,3\rangle, \langle 2,2\rangle, \langle 3,1\rangle, \langle 1,4\rangle, \langle 2,3\rangle, \langle 3,2\rangle, \langle 4,1\rangle, \dots,$$

in which $\langle m, n\rangle$ precedes $\langle m', n'\rangle$ if either $m + n < m' + n'$ or $m + n = m' + n'$ and $m < m'$. Show that in the above enumeration the pair $\langle m, n\rangle$ comes at position $\frac{1}{2}(m+n)(m+n-1) + m$, and therefore the mapping

$$\langle m, n\rangle \mapsto \frac{(m+n)(m+n-1)}{2} + m,$$

gives us another effective bijection from $\mathbf{N} \times \mathbf{N}$ onto $\mathbf{N}$.

We do not prefer any specific effective bijection from $\mathbf{N} \times \mathbf{N}$ onto $\mathbf{N}$ over any other, but instead record their useful properties in the following form, obtained by considering the inverse of any such effective bijection.

Problem 314 (Effective Pairing Functions). *There are effective "pairing functions" $\pi: \mathbf{N} \times \mathbf{N} \to \mathbf{N}$ (bijective) and $\lambda, \rho: \mathbf{N} \to \mathbf{N}$ (surjective) such that for all $m, n, k \in \mathbf{N}$:*

$$\lambda(\pi(m,n)) = m, \ \rho(\pi(m,n)) = n, \quad \text{and} \quad \pi(\lambda(k), \rho(k)) = k.$$

In particular, for all $m, n \in \mathbf{N}$ there is $k \in \mathbf{N}$ with $m = \lambda(k)$ and $n = \rho(k)$.

Problem 315. *Let π be the pairing function of Problem 313 and λ, ρ be as in Problem 314. Show that*

1. *$m < m' \Rightarrow \pi(m,n) < \pi(m',n)$ and $n < n' \Rightarrow \pi(m,n) < \pi(m,n')$.*
2. *$\lambda(n) \leq n$ and $\rho(n) \leq n$ for all n.*
3. *For all k there are infinitely many n with $\lambda(n) = k$ (and similarly for ρ).*

Suppose that $\langle f_m \mid m \in \mathbf{N} \rangle$ is a sequence of sequences, where each f_m is a function with domain $\mathbf{N}$ which enumerates the set $A_m := \operatorname{ran}(f_m)$. Fix effective functions π, λ, and ρ as in Problem 314, and define a sequence g by setting

$$g(k) := f_{\lambda(k)}(\rho(k)) \qquad (k \in \mathbf{N}).$$

Note that g effectively "combines" the sequence of sequences $\langle f_m \mid m \in \mathbf{N} \rangle$ into a single sequence in the following sense:

1. We have $f_m(n) = g(\pi(m,n))$, and so each f_m can be recovered from g as a "subsequence of g" given by the mapping $n \mapsto g(\pi(m.n))$.
2. We have $\operatorname{ran}(g) = \bigcup_m \operatorname{ran}(f_m)$, and so the set enumerated by g is the union of the sets enumerated by the functions f_m.

We summarize this as follows.

Proposition 316. *A given sequence $\langle f_n \mid n \in \mathbf{N} \rangle$ of enumerations of sets $\langle A_n \mid n \in \mathbf{N} \rangle$ (with $\operatorname{ran}(f_n) = A_n$) can be effectively combined into a single enumeration of the union $A := \bigcup_n A_n$ of the sets.*

Countable Union of Countable Sets

We want to use the last result to establish the useful fact that *the union of a countable family of countable sets is countable.* We can attempt to reason as follows. Let $\langle A_n \mid n \in \mathbf{N} \rangle$ be a sequence of nonempty countable sets, and enumerate each set A_m as

$$A_m = \{a_{m,1}, a_{m,2}, \ldots, a_{m,n}, \ldots\} \qquad (m \in \mathbf{N}).$$

Then the union $\bigcup_{m \in \mathbf{N}} A_m$ can be enumerated as

$$\bigcup_{m \in \mathbf{N}} A_m = \{a_{\lambda(k),\rho(k)} \mid k \in \mathbf{N}\},$$

and so $\bigcup_{m \in \mathbf{N}} A_m$ is countable.

But this proof is not effective. It makes a subtle use of the Axiom of Choice, since (unlike Proposition 316) no sequence of enumerations of the sets A_m is given beforehand. Each set A_m can be enumerated in many different ways and saying "A_m is enumerated as $A_m = \{a_{m,1}, a_{m,2}, \dots, a_{m,n}, \dots\}$" involves *implicitly choosing and fixing* one such enumeration. Since we have an infinite sequence of sets, this results in an infinite number of choices, requiring AC.[5]

The proof given below (for Proposition 317) makes this use of AC explicit. However, since we will be making "at most countably many" choices, the full general version of AC will not be used in the proof. Instead, the following special case of the Axiom of Choice, known as the *Countable Axiom of Choice* or CAC, will suffice.

5.7 The Countable and Dependent Axioms of Choice

CAC (The Countable Axiom of Choice). Every countable family of nonempty sets has a choice function: If I is countable and $\langle A_i \mid i \in I \rangle$ is a family of sets with $A_i \neq \emptyset$ for all $i \in I$ then there is a choice function $\varphi : I \to \bigcup_{i \in I} A_i$ such that $\varphi(i) \in A_i$ for all $i \in I$.

We then have:

Proposition 317 (CAC). *A countable union of countable sets is countable: If I is countable and A_i is countable for each $i \in I$, then $\bigcup_{i \in I} A_i$ is countable.*

Proof. Without loss of generality, we may assume that I and the sets A_i, for all $i \in I$, are nonempty. Fix effective pairing functions λ and ρ as in Problem 314, so that for all m, n there is k with $m = \lambda(k)$ and $n = \rho(k)$. Since I is nonempty countable, it can be enumerated by some function $h : \mathbf{N} \to I$ with $\operatorname{ran}(h) = I$. For each $i \in I$, define

$$\begin{aligned} \mathsf{E}_i &:= \{f \mid f \text{ is an enumeration of } A_i\} \\ &= \{f \mid f : \mathbf{N} \to A_i \text{ with } \operatorname{ran}(f) = A_i\}. \end{aligned}$$

[5] AC is needed even if the sets A_m are all finite, as illustrated by Russell's example: Given $\aleph_0$ pairs of socks and $\aleph_0$ pairs of boots, how many socks do we have in total, and how many boots? With boots, the answer is $\aleph_0$, but the socks may form an infinite Dedekind finite set and the answer may be a non-reflexive cardinal.

Each E_i is nonempty since A_i is nonempty and countable. Hence by CAC, there is a choice function φ with $\mathrm{dom}(\varphi) = I$ such that $\varphi(i) \in \mathsf{E}_i$ for each $i \in I$. Thus $\varphi(i)$ is a function enumerating A_i and let us abbreviate $\varphi(i)$ as ϕ_i (for each $i \in I$). Finally, define g by

$$g(k) := \phi_{h(\lambda(k))}(\rho(k)).$$

Then it is routine to verify that g enumerates $\bigcup_{i \in I} A_i$. □

The reader may once again compare Proposition 316 with Proposition 317 and note how the former can be proved effectively while the latter requires the use of CAC.

Problem 318. *Prove that if Σ is any nonempty countable alphabet (= set), then the set Σ^* of all words over Σ (= finite sequences from Σ) is countable. Moreover, if Σ is effectively countable, then so is Σ^*.*

Problem 319 (CAC). *Let $A_1, A_2, \ldots$ be an infinite sequence of pairwise disjoint sets such that $A_m \sim A_n$ for all $m, n \in \mathbf{N}$. Let $\kappa = |A_1|$. If $I \subseteq \mathbf{N}$ and I is infinite, then*

$$\left| \bigcup_{n \in I} A_n \right| = \kappa \cdot \aleph_0 = \left| \bigcup_{n \in \mathbf{N}} A_n \right|.$$

The problem below is relevant to the proof of the Cantor–Bernstein theorem.

Problem 320. *Suppose that we have a one-to-one reflection $f : C \to C$, and let A be a subset of C disjoint from $f[C]$, that is, with $A \subseteq C \smallsetminus f[C]$. Define the sets $A_1, A_2, A_3, \ldots$ recursively as follows:*

$$A_1 := f[A] \quad \textit{and} \quad A_{n+1} := f[A_n].$$

1. *Show that the sets $A, A_1, A_2, \ldots$ are pairwise disjoint.*
2. *Put $A^* := A_1 \cup A_2 \cup \cdots = \bigcup_{n \geq 1} A_n$. Prove effectively that $A \cup A^* \sim A^*$.*

Theorem 321 (CAC). *Every infinite set is Dedekind infinite.*

Proof. For each $n \in \mathbf{N}$, the following set is nonempty since A is infinite:

$$\mathsf{F}_n := \{f \mid f : \mathbf{N} \to A \text{ and } |f[\mathbf{N}]| \geqslant n\}.$$

By CAC, select a sequence $\langle f_n \mid n \in \mathbf{N} \rangle$ with $f_n \in \mathsf{F}_n$ for all $n \in \mathbf{N}$.

As in Proposition 316, combine the f_n's into a single $g : \mathbf{N} \to A$ where $g(n) := f_{\lambda(n)}(\rho(n))$. Then $g[\mathbf{N}] = \bigcup_n f_n[\mathbf{N}]$ is infinite (since $|f_n[\mathbf{N}]| \geqslant n$) and countable, so $g[\mathbf{N}] \sim \mathbf{N}$. Hence A is Dedekind infinite.

(Or, directly define an injection $h : \mathbf{N} \to A$ by recursion: $h(1) := f_1(1)$ and $h(n+1) := f_{n+1}(k)$ where k is the least number such that $f_{n+1}(k) \notin \{h(1), h(2), \ldots, h(n)\}$, which is well defined since $|\mathrm{ran}(f_{n+1})| \geqslant n + 1$.) □

So *under countable choice, the two notions of finite sets coincide: A set A is infinite (in either sense) if and only if* $\mathbf{N} \preccurlyeq A$. But note that if A is Dedekind infinite, we get $\mathbf{N} \preccurlyeq A$ *effectively* (why?), while for a general infinite set A, we get $\mathbf{N} \preccurlyeq A$ only by appealing to choice (so non-effectively).

Problem 322. *Show without using any form of AC that if A is infinite then* $\mathrm{P}(\mathrm{P}(A))$ *is Dedekind infinite.*

It follows that A is finite if and only if $\mathrm{P}(\mathrm{P}(A))$ is Dedekind finite, giving a characterization of (inductive) finiteness in terms of Dedekind finiteness.

The Axiom of Dependent Choice

The *Axiom of Dependent Choice* (DC) allows one to make a sequence of choices where each choice may depend on the previous one. We will use it to derive later that an order which is not well-ordered (or a relation which is not well-founded) contains a strictly decreasing sequence of elements.

DC (The Axiom of Dependent Choice). Let R be a relation on A such that for all $x \in A$ there is $y \in A$ with xRy, and let $a \in A$. Then there is a sequence $\langle a_n \mid n \in \mathbf{N} \rangle \in A^{\mathbf{N}}$ such that $a_1 = a$ and $a_n R a_{n+1}$ for all $n \in \mathbf{N}$.

Proof (AC). Put $\mathrm{P}^*(A) := \mathrm{P}(A) \smallsetminus \{\emptyset\}$ and fix a choice function $\phi\colon \mathrm{P}^*(A) \to A$ such that $\phi(E) \in E$ for all $E \in \mathrm{P}^*(A)$. Define $g\colon A \to A$ by setting $g(x) := \phi(\{y \in A \mid xRy\})$. Then g is well defined by the given condition for the relation R. Hence by the principle of recursive definition there exists a function $f\colon \mathbf{N} \to A$ such that $f(1) = a$ and $f(n+1) = g(f(n))$ for all $n \in \mathbf{N}$. Finally, put $a_n := f(n)$. Then $a_1 = a$, and for all n we have

$$a_{n+1} = g(a_n) = \phi(\{y \in A \mid a_n R y\}) \in \{y \in A \mid a_n R y\},$$

hence $a_n R a_{n+1}$ for all n. □

DC is weaker than the full Axiom of Choice, but it is stronger than CAC.

Problem 323. *Show (without using any form of AC) that DC implies CAC.*

Problem 324. *Use DC to formalize the argument at the end of Sect. 5.4 and give a proof of Theorem 321 using DC instead of CAC.*

5.8 $\aleph_0 < \mathfrak{c}$: The Cardinality of the Continuum

Definition 325. An *interval* in $\mathbf{R}$ is a subset of $\mathbf{R}$ having one of the forms:

$$(a,b),\ (a,b],\ [a,b),\ [a,b];\ \ (-\infty,a),\ (a,\infty),\ (-\infty,a],\ [a,\infty);\ \text{ or } (-\infty,\infty),$$

The first four forms above are called *bounded* intervals, the next four are called *half-infinite* intervals, and the last interval $(-\infty, \infty)$ equals $\mathbf{R}$ itself.

An interval is *proper* if it contains at least two points, while the empty set $\emptyset = (a, a)$ and singleton sets $\{a\} = [a, a]$ are *improper* intervals.

Problem 326. *Prove that if $a < b$ are real numbers, then the interval (a, b) is effectively equinumerous with $(0, 1)$, the interval $[a, b]$ is effectively equinumerous with $[0, 1]$, and each of the intervals $(a, b]$ and $[a, b)$ is effectively equinumerous with $(0, 1]$, all via suitable linear mappings.*

The figure below shows the geometric view of a one-to-one correspondence between the line segment $\overline{AB}$ and the line segment $\overline{CD}$.

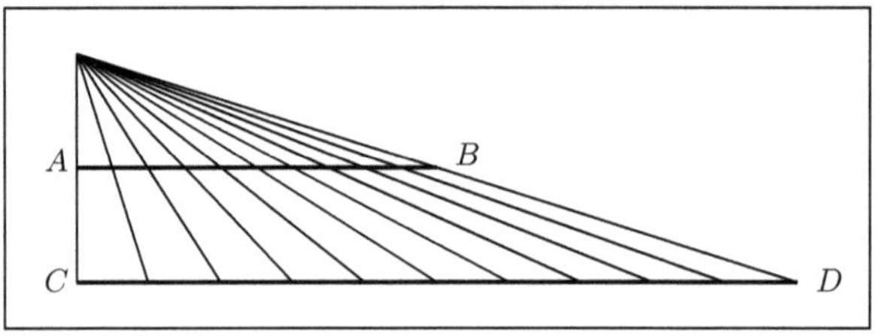

More interestingly, we have the following result.

Problem 327. $(0, 1] \sim (0, 1)$, $[0, 1] \sim [0, 1)$, *and so* $[0, 1] \sim (0, 1)$, *effectively.*

[Hint: For $(0, 1] \sim (0, 1)$, remove from $(0, 1]$ the set $\{1, \frac{1}{2}, \frac{1}{4}, \frac{1}{8}, \ldots\}$ and remove from $(0, 1)$ the set $\{\frac{1}{2}, \frac{1}{4}, \frac{1}{8}, \ldots\}$. Note that if $A \cap B = \emptyset = A \cap C$ then $B \sim C \Rightarrow A \cup B \sim A \cup C$.]

Corollary 328. *Any two bounded proper intervals in* $\mathbf{R}$, *whether open, half-open, or closed, are effectively equinumerous with each other.*

Problem 329. *For any $a, b \in \mathbf{R}$, $[a, \infty) \sim [0, \infty) \sim (-\infty, b]$, effectively.*

[Hint: Use the maps $x \mapsto x + a$ and $x \mapsto b - x$ defined on $[0, \infty)$.]

Problem 330. $[0, \infty) \sim (0, \infty)$, *effectively..*

[Hint: Remove $\mathbf{J}$ and $\mathbf{N}$ from the intervals $[0, \infty)$ and $(0, \infty)$, respectively.]

Corollary 331. *Any two half-infinite intervals, whether open or closed, are effectively equinumerous with each other.*

The following result now implies that any half-infinite interval is effectively equinumerous with any bounded interval.

Problem 332. *Show that* $x \mapsto y = \dfrac{x}{x + 1}$ *maps the interval $(0, \infty)$ bijectively onto $(0, 1)$.*

The next figure geometrically illustrates how the ray $\overrightarrow{OA} \equiv (0, \infty)$ gets mapped onto the line segment $\overline{OB} \equiv (0, 1)$.

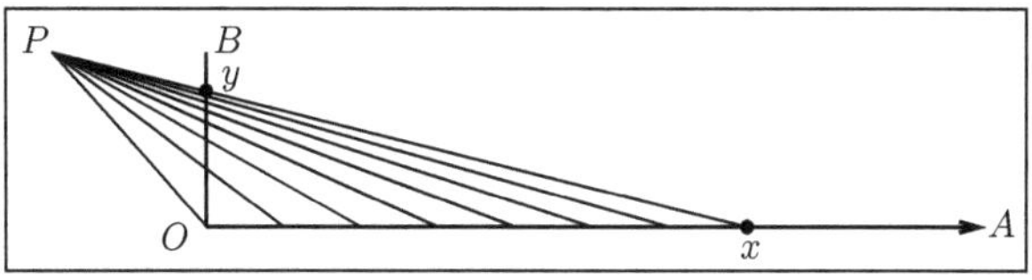

From what we have obtained so far, we see that

$$\mathbf{R} = (-\infty, 0) \cup \{0\} \cup (0, \infty) \sim (-1, 0) \cup \{0\} \cup (0, 1) = (-1, 1),$$

and so $\mathbf{R} = (-\infty, \infty)$ is effectively equinumerous with any bounded interval and so also with any half-infinite interval. We record this important result as

Corollary 333. *Any two proper intervals in* $\mathbf{R}$*, any of which may be bounded, half-infinite, or* $\mathbf{R}$ *itself, are effectively equinumerous with each other.*

[Note: The above results all hold for any ordered field, not just $\mathbf{R}$.]

By the last result, all proper real intervals have the same cardinal number which is denoted by $\mathfrak{c}$, and is called the *cardinality of the continuum.*

Definition 334. $\mathfrak{c} := |(0, 1]| = |[0, 1]| = |(0, 1)| = |\mathbf{R}|$.

Since $\mathbf{N} \subseteq \mathbf{R}$, it follows that

Problem 335. $\aleph_0 \leqslant \mathfrak{c}$.

Problem 336. *Using the earlier results of this section establish each of the following results* effectively*:*

1. $\mathfrak{c} + \mathfrak{c} = \mathfrak{c}$*, i.e.,* $2\mathfrak{c} = \mathfrak{c}$*; and so by induction we also have* $n \cdot \mathfrak{c} = \mathfrak{c}$.
2. $\aleph_0 \cdot \mathfrak{c} = \mathfrak{c}$.
3. $\mathfrak{c} + 1 = \mathfrak{c}$*; and so by induction* $\mathfrak{c} + n = \mathfrak{c}$.
4. $\mathfrak{c} + \aleph_0 = \mathfrak{c}$.

We now turn to the following remarkable result of Cantor, often expressed by the statement "$\mathbf{R}$ is uncountable." Unlike the set of rational numbers, the reals numbers cannot be exhaustively listed as a sequence.

Theorem 337 (Uncountability of R, Cantor). *No proper interval is countable. Hence* $\mathbf{R}$ *is not countable, that is,* $\mathbf{R} \not\preccurlyeq \mathbf{N}$*, and so* $\mathbf{N} \prec \mathbf{R}$.

Proof. Since all proper intervals are equinumerous to each other and to $\mathbf{R}$, it suffices to show that $[0, 1]$ is not countable, that is, $[0, 1]$ cannot be the range of any function with domain $\mathbf{N}$. We will show, following Cantor, that if $f : \mathbf{N} \to \mathbf{R}$ then its range

$f[\mathbf{N}] = \text{ran}(f)$ cannot include all of $[0, 1]$, i.e., there is $p \in [0, 1]$ which is not in the range of f, and so, in particular, f certainly cannot be a bijection between $\mathbf{N}$ and $[0, 1]$.

Given a bounded closed interval $I = [a, b]$ with $a < b$, we trisect $[a, b]$ and subdivide it into three equal subintervals each of length $\ell := \frac{1}{3}\,\text{len}(I) = \frac{b-a}{3}$, so that $a < a + \ell < a + 2\ell < b$. By removing the middle-third open interval, $I = [a, b]$ splits into two disjoint closed subintervals, called the *left-third* and *right-third* subintervals, as

$$L[I] := [a, a + \ell] = \text{left-third of } [a, b], \quad \text{and,}$$
$$R[I] := [a + 2\ell, b] = \text{right-third of } [a, b].$$

The figure below illustrates how this is done.

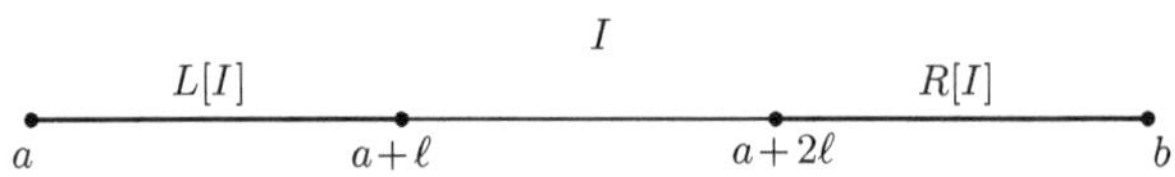

Trisecting the interval $I = [a,b]$, where $\ell := \frac{b-a}{3}$

Now let $f : \mathbf{N} \to \mathbf{R}$ be an arbitrary mapping, and put $a_n = f(n)$ for each $n \in \mathbf{N}$, which gives the sequence $\langle a_1, a_2, \ldots, a_n, \ldots \rangle$ enumerating the range of f. We will find an element $p \in [0, 1]$ which is not in the range of f, showing that f is not onto.

We say that *a set A avoids the real x* if and only if $x \notin A$. The crucial fact used in this proof is that *for any given interval I and any real x, one of the subintervals $L[I]$ or $R[I]$ will avoid x.*

Let $I = [0, 1]$. Then a_1 cannot be both in $L[I] = [0, \frac{1}{3}]$ and in $R[I] = [\frac{2}{3}, 1]$. We will take I_1 to be one of these two subintervals, making sure that I_1 avoids a_1. To be definite, let

$$I_1 = \begin{cases} L[I] & \text{if } a_1 \notin L[I], \\ R[I] & \text{otherwise.} \end{cases}$$

The point is that I_1 is a closed subinterval of I such that I_1 avoids a_1.

We continue in this fashion in stages, choosing closed intervals $I_1 \supseteq I_2 \supseteq \cdots \supseteq I_n$ such that I_n avoids a_n. To be definite, we specify how to go from stage-n to stage-$(n + 1)$. Given that I_n has been constructed in stage-n, put

$$I_{n+1} = \begin{cases} L[I_n] & \text{if } a_{n+1} \notin L[I_n], \\ R[I_n] & \text{otherwise.} \end{cases}$$

Then I_{n+1} is a closed subinterval of I_n such that I_{n+1} avoids a_{n+1}.

This gives us a sequence of nested and bounded closed intervals

$$I_1 \supseteq I_2 \supseteq \cdots \supseteq I_n \supseteq I_{n+1} \supseteq \cdots \quad \text{with } a_n \notin I_n \text{ for all } n,$$

and so by the Nested Interval Property there is $p \in \bigcap_{n=1}^{\infty} I_n$. But $p \neq a_n$ for any $n \in \mathbf{N}$, since $p \in I_n$ while $a_n \notin I_n$. So p is not in the range of f. □

Corollary 338. $\aleph_0 < \mathfrak{c}$.

The Cantor Machine

Note the effective nature of Cantor's proof: Given any sequence of real numbers $\langle a_n \mid n \in \mathbf{N}\rangle \in \mathbf{R}^{\mathbf{N}}$, the procedure in the above proof effectively and uniquely produces a point $p \in [0, 1]$ with $p \neq a_n$ for all n. This effective mapping $\langle a_n \rangle \mapsto p$ will be denoted by M, so that $\mathsf{M}\colon \mathbf{R}^{\mathbf{N}} \to [0, 1]$ with $\mathsf{M}(\langle a_n \rangle) \notin \{a_n \mid n \in \mathbf{N}\}$ for all sequences $\langle a_n \rangle \in \mathbf{R}^{\mathbf{N}}$. The mapping M, which is pictured below, will be referred to as the *Cantor Machine.*

$$\langle a_n \rangle \quad \longrightarrow \quad \boxed{\mathsf{M}} \quad \longrightarrow \mathsf{M}(\langle a_n \rangle) \in [0, 1] \setminus \{a_n \mid n \in \mathbf{N}\}$$

Input sequence of reals | Cantor Machine | Output real

Thus the phenomenon in Cantor's proof is summarized as follows: *Given a sequence of reals $\langle a_n \rangle$ as input, the Cantor Machine M responds by producing an output real $p = \mathsf{M}(\langle a_n \rangle) \in [0, 1]$ which differs from every term of the given sequence.* We informally express this by saying that the point p is *diagonalized away* from the given list of rcals $a_1, a_2, \ldots, u_n, \ldots$.

5.9 CH: The Continuum Hypothesis

We now have the following examples of distinct cardinal numbers:

$$0 < 1 < 2 < \cdots < n < n + 1 < \cdots \cdots < \aleph_0 < \mathfrak{c}.$$

By Corollary 300, we know that the sequence of cardinals from 0 to $\aleph_0$ is complete in the sense that other than the finite cardinals and $\aleph_0$ there is no cardinal which can be "placed in between them." The question now arises if the sequence of cardinals from 0 to $\mathfrak{c}$ displayed above is complete in the above sense, and it reduces to the question:

Is there a cardinal κ such that $\aleph_0 < \kappa < \mathfrak{c}$?

The *Continuum Hypothesis,* or *CH,* is the assertion that there is no such cardinal. CH is equivalent to the statement that for every subset A of $\mathbf{R}$, either A is countable (i.e., $|A| \leqslant \aleph_0$) or $A \sim \mathbf{R}$ (i.e., $|A| = \mathfrak{c}$). If CH is false, there would exist uncountable subsets of $\mathbf{R}$ not equinumerous to $\mathbf{R}$.

Cantor tried to decide if CH is true or not, but failed. Other mathematicians in early twentieth century also tried, but the question remained open. Much of research in set theory in the twentieth century was dominated by this question. We will return to the topic later.

5.10 More Countable Sets and Enumerations

Recall that $\mathbf{J} = \{0, 1, 2, \dots\}$ denotes the set of all finite (inductive) cardinals, $\mathbf{N} = \{1, 2, 3, \dots\}$ is the set of all natural numbers (nonzero finite cardinals), and $\mathbf{Z}$ the set of all integers (positive, negative, or zero). Given any set A, we let A^* denote the set of all finite sequences from A. The members of A^* are also called the *strings* or *words* over the *alphabet* A.

Problem 339. *Find an effective bijection between* $\mathbf{J}$ *and the collection of all finite subsets of* $\mathbf{J}$. *Conclude that the collection of all finite subsets of* $\mathbf{N}$ *is effectively denumerable, and more generally that the collection of all finite subsets of an effectively denumerable set is effectively denumerable.*

[Hint: Given $m \in \mathbf{J}$, consider the set A_m of all $k \in \mathbf{J}$ such that the bit at position k in the binary representation of m is 1, where the least significant bit is defined as position 0. In other words, $A_m = \{k \in \mathbf{J} \mid \lfloor m/2^k \rfloor \text{ is odd}\}$, where $\lfloor x \rfloor$ denotes the greatest integer not greater than x.]

Problem 340. *Show that the set* $\mathbf{N}^*$ *of all finite sequences of natural numbers is effectively bijective with the collection of all finite subsets of* $\mathbf{N}^*$. *Conclude that* $\mathbf{N}^*$ *is effectively denumerable.*

[Hint: Consider the mapping which sends a finite sequence $\langle n_1, n_2, \dots, n_k \rangle$ in $\mathbf{N}^*$ to the finite set $\{n_1,\ n_1 + n_2,\ \dots,\ n_1 + n_2 + \dots + n_k\}$.]

Problem 341. *Prove that the set of words over any nonempty effectively countable alphabet is effectively denumerable. In particular, the set of all computer programs in any programming language is effectively denumerable.*

Problem 342. *Prove that the set* $\mathbf{Z}[x]$ *of all polynomials with integer coefficients is effectively countable.*

Definition 343. A real number is *algebraic* if it is a root of some nonzero polynomial with integer coefficients. Otherwise, it is *transcendental.*

Problem 344. *Every rational number is algebraic, but there are infinitely many algebraic numbers which are not rational.*

Problem 345 (Cantor). *The set of algebraic numbers is countable.*

Corollary 346 (Cantor). *There exist transcendental numbers.*

Since every nonzero polynomial can have at most finitely many roots, one can establish Problem 345 using the result that a countable union of finite sets is countable (Proposition 317), but such a proof requires the use of CAC (the Countable Axiom of Choice) and so is not effective. However, it is in fact possible to prove that the set of algebraic numbers is effectively countable.

Problem 347. *Show that given an effectively countable family of finite subsets of* **R**, *their union is effectively countable. Conclude that the set of algebraic numbers is effectively countable.*

By the last problem, the algebraic numbers can be effectively enumerated in a specific sequence $a_1, a_2, \ldots, a_n, \ldots$. Since the Cantor Machine effectively "diagonalizes out" a real p different from all the a_n's, so Cantor's method is able to effectively specify a particular transcendental number.

Long before Cantor, Liouville had given examples of transcendental numbers which are even more effective. For example, he proved that the *Liouville Constant*

$$\sum_{n=1}^{\infty} \frac{1}{10^{n!}} = \frac{1}{10} + \frac{1}{10^2} + \frac{1}{10^6} + \frac{1}{10^{24}} + \cdots = 0.110001000000000000000001000\cdots$$

whose decimal expansion has the digit one at position $n!$ for every n with all other digits being zero, is a transcendental number.

Liouville's proof method was specific to number theory. Cantor's new method of proof, on the other hand, is applicable to much wider contexts beyond the theory of algebraic and transcendental numbers.

Problem 348. *Show that the rules*

$$f(1) := 1, \quad f(2n) := f(n) + 1, \quad \text{and} \quad f(2n+1) := 1/f(2n) \quad (n \in \mathbf{N})$$

define a unique function $f : \mathbf{N} \to \mathbf{Q}^+$ *which is in fact a one-to-one correspondence between* **N** *and the set* $\mathbf{Q}^+$ *of positive rational numbers.*

Chapter 6
Cardinal Arithmetic and the Cantor Set

Abstract We continue the basic theory of cardinals, covering the Cantor–Bernstein Theorem, arbitrary cardinal products and cardinal arithmetic, binary trees and the construction of the Cantor set, the identity $2^{\aleph_0} = \mathfrak{c}$ and effective bijections between familiar sets of cardinality $\mathfrak{c}$, Cantor's theorem and König's inequality, and the behavior of $\kappa^{\aleph_0}$ for various cardinals κ.

6.1 The Cantor–Bernstein Theorem

The following basic result says that for cardinals α and β, if $\alpha \leqslant \beta$ and $\beta \leqslant \alpha$ then $\alpha = \beta$. Among other things, it greatly facilitates cardinal arithmetic.

Theorem 349 (Cantor–Bernstein). *If $C \supseteq B$ and $f : C \to B$ is a one-to-one function "reflecting" C into the subset $f[C]$ of B so that $C \supseteq B \supseteq f[C]$, then $B \sim C$.*

Discussion and proof. Figure 6.1 shows the analogy with Royce's illustration of map.

Put $A := C \setminus B$. The set C is then viewed as a "country" with "provinces" A and B, and f is viewed as a "mapping" in the sense of cartography: *Country C has just two provinces A and B (Fig. 6.1a), and a perfect map C_1 of Country C is made upon the surface of Province B,* so that C_1 consists of a map A_1 of A and a map B_1 of B (Fig. 6.1b). Since the map is correct, B_1 must contain a "map of the map," C_2, consisting of A_2 and B_2; and B_2 must contain a "map of the map of the map"; and so on (Fig. 6.1c).

The one-to-one mapping f maps A onto its map $A_1 = f[A]$, so $A \sim A_1$, with A_1 disjoint from A (since $A_1 \subseteq B$). Similarly $A_2 = f[A_1]$ is similar (equinumerous) to both A and A_1, and is disjoint from both of them. So the "iterated maps" of Province A, shown shaded in Fig. 6.1c as $A_1, A_2, \ldots$, form an infinite sequence of pairwise disjoint "copies" of Province A:

$$A \sim A_1 \sim A_2 \sim A_3 \sim \cdots \qquad \text{(all similar and pairwise disjoint).}$$

A. Dasgupta, *Set Theory: With an Introduction to Real Point Sets*,
DOI 10.1007/978-1-4614-8854-5_6,

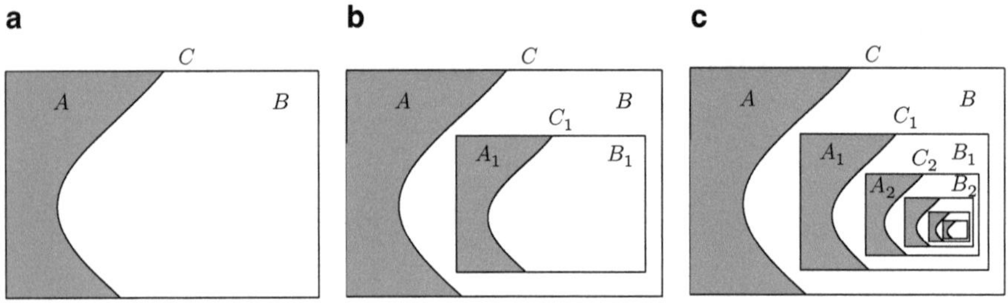

Fig. 6.1 Royce's illustration of map for the proof of the theorem. (**a**) Country C consists of two provinces: A (shown *shaded*) and B (shown *unshaded*); (**b**) A *map* C_1 of Country C is placed within Province B, so C_1 itself contains maps A_1 and B_1 for Provinces A and B; (**c**) The map C_1 in turn must contain a *"map of the map"* C_2, consisting of A_2 and B_2, and so on

Put $A^* := A_1 \cup A_2 \cup A_3 \cup \cdots$. Now a key observation is that $(A \cup A^*) \sim A^*$, since f "shifts" the pairwise disjoint sets $A, A_1, A_2, \ldots$ (shaded in the figure) to the "next" sets $A_1, A_2, A_3, \ldots$. More precisely, $f \restriction_{A \cup A^*}: A \cup A^* \to A^*$ is a bijection from $A \cup A^*$ onto A^*, since f is injective and

$$\begin{aligned} f[A \cup A^*] = f[A] \cup f[A^*] &= f[A] \cup f[A_1 \cup A_2 \cup \cdots] \\ &= A_1 \cup f[A_1] \cup f[A_2] \cup \cdots \\ &= A_1 \cup A_2 \cup A_3 \cup \cdots \;=\; A^*. \end{aligned}$$

Finally, let $E := C \smallsetminus (A \cup A^*)$, the entire remaining unshaded part in Fig. 6.1c. Then $C = (A \cup A^*) \cup E$ and $B = A^* \cup E$. Since $(A \cup A^*) \sim A^*$ and since E is disjoint from both $A \cup A^*$ and A^*, it follows that:

$$(A \cup A^*) \cup E \sim A^* \cup E, \qquad \text{or:} \qquad C \sim B. \qquad \square$$

It should be noted in the above proof that the final one-to-one correspondence, say g, between C and B is described as the mapping on C which fixes every point outside $A \cup A^*$ and sends each point x in $A \cup A^*$ to $f(x)$. Formally, the bijection $g: C \to B$ is the union of f restricted to $A \cup A^*$ with the identity map restricted to $C \smallsetminus (A \cup A^*)$, that is:

$$g(x) = \begin{cases} f(x) & \text{if } x \in (A \cup A^*) \\ x & \text{if } x \in C \smallsetminus (A \cup A^*) \end{cases}.$$

Therefore the Cantor–Bernstein theorem is an effective theorem: *A bijection $g: C \to B$ can be effectively specified in terms of the given function and sets.*

The theorem is often stated in the following equivalent "symmetric" form.

Theorem 350 (Cantor–Bernstein, Symmetric Version). *If $A \preccurlyeq B$ and $B \preccurlyeq A$ then $A \sim B$. Therefore, the relation $\leqslant$ defined on the cardinals is antisymmetric,* *i.e., if $\alpha \leqslant \beta$ and $\beta \leqslant \alpha$, or equivalently if $\alpha =^* \beta$, then $\alpha = \beta$. It follows that $\alpha \leqslant \beta$ if and only if $\alpha < \beta$ or $\alpha = \beta$.*

Problem 351. *Show that this symmetric version of the Cantor–Bernstein Theorem is equivalent to the earlier version.*

Problem 352. *The Cantor–Bernstein Theorem is equivalent to the assertion that if α, β, γ are cardinals with $\alpha + \beta + \gamma = \alpha$, then $\alpha + \beta = \alpha$.*

Historical note. The Cantor–Bernstein Theorem was conjectured by Cantor, partially proved by Schröder, and a full proof was published by Bernstein. But earlier than all of these, Dedekind had actually proved the theorem, but he never published his proof. The theorem is most often called the *Schröder–Bernstein Theorem,* and sometimes simply *Bernstein's Theorem.*

6.2 Arbitrary Sums and Products of Cardinals

Given an indexed family $\langle \alpha_i \mid i \in I \rangle$ of cardinal numbers, we wish to define the "general sum" $\sum_{i \in I} \alpha_i$ as follows. First use AC to choose representative sets A_i of cardinality α_i so that $|A_i| = \alpha_i$ for each $i \in I$. Then the sets $\{i\} \times A_i$ are pairwise disjoint, with $|\{i\} \times A_i| = \alpha_i$ $(i \in I)$. So we may define:

$$\sum_{i \in I} \alpha_i := \left| \bigcup_{i \in I} \{i\} \times A_i \right|.$$

For this definition to work properly, we need one more application of AC:

Problem 353 (Uniqueness of General Sum, AC). *Let $\langle A_i \mid i \in I \rangle$ and $\langle B_i \mid i \in I \rangle$ be indexed families of pairwise disjoint sets (i.e., for $i, j \in I$, $i \neq j \Rightarrow A_i \cap A_j = \emptyset = B_i \cap B_j$). If $A_i \sim B_i$ for each $i \in I$, then*

$$\bigcup_{i \in I} A_i \sim \bigcup_{i \in I} B_i.$$

Let $\langle \alpha_i \mid i \in I \rangle$ be an indexed family of cardinals. Without AC, we cannot guarantee the existence of a representative family of sets $\langle A_i \mid i \in I \rangle$ with $|A_i| = \alpha_i$ for $i \in I$. Even if such a family exists, we cannot assume, without AC, that if $|A_i| = |A'_i| = \alpha_i$ for all $i \in I$ then $\bigcup_{i \in I} \{i\} \times A_i \sim \bigcup_{i \in I} \{i\} \times A'_i$.

Definition 354 (Sum-Adequate Families of Cardinals). An indexed family of cardinals $\langle \alpha_i \mid i \in I \rangle$ is *sum-adequate* if there is a representative family of sets $\langle A_i \mid i \in I \rangle$ such that $|A_i| = \alpha_i$ for all $i \in I$ and whenever $A'_i \sim A_i$ for all $i \in I$, we have $\bigcup_{i \in I} \{i\} \times A_i \sim \bigcup_{i \in I} \{i\} \times A'_i$.

Under AC, every family of cardinals is sum-adequate. If AC is not assumed, arbitrary cardinal sums are significant only for sum-adequate families.[1]

Definition 355 (General Cardinal Sum). For a sum-adequate family of cardinal numbers $\langle \alpha_i \mid i \in I \rangle$, we define the *general cardinal sum* as:

$$\sum_{i \in I} \alpha_i := \left| \bigcup_{i \in I} \{i\} \times A_i \right|,$$

where each A_i is a representative set of cardinality α_i, i.e., $|A_i| = \alpha_i$ for $i \in I$. If $\langle \alpha_i \mid i \in I \rangle$ is not sum-adequate, we define the sum $\sum_{i \in I} \alpha_i$ to be zero.

Problem 356 (AC). *Assuming AC, show that the general cardinal sum of any indexed family of cardinals is well defined and unique: For any indexed family of cardinal numbers $\langle \alpha_i \mid i \in I \rangle$, there is a unique cardinal number α and a pairwise disjoint family $\langle A_i \mid i \in I \rangle$ of sets such that*

$$|A_i| = \alpha_i \text{ for all } i \in I, \qquad \text{and} \qquad \alpha = \left| \bigcup_{i \in I} A_i \right|.$$

Problem 357 (AC). *If $|A| = \alpha$, $|B| = \beta$, and $\alpha_b = \alpha$ for each $b \in B$, then*

$$\alpha\beta = \sum_{b \in B} \alpha_b = \sum_{b \in B} \alpha.$$

[Hint: $A \times B$ partitions as $\cup_{b \in B} A \times \{b\}$ with $A \sim A \times \{b\}$ for all $b \in B$.]

Problem 358 (AC). *Prove the distributive law for arbitrary cardinal sums:*

$$\alpha\Big(\sum_{i \in I} \beta_i\Big) = \sum_{i \in I} \alpha\beta_i.$$

Definition 359. Let I be any set. We say that *a is an I-tuple* if a is a function with domain $\text{dom}(a) = I$. If a is an I-tuple, then the value $a(i)$ is called *the i-th coordinate of a* and is denoted by a_i, so that $a = \langle a_i \mid i \in I \rangle$.

Definition 360. The *Cartesian product* of an indexed family $\langle A_i \mid i \in I \rangle$ of sets is defined as the set of all I-tuples $\langle a_i \mid i \in I \rangle$ whose i-th coordinate ranges over the set A_i, for each $i \in I$. In notation:

---

[1] As Russell's socks-and-boots example noted, without AC, the family $\langle 2, 2, 2, \dots \rangle$ may fail to be sum-adequate: The ambiguous infinite sum $2 + 2 + 2 + \cdots$ may be $\aleph_0$, non-reflexive, or $> \aleph_0$. The failure for the sequence $\langle \aleph_0, \aleph_0, \aleph_0, \dots \rangle$ is striking too: $\mathbf{R}$ may be partitioned into countably many countable sets (Feferman–Levy, see [33]). If each α_i is well-orderable, then one may use canonical representatives to define a "principal value" for $\sum_i \alpha_i$ even if $\langle \alpha_i \rangle$ is not sum-adequate (Whitehead, see [48]).

$$\prod_{i\in I} A_i := \{\langle a_i \mid i \in I\rangle \mid \forall i \in I,\ a_i \in A_i\}.$$

Problem 361 (AC). *If $A_i \sim A_i'$ for all $i \in I$, then* $\prod_{i\in I} A_i \sim \prod_{i\in I} A_i'$.

Let $\langle \kappa_i \mid i \in I\rangle$ be an indexed family of cardinals. Without AC, if $|A_i| = |A_i'| = \kappa_i$ for all $i \in I$, we cannot conclude that $\prod_{i\in I} A_i \sim \prod_{i\in I} A_i'$. In fact, without AC we cannot even conclude that a representative family of sets $\langle A_i \mid i \in I\rangle$ with $|A_i| = \kappa_i$ exists.

Definition 362 (Product-Adequate Families of Cardinals). An indexed family of cardinals $\langle \kappa_i \mid i \in I\rangle$ is *product-adequate* if there is a representative family of sets $\langle A_i \mid i \in I\rangle$ such that $|A_i| = \kappa_i$ for all $i \in I$ and whenever $A_i' \sim A_i$ for all $i \in I$, we have $\prod_{i\in I} A_i \sim \prod_{i\in I} A_i'$.

If AC is not assumed, arbitrary cardinal products are significant only for product-adequate families.

Definition 363 (Arbitrary Cardinal Products). For a product-adequate family $\langle \kappa_i \mid i \in I\rangle$ of cardinals, define the *cardinal product* of the family as:

$$\prod_{i\in I} \kappa_i := \left|\prod_{i\in I} A_i\right|,$$

where A_i is a set with $|A_i| = \kappa_i$ for each $i \in I$. If $\langle \kappa_i \mid i \in I\rangle$ is not product-adequate, we define the product $\prod_{i\in I} \kappa_i$ to be zero.

Under AC, every family of cardinals is product-adequate, and every family of nonzero cardinals has a unique nonzero cardinal as the product.

Alternative View of Products

Suppose that $\langle B_i \mid i \in I\rangle$ is an indexed family of pairwise disjoint nonempty sets (i.e., $B_i \neq \emptyset$ and $B_i \cap B_j = \emptyset$ whenever $i \neq j$), so that $\{B_i \mid i \in I\}$ forms a partition. We would expect the number of choice sets from this partition $\{B_i \mid i \in I\}$ to be equal to the product of the size of the B_i's. The following result confirms this formally.

Problem 364. *Given an indexed family $\langle A_i \mid i \in I\rangle$ of nonempty sets, put $B_i = \{i\} \times A_i$ so that each $B_i \sim A_i$, but $B_i \cap B_j = \emptyset$ for $i \neq j$. Let $\mathbf{C}$ be the collection of all choice sets from the partition $\{B_i \mid i \in I\}$. Then $\mathbf{C} \sim \prod_{i\in I} A_i$. In fact, $\mathbf{C} = \prod_{i\in I} A_i$.*

Thus the alternative definition of product (using choice sets from a partition) coincides with the original Cartesian product!

Here is why Russell called the Axiom of Choice the *Multiplicative Axiom.*

Problem 365. *The Axiom of Choice is equivalent to the assertion that an arbitrary product of nonzero cardinals is nonzero: If $\langle \kappa_i \mid i \in I \rangle$ is an indexed family of cardinals with $\kappa_i \neq 0$ for all $i \in I$, then $\prod_{i\in I} \kappa_i \neq 0$.*

6.3 Cardinal Exponentiation: $|\,\mathrm{P}(A)| = 2^{|A|}$

If $A_i = A$ for all $i \in I$, so that $\kappa_i = |A_i| = |A| = \kappa$ (say) for all i, then the cardinal product $\prod_{i\in I} \kappa_i$ reduces to exponentiation, as in (informally):

$$\prod_{i\in I} |A_i| = \prod_{i\in I} \kappa_i = \prod_{i\in I} |A| = \prod_{i\in I} \kappa = |A|^{|I|} = \kappa^{|I|}.$$

Problem 366. *If $A_i = A$ for all $i \in I$, then the Cartesian product $\prod_{i\in I} A_i = \prod_{i\in I} A$ equals the collection of all functions from I to A, i.e., $\prod_{i\in I} A = \{F \mid F\colon B \to A\}$.*

The last fact simplifies the definition of cardinal exponentiation.

Definition 367 (Exponential Sets). Given sets A and B, we define A^B to be the set of all functions from B to A, i.e.,

$$A^B := \{F \mid F\colon B \to A\}\,.$$

The following is similar to Problem 361, but here we do not need AC.

Problem 368 (Invariance of Exponentiation). *If $A \sim C$ and $B \sim D$ then $A^B \sim C^D$.*

The following is therefore well defined:

Definition 369 (Cardinal Exponentiation, Cantor). Given cardinals α and β, define

$$\alpha^\beta := \left|A^B\right|, \quad \text{where } A \text{ and } B \text{ are representative sets with } |A| = \alpha,\ |B| = \beta.$$

Problem 370. *Let κ be a cardinal. Using the last definition of exponentiation show that $\kappa^1 = \kappa$, and $\kappa^{n+1} = \kappa^n{\cdot}\kappa$ for any $n \in \mathbf{J}$. Informally: $\kappa^n = \kappa{\cdot}\kappa{\cdot}\cdots{\cdot}\kappa$ (n times).*

Definition 371 (Characteristic Functions). We say that *f is a characteristic function on I* if $f \in \{0,1\}^I$, that is, if $f\colon I \to \{0,1\}$.

If $E \subseteq I$, we let χ_E denote the characteristic function on I defined by $\chi_E(i) = 1$ if $i \in E$ and $\chi_E(i) = 0$ if $i \notin E$.

Definition 372 (Binary I-tuples). An I-tuple $a = \langle a_i \mid i \in I \rangle$ is called a *binary I-tuple* if and only if $\forall i \in I,\ a_i = 0$ or $a_i = 1$. Thus a binary I-tuple is simply a characteristic function on I. Given a binary I-tuple $a = \langle a_i \mid i \in I \rangle$, the value of a at i, i.e., $a(i) = a_i$, is called *the i-th bit of a*.

Recall the definitions of infinite sequences and strings from Sect. 1.7.

Definition 373 (Infinite Binary Sequences). When $I = \mathbf{N}$, a binary $\mathbf{N}$-tuple is called an *infinite binary sequence,* so that $\{0, 1\}^{\mathbf{N}}$ is the set of all infinite binary sequences. An infinite binary sequence $a = \langle a_n \mid n \in \mathbf{N} \rangle$ is written by simply writing out its bits in order, i.e., as $a = a_1 a_2 a_3 \cdots a_n \cdots$.

For any set I, *there is a very natural effective one-to-one correspondence between its power set* $\mathrm{P}(I)$ *and the set* $\{0, 1\}^I$ *of all characteristic functions on* I*, making the two collections* $\mathrm{P}(I)$ *and* $\{0, 1\}^I$ *virtually interchangeable.*

Problem 374 (Cantor). $\mathrm{P}(I) \sim \{0, 1\}^I$*, via an effective natural bijection.*

In particular, $\mathrm{P}(\mathbf{N}) \sim \{0, 1\}^{\mathbf{N}}$ *via a natural effective bijection between the set of all subsets of* $\mathbf{N}$ *and the set of all infinite binary sequences.*

[Hint: Consider the mapping $E \mapsto \chi_E$ from $\mathrm{P}(I)$ to $\{0, 1\}^I$.]

Corollary 375. *If* $|A| = \kappa$ *then* $|\,\mathrm{P}(A)| = 2^\kappa$*. In other words,* $|\,\mathrm{P}(A)| = 2^{|A|}$.

Corollary 376. *If A is denumerable, then* $|\,\mathrm{P}(A)| = 2^{\aleph_0}$*, and thus every denumerable set has exactly* $2^{\aleph_0}$ *subsets. In particular* $|\,\mathrm{P}(\mathbf{N})| = |\,\mathrm{P}(\mathbf{Q})| = 2^{\aleph_0}$*, i.e., each of* $\mathbf{N}$ *and* $\mathbf{Q}$ *has exactly* $2^{\aleph_0}$ *subsets.*

Problem 377. $A \preccurlyeq \mathrm{P}(A) \sim \{0, 1\}^A$ *for any set A, and* $\kappa \leqslant 2^\kappa$ *for any cardinal* κ.

Problem 378. $\mathfrak{c} \leqslant 2^{\aleph_0}$.

[Hint: Find an injection from $\mathbf{R}$ into $\mathrm{P}(\mathbf{Q})$.]

Since we have earlier proved that $\aleph_0 < \mathfrak{c}$, we now get:

Corollary 379. $\aleph_0 < 2^{\aleph_0}$. *Thus there are uncountably many infinite binary sequences, and so uncountably many subsets of* $\mathbf{N}$.

This last fact is a special case of *Cantor's Theorem,* and we will revisit this in a later section where we will also prove the general case.

6.4 Cardinal Arithmetic

Since we have already defined sum, product, and power of cardinals, we can look for their algebraic properties. *Throughout this section we assume AC.*

Definition 380 (Rearrangement or Permutation). If $\langle \kappa_i \mid i \in I \rangle$ and $\langle \kappa'_i \mid i \in I \rangle$ are two families of cardinals indexed by the same index set I, we say that $\langle \kappa'_i \mid i \in I \rangle$

is a *rearrangement* (or *permutation*) of $\langle\kappa_i \mid i \in I\rangle$ if there exists a permutation σ of I (that is $\sigma: I \to I$ a bijective transformation of I onto I) such that

$$\kappa'_i = \kappa_{\sigma(i)} \qquad \text{for all } i \in I.$$

Problem 381 (The Generalized Commutative Laws). *If $\langle\kappa'_i \mid i \in I\rangle$ is a rearrangement of the cardinals $\langle\kappa_i \mid i \in I\rangle$, then*

$$\sum_{i\in I}\kappa'_i = \sum_{i\in I}\kappa_i, \qquad \text{and} \qquad \prod_{i\in I}\kappa'_i = \prod_{i\in I}\kappa_i.$$

Problem 382. *Formulate and prove "Generalized Associative Laws" for arbitrary sums and products of cardinal numbers.*

Problem 383 (Monotonicity of Sum and Product). *Prove that if $\alpha_i \leqslant \beta_i$ for all $i \in I$, then*

$$\sum_{i\in I}\alpha_i \leqslant \sum_{i\in I}\beta_i, \qquad \text{and} \qquad \prod_{i\in I}\alpha_i \leqslant \prod_{i\in I}\beta_i.$$

Problem 384 (Laws of Exponents). *Prove that if α, β, and κ are cardinals, then*

$$\kappa^\alpha\kappa^\beta = \kappa^{\alpha+\beta}, \qquad \alpha^\kappa\beta^\kappa = (\alpha\beta)^\kappa, \qquad \text{and} \qquad (\kappa^\alpha)^\beta = \kappa^{\alpha\beta}.$$

Problem 385 (Generalized Laws of Exponents).

$$\prod_{i\in I}\kappa^{\alpha_i} = \kappa^{\sum_{i\in I}\alpha_i}, \qquad \text{and} \qquad \prod_{i\in I}\kappa_i^\alpha = \left(\prod_{i\in I}\kappa_i\right)^\alpha.$$

Problem 386. *Are the following strict inequalities true?*

$$\alpha < \beta \Rightarrow \alpha^\kappa < \beta^\kappa, \quad \text{and} \quad \alpha < \beta \Rightarrow \kappa^\alpha < \kappa^\beta.$$

Here are some examples of computation using cardinal arithmetic.

Example 387. Find $\sum\limits_{n\in\mathbf{N}} n = 1 + 2 + 3 + \cdots$.

Proof (Solution). Put $\kappa = \sum\limits_{n\in\mathbf{N}} n = 1 + 2 + 3 + \cdots$. Since $n \geqslant 1$ for all $n \in \mathbf{N}$, we get

$$\kappa = \sum_{n\in\mathbf{N}} n = 1 + 2 + 3 + \cdots \geqslant 1 + 1 + 1 + \cdots = 1 \cdot \aleph_0 = \aleph_0,$$

but also $n \leqslant \aleph_0$ for all $n \in \mathbf{N}$, and so

$$\kappa = \sum_{n \in \mathbf{N}} n = 1 + 2 + 3 + \cdots \leqslant \aleph_0 + \aleph_0 + \aleph_0 + \cdots = \aleph_0 \cdot \aleph_0 = \aleph_0^2 = \aleph_9,$$

Combining the inequalities $\kappa \geqslant \aleph_0$ and $\kappa \leqslant \aleph_0$ we get $\kappa = \aleph_0$. □

Example 388. Find $\prod_{n \in \mathbf{N}} n = 1 \cdot 2 \cdot 3 \cdots$.

Proof (Solution). Put $\kappa = \prod_{n \in \mathbf{N}} n = 1 \cdot 2 \cdot 3 \cdots = 2 \cdot 3 \cdot 4 \cdots = \prod_{n \in \mathbf{N}} (n+1)$. Since $n \geqslant 2$ for all $n \in \mathbf{N}$, we get

$$\kappa = \prod_{n \in \mathbf{N}} (n+1) = 2 \cdot 3 \cdot 4 \cdots \geqslant 2 \cdot 2 \cdots 2 \cdots = 2^{\aleph_0},$$

but also $n \leqslant \aleph_0$ for all $n \in \mathbf{N}$, and so

$$\kappa = \prod_{n \in \mathbf{N}} n = 1 \cdot 2 \cdot 3 \cdots \leqslant 2^{\aleph_0} \cdot 2^{\aleph_0} \cdot 2^{\aleph_0} \cdots = \left(2^{\aleph_0}\right)^{\aleph_0} = 2^{\aleph_0^2} = 2^{\aleph_0}.$$

As $\kappa \geqslant 2^{\aleph_0}$ and $\kappa \leqslant 2^{\aleph_0}$ we get $\kappa = 2^{\aleph_0}$. □

6.5 The Binary Tree

Recall that a *binary word* (or a *finite binary sequence*) is a finite word made only of 0 and 1, and the set of all binary words is:

$$\{0,1\}^* := \{\varepsilon, 0, 1, 00, 01, 10, 11, 000, 001, 010, 011, 100, 101, \ldots\}.$$

The unique word of length zero is the *empty word,* denoted by ε, and for any n, there are exactly 2^n binary words of length n.

We can arrange the binary words in a tree structure by regarding prefixes of words as "ancestors" and extensions as "descendents." This will be called *the tree of binary words,* or simply *the binary tree.* The binary words are also called the *nodes* of the tree. For each n, there are 2^n binary words of length n forming the nodes of the tree at *level n,* and each node u of length n has two immediate descendents (extensions) of length $n+1$, namely $u^\frown 0$ and $u^\frown 1$. Since the empty word ε is a prefix of every word, it will be an ancestor of every word, and hence will form the *root* node of the tree.

A picture of the first few levels of the binary tree is shown below.

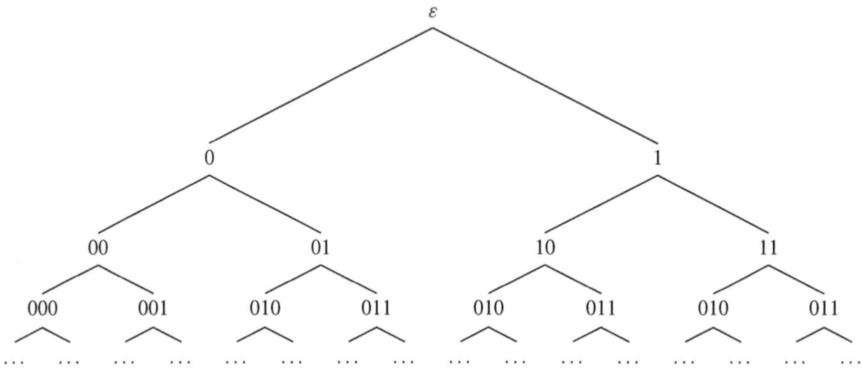

Problem 389. *Find an effective bijection between* $\mathbf{N}$ *and the set of nodes in the binary tree. Conclude that there are* $\aleph_0$ *nodes in the binary tree.*

Starting from the root node ε, one can descend down the tree to obtain an *infinite branch through the binary tree,* that is an infinite set of nodes of the form $\{u_0, u_1, u_2, \ldots, u_n, \ldots\}$, where $u_0 = \varepsilon$ and each u_{n+1} is one of the two possible immediate extensions of u_n, so that u_{n+1} equals either $u_n^\frown 0$ or $u_n^\frown 1$, and we get $\operatorname{len}(u_n) = n$ by induction.

There are an infinite (in fact, uncountable) number of *infinite branches* through this tree, where *each branch is represented uniquely by an infinite binary sequence: For any given infinite binary sequence* $x \in \{0,1\}^{\mathbf{N}}$, *the set* $\{x|n \mid n = 0, 1, 2, \ldots\}$ *of all prefixes of* x *forms a branch through the tree.* E.g., the infinite binary sequence $000000\cdots$ represents the leftmost branch, $111111\cdots$ the rightmost branch, and the sequence $010101\cdots$ represents a "left-right-left-right-$\cdots$" zigzag branch: $\varepsilon, 0, 01, 010, 0101, \ldots$.

Problem 390. *An infinite branch* B *through the binary tree is a set of binary words which contains exactly one node of length* n *for each* $n = 0, 1, 2, \ldots$ *and is linearly ordered by the "prefix" relation (i.e., for any two nodes in* B*, one is a prefix of the other). Thus* $B = \{u_0, u_1, u_2, \ldots, u_n, \ldots\}$ *where each* u_n *has length* n *and* u_{n+1} *extends* u_n *by postfixing a single bit to it. Let* $\mathbf{B}$ *be the set of all branches through the binary tree. Prove that* $\mathbf{B} \sim \{0,1\}^{\mathbf{N}}$*, via a very effective natural correspondence.*

Problem 391. *Prove that there are* $2^{\aleph_0}$ *branches through the binary tree.*

Almost Disjoint Families of Subsets of N

Definition 392. A family $\mathrm{D} \subseteq \mathbf{P}(\mathbf{N})$ of sets of natural numbers is said to be an *almost disjoint family* if every set in D is infinite and the intersection of any two distinct sets in D is finite.

While any *pairwise disjoint family* of subsets of **N** is effectively countable (why?), there are almost disjoint families which are uncountable (of size $\geqslant \mathfrak{c}$).

Problem 393. *Show that there is an almost disjoint family* D *of subsets of* **N** *with* $|\mathsf{D}| = 2^{\aleph_0} \geqslant \mathfrak{c}$.

[Hint: Replace **N** by $\{0,1\}^*$ and note that the intersection of two distinct infinite branches (as defined in Problem 390) through the binary tree is finite.]

6.6 The Cantor Set K

In this section we define the important *Cantor set* **K**, a set of reals which naturally has cardinality $2^{\aleph_0}$. The Cantor set is constructed using a very special "binary tree of intervals" called the *Cantor system of intervals.*

More precisely, we will map the nodes of the tree of binary words into a collection of intervals, and assign a closed interval $I[u]$ to every binary word u in such a way that[2]

> *For every binary word u, the two intervals $I[u^\frown 0]$ and $I[u^\frown 1]$ will be disjoint subintervals contained within $I[u]$.*

The definition below proceeds by induction on node depth (= word length).

Recall that any closed interval $I = [a,b]$ (with $a < b$) can be trisected into three equal subintervals each of length $\ell = \frac{1}{3}(b-a)$, so that $a < a+\ell < a+2\ell < b$. If we remove the *middle-third open interval* $(a+\ell, a+2\ell)$ from $I = [a,b]$, then I splits into two closed subintervals: $[a, a+\ell] =$ *the left-third of* I, and $[a+2\ell, b] =$ *the right-third of* I.

Definition 394 (The Cantor System of Intervals). Let $I[\varepsilon] = [0,1]$, and having defined $I[u]$ for all words of length n, define $I[v]$ for words v of length $n+1$ by the rule

$$I[u^\frown 0] = \text{Left-third of } I[u], \qquad \text{and} \qquad I[u^\frown 1] = \text{Right-third of } I[u].$$

Since each binary word v of length $n+1$ has one of the forms $v = u^\frown 0$ or $v = u^\frown 1$ where u has length n, by induction this completely defines $I[u]$ for all binary words u.

For example, we have $I[0] =$ left-third of $[0,1] = [0, \frac{1}{3}]$, and $I[1] =$ right-third of $[0,1] = [\frac{2}{3}, 1]$. Also $I[00] =$ left-third of $I[0] = [0, \frac{1}{9}]$, and $I[01] =$ right-third of $I[0] = [\frac{2}{9}, \frac{1}{3}]$, Similarly, $I[10] = [\frac{2}{3}, \frac{7}{9}]$, $I[11] = [\frac{8}{9}, 1]$, etc.

---

[2]In fact, we have already seen a form of this construction in the previous chapter as part of the proof of Cantor's theorem that **R** is uncountable. However, in that proof only a specific "infinite branch of intervals" was "diagonalized out" to produce a real number distinct from the given sequence of reals. Here we will deal with the full tree of intervals.

This gives the following "binary tree of intervals," which we call *the Cantor System of Intervals,* or *the Cantor Tree of Intervals.*

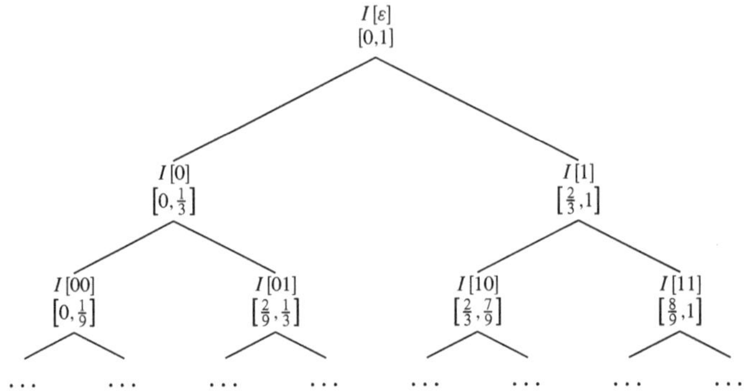

If $x \in \{0,1\}^{\mathbf{N}}$ is an infinite binary sequence, say

$$x = x_1 x_2 x_3 \cdots x_n \cdots, \qquad \text{each } x_n \text{ equals 0 or 1.}$$

then let $x|n := x_1 x_2 \ldots x_n$ denote the prefix word of x of length n. Thus x represents the infinite branch through the binary tree given by its prefix words $x|0 = \varepsilon$, $x|1 = x_1$, $x|2 = x_1x_2$, $x|3 = x_1x_2x_3$, etc.

Now note that given any $x \in \{0,1\}^{\mathbf{N}}$, the infinite branch of its prefix words $x|0$, $x|1$, $x|2$, ..., $x|n$, ..., etc., *also determines an infinite branch through the above Cantor Tree of Intervals, giving the corresponding nested sequence*

$$I[\varepsilon] \supseteq I[x|1] \supseteq I[x|2] \supseteq I[x|3] \supseteq \cdots \supseteq I[x|n] \supseteq \cdots$$

of closed intervals, where the n-th interval $I[x|n]$ *has length* $1/3^n$. Therefore, by the Nested Interval Property, this nested sequence of intervals *must contain a unique real number,* which we denote by $\mathbf{F}(x)$.

So each $x \in \{0,1\}^{\mathbf{N}}$, via the nested branch $I[x|0] \supseteq I[x|1] \supseteq I[x|2] \supseteq \cdots$ through the Cantor Tree of Intervals, determines the unique point $\mathbf{F}(x)$ in their intersection. We thus have a function $\mathbf{F}\colon \{0,1\}^{\mathbf{N}} \to [0,1]$ which maps the set of infinite binary sequences into the interval $[0,1]$. Officially, $\mathbf{F}$ is defined by setting $\mathbf{F}(x) :=$ The unique member of $\bigcap_{n=1}^{\infty} I[x|n]$.

Theorem 395. *Let* $I[u]$*, where u ranges over all binary words, be the Cantor System of Intervals, so that* $I[\varepsilon] = [0,1]$, $I[0] = [0,\frac{1}{3}]$, $I[1] = [\frac{2}{3},1]$, *etc. The Cantor System of Intervals then naturally determines a unique injective function* $\mathbf{F}\colon \{0,1\}^{\mathbf{N}} \to [0,1]$ *such that for every* $x \in \{0,1\}^{\mathbf{N}}$,

$$\mathbf{F}(x) = \textit{the unique member of } \bigcap_{n=1}^{\infty} I[x|n].$$

Problem 396. *Prove Theorem 395 by showing that* **F** *is one-to-one.*

[Hint: Distinct branches through the Cantor Tree of Intervals determine distinct nested sequences of intervals which eventually become disjoint.]

Problem 397. *For each of the following infinite binary sequences x, find the first four of the nested intervals determined by x, and then compute* $\mathbf{F}(x)$. *1.* $x = 000000\cdots$ *2.* $x = 111111\cdots$ *3.* $x = 010101\cdots$ *4.* $x = 101010\cdots$

[Hint: $\mathbf{F}(x)$ is a limit of the endpoints of the nested intervals $I[x|n]$, $n \in \mathbf{N}$.]

Definition 398 (The Cantor Set K). The *Cantor Set* **K** is defined as the subset of $[0, 1]$ which equals the range of the function **F** of Theorem 395, i.e., $\mathbf{K} := \text{ran}(\mathbf{F}) = \{\mathbf{F}(x) \mid x \in \{0, 1\}^{\mathbf{N}}\}$.

The bijection $\mathbf{F}\colon \{0, 1\}^{\mathbf{N}} \to \mathbf{K}$ *establishes a natural identification of infinite binary sequences with the points of the Cantor set* **K**.

Corollary 399. $\mathbf{K} \sim \{0, 1\}^{\mathbf{N}}$, *so* $|\mathbf{K}| = 2^{\aleph_0}$*: The Cantor Set contains exactly* $2^{\aleph_0}$ *elements.*

Corollary 400. $2^{\aleph_0} \leqslant \mathfrak{c}$.

Since we have earlier established that $\mathfrak{c} \leqslant 2^{\aleph_0}$, an application of the Cantor–Bernstein Theorem yields the following important result.

Corollary 401 (Cantor). $2^{\aleph_0} = \mathfrak{c}$.

In the next section we will indicate how to build an *explicit effective bijection* between **R** and $\{0, 1\}^{\mathbf{N}}$.

Let $\mathbf{K}_n$ be the set formed by taking the union of the 2^n intervals at level (depth) n of the Cantor Tree of Intervals. So $\mathbf{K}_0 = [0, 1]$, $\mathbf{K}_1 = [0, \frac{1}{3}] \cup [\frac{2}{3}, 1]$, $\mathbf{K}_2 = [0, \frac{1}{9}] \cup [\frac{2}{9}, \frac{1}{3}] \cup [\frac{2}{3}, \frac{7}{9}] \cup [\frac{8}{9}, 1]$, etc., and in general $\mathbf{K}_n$ is the union of the 2^n intervals $I[u]$ where u ranges over the 2^n binary words of length n:

$$\mathbf{K}_n := \bigcup \{I[u] \mid u \text{ is a binary word of length } n\}.$$

Note also that the intervals of $\mathbf{K}_{n+1}$ are obtained by removing the middle-third open intervals from each of the intervals of $\mathbf{K}_n$, thus doubling the number of intervals as we go from $\mathbf{K}_n$ to $\mathbf{K}_{n+1}$. The first few of the sets $\mathbf{K}_n$ are shown below.

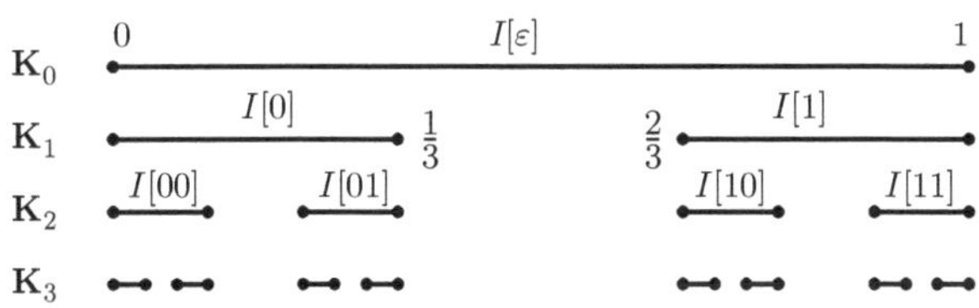

The sets $\mathbf{K}_n$

The following problem is instructive.

Problem 402 (Alternative Characterization of the Cantor Set). *Show that* $\mathbf{K} = \bigcap_{n=1}^{\infty} \mathbf{K}_n$.

The Cantor set is uncountable (with cardinality $\mathfrak{c} = 2^{\aleph_0} > \aleph_0$), but note that each set $\mathbf{K}_n$ consists of 2^n disjoint closed intervals each of length $1/3^n$, so the total length or *measure* of the set $\mathbf{K}_n$ is $(2/3)^n$. Since $\mathbf{K} \subseteq \mathbf{K}_n$ for all n, and since $\lim_{n\to\infty} (2/3)^n = 0$, this seems to indicate that the Cantor Set $\mathbf{K}$ has *zero measure,* a notion that will be officially defined in Chap. 15.

Problem 403. *Prove that* $[0, 1] \smallsetminus \mathbf{K}$*, the complement of the Cantor Set within* $[0, 1]$*, can be partitioned into a pairwise disjoint countable sequence of open intervals whose lengths add up to* 1.

Thus one can also view the Cantor set being constructed *by removing middle-third open intervals in stages,* where at each stage we have a finite disjoint union of closed intervals, and proceed to the next stage by removing the middle-thirds of all the closed intervals of the current stage—which splits every interval into two and in effect doubles the number of intervals. We start at stage 0 with the unit interval $\mathbf{K}_0 = [0, 1]$, and having the set $\mathbf{K}_n$ at stage n, remove the middle-thirds of each of the 2^n disjoint closed intervals of $\mathbf{K}_n$ to obtain the set $\mathbf{K}_{n+1}$ of stage $n + 1$ consisting of 2^{n+1} disjoint closed intervals. Once all the middle-third open intervals are removed, what remains is the Cantor set.

While this "top-down" definition of the Cantor set is often useful, we re-emphasize the original "bottom up" construction via the Cantor Tree of Intervals where the points of the Cantor set are represented uniquely by infinite binary sequences: Since each sequence of nested intervals $\langle I_n \rangle$ given by an infinite branch through the binary tree produces a unique real in its intersection $\cap_n I_n$, we get *a natural effective one-to-one correspondence between the infinite binary sequences in* $\{0, 1\}^{\mathbf{N}}$ *and the points of the Cantor set* $\mathbf{K}$.

Problem 404. *For any infinite binary sequence* $x = x_1x_2x_3 \ldots x_n \ldots$,

$$\mathbf{F}(x) = 2\sum_{n=1}^{\infty} \frac{x_n}{3^n} \qquad \textit{(where } \mathbf{F} \textit{ is as in Theorem 395).}$$

Problem 405. *Does* $1/4 \in \mathbf{K}$*? Does* $1/e \in \mathbf{K}$*?*

Problem 406. *Recall the Cantor Machine* $\mathsf{M}\colon \mathbf{R}^{\mathbf{N}} \to [0, 1]$ *used in the proof of Cantor's theorem, which satisfies* $\mathsf{M}(\langle a_n \rangle) \in [0, 1] \smallsetminus \{a_n \mid n \in \mathbf{N}\}$ *for all sequences* $\langle a_n \rangle \in \mathbf{R}^{\mathbf{N}}$*. Show that*

1. *The mapping* M *is highly non-injective in the sense that for any* $\langle a_n \rangle \in \mathbf{R}^{\mathbf{N}}$ *the set* $\{\langle x_n \rangle \in \mathbf{R}^{\mathbf{N}} \mid \mathsf{M}(\langle x_n \rangle) = \mathsf{M}(\langle a_n \rangle)\}$ *has cardinality* $\mathfrak{c} = 2^{\aleph_0}$.
2. *The set of all possible responses produced by the Cantor Machine* M *equals the Cantor set, that is,* $\operatorname{ran}(\mathsf{M}) = \mathbf{K}$.

Problem 407. *Exhibit a natural effective bijection between the Cartesian product* $\{0,1\}^{\mathbf{N}} \times \{0,1\}^{\mathbf{N}}$ *and* $\{0,1\}^{\mathbf{N}}$.

[Hint: Intertwine two sequences $x_1 x_2 \cdots$ and $y_1 y_2 \cdots$ into one: $x_1 y_1 x_2 y_2 \cdots$.]

Problem 408. *Let* $\mathbf{K} \times \mathbf{K}$ *be the "planar Cantor set." Show that there is a natural effective bijection between* $\mathbf{K} \times \mathbf{K}$ *and* $\mathbf{K}$.

6.7 The Identity $2^{\aleph_0} = \mathfrak{c}$

The identity $2^{\aleph_0} = \mathfrak{c}$ can be used to obtain the following results.

Problem 409. *Let* $\mathbf{C} = \mathbf{R}^2$ *be the complex plane. Prove that* $\mathfrak{c}^2 = \mathfrak{c}$ *and so* $\mathbf{C} \sim \mathbf{R}$.

By induction, this can be generalized to any dimensions.

Problem 410. $\mathfrak{c}^n = \mathfrak{c}$, *and so* $\mathbf{R}^n \sim \mathbf{R}$, *for any* $n \in \mathbf{N}$.

Problem 411. *Prove that any subset of an Euclidean space (=* $\mathbf{R}^n$ *for some* n*) containing a line segment (and in particular any subset with nonempty interior) must be equinumerous with the entire space.*

Cantor's proofs of these facts initially resulted in controversy, since they seemed to contradict the familiar principle of "invariance of dimensions" which says that there cannot be a continuous one-to-one correspondence between two Euclidean spaces of different dimensions. For example, Cantor's result $\mathbf{R}^2 \sim \mathbf{R}$ implies that one can represent the points of the Cartesian plane using a *single* real coordinate in a one-to-one fashion (as opposed to the usual form which uses a pair of real coordinates)! However, it soon became clear that the confusion resulted from a failure to recognize the requirement of continuity in invariance of dimensions. As the field of topology developed, it was firmly established that the bijections obtained by Cantor's method, while being effective, cannot be continuous, and so the principle of invariance of dimensions remains intact.

Definition 412. The mapping $\mathbf{h}\colon \{0,1\}^{\mathbf{N}} \to [0,1]$ is defined by setting, for each $x = \langle x_n \mid n \in \mathbf{N} \rangle \in \{0,1\}^{\mathbf{N}}$,

$$\mathbf{h}(x) := \sum_{n=1}^{\infty} \frac{x_n}{2^n},$$

so that $\mathbf{h}(x)$ is the real number in $[0,1]$ having an infinite binary representation $0 \cdot x_1 x_2 x_3 \cdots x_n \cdots$.

Problem 413. *The map* $\mathbf{h}\colon \{0,1\}^{\mathbf{N}} \to [0,1]$ *is surjective but not injective. For which* $x \in \{0,1\}^{\mathbf{N}}$ *can you find* $y \in \{0,1\}^{\mathbf{N}}$, $y \neq x$, *with* $\mathbf{h}(y) = \mathbf{h}(x)$?

Definition 414. Let $\{0,1\}^{\mathbf{N}}_{\infty}$ be the set of infinite binary sequences which are not eventually zero, that is those which have infinitely many entries of 1:

$$\{0,1\}^{\mathbf{N}}_{\infty} := \{x \in \{0,1\}^{\mathbf{N}} \mid x(n) = 1 \text{ for infinitely many } n\}.$$

Problem 415. *Show that the restriction of* $\mathbf{h}$ *to* $\{0,1\}^{\mathbf{N}}_{\infty}$ *is injective function with range* $(0,1]$. *Hence there is an effective bijection from* $\{0,1\}^{\mathbf{N}}_{\infty}$ *onto* $(0,1]$.

Problem 416. *Show that there is an effective bijection from* $\{0,1\}^{\mathbf{N}}$ *onto* $\{0,1\}^{\mathbf{N}}_{\infty}$.

[Hint: Given $x = \langle x_n \rangle \in \{0,1\}^{\mathbf{N}}$, put $h(x) := \langle 1, 1-x_1, 1-x_2, \dots \rangle$ if x is eventually zero, put $h(x) := \langle 0, x_1, x_2, \dots \rangle$ if x is eventually one, and $h(x) := x$ otherwise.]

Combining the results of the last two problems, we get the following.

Corollary 417. *There is an explicit effective bijection from the set* $\{0,1\}^{\mathbf{N}}$ *of all infinite binary sequences onto the interval* $(0,1]$.

One could also express this explicit effective bijection in the following alternative form.

Problem 418. *Define* $\Phi\colon \{0,1\}^{\mathbf{N}} \to (0,1]$ *by*

$$\Phi(x) := \begin{cases} 1 - \frac{1}{2}\mathbf{h}(x) & \textit{if } x \textit{ is eventually zero,} \\ \frac{1}{2}\mathbf{h}(x) & \textit{if } x \textit{ is eventually one,} \\ x & \textit{otherwise.} \end{cases}$$

Then Φ *is an effective bijection from* $\{0,1\}^{\mathbf{N}}$ *onto the interval* $(0,1]$.

In the last chapter we had seen effective bijections between $(0,1]$ and $(0,1)$, and between $(0,1)$ and $\mathbf{R}$. Therefore we have:

Corollary 419. *There is an explicit effective bijection from the set* $\{0,1\}^{\mathbf{N}}$ *onto* $\mathbf{R}$. *Hence we get* $2^{\aleph_0} = \mathfrak{c}$ *in an especially direct and effective way.*

Problem 420. *Find a natural effective bijection between the set* $\mathbf{N}^{\mathbf{N}}$ *of all infinite sequences of natural numbers and the set* $\{0,1\}^{\mathbf{N}}_{\infty}$ *of all infinite binary sequences which are not eventually zero.*

[Hint: Given an infinite sequence $\langle n_1, n_2, \dots, n_k, \dots \rangle$ of natural numbers, consider the infinite binary sequence which has a 1 at position n_1, position $n_1 + n_2$, position $n_1 + n_2 + n_3$, and so on, and has a 0 at every other place.]

Problem 421. *Prove that* $\aleph_0^{\aleph_0} = \mathfrak{c}$ *effectively, by exhibiting an explicit bijection* $\mathbf{H}$ *from* $\mathbf{N}^{\mathbf{N}}$ *onto* $(0,1]$.

[Hint: Consider the mapping $\mathbf{H}\colon \mathbf{N}^{\mathbf{N}} \to (0,1]$

$$\mathbf{H}(\langle n_1, n_2, n_3, \ldots, n_k, \ldots \rangle) := \frac{1}{2^{n_1}} + \frac{1}{2^{n_1+n_2}} + \frac{1}{2^{n_1+n_2+n_3}} + \cdots$$

which maps $\mathbf{N}^{\mathbf{N}}$ bijectively onto $(0, 1]$.]

Problem 422. *Show effectively that* $\left(\{0,1\}^{\mathbf{N}}\right)^{\mathbf{N}} \sim \{0,1\}^{\mathbf{N}}$.

[Hint: Use the method of Proposition 316 in which a sequence of sequences is effectively combined into a single sequence.]

Corollary 423. *We have* $\mathfrak{c}^{\aleph_0} = \mathfrak{c}$ *and* $\mathbf{R}^{\mathbf{N}} \sim \mathbf{R}$ *effectively. In particular, the set of all real sequences is effectively equinumerous with the real line.*

The results established so far can be summarized as follows.

Theorem 424. *One can explicitly construct effective bijections between any two of the following sets:*

$$\{0,1\}^{\mathbf{N}}, \ \mathbf{N}^{\mathbf{N}}, \ \mathbf{R}, \ \mathbf{R}^2, \ \mathbf{R}^n, \ \mathbf{R}^{\mathbf{N}}, \ (0,1), \ (0,1], \ [0,1], \ [0,1]^2, \ [0,1]^{\mathbf{N}}.$$

6.8 Cantor's Theorem: The Diagonal Method

In this section we will generalize the inequality $\aleph_0 < 2^{\aleph_0}$ to arbitrary cardinals, an important result known as *Cantor's Theorem.* An even more general (but non-effective) result called *König's Inequality* will also be proved. We start with a proof of $\aleph_0 < 2^{\aleph_0}$ in terms of binary sequences which readily generalizes to arbitrary cardinals.

Problem 425 (Cantor Diagonalization of Binary Sequences). $\{0,1\}^{\mathbf{N}} \not\preccurlyeq \mathbf{N}$. *More specifically, given any sequence of infinite binary sequences, one can effectively find another infinite binary sequences different from all the given ones.*

[Hint: Write out the given sequence of infinite binary sequences as an infinite array (matrix) of bits whose first row is the first given sequence, the second row is the second given sequence, etc. Now let d_1 be the complement of the first bit of the first row, d_2 be the complement of the second bit of the second row, etc. Notice that $d_1 d_2 \cdots d_n \cdots$ is the sequence obtained by taking the *diagonal* of the given array and then inverting every one of its bits.]

The method outlined above, known as *Cantor diagonalization,* thus gives us a direct proof that there are uncountably many infinite binary sequences.

Corollary 426. $\mathbf{N} \prec \{0,1\}^{\mathbf{N}}$, *and so* $\aleph_0 < 2^{\aleph_0}$.

Since $\{0,1\}^{\mathbf{N}} \sim \mathrm{P}(\mathbf{N})$, so we also get:

Corollary 427. $\mathbf{N} \prec \mathrm{P}(\mathbf{N})$, *i.e.,* $\aleph_0 < |\mathrm{P}(\mathbf{N})|$. *So* $\mathrm{P}(\mathbf{N})$ *is uncountable, i.e.,* $\mathbf{N}$ *has an uncountable number of subsets.*

Cantor diagonlization can be readily generalized to arbitrary cardinalities:

Theorem 428 (Cantor's Theorem). *$\kappa < 2^\kappa$ for any cardinal κ, i.e., the cardinal 2^κ is strictly greater than κ. It follows that $\mathbf{P}(A) \not\preccurlyeq A$ and so $A \prec \mathbf{P}(A)$, for any set A, i.e., $|A| < |\mathbf{P}(A)|$ and so the number subsets of A is strictly greater than $|A|$.*

Problem 429. *Prove Cantor's Theorem.*

[Hint: Since $\mathbf{P}(A) \sim \{0,1\}^A$ one can work with $\{0,1\}^A$ instead of $\mathbf{P}(A)$.]

Problem 430. *Give a direct effective proof of Cantor's Theorem by showing that if $F\colon A \to \mathbf{P}(A)$ then the "anti-diagonal set"*

$$D_F := \{x \in A \mid x \notin F(x)\}$$

is not in the range of F, i.e., $D_F \in \mathbf{P}(A) \smallsetminus \operatorname{ran}(F)$, and so F is not onto.

The following beautiful result of König generalizes Cantor's Theorem.

Theorem 431 (König's Inequality (AC)). *Let K be any set and let α_k and β_k be cardinals for each $k \in K$.*

$$\textit{If } \alpha_k < \beta_k \textit{ for all } k \in K, \textit{ then } \sum_{k \in K} \alpha_k < \prod_{k \in K} \beta_k.$$

Proof. Assume $\alpha_k < \beta_k$ for each $k \in K$, and let $\alpha := \sum_{k\in K} \alpha_k$ and $\beta := \prod_{k \in K} \beta_k$. We will be using the Axiom of Choice several times in this proof to choose certain sets and elements. For each $k \in K$, fix a set B_k such that $|B_k| = \beta_k$. By replacing each B_k with $\{k\} \times B_k$ if necessary, we can assume that the sets B_k, $k \in K$, are pairwise disjoint. Also since $\alpha_k < \beta_k$, we can fix, for each $k \in K$, a subset $A_k \subsetneq B_k$ with $|A_k| = \alpha_k$ and an element $b_k \in B_k \smallsetminus A_k$. Put $A := \cup_k A_k$ and $B := \prod_{k \in K} B_k$, so that $|A| = \alpha$ and $|B| = \beta$.

Define $F\colon A \to B$ by setting, for each $x \in A$, $F(x) = y$, where $y = \langle y_k \mid k \in K\rangle$ is defined as:

$$y_k = \begin{cases} x & \text{if } x \in A_k \\ b_k & \text{otherwise} \end{cases}.$$

Then F is a one-to-one function, for if $x \neq x'$ are in A, $y = F(x)$ and $y' = F(x')$, then either there is k such that $x, x' \in A_k$, so that $y_k = x \neq x' = y'_k$, or else there are distinct $k, k' \in K$ with $x \in A_k$ and $x' \in A_{k'}$ giving $y_k = x \neq b_k = y'_k$, hence $y \neq y'$ in both cases. It follows that F is injective, and so $\alpha \leqslant \beta$.

Finally, let $G\colon A \to B$ be an arbitrary function from A to B. We show that G cannot be surjective. For each $k \in K$, the set $D_k := \{\pi_k(G(x)) \mid x \in A_k\}$ is a subset of B_k of cardinality at most α_k (where $\pi_k\colon B \to B_k$ is the k-th coordinate projection function), and so we can fix $d_k \in B_k \smallsetminus D_k$. Then the element $d := \langle d_k \mid k \in K\rangle \in B$

is not in the range of G, for if $x \in A$, then $x \in A_k$ for some $k \in K$, so $\pi_k(G(x)) \in D_k$ while $\pi_K(d) = d_k \notin D_k$, hence $G(x) \neq d$. Thus G is not surjective. Hence $\alpha < \beta$. □

Problem 432. *Derive Cantor's Theorem as a direct corollary of König's Inequality.*

6.9 The Cardinal $\mathfrak{f} = 2^{\mathfrak{c}}$ and Beyond

By Cantor's Theorem, there is no largest cardinal number. For any cardinal κ, the cardinal 2^{κ} is still bigger.

Definition 433. $\mathfrak{f} = 2^{\mathfrak{c}}$.

Thus we have: $0 < 1 < 2 < \cdots < \aleph_0 < \mathfrak{c} < \mathfrak{f}$.

Problem 434. *Prove that* $\mathfrak{c}^{\mathfrak{c}} = \mathfrak{f}$. *Prove also that* $\mathfrak{f}^{\mathfrak{c}} = \mathfrak{f} = \mathfrak{f}^{\aleph_0}$.

Thus it follows that the set of all functions from $\mathbf{R}$ to $\mathbf{R}$ has cardinality $\mathfrak{f}$.

Problem 435. *Prove that*

1. *The collection of all bounded real-valued Riemann Integrable functions defined on the unit interval* $[0, 1]$ *has cardinality* $\mathfrak{f}$.
2. *On the other hand, that the set of all continuous real-valued functions with domain* $\mathbf{R}$ *has cardinality* $\mathfrak{c}$.

[Hint: For the first result, use the fact that any bounded function defined on the unit interval which is constant on the complement of the Cantor set must be Riemann integrable. For the second result, note that two continuous functions which agree on all rational points must agree on all real numbers.]

Cantor's theorem enables us to obtain larger and larger infinite cardinals in an endless fashion. Starting from $\aleph_0$ and repeatedly applying Cantor's theorem we get:

$$\aleph_0 < 2^{\aleph_0} < 2^{2^{\aleph_0}} < \cdots .$$

We can then get a cardinal β larger than all the cardinals above by taking β to be the sum of all these cardinals:

$$\beta := \aleph_0 + 2^{\aleph_0} + 2^{2^{\aleph_0}} + \cdots .$$

Then we can start again from β and keep applying Cantor's Theorem to get

$$\beta < 2^{\beta} < 2^{2^{\beta}} < \cdots$$

and so on. Iterating the process endlessly into the transfinite needs the notion of *ordinals*, which will be defined later.

We have already seen that

$$\mathfrak{c}^{\aleph_0} = \mathfrak{c} \quad \text{and} \quad \mathfrak{f}^{\aleph_0} = \mathfrak{f}.$$

More generally, we have $\kappa^{\aleph_0} = \kappa$ if κ is any of $\mathfrak{c}, 2^{\mathfrak{c}}, 2^{2^{\mathfrak{c}}}, \ldots$, etc.:

Problem 436. *Put $\kappa_1 := 2^{\aleph_0}$ and $\kappa_{n+1} = 2^{\kappa_n}$ for $n = 1, 2, \ldots$. Show that for any $n \geq 1$, $\kappa_n^{\aleph_0} = \kappa_n$.*

In view of these last facts, one may think that the cardinals such as $\mathfrak{c} = 2^{\aleph_0}$, $\mathfrak{f} = 2^{\mathfrak{c}}$, and $2^{\mathfrak{f}}$ are "too large" to be increased by raising to the power $\aleph_0$, and one may conjecture that if κ is a sufficiently large cardinal, say if $\kappa \geq 2^{\aleph_0}$, then $\kappa^{\aleph_0} = \kappa$. But this is false, since the cardinal β mentioned above is larger than all of $\mathfrak{c}$, $2^{\mathfrak{c}}$, $2^{2^{\mathfrak{c}}}$, etc., yet we have $\beta^{\aleph_0} > \beta$.

Problem 437. *Prove that $\beta^{\aleph_0} > \beta$.*

[Hint: Use König's inequality.]

Problem 438. *Prove that for any cardinal κ there are cardinals $\mu > \kappa$ and $\nu > \kappa$ such that $\mu^{\aleph_0} = \mu$ and $\nu^{\aleph_0} > \nu$.*

The phenomenon in the last problem can be illustrated in a more general way using the concept of *cofinality*. We will define cofinality in a later chapter where the "cofinality version" of König's result will be proved.

6.10 Additional Problems

Problem 439. *Show that the set of all monotone real functions has cardinality* $\mathfrak{c}$. *(A function $f : \mathbf{R} \to \mathbf{R}$ is* monotone *if either $f(x) \leq f(y)$ for all $x < y$ or $f(x) \geq f(y)$ for all $x < y$.)*

Problem 440. *Given a set $A \subseteq \mathbf{N}$, define a real number x_A as*

$$x_A := \sum_{k=1}^{\infty} \frac{\chi_A(k)}{10^{k^2}},$$

where we write $\chi_A(n) := 1$ if $n \in A$ and $\chi_A(n) = 0$ if $n \notin A$ (thus χ_A is the characteristic function of A). Prove that

1. *$A \cap B = \emptyset \Rightarrow x_{A \cup B} = x_A + x_B$, and*
2. *x_A is irrational if and only if A is infinite.*

Problem 441. *Find a specific mapping*

$$F : \mathbf{R} \to \mathbf{P}(\mathbf{N})$$

such that for all x, y, if $x < y$ then $F(x) \subsetneq F(y)$ with $F(y) \setminus F(x)$ infinite.

[Hint: It may be easier to first obtain such a function F from $\mathbf{R}$ to $\mathrm{P}(\mathbf{Q})$.]

Problem 442. *Find a specific function $g: \mathbf{R} \to \mathbf{R}$ such that $x \neq y \Rightarrow g(x) - g(y) \in \mathbf{R} \setminus \mathbf{Q}$. (This gives an effective one-to-one map from $\mathbf{R}$ into the partition $\mathbf{R}/\mathbf{Q}$.)*

Problem 443. *A nonempty subset S of the set $\mathbf{Q}$ of rational numbers is called a* subring of $\mathbf{Q}$ *if it is closed under addition, subtraction, and multiplication, that is, $x, y \in S \Rightarrow x + y, x - y, xy \in S$. How many subsets of $\mathbf{Q}$ are subrings? Describe the subrings of $\mathbf{Q}$ completely, and exhibit an effective one-to-one correspondence between the collection of all subrings of $\mathbf{Q}$ and a familiar set.*

Chapter 7
Orders and Order Types

Abstract This chapter introduces order isomorphisms and order types, as well as the basic operations of sums and product of order types.

7.1 Orders, Terminology, and Notation

Consider the natural numbers arranged in ascending order of magnitude:

$$1,\ 2,\ 3,\ \ldots,\ n,\ n+1,\ \ldots$$

This arrangement is determined by the binary relation $<$ (*less than*), and a smaller number is always placed to the left of any larger number: If m and n are two distinct numbers, then *m precedes n* in the above arrangement if and only if $m < n$. Here, the precedence relation $<$ is a transitive relation on $\mathbf{N}$ with the additional *trichotomy* property that given any two distinct natural numbers exactly one of them precedes the other.

Another familiar linear ordering is on the "points of the real number line"—the set of real numbers ordered by magnitude—where the precedence relation is again a transitive relation with the trichotomy property.

Intuitively, by a *linear ordering* we mean a set whose members are viewed as points arranged in a single line by a transitive relation of precedence where for any two distinct points exactly one precedes the other. If P denotes this relation on (say) the set X of points, "xPy" stands for "x precedes y."

We will first review some basic definitions from Sect. 1.9.

Definition 444. We say that *P is an order on A*, or that *P orders A*, or equivalently, that *$\langle A, P\rangle$ is an order* (or a *total* or *linear* order) if P is a transitive relation on the set A which satisfies trichotomy on A, i.e., for all $x, y \in A$ exactly one of the conditions

A. Dasgupta, *Set Theory: With an Introduction to Real Point Sets*,
DOI 10.1007/978-1-4614-8854-5_7, © Springer Science+Business Media New York 2014

$$xPy, \qquad x = y, \qquad yPx$$

holds. If A has more than one element, the order is called *nontrivial*.

Recall also that a relation P is called *connected* on a set A if for all distinct $x, y \in A$ at least one of xPy or yPx holds.

Variants of the following basic problems were given in Chap. 1.

Problem 445. *Let P be a relation on a set A. If P is asymmetric on A then it is irreflexive on A, but the converse implication may fail. If P is transitive, then P is asymmetric on A if and only if it is irreflexive on A.*

Problem 446. *Let P be a transitive relation on the set A. Then each of the following conditions is equivalent to the others:*

1. *P orders A.*
2. *P is asymmetric and connected on A.*
3. *P is irreflexive and connected on A.*

Remark. If P is an order on A, then by trichotomy the Cartesian product $A \times A$ is partitioned into three pairwise disjoint sets:

$$\{\langle x, y\rangle \mid xPy\}, \qquad \{\langle x, y\rangle \mid x = y\}, \qquad \text{and} \qquad \{\langle x, y\rangle \mid yPx\}.$$

In particular for any $a \in A$, the three sets

$$\{x \in A \mid xPa\}, \qquad \{a\}, \qquad \text{and} \qquad \{x \in A \mid aPx\}$$

are pairwise disjoint with union equal to A.

Remark. We are defining order in the *strict* sense, i.e., as an irreflexive relation. One could also define an ordering relation as a reflexive relation without any essential changes.

Terminology and Notation

If P orders A, we will write $x <_P y$ to denote xPy. When there is no chance of confusion, we even drop the subscript P and simply write $x < y$ for xPy, a notation that will be used routinely. We will further abuse terminology and usually say "A is an order" in place of "$\langle A, <\rangle$ is an order." The informal phrase "A is an order" is really an abbreviation for "A is a set with an associated ordering which will be denoted by $<$."

Of course, if there are multiple orderings, say P and Q, on the same set A, then the notation $x < y$ may be ambiguous and we may need to explicitly distinguish between $x <_P y$ and $x <_Q y$. Also, two orderings on two different sets X and Y will sometimes be denoted by $<_X$ and $<_Y$, respectively.

In addition, the usual notational enhancements for the symbol $<$ will be used. For example, "$x \leq y$" stands for "$x < y$ or $x = y$," "$x > y$" means "$y < x$," "$x < y < z$" is an abbreviation for "$x < y$ and $y < z$," etc.

7.2 Some Basic Definitions: Suborders

Definition 447. Let X be an order.

Given $a \in X$, the elements of the set $\{x \in X \mid x < a\}$ are called *the predecessors of* a, and the elements of the set $\{x \in X \mid a < x\}$ are called *the successors of* a.

If $x, y \in X$, x *is an immediate predecessor of* y *in* X, or equivalently y *is an immediate successor of* x *in* X, if $x < y$ and there is no $z \in X$ with $x < z < y$. We also say that x and y are *consecutive elements* or *immediate neighbors* to mean that one of them is an immediate successor of the other.

It is easily seen that each element has at most one immediate successor or immediate predecessor.

Example 448. The sets **N** and **R** can each be equipped with the usual order of magnitude among the elements, but these are two separate orderings. In **N**, the set of predecessors of the element 4 is the finite set $\{1, 2, 3\}$, while in **R** the set of predecessors of 4 is the entire open interval $(-\infty, 4)$. The set of successors of 4 in **N** is the infinite but countable set $\{5, 6, 7, 8, \ldots\}$, while in **R** the set of successors of 4 is the uncountable open interval $(4, \infty)$.

In **N**, 7 is an immediate successor of 6, so 6 and 7 are consecutive elements of **N**, while in **R** the same elements 6 and 7 are not consecutive. In fact, in **N**, every element has a (unique) immediate successor and every element other than 1 has an immediate predecessor (1 has no predecessor at all in **N**). On the other hand, in **R**, no element has an immediate successor or immediate predecessor, and so there are no consecutive elements in **R**.

Definition 449. Let X be an order, $A \subseteq X$, and $a \in X$.

a is a *lower bound of* A, written $a < A$, if $a < x$ for all $x \in A$, and a is a *first (or least) element of* A if $a \in A$ and $a \leq A$.

Upper bounds and *last* or *greatest* elements are defined similarly.

An element $a \in X$ is called an *endpoint* of the order X if a is either a first or a last element of the whole set X. An ordering which does not have either a first or last element is called an *ordering without endpoints.*

The subset A is called *bounded below* if there is some $a \in X$ which is a lower bound of A. Similarly we define *bounded above.* A is *bounded* if it is both bounded above and bounded below.

If $A, B \subseteq X$, we write $A < B$ to mean $(\forall x \in A)(\forall y \in B)(x < y)$.

By trichotomy, a first (or last) element of a set, if it exists, is unique, and we write $a = \min A$ for "a is the first element of A," and $a = \max A$ for "a is the last element of A."

Example 450. Let each of **N** and **R** be ordered as before (by usual order of magnitude). The ordering **N** has a first element, but no last element. **R** is an ordering without endpoints. If $A = [0, \infty)$ is the subset of **R** consisting of the nonnegative reals, then A has a least element, 0. Also A is bounded below in **R** but not bounded above.

Problem 451. *For each of the following, give an example of a set X and a nontrivial order on X satisfying the given condition.*

1. *X is countable, has a first and a last element, but there are no consecutive elements in X,*
2. *X has a last element but no first element.*
3. *X is infinite, X has a first and a last element, and each element except the last has an immediate successor and each element except the first has an immediate predecessor.*
4. *X has a unique element which has neither an immediate successor nor an immediate predecessor while every other element has both an immediate successor and an immediate predecessor.*

Problem 452. *If A is a nonempty finite subset of an order X, show that* $\min A$ *and* $\max A$ *both exist, that is A contains a least element and a greatest element.*

Suborders

The sets **N**, **Z**, **Q**, and **R**, each ordered by the natural order of magnitude among its elements, are familiar examples of orders. One can obtain many more examples of orders either by rearranging the elements of the set (next section), or by passing to a subset and regarding the subset as a new order with the ordering on the subset inherited from the original order.

More specifically, given an ordering X and a subset $Y \subseteq X$, Y becomes an order on its own right by restricting the order on X to the elements of Y. The resulting order on Y is said to be the *suborder induced by* (or the *suborder inherited from*) the ordering of X.

For example, let X be the set of real numbers with the usual order and let $Y \subseteq X$ be the subset $Y := \{\frac{n}{n+1} \mid n \in \mathbf{N}\}$. Then the suborder Y has a first element $(1/2)$, but Y does not have any last element. In Y, the elements $2/3$ and $3/4$ are consecutive, with $3/4$ being the immediate successor of $2/3$.

Order properties or relations for points and subsets may not be preserved between the suborder and the original parent order. In $\mathbf{N} \subseteq \mathbf{R}$, the suborder on **N** inherited from the usual order of **R** is same as the usual order on **N**, but we had noted earlier that 6 and 7 are consecutive elements in the suborder **N**, while the parent order **R** has no consecutive elements at all.

As another example, consider the interval $[0, 1)$ as a suborder of **R**, and let $A{:=}\{\frac{n}{n+1} \mid n \in \mathbf{N}\}$, so that $A \subseteq [0, 1) \subseteq \mathbf{R}$. Then A is bounded in the parent order **R**, but it is not bounded in the suborder $[0, 1)$.

Definition 453 (Intervals, segments, cofinal sets). Let X be an order.

An *open interval* in X is any subset which can be expressed (for some $a, b \in X$) in one of the four forms

$$\{x \mid x < a\} \quad \text{or} \quad \{x \mid a < x\} \quad \text{or} \quad \{x \mid a < x < b\} \quad \text{or the entire order } X.$$

A *closed interval* in X is a subset of X of the form $\{x \mid x \leq a\}$ or $\{x \mid a \leq x\}$ or $\{x \mid a \leq x \leq b\}$ (for some $a, b \in X$) or X or $\emptyset$. An *interval* is a subset which is either an open or a closed interval.

A subset A is an *initial segment of* X if it is "closed under precedence," that is, if $a \in A$ and $x < a \Rightarrow x \in A$ (any predecessor of any element of A is also in A). *Final segments* are similarly defined. A subset A is a *segment in* X if whenever $x, y \in A$ and $x < z < y$ then $z \in A$.

A subset A is *cofinal in* X if for all $x \in X$ there is some $a \in A$ with $x \leq a$. *Coinitial* subsets are similarly defined.

Note that every initial (or final) segment is a segment, and that every interval is a segment, but in some orders there are segments which are not intervals. For example, the subset $\{x \in \mathbf{Q} \mid x^2 < 2\}$ is a segment, but not an interval, in $\mathbf{Q}$ (with usual order).

The complement of an initial segment is a final segment, and vice versa. For an order without a last element a subset is cofinal if and only if it is unbounded above. For an order with a last element a subset is cofinal if and only if it contains the last element.

7.3 Isomorphisms, Similarity, and Rearrangements

Let $A = \{1, 3, 5, 7, \ldots\}$ be the set of odd positive integers and $B = \{2, 4, 6, 8, \ldots\}$ be the set of even positive integers, with each of them ordered by the usual order of magnitude. Consider the correspondence between them as displayed below:

$$\begin{array}{lcccccccc} A: & 1 < & 3 < & 5 < & 7 < & \cdots < & 2n-1 & < \cdots \\ & \updownarrow & \updownarrow & \updownarrow & \updownarrow & & \updownarrow & \\ B: & 2 < & 4 < & 6 < & 8 < & \cdots < & 2n & < \cdots \end{array}$$

If f denotes this mapping from A to B so that $f(n) = n + 1$, then note that $f: A \to B$ not only is a bijection, but also *preserves order* in the sense that for all $m, n \in A$, $m < n$ in A if and only if $f(m) < f(n)$ in B. Such an order-preserving bijection between two orders is called an *order isomorphism,* and two orders are said to be *similar* or *order isomorphic* if there is an order-preserving bijection between them.

Definition 454. Let A and B be orders. A mapping f is an *order isomorphism from A to B* if $f: A \to B$ is a bijection which also preserves order, that is, for all $x, y \in A$, $x < y$ in A if and only if $f(x) < f(y)$ in B.

Two orders A and B are *similar* or *order isomorphic*, written $A \cong B$, if there is some order isomorphism between them.

Problem 455. *If A and B are orders, then $f: A \to B$ is said to be* strictly increasing *if whenever $x < y$ in A then $f(x) < f(y)$ in B. Show that a mapping from one order to another is an order isomorphism if and only if it is strictly increasing and onto.*

Example 456. Let $A := \mathbf{N}$ and $B := \{\frac{n}{n+1} \mid n \in \mathbf{N}\}$, where both sets are ordered by the usual order of magnitude. Then A and B are isomorphic via the order-preserving correspondence $n \leftrightarrow \frac{n}{n+1}$.

In fact, we had already seen some examples of order isomorphisms in the chapter on cardinals, where we saw that there is a bijective order-preserving correspondence between any two proper closed intervals of the real line. Furthermore, the mapping $x \mapsto \frac{x}{x+1}$ was seen to be an order isomorphism between $[0, \infty)$ and $[0, 1)$.

When two orders X and Y are isomorphic they share all order properties which do not mention actual elements or subsets of the orders: X has a first element if and only if Y has a first element, X has consecutive elements if and only if Y does too, and so on.

Even when specific points and subsets of the orders are mentioned, the isomorphism function will preserve all order properties and relations between them so long as those points and subsets are replaced by their appropriate images under the function when moving between the two orders: If $f: X \to Y$ is an order isomorphism between the orders X and Y, $a \in X$, and $A \subseteq X$, then a is the first element of A in X if and only if $f(a)$ is the first element of $f[A]$ in Y, a is an upper bound of A in X if and only if $f(a)$ is an upper bound of $f[A]$ in Y, A is bounded above in X if and only if $f[A]$ is bounded above in Y, and so on.

Informally, *two orders are isomorphic if one can be obtained from the other by renaming or replacing its points while preserving the order.*

Rearrangements

Distinct orders on the same set will be called *rearrangements*. For example, the finite set $\{a, b, c\}$, where a, b, c are three distinct elements, can be ordered in six different ways:

$$a < b < c; \quad a < c < b; \quad b < a < c; \quad b < c < a; \quad c < a < b; \quad c < b < a,$$

where each order is a rearrangement of the others. Here all six orders are similar.

Problem 457. *If A is a finite set with n elements, then show that there are exactly $n!$ distinct orders on A all of which are similar.*

If we start with the usual order of magnitude on the infinite set **N**,

$$1 < 2 < 3 < 4 < 5 < 6 < 7 < 8 < \cdots,$$

we can rearrange it to obtain new orders, such as the order

$$2 < 1 < 4 < 3 < 6 < 5 < 8 < 7 < \cdots$$

which is distinct from the original order since in this new rearrangement of **N** the first element is 2 and the immediate successor of 1 is 4. However, note that this rearrangement is still similar to the usual order on **N** (under the bijection f defined by $f(n) = n + 1$ if n is odd and $f(n) = n - 1$ if n is even).

We can get other rearrangements of **N** like P, Q, or R below

$$\begin{aligned} P: &\quad \cdots < 8 < 7 < 6 < 5 < 4 < 3 < 2 < 1 \\ Q: &\quad \cdots < 7 < 5 < 3 < 1 < 2 < 4 < 6 < 8 < \cdots \\ R: &\quad 1 < 3 < 5 < 7 < \cdots \cdots < 8 < 6 < 4 < 2, \end{aligned}$$

none of which is isomorphic to any other or to the usual order on **N**: The order P is the *reverse order* of the usual order and has a last element but no first; it is isomorphic to the set of negative integers with the usual ordering. The ordering Q has neither a first nor a last element, and is isomorphic to the set **Z** of all integers with the usual ordering. The ordering R has both a first element and a last element. These differences in the presence of first and last these orders show that none of the orders P, Q, R is isomorphic to any other or to the usual order on **N**.

The usual method for showing that two orders are not isomorphic is to find an order property which holds in one order but not the other. Here is one more example. Consider the rearrangement S of **N** shown as

$$S: \quad 2 < 3 < 4 < 5 < 6 < \cdots < 1.$$

This rearrangement S of **N** has both a first element (2) and a last element (1), making it an ordering having both endpoints. Therefore it is not similar to any of the orders above—except possibly R, which also has both endpoints. But note that in R the last element has an immediate predecessor, while in S the last element has no immediate predecessor, so S and R cannot be isomorphic. Thus S is not isomorphic to any of other order mentioned above.

Problem 458. *Show that the three rearrangements of* **N** *shown by*

$$\begin{aligned} &3 < 4 < 5 < 6 < 7 < 8 < 9 < 10 < \cdots 1 < 2, \\ &1 < 3 < 5 < 7 < \cdots < 2 < 4 < 6 < 8 < \cdots, \\ \textit{and} \quad &3 < 5 < 7 < 9 < \cdots < 2 < 4 < 6 < 8 < \cdots 1 \end{aligned}$$

are not isomorphic to each other or to any other rearrangement of **N** *mentioned earlier in this section.*

7.4 Order Types and Operations

Problem 459. *Similarity of orders is an equivalence relation.*

We can therefore appeal to the principle of abstraction to fix a complete invariant "OrdTyp" for the equivalence relation of similarity between orders.[1]

Definition 460 (Order Types, Cantor). For each order X, the *order type of* X is denoted by OrdTyp(X). It is a complete invariant for similarity of orders, so that for all orders X and Y, $X \cong Y \Leftrightarrow \mathrm{OrdTyp}(X) = \mathrm{OrdTyp}(Y)$. We will sometimes write $\mathrm{OrdTyp}_{<}(X)$ to make the ordering $<$ explicit.

Problem 461. *Two finite orders are isomorphic if and only if they have the same number of elements.*

Thus for each finite cardinal number n, there is a unique order type for orders on n-element sets, and we denote this order type by $\mathbf{n}$.

We now introduce special notation for important basic order types. All orders in the definition below are assumed to be the usual order of magnitude.

Definition 462 (Notation for Standard Order Types).

1. $\mathbf{n} := \mathrm{OrdTyp}(\{1 < 2 < \cdots < n\})$.
2. $\omega := \mathrm{OrdTyp}(\mathbf{N})$.
3. $\zeta := \mathrm{OrdTyp}(\mathbf{Z})$.
4. $\eta := \mathrm{OrdTyp}(\mathbf{Q})$.
5. $\lambda := \mathrm{OrdTyp}(\mathbf{R})$.

Definition 463. If $<$ is an order on X, its *reverse order* $^*<$ is the order on the same set X defined by $x \; ^*< y \Leftrightarrow y < x$. When X is equipped with the reverse order $^*<$, we will refer to it as $^*X$. An order is *symmetric* if $X \cong {}^*X$.

Since $X \cong X' \Rightarrow {}^*X \cong {}^*X'$, we can make the following definition.

Definition 464. Given an order type α, its *reverse order type* $^*\alpha$ is defined as the order type of the reverse order of any order of type α.

An order type α is called *symmetric* if $^*\alpha = \alpha$.

Problem 465. *Which of the order types* $\mathbf{n}$, ω, ζ, η, *and* λ, *are symmetric?*

---

[1]Definition 1299 gives a formal definition of order types in ZF set theory.

Sum of Order Types

Informally, to obtain the sum $\alpha + \beta$ of two order types α and β we take disjoint representative orders A and B of type α and β respectively, and then form a single order by "placing A before B." More precisely, given order types α and β, one can construct an order X which can be partitioned into disjoint sets L and U such that the suborder L has order type α, the suborder U has order type β, and all elements of L precede all elements of U (so that L is an initial segment in X whose complement is the final segment U):

$$X = L \cup U, \quad L < U, \quad \mathrm{OrdTyp}(L) = \alpha, \quad \text{and} \quad \mathrm{OrdTyp}(U) = \beta.$$

(This is very much like a Dedekind partition except that here we are allowing L and U to be empty.) Such an order X consists of "an order of type α followed by an order of type β," and X is easily seen to be uniquely determined up to order isomorphism by only the order types α and β.

Theorem 466. *Given any pair of order types α and β, there is a unique order type γ such that any order X of type γ consists of an initial segment L having type α and a (complimentary) final segment $U = X \smallsetminus L$ having type β.*

Proof. Uniqueness is routine. For the existence part, let α and β be given order types. Fix orders L and U with $\mathrm{OrdTyp}(L) = \alpha$ and $\mathrm{OrdTyp}(U) = \beta$. By replacing L with $L' := \{0\} \times L$ and U with $U' := \{1\} \times U$ (and transferring the orders on L and U to L' and U'), we may assume that $L \cap U = \emptyset$. Now put $X = L \cup U$, and order X by the rule

$$\begin{aligned} x < y \;\Leftrightarrow\; & \text{Either } x, y \in L \text{ and } x < y \text{ in } L, \\ & \text{or } x, y \in U \text{ and } x < y \text{ in } U, \\ & \text{or } x \in L \text{ and } y \in U. \end{aligned}$$

Then it is easy to see that X is a well-defined order satisfying the conditions of the theorem. □

Definition 467 (Sum of two order types). If α and β are any two order types, then $\alpha + \beta$ denotes the order type γ of the preceding theorem. In other words, $\alpha + \beta$ is the unique order type of any order X which can be partitioned into an initial segment L of type α and a final segment $U = X \smallsetminus L$ of type β.

Example 468. Take $L = \{1 < 2 < \cdots < n\}$ and $U = \{n+1 < n+2 < \cdots\}$ in $\mathbf{N}$ with the usual order:

$$\underbrace{1 < 2 < \cdots < n}_{L} < \underbrace{n+1 < n+2 < \cdots}_{U}$$

Since L has order type n, U has order type ω, and the entire order $\mathbf{N}$ has order type ω, it follows that (for any $n \in \mathbf{N}$):

$$n + \omega = \omega.$$

In particular

$$1 + \omega = \omega.$$

On the other hand, consider the rearrangement of $\mathbf{N}$ displayed by

$$2 < 3 < 4 < 5 < \cdots < 1,$$

which has order type $\omega + 1$ (take $L = \{2 < 3 < 4 < \cdots\}$ and $U = \{1\}$). This ordering has a last element, and so is not similar to the usual order on $\mathbf{N}$. Hence:

$$1 + \omega \neq \omega + 1,$$

which shows that the commutative law fails for addition of order types.

On the other hand, the associative law holds, allowing us to write expressions such as $\alpha + \beta + \gamma$ unambiguously without using parentheses. In particular, if the order A has order type α and A is partitioned into segments $A_1, A_2, \cdots, A_n$ with $A_1 < A_2 < \cdots < A_n$ and α_k = order type of A_k ($k = 1, 2, \ldots, n$), then we have:

$$\alpha = \alpha_1 + \alpha_2 + \cdots + \alpha_n.$$

Problem 469. *Show that* ${}^*(\alpha + \beta) = {}^*\beta + {}^*\alpha$.

Problem 470. *Verify which of the following equations involving order types are correct:*

1. $\omega + {}^*\omega = \zeta$
2. ${}^*\omega + \omega = \zeta$
3. ${}^*\omega + 3 + \omega = \zeta$
4. $\lambda + 1 + \lambda = \lambda$
5. $\lambda + \lambda = \lambda$
6. $\eta + 1 + \eta = \eta$
7. $\eta + \eta = \eta$
8. $\zeta + \omega = {}^*\omega + \zeta$

The cancellation laws for addition fail. For example,

$$1 + \omega = 2 + \omega, \quad \text{but} \quad 1 \neq 2.$$

However, certain special forms of cancellation work.

Problem 471. *Show that if n is a finite order type then n can be cancelled both from the left and from the right:*

$$n + \alpha = n + \beta \Rightarrow \alpha = \beta, \quad \textit{and} \quad \alpha + n = \beta + n \Rightarrow \alpha = \beta.$$

Show also that if m, n are finite order types and α is the order type of an order without a first element, then

$$m + \alpha = n + \alpha \Rightarrow m = n.$$

Problem 472. *Show that the order type ζ of the integers can be cancelled both from the left and from the right:*

$$\zeta + \alpha = \zeta + \beta \Rightarrow \alpha = \beta, \quad \textit{and} \quad \alpha + \zeta = \beta + \zeta \Rightarrow \alpha = \beta.$$

Ordered Sum of a Family of Order Types

We now define the ordered sum of any family of order types indexed by an ordered set.

Definition 473 (Ordered sum of order-indexed order types (AC)). Let I be an order (which is to be the index set) and for each $i \in I$ let α_i be an order type. The ordered sum

$$\sum_{i \in I} \alpha_i$$

is defined as the order type of any order X which can be partitioned into pairwise disjoint segments $X_i, i \in I$ such that the suborder X_i has order type α_i for each $i \in I$ and $X_i < X_j$ in X whenever $i < j$ in I.

The proof of existence and uniqueness of the ordered sum $\sum_{i \in I} X_i$ is similar to the previous proof, but the Axiom of Choice is used as needed to fix representative orders (or order isomorphisms). We outline the proof of existence. Given a family of order types $\langle \alpha_i \mid i \in I \rangle$, where I is an order, first fix (using AC) an order X_i of type α_i for each $i \in I$. By replacing X_i by $X_i' := \{i\} \times X_i$ (and transferring the order on X_i to X_i') we may assume that the orders X_i $(i \in I)$ are pairwise disjoint. Put $X := \bigcup_{i \in I} X_i$, and define an order on X by

$$\begin{aligned} x < y \quad \Leftrightarrow \quad & \text{either for some } i \in I\colon\ x, y \in X_i \text{ and } x < y \text{ in } X_i, \\ & \text{or, for some } i, j \in I\colon\ i < j \text{ in } I, x \in X_i, \text{ and } y \in X_j. \end{aligned}$$

Then X becomes an order satisfying the condition of the definition. □

Problem 474. *For $n \in \mathbf{N}$, let $\alpha_n = 1 + {}^*\omega$ if n is odd and $\alpha_n = \omega + 1$ if n is even. Show that*

$$\sum_{n\in\mathbf{N}} \alpha_n = 1 + (\zeta + 2)\omega.$$

For simple index sets I, it may be possible to write the ordered sum $\sum_{i\in I}$ as an informal expanded notation. For example if $I = \mathbf{N}$ (with the usual order) and for each $n \in \mathbf{N}$, $\alpha_n = \lambda + 1$, then

$$\begin{aligned}\sum_{n\in\mathbf{N}} \alpha_n &= \alpha_1 + \alpha_2 + \alpha_3 + \cdots \\ &= (\lambda + 1) + (\lambda + 1) + (\lambda + 1) + \cdots \\ &= \mathrm{OrdTyp}((0,1]) + \mathrm{OrdTyp}((1,2]) + \mathrm{OrdTyp}((2,3]) + \cdots \\ &= \mathrm{OrdTyp}((0,1] \cup (1,2] \cup (2,3] \cup \cdots) \;=\; \mathrm{OrdTyp}((0,\infty)) \;=\; \lambda.\end{aligned}$$

Similarly we can write:

$$\begin{aligned}1 + 1 + 1 + \cdots &= \omega \\ 1 + 2 + 3 + \cdots &= \omega \\ \cdots + 3 + 2 + 1 &= {}^*\omega,\end{aligned}$$

etc.

The above informal notation assumes a generalized version of the associative law where all groupings of the summands yield identical sums so long as the overall order of the summands is preserved.

Problem 475. *Formulate and prove the generalized associative law for ordered sums of families of order types.*

Lexicographic and Anti-lexicographic Ordering

By *lexicographic order* we mean the "left-to-right dictionary order" or "ordering by first differences," where two words x and y of same length are compared by reading their letters from left to right until the first place where the words differ is located, and we declare $x < y$ if the letter of x at that position alphabetically precedes the corresponding letter of y. Thus in lexicographic order, letters are more significant to the left. The *anti-lexicographic order* is the opposite "right-to-left dictionary order" (or "ordering by last differences") where letters on the right are more significant.

Definition 476 (Lexicographic and Anti-lexicographic orders). Let A and B be orders. The *lexicographic order on* $A \times B$ is defined by the rule (for $\langle a, b \rangle, \langle a', b' \rangle \in A \times B$):

$$(a, b) < (a', b') \Leftrightarrow a < a' \text{ in } A \text{ or } a = a' \text{ and } b < b' \text{ in } B.$$

The *anti-lexicographic order on* $A \times B$ is defined by the rule (for $\langle a, b \rangle, \langle a', b' \rangle \in A \times B$):

$$(a, b) < (a', b') \Leftrightarrow b < b' \text{ in } B \text{ or } b = b' \text{ and } a < a' \text{ in } A.$$

Problem 477. *Show that the anti-lexicographic order on $A \times B$ is similar to the lexicographic order on $B \times A$.*

Product of Order Types

It is easily verified that if $A \cong A'$ and $B \cong B'$ then under lexicographic orders, $A \times B \cong A' \times B'$. This allows us to define the product of two order types, but the standard convention is to define it with the order of the factors reversed as follows.

Definition 478 (Product of two order types). If α and β are order types, then the product $\alpha\beta$ is defined to be the order type of the Cartesian product $B \times A$ under the lexicographic order (or the order type of $A \times B$ under the anti-lexicographic order) where A and B are orders of type α and β, respectively. *Notice the reversal of the order of the factors.*

In particular, note that

> *If A has type α and B has type β, then the lexicographic product $A \times B$ has type $\beta\alpha$, not $\alpha\beta$.*

Product as Repeated Sum

It is often more convenient to view the product defined above as a "repeated sum" of the first factor. The following useful result, whose proof is routine, allows us to view the product as a "repeated sum":

Theorem 479 (Product as Repeated Sum, AC). *If α and β are two order types, then their product $\alpha\beta$ equals the ordered sum of an indexed family of types where the index set has type β and where each summand is α:*

$$\alpha\beta := \sum_{i \in B} \alpha_i, \qquad \text{where } B \text{ has order type } \beta \text{ and } \alpha_i = \alpha \text{ for all } i \in B.$$

This sum is independent of the choice of the representative order B.

Notice that order of the factors matters heavily in the product, and $\alpha\beta$ is viewed as "α repeated β times."

Thus 2ω is "2 repeated ω times" and so equals ω:

$$2\omega = 2 + 2 + 2 + \cdots \; = \; \omega,$$

while $\omega 2$ is "ω repeated 2 times," or:

$$\omega 2 = \omega + \omega.$$

But $\omega + \omega \neq \omega$ because an order of type $\omega + \omega$ contains elements with infinitely many predecessors, whereas any element in an order of type ω has only finitely many predecessors. Thus

$$\omega 2 \neq 2\omega,$$

and so *multiplication of order types is not commutative.*

Also, both the right and left cancellation laws for multiplication fail, since

$$\begin{array}{llll} & (1+\lambda)\zeta = (\lambda+1)\zeta = \lambda & \text{but} & 1+\lambda \neq \lambda+1, \\ \text{and} & (1+\lambda)1 = (1+\lambda)2 & \text{but} & 1 \neq 2. \end{array}$$

On the other hand, the associative law for multiplication and the left distributive law (but not the right) hold.

Problem 480. *Prove the associative law for multiplication of order types.*

Problem 481. *Show that the left distributive law* $\alpha(\beta + \gamma) = \alpha\beta + \alpha\gamma$ *holds, but the right distributive law* $(\alpha + \beta)\gamma = \alpha\gamma + \beta\gamma$ *fails.*

Problem 482. *Show that* ${}^*(\alpha\beta) = {}^*\alpha{}^*\beta$.

We write α^2 for $\alpha\alpha$, α^3 for $\alpha\alpha\alpha$, etc. We also define $\alpha^0 = 1$ and $\alpha^1 = \alpha$.

In particular, the lexicographic ordering on $\mathbf{N} \times \mathbf{N}$ has order type ω^2, but let us examine the following example of rearranging the natural numbers into an ordering of type ω^2.

Example 483. Consider the following rearrangement of $\mathbf{N}$ in which we first put the odd numbers in increasing order of magnitude, followed by "the doubles of the odd numbers" (numbers divisible by 2 but not by 4), followed by the "doubles of the doubles" (numbers divisible by 4 but not by 8), and so on:

$$1 < 3 < 5 < \cdots \; 2 < 6 < 10 < \cdots \; 4 < 12 < 20 < \cdots \; 8 < 24 < 40 < \cdots \; \cdots$$

This order starts with an order of type ω, followed by another order of type ω, and so on in an infinite sequence of successive orders each having type ω. This ordering thus has order type $\omega\omega = \omega^2$. We can formally define the rearrangement displayed above in terms of the usual order by the rule that x precedes y in this order if and only if

$$\nu_2(x) < \nu_2(y) \text{ or } \nu_2(x) = \nu_2(y) \text{ and } x < y,$$

where $\nu_2(x)$ denotes the highest power of 2 that divides x ($\nu_2(x) = 0$ if x is odd). Of course we could have used any other prime in place of 2 and obtained a different rearrangement of $\mathbf{N}$ of order type ω^2.

Each segment of type ω in the above order can itself be rearranged into an order of type ω^2 making the order type of the overall arrangement ω^3. Evidently, the process can be iterated to generate orders of type ω^4, ω^5, etc.

Problem 484. *Consider the order on* $\mathbf{N}$ *in which* x *precedes* y *if and only if either* $\nu_2(x) < \nu_2(y)$, *or* $\nu_2(x) = \nu_2(y)$ *and* $\nu_3(x) < \nu_3(y)$, *or* $\nu_2(x) = \nu_2(y)$, $\nu_3(x) = \nu_3(y)$ *and* $x < y$.

1. *What is the order type of this order?*
2. *What is the order type of the suborder consisting of all predecessors of the element* 600 *in this order?*
3. *What is the order type of the suborder consisting of all successors of the element* 600 *in this order?*

Problem 485. *Find a rearrangement of* $\mathbf{N}$ *of order type* ω^4.
Find a rearrangement of $\mathbf{N}$ *of order type* $\sum_n \omega^n = \omega + \omega^2 + \cdots + \omega^n + \cdots$.

The left distributive law can be used to simplify expressions involving powers. For example, $\omega + \omega^2 = \omega(1 + \omega) = \omega\omega = \omega^2$. More generally, any integral power of ω can "absorb" any lower power of ω from the left, but not from the right:

Problem 486. *If* $0 \leq m < n$ *are nonnegative integers, then* $\omega^m + \omega^n = \omega^n$ *but* $\omega^n + \omega^m \neq \omega^n$.

Lexicographic Orders with Many Factors

One can readily extend the definitions of lexicographic ordering to n factors ($n \in \mathbf{N}$) instead of just two factors. For example, the lexicographic order on $\mathbf{N}^3 = \mathbf{N}\times\mathbf{N}\times\mathbf{N}$ has order type ω^3. In general, given $u = \langle u_1, u_2, \ldots, u_k\rangle$ and $v = \langle v_1, v_2, \ldots, v_k\rangle$ in $\mathbf{N}^k$, we say that u precedes v in the lexicographic order on $\mathbf{N}^k$ if and only if $u_j < v_j$ for the least index j with $u_j \neq v_j$.

It is also possible to extend this notion to infinitely many factors indexed by an index set provided that the index set is *well-ordered,* an order-property that will be studied in a later chapter. Here we introduce an important special case, namely the *lexicographic order on the "power"* $A^{\mathbf{N}}$, *the set of sequences from the order* A.

Definition 487. Let A be an order. For sequences $a = \langle a_n \mid n \in \mathbf{N} \rangle$ and $b = \langle b_n \mid n \in \mathbf{N} \rangle$ in $A^{\mathbf{N}}$, we say that *a precedes b in the lexicographic order on* $A^{\mathbf{N}}$ if $a_n < b_n$ for the least n at which $a_n \neq b_n$. That is,

$$a < b \Leftrightarrow \quad \text{for some } n,\ a_n < b_n \text{ but } a_k = b_k \text{ for all } k < n.$$

Recall that the set $\mathbf{2^N}$ of binary sequences is equinumerous with the Cantor set via the bijection $\mathbf{F}$ given by

$$\mathbf{F}(a) = \sum_{n=1}^{\infty} \frac{2a_n}{3^n} \qquad (a \in \mathbf{2^N}).$$

It turns out that with the lexicographic order on $\mathbf{2^N}$, the bijection $\mathbf{F}$ is also order-preserving:

Problem 488. *Show that the map* $\mathbf{F}$ *above from* $\mathbf{2^N}$ *onto the Cantor set is an order isomorphism, and so the Cantor set (with the usual order) is similar to the set* $\mathbf{2^N}$ *of binary sequences ordered lexicographically.*

Regarding elements of $\mathbf{2^N}$ as binary expansions of reals in $[0, 1)$ we get another order-preserving map, provided that we discard duplicate binary expansions.

Problem 489. *Let D be the suborder of* $\mathbf{2^N}$ *(ordered lexicographically) consisting of binary sequences with infinitely many zeros, that is*

$$D := \{a \in \mathbf{2^N} \mid a_n = 0 \textit{ for infinitely many } n\}.$$

Show that D is similar to the real interval $[0, 1)$ *with the usual order.*
[Hint: Use the map $\phi\colon D \to [0, 1)$ *defined by* $\phi(a) = \sum_{n=1}^{\infty} a_n/2^n$.*]*

Problem 490. *Show that the Cantor set has a subset of order type* λ.

Problem 491. *Suppose that the set* $\mathbf{N^N}$ *of all sequences of positive integers is ordered lexicographically, and let* $\mathbf{N}^{\mathbf{N}}_{\uparrow}$ *be the suborder of* $\mathbf{N^N}$ *consisting of* strictly increasing *sequences. Show that, under the lexicographic ordering,*

1. $\mathbf{N^N}$ *is isomorphic to the suborder* $\mathbf{N}^{\mathbf{N}}_{\uparrow}$.
2. *Each of* $\mathbf{N^N}$ *and* $\mathbf{N}^{\mathbf{N}}_{\uparrow}$ *has order type* $1 + \lambda$.

[Hint: For the first part, consider the mapping from $\mathbf{N^N}$ to $\mathbf{N}^{\mathbf{N}}_{\uparrow}$ given by $\langle n_1, n_2, n_3, \dots \rangle \mapsto \langle n_1,\ n_1 + n_2,\ n_1 + n_2 + n_3,\ \dots \rangle$. For the second part, show that the mapping

$$\langle n_1, n_2, n_3, \dots \rangle \mapsto \frac{1}{2^{n_1}} + \frac{1}{2^{n_2}} + \frac{1}{2^{n_3}} + \cdots \qquad (n_1 < n_2 < n_3 < \cdots)$$

is an order-reversing bijection from $\mathbf{N}^{\mathbf{N}}_{\uparrow}$ onto $(0, 1]$.]

The definitions of lexicographic and anti-lexicographic orderings given above can only compare words of the same length, and the question of how to compare two words of different length is left open. The problem below considers several possibilities for extending the definition of lexicographic order to the collection $\mathbf{N}^*$ of finite sequences of positive integers of *all* possible lengths.

Problem 492. *Let* $\mathbf{N}^*$ *be the set of all finite sequences (strings) of natural numbers. Try to find the order type for each of the following orders defined on* $\mathbf{N}^*$*, all of which extend the lexicographic orders on* $\mathbf{N}^k$*,* $k = 1, 2, \ldots$.

1. *The order on* $\mathbf{N}^*$ *defined by the rule that* $u = \langle u_1, u_2, \ldots, u_m \rangle$ *precedes* $v = \langle v_1, v_2, \ldots, v_n \rangle$ *in* $\mathbf{N}^*$ *if and only if* $m < n$ *or* $m = n$ *and* u *precedes* v *lexicographically in* $\mathbf{N}^m (= \mathbf{N}^n)$.
2. *The order on* $\mathbf{N}^*$ *defined by the rule that* $u = \langle u_1, u_2, \ldots, u_m \rangle$ *precedes* $v = \langle v_1, v_2, \ldots, v_n \rangle$ *in* $\mathbf{N}^*$ *if and only if either* u *is a proper initial segment of* v*, or there is* $k \le \min(m, n)$ *with* $u_k < v_k$ *and* $u_j = v_j$ *for all* $j < k$.
3. *The order on* $\mathbf{N}^*$ *defined by the rule that* $u = \langle u_1, u_2, \ldots, u_m \rangle$ *precedes* $v = \langle v_1, v_2, \ldots, v_n \rangle$ *in* $\mathbf{N}^*$ *if and only if either* v *is a proper initial segment of* u*, or there is* $k \le \min(m, n)$ *with* $u_k < v_k$ *and* $u_j = v_j$ *for all* $j < k$.

The order in the last part of the problem is known as the *Kleene–Brouwer order* and will reappear as Problem 550.

Chapter 8
Dense and Complete Orders

Abstract This chapter introduces some basic topological notions in the context of orders, and then develops the theories of dense orders and complete orders in a general setting. We cover Cantor's theorem on countable dense linear orders, Dedekind completeness and completions, order characterization of **R**, connectedness and the intermediate value theorems for linear continuums, and the perfect set theorem for complete orders.

8.1 Limit Points, Derivatives, and Density

Definition 493. Let X be an order, let $a \in X$, and let $A \subseteq X$.

We say that a is an *upper limit point of A in X* if a is not the first element of X, and for every $x < a$ there is some $p \in A$ such that $x < p < a$. *Lower limit points of A in X* are similarly defined.

We say that a is a *limit point of A in X* if a is a lower or an upper limit point of A in X, and a is a *two sided limit point of A in X* if a is both a lower and an upper limit point of A in X.

If the parent order X is clear, we simply use the phrase "limit point of A" without the qualifier "in X."

The set of all limit points of A (in X) is called the *derived set of A* or the *derivative of A*, and will be denoted by $D(A)$.

Example 494. Let X be the order **R**, and let

$$A := \left\{ \frac{n}{n+1} \mid n \in \mathbf{N} \right\} \cup \{2\}.$$

Then 1 is an upper limit point of A in X, and A has no other limit points (upper or lower) in X. So $D(A) = \{1\}$ in X. Thus in general the notions of limit point and derived set are to be understood *relative to the parent order.*

A. Dasgupta, *Set Theory: With an Introduction to Real Point Sets*,
DOI 10.1007/978-1-4614-8854-5_8,

On the other hand, if we consider the suborder A as an order by itself, without bringing back the parent order X, then 2 will be an upper limit point of A *in the order* A, and so we will have $D(A) = \{2\}$ *in the order* A.

An order of type ω, such as **N**, by itself has no limit points, upper or lower. The same applies to any order of type ζ. On the other hand, in an order of type η (such as **Q**) or type λ (such as **R**) every point is both an upper and a lower limit point (of the entire order).

An upper limit point cannot have an immediate predecessor. If X is an order and $a \in X$ is not the first element of X, then a is an upper limit point in X if and only if a has no immediate predecessor in X.

Problem 495. *With the lexicographic order, which points are the limit points in* $\mathbf{Z} \times \mathbf{N}$*? In* $\mathbf{N} \times \mathbf{Z}$*?*

Problem 496. *Show that an order X is without endpoints and without any limit point (i.e., every element of X has both an immediate predecessor and an immediate successor) if and only if the order type of X has the form $\zeta\alpha$ for some order type α.*

Recall the order X from Example 483 with order type ω^2:

$$1 < 3 < 5 < \cdots \; 2 < 6 < 10 < \cdots \; 4 < 12 < 20 < \cdots \; 8 < 24 < 40 < \cdots \; \cdots$$

In X the elements $2, 4, 8, \ldots$ are the upper limit points, but there are no lower limit points. If A denotes the subset of X consisting of the odd numbers, then

$$D(A) = \{2\}, \quad \text{while} \quad D(X) = \{2, 4, 8, 16, \ldots\}.$$

Example 497. We slightly modify the order of Example 483 by moving 1 to the last position to get an order of type $\omega^2 + 1$:

$$3 < 5 < 7 < \cdots \; 2 < 6 < 10 < \cdots \; 4 < 12 < 20 < \cdots \; 8 < 24 < 40 < \cdots \; 1.$$

If Y denotes this order, note that in Y the elements $2, 4, 8, \ldots$, as well as the last element 1 are all limit points. Hence:

$$D(Y) = \{2 < 4 < 8 < \cdots \; 1\}.$$

If we regard the suborder $D(Y)$ as an order by itself (of order type $\omega + 1$), then note that the elements $2, 4, 8, \ldots$ are no longer limit points in $D(Y)$, but 1 is still a limit point in $D(Y)$. This is because in the original order Y, 1 was a *limit point of limit points.* Thus

$$D(D(Y)) = \{1\}.$$

Problem 498. *Let X be an order and let $A \subseteq X$.*

1. *Show that if A has a limit point in X then A must be infinite. Hence $D(A) = \emptyset$ if A is finite.*
2. *If $A, B \subseteq X$ then show that $D(A \cup B) = D(A) \cup D(B)$.*
3. *Show that*

$$D(D(A)) \subseteq D(A),$$

that is, "a limit point of limit points of A is a limit point of A."

Thus the elements of $D(A)$ are limit points of A, elements of $D(D(A))$ are limit points of limit points—or *second order limit points* of A, the elements of $D(D(D(A)))$ are the *third order limit points* of A, and so on.

Let us write $D^{(0)}(A) := A$, $D^{(1)}(A) := D(A)$, $D^{(2)}(A) := D(D(A))$, etc., so that the elements of $D^{(k)}(A)$ are the *limit points of A of order k*.

Problem 499. *Let $X = \mathbf{R}$ and let*

$$A := \left\{ \frac{1}{2^m} + \frac{1}{2^{m+n}} \mid m, n \in \mathbf{N} \right\}$$

What is the order type of the suborder A? Compute $D^{(k)}(A)$ for $k = 1, 2, 3$.

Problem 500. *Give an example of a subset A of $\mathbf{R}$ such that $D^{(3)}(A) \neq \emptyset$ but $D^{(4)}(A) = \emptyset$.*

Problem 501. *If X is an order of type $\omega^n + 1$ (where $n \in \mathbf{N}$), what are the order types of $D^{(k)}(X)$ for various $k \in \mathbf{N}$?*

Consider an order of type

$$\begin{aligned}\left(\sum_{n \in \mathbf{N}} (\omega^n + 1)\right) + 1 &= \omega + 1 + \omega^2 + 1 + \cdots + \omega^n + 1 + \cdots + 1 \\ &= \omega + \omega^2 + \cdots + \omega^n + \cdots + 1 = \left(\sum_{n \in \mathbf{N}} \omega^n\right) + 1\end{aligned}$$

where the last element is a limit point of limit points of order k for arbitrarily large $k \in \mathbf{N}$, and is called a *limit point of order ω*. In fact, we can define

$$D^{(\omega)}(A) := \bigcap_{k \in \mathbf{N}} D^{(k)}(A), \quad D^{(\omega+1)}(A) := D(D^{(\omega)}(A)), \quad \ldots, \text{ etc.,}$$

and keep *iterating the derivative operator D* indefinitely without end—a process that leads to the notion of *ordinal numbers* that will be studied later.

Problem 502. *For each $k \in \mathbf{N}$, put*

$$A_k := \left\{ \frac{1}{2^k} + \frac{1}{2^{k+n_1}} + \cdots + \frac{1}{2^{k+n_1+\cdots+n_k}} \mid n_1, n_2, \ldots, n_k \in \mathbf{N} \right\}.$$

In terms of fractional binary expansion, the set A_1 consists of binary fractions of the form $0{\cdot}10^{\star}1$, where "$0^{\star}$" stands for a string of zero or more 0s. Thus $A_1 \subseteq (\frac{1}{2}, 1)$. The set A_2 consists of binary fractions of the form $0.010^{\star}10^{\star}1$, with $A_2 \subseteq (\frac{1}{4}, \frac{1}{2})$, etc. Finally, put:

$$A = \bigcup_k A_k.$$

1. *Show that each A_k has infinitely many limit points of order $< k$, exactly one limit point of order k (namely, $\frac{1}{2^k}$), but no limit points of order $> k$.*
2. *Show that $D^{(\omega)}(A) \neq \emptyset$ but $D^{(\omega+1)}(A) = \emptyset$.*

Dense Orders and Dense Subsets

Definition 503 (Order Density: Dense Orders). A nontrivial order X is said to be a *dense order* or *order-dense* if for all $x, y \in X$ with $x < y$ there is $z \in X$ such that $x < z < y$, that is, if X does not contain any consecutive elements.

Q, **R**, and any nontrivial rational or real interval are familiar examples of dense orders. Finite orders and orders of type ω or ζ are example of orders which are not order-dense. The following problem illustrates the relationship between order density and limit points.

Problem 504. *Let X be a nontrivial order. Show that*

1. *Every element of X is an upper limit point if and only if X is order-dense and has no first element.*
2. *Every element of X is a lower limit point if and only if X is order-dense and has no last element.*
3. *Every element of X is a two-sided limit point (both an upper and a lower limit point) if and only if X is order-dense and without endpoints.*

Conclude that the following conditions are all equivalent:

1. *X is order-dense.*
2. *Every element of X except the first element (if present) is an upper limit point.*
3. *Every element of X except the last element (if present) is a lower limit point.*

If X has more than two points, the above conditions are also equivalent to:

4. *Every element of X except the endpoint(s) (if present) are two-sided limit points.*

Definition 505 (Relative Density: Dense Subsets). Let X be an order and suppose that $A \subseteq X$. We say that *A is dense in X* or that *A is a dense subset of X* if every element of $X \smallsetminus A$ is a limit point of A, that is, if $X \smallsetminus A \subseteq D(A)$ (or equivalently, if $X = A \cup D(A)$).

For example, **Q** is dense in X. This follows from the fact that between any two real numbers there is a rational number.

Problem 506. *Let X be a dense order and $A \subseteq X$. Then A is dense in X if and only if for all $x, y \in X$ with $x < y$ there is $a \in A$ with $x < a < y$.*

Problem 507. *Let A be a dense subset of a dense order X. Show that the suborder A as an order by itself is a dense order.*

Problem 508. *Assume that each of the sets* **N**, **Z**, **Q**, *and* **R** *is ordered by the natural order of magnitude among its elements.*

1. *Give examples of two disjoint subsets of* **R** *both of which are dense in* **R**.
2. *Prove rigorously that* **Q** *is a dense subset of* **R**, *but that* **Z** *is not dense in* **R**.
3. *Which subsets of* **N** *are dense in* **N**?
4. *If X is an order of type $\zeta + 1 + \zeta$, which subsets of X are dense in X?*
5. *If X is an order of type $\omega^2 + 1$, which subsets of X are dense in X?*

Problem 509. *Recall that an open interval in an order X is a subset which has one of the following four forms:*

$$\{x \in X \mid a < x < b\}, \quad \{x \in X \mid x < a\}, \quad \{x \in X \mid x > a\}, \quad \textit{or} \quad X.$$

Show that a subset A of X is dense in X if and only if A has nonempty intersection with every nonempty open interval of X.

Problem 510. *For $A \subseteq$* **R** *if* **R**$\smallsetminus A$ *is countable, show that A is dense in* **R**. *Conclude that the set of irrational numbers is dense in* **R**.

Problem 511. *Let X be an order. Given $A, B \subseteq X$, we say that A* is dense in *B if $B \smallsetminus A \subseteq D(A)$. Suppose that $A \subseteq B \subseteq C \subset X$. Show that if A is dense in B and B is dense in C then A is dense in C.*

A Note on Terminology

The notion of *dense order* (order-density) should be carefully distinguished from the notion *dense subset* (relative density), as they are of totally different category: Order-density is a property of *entire orders,* while relative density is a property of *subsets of orders.* Thus saying "Y is dense" may be ambiguous. To avoid this ambiguity, we can explicitly indicate order-density by saying "Y is a dense order" (or "Y is order dense"), and explicitly indicate relative density by saying "Y is a dense subset" (or "Y is dense in its parent order").

8.2 Continuums, Completeness, Sup, and Inf

Recall Dedekind's method, in which we partition or "cut" a given order into two nonempty pieces with one piece completely preceding the other:

Definition 512. A *Dedekind partition* or *Dedekind cut* for an order X is a partition of X consisting two nonempty disjoint sets L and U such that $x \in L, y \in U \Rightarrow x < y$, i.e., every member of L precedes every member of U, as pictured below.

In other words, all elements of U are upper bounds for L, all elements of L are lower bounds for U, as well as $L \neq \emptyset \neq U$, $L \cap U = \emptyset$, and $L \cup U = X$.

A Dedekind partition L, U in an order X can be classified as exactly one of the following four types:

1. *Jump*: Both L has a largest element and U has a smallest element.
2. *Upper limit point cut*: L does not have a largest element, but U has a smallest element which therefore is an upper limit point of L.
3. *Lower limit point cut*: U does not have a smallest element, but L has a largest element which therefore is a lower limit point of U.
4. *Gap*: Neither L has a largest element nor U has a smallest element.

Note that an order is order-dense if and only if it has no jumps (i.e., if no Dedekind partition is a jump).

In a cut of type (2) or (3), which we call a *limit point cut* or a *boundary cut*, the two halves of the partition are "connected together" in the sense that one of them contains a limit point of the other.

On the other hand, a cut of type (1) or (4) (a jump or a gap) is a "separation" of the order into two "disconnected pieces" none of which contains a limit point of the other. A continuum is an order which does not admit any such disconnection, that is one which has no jump or gap.

Definition 513 (Dedekind Continuity). A nontrivial order is said to be *a (linear) continuum* or *Dedekind continuous* if no Dedekind cut in it is a jump or a gap, or equivalently if every Dedekind cut in it is a limit point cut.

A more general notion is order completeness (Dedekind completeness).

Definition 514 (Dedekind Completeness). An order X is said to be *complete* or *Dedekind complete* if it has no gaps, i.e., if no Dedekind cut for it is a gap.

So *an order is a continuum if and only if it is both order-dense and complete.*

We had already seen that the real line $\mathbf{R}$ and all intervals in it are continuums (being both order-dense and complete). On the other hand neither the rationals $\mathbf{Q}$ nor the integers $\mathbf{Z}$ form a continuum: $\mathbf{Q}$ is order-dense but not complete, while $\mathbf{Z}$ is complete but not order dense.

Since each of the properties of being order-dense, being a continuum, and being complete is preserved under order isomorphisms, we can meaningfully speak of an *order type* being order-dense, or a continuum, or complete. Thus η is not complete, and neither ω nor ζ is dense, while λ, $1 + \lambda$, $\lambda + 1$, and $1 + \lambda + 1$ are continuums (corresponding to the various kinds of intervals). Later we will see several examples of continuums which are essentially different from these.

Definition 515 (Supremum and Infimum). Let X be an order and let $A \subseteq X$.

We say that $a \in X$ is a *least upper bound of* A or a *supremum of* A if a is the least element of the set of all upper bounds of A in X, that is, if a is an upper bound of A and $a \leq b$ for every upper bound b of A. *Greatest lower bounds* or *infimums* are similarly defined.

It is easily seen that a least upper bound or supremum of a set, if it exists, must be unique. Therefore we make the following definition.

Definition 516 (sup A **and** inf A)**.** If a set A has a supremum, we denote it by sup A, and thus the statement "$a = \sup A$" stands for "a is the least upper bound of A." Similarly the infimum of A is denoted by inf A.

For a set with a largest element, the supremum coincides with the maximum. For a nonempty set without a maximum, the supremum, if it exists, is an upper bound which is also a limit point of the set.

Problem 517. *Let A be a nonempty subset of an order X and let a be an upper bound of A in X. Prove that a is the least upper bound of A if and only if either a is the maximum element of A, or $a \notin A$ and a is an upper limit point of A.*

Problem 518 (Completeness as the Least Upper Bound Property). *Given an order X, prove that the following conditions are equivalent.*

1. *X is complete.*
2. *X has the* least-upper-bound property*: Every nonempty subset of X which is bounded above has a supremum (least upper bound)*
3. *X has the* greatest lower bound property*: Every nonempty subset of X which is bounded below has an infimum (greatest lower bound).*

Problem 519. *An order is complete if and only if every segment is an interval.*

Problem 520. *For each of the following orders, express its order type in terms of familiar order types, and determine if it is order dense, if it is complete, and if it is a continuum.*

1. **N** *with usual order.*
2. **Z** *with usual order.*
3. **Q** *with usual order.*
4. **R** *with usual order.*
5. *The real interval* $(0, 1]$ *with usual order.*
6. *The set* $\{0\} \cup \{\frac{1}{n} \mid n \in \mathbf{N}\}$

7. *The set* $\{-\frac{1}{n} \mid n \in \mathbf{N}\} \cup \{0\} \cup \{\frac{1}{n} \mid n \in \mathbf{N}\}$.
8. *The set* $\{-\frac{1}{n} \mid n \in \mathbf{N}\} \cup \{\frac{1}{n} \mid n \in \mathbf{N}\}$.
9. *The closed unit square* $[0, 1] \times [0, 1]$*, with lexicographic order.*
10. *The half-open unit square* $[0, 1) \times [0, 1)$*, with lexicographic order.*
11. *The subset of the plane* $[0, 1] \times \{0, 1\}$*, with lexicographic order.*
12. $\mathbf{N} \times \mathbf{Z}$*, with lexicographic order.*
13. $\mathbf{Z} \times \mathbf{N}$*, with lexicographic order.*

Problem 521. *An order is called* totally discrete *if every Dedekind cut for it is a jump.*

1. *An order is totally discrete if and only if it is complete and has no limit points.*
2. *Give example of an order which has no limit points yet is not totally discrete.*
3. *Give examples of three pairwise non-isomorphic infinite totally discrete orderings.*
4. *Prove that there does not exist four pairwise non-isomorphic infinite totally discrete orderings.*
5. *List the possible order types of totally discrete orders.*

8.3 Embeddings and Continuity

Definition 522. Let X and Y be orders. An order isomorphism between X and a suborder of Y is called an *order embedding*. If $f: X \to Y$ is an embedding, we say that f *embeds* X *into* Y. We say that X *is embeddable in* Y if there is some embedding of X into Y.

If f embeds X into Y, then the suborder $f[X] = \operatorname{ran}(f)$ of Y is an "isomorphic copy" of X sitting inside Y.

Any strictly increasing map from one order into another is an embedding:

Problem 523. *Let X and Y be orders and let $f: X \to Y$ be a function which is strictly increasing: If $x < y$ in X then $f(x) < f(y)$ in Y. Then f is an embedding of X into Y, and so X is isomorphic to the suborder $f[X] = \operatorname{ran}(f)$ of Y via f.*

This makes it easy to find examples of embeddings. For example, the map $n \mapsto n^2$ is an embedding of $\mathbf{N}$ into itself, and $n \mapsto n/(n + 1)$ is an embedding of $\mathbf{N}$ into $\mathbf{R}$.

If X is a suborder of Y, then the identity map on X is an embedding of X into Y, called the *inclusion map*. Thus *every suborder is embedded in the parent order via the inclusion map.*

The analogue of the Cantor–Bernstein Theorem fails for orders:

Problem 524. *Give examples of two non-isomorphic orders each of which is isomorphic to a suborder of the other.*

Unlike order isomorphisms, order embeddings fail to preserve many order notions. In particular, limit points of subsets are not preserved. Let

$$X := \left\{ \frac{n}{n+1} \mid n \in \mathbf{N} \right\} \cup \{1\},$$

so that X has order type $\omega+1$ and the point 1 is a limit point in X. Define $f: X \to \mathbf{R}$ by setting $f(1) = 2$ and $f(x) = x$ otherwise. Then f is an embedding of X into $\mathbf{R}$, but f does not preserve limit points: If $A \subseteq X$ is the subset

$$A := \left\{ \frac{n}{n+1} \mid n \in \mathbf{N} \right\},$$

then 1 is a limit point of A in X but $f(1) = 2$ is not a limit point of $f[A] = A$ in $\mathbf{R}$.

We say that the reason for this failure is the "discontinuity" of the map f, and a map which preserves limit points is called a continuous map:

Definition 525 (Continuous Maps). Let X and Y be orders. A map $f: X \to Y$ is called *continuous* if whenever $A \subseteq X$ and $a \in X$ is a limit point of A in X, then either $f(a) \in f[A]$ or $f(a)$ is a limit point of $f[A]$ in Y.

When an order embedding is onto, it becomes an isomorphism and hence continuous (an isomorphism preserves all order notions, including limit points).

Example 526. Consider the rearrangement of $\mathbf{N}$

$$1 < 3 < 5 < \cdots 2 < 6 < 10 < \cdots 4 < 12 < 20 < \cdots \cdots$$

of order type ω^2 that we have encountered before, and define f and g by

$$f\left(2^{m-1}(2n-1)\right) = m + \frac{n}{n+1}, \qquad g\left(2^{m-1}(2n-1)\right) = 2m + \frac{n}{n+1}.$$

Then both f and g embed the above rearrangement of $\mathbf{N}$ into $\mathbf{R}$, but while f is continuous, g is not.

Problem 527. *Let X and Y be nontrivial orders without endpoints, and $f: X \to Y$. Show that f is continuous if and only if whenever $p \in X$ and $c < f(p) < d$ in Y, there exist $a, b \in X$ with $a < p < b$ such that we have $c < f(x) < d$ for all x satisfying $a < x < b$.*

Limit Points and Suborders: Continuously Embedded Suborders

We now look at the above phenomenon (where an embedding fails to preserve limit points) for the special case of suborders. Consider $X := (0,1) \cup [2,3)$ as a suborder of $\mathbf{R}$. When X is considered as a suborder by itself, it has order type λ and 2 is a limit point of $(0,1)$ in the suborder X. But 2 is not a limit point of $(0,1)$ in the parent order $\mathbf{R}$.

Definition 528. Let Y be an order and $X \subseteq Y$ be a suborder. We say that the suborder X is *continuously order embedded* in the parent order Y if for any $A \subseteq X$ and $a \in X$, if a is a limit point of A in X, then a is a limit point of A in the parent order Y as well.

For example, both the suborders $A := \{n/(n+1) \mid n \in \mathbf{N}\} \cup \{1\}$ and $B := \{n/(n+1) \mid n \in \mathbf{N}\} \cup \{2\}$ of $\mathbf{R}$ have order type $\omega + 1$, but while A is continuously embedded in $\mathbf{R}$, B is not.

Problem 529. *Let Y be an order and $X \subseteq Y$ be a suborder. Show that X is continuously embedded in Y if and only if the inclusion map from X to Y is continuous, where the inclusion map $\iota_X : X \to Y$ is defined as $\iota_X(x) = x$ (for all $x \in X$).*

Problem 530. *Show that if $f : X \to Y$ is an order embedding of the order X into the order Y, then f is continuous if and only if $f[X] = \operatorname{ran}(f)$ is continuously embedded in Y.*

Problem 531. *Let X be a suborder of the order Y. Show that X is continuously embedded in Y if and only if for any $A \subseteq X$ and any $a \in X$, if $a = \sup A$ in X then $a = \sup A$ in Y and if $a = \inf A$ in X then $a = \inf A$ in Y.*

The following theorem, whose proof is left as an exercise, gives sufficient conditions for a suborder to be continuously embedded in the parent order.

Theorem 532. *Let X be a suborder of the order Y.*

1. *If every point of X is a two sided limit point of X in Y, then X is continuously embedded in Y.*
2. *If every point of $Y \smallsetminus X$ is a two sided limit point of X in Y, then X is continuously embedded in Y.*

Problem 533. *Let X be an order, $A \subseteq X$, and $a \in A$.*

1. *Show that if a is a limit point of A in the parent order X then a is a limit point of A in the suborder A.*
2. *Give an example to show that if a is a limit point of X (in the parent order X), then a may not be a limit point of A either in the suborder A or in the parent order X.*
3. *Give an example to show that if a is a limit point of A in the suborder A, then a may not be a limit point of X (in the parent order X).*

Problem 534*. *Give an example of an order X and a suborder $Y \subseteq X$ such that in the suborder Y every point of Y is a limit point, but in the original order X no point of Y is a limit point.*

Problem 535. *Give an example of an order which contains a suborder of type η, but in which every point has an immediate successor and an immediate predecessor.*

Example 536. Let X be a complete order with endpoints in which the only limit point is the last element. Then the order type of X is $\omega + 1$.

Proof. Note that every element (other than the last) has an immediate successor, while the last element (being an upper limit) is not the immediate successor of any element. Let a_1 be the first element, a_2 be its immediate successor (which, being an immediate successor, cannot be the last element), a_3 be the immediate successor of a_2 (which again cannot be the last element), and so on. Put $A = \{a_1 < a_2 < \dots\}$, which has order type ω. A is bounded above (by the last element of X), and so $a = \sup A$ exists and is a limit point in X. Hence a must be the last element. Hence the order type of X must be $\omega + 1$. □

Problem 537. *Let X be a complete order with endpoints and exactly one limit point. What are the possible order types that X can have?*

Theorem 538 (Hausdorff). *There are exactly $2^{\aleph_0} = \mathfrak{c}$ distinct order types for countable orders.*

Proof. Every countable order is isomorphic to an order defined on some subset of **N**. Since every order defined on a subset of **N** equals $\langle A, R\rangle$ with $A \subseteq \mathbf{N}$ and $R \subseteq \mathbf{N}^2$ (so that $\langle A, R\rangle \in \mathbf{P}(\mathbf{N}) \times \mathbf{P}(\mathbf{N}^2)$), there are at most

$$\left|\mathbf{P}(\mathbf{N}) \times \mathbf{P}(\mathbf{N}^2)\right| = 2^{\aleph_0} \cdot 2^{\aleph_0} = 2^{\aleph_0} = \mathfrak{c}$$

such orders. Hence there are at most $2^{\aleph_0} = \mathfrak{c}$ distinct order types for countable orders.

By the Cantor–Bernstein theorem, it suffices to exhibit $2^{\aleph_0} = \mathfrak{c}$ pairwise non-isomorphic countable orders. Consider order types of the form:

$$\sum_{n \in \mathbf{N}} \alpha_n = \alpha_1 + \alpha_2 + \cdots + \alpha_n + \cdots,$$

where each α_n is either $\omega + 1$ or $1 + {}^*\omega$. Any order X with order type of this form is countable and the set of limit points in X will form a suborder of X of type ω, hence the limit points of X can be listed in a sequence as "the first limit point," "the second limit point," and so on.

Now note that in an order of the above type there is no two-sided limit point: The n-th limit point will be a one-sided upper limit point if $\alpha_n = \omega + 1$ and will be a one-sided lower limit point if $\alpha_n = 1 + {}^*\omega$.

Thus given an infinite binary sequence $\langle b_1, b_2, \dots, b_n, \dots\rangle$ in $2^{\mathbf{N}}$, we can set $\alpha_n = 1 + {}^*\omega$ if $b_n = 0$ and $\alpha_n = \omega + 1$ if $b_n = 1$ to get an order of the above type in which the n-th limit point is a lower limit point if $b_n = 0$ and is an upper limit point if $b_n = 1$. Distinct binary sequences will give non-isomorphic orders, since in two isomorphic orders the n-th limit points (if they exist) will be of the same type. This gives $2^{\aleph_0} = \mathfrak{c}$ non-isomorphic countable orders. □

8.4 Cantor's Theorem on Countable Dense Orders

The rational numbers with the usual ordering form an order which is countable, order dense, and without endpoints. A remarkable result of Cantor asserts that, up to isomorphism, it is the only one with those properties.

Theorem (Cantor). If X and Y are orders both of which are countable, order dense, and without endpoints, then X and Y are similar orders. Hence any order which is countable, order dense, and without endpoints is isomorphic to the rational numbers $\mathbf{Q}$ with the usual ordering, and so must have order type η.

This gives a characterization of the order type η in terms of its structural properties.

Definition 539 (Finite Partial Isomorphisms). Let A and B be orders. By a *finite partial isomorphism* between A and B we mean an order isomorphism (order preserving bijection) between a finite suborder of A and a finite suborder of B. In other words, a finite partial isomorphism between A and B is a bijection $f: E \to F$ where E is a finite subset of A, F is a finite subset of B, and for all $x, y \in E$, $x <_A y \Leftrightarrow f(x) <_B f(y)$.

Problem 540 (Extension Lemma for Finite Partial Isomorphisms). *Let A and B be orders, E be a finite subset of A, F be a finite subset of B, and $f: E \to F$ be a finite partial isomorphism between A and B. Prove that*

1. *If B is order-dense without endpoints and $a \in A$ then there is a finite partial isomorphism g extending f with $a \in \operatorname{dom}(g)$, i.e., there are finite subsets $P \subseteq A$, $Q \subseteq B$, and an order-preserving bijection $g: P \to Q$ such that $g|_E = f$ and $a \in P$.*
2. *Similarly, if A is order-dense without endpoints and $b \in B$ then there is a finite partial isomorphism h extending f with $b \in \operatorname{ran}(h)$, i.e., there are finite subsets $P \subseteq A$, $Q \subseteq B$, and an order-preserving bijection $h: P \to Q$ such that $h|_E = f$ and $b \in Q$.*

Theorem 541 (Cantor's Theorem on Countable Dense Orders). *Any two countable dense orders without endpoints are isomorphic to each other.*

Proof. Let A and B be countable dense orders without endpoints, and enumerate the elements of A and B as

$$A = \{a_1, a_2, \dots, a_n, \dots\}, \qquad B = \{b_1, b_2, \dots, b_n, \dots\}.$$

We will inductively define a sequence $f_1 \subseteq f_2 \subseteq \cdots \subseteq f_n \subseteq \cdots$ of finite partial isomorphisms between A and B with each f_n extending the preceding ones, $\operatorname{dom}(f_n) \supseteq \{a_1, \dots, a_n\}$, and $\operatorname{ran}(f_n) \supseteq \{b_1, \dots, b_n\}$.

Let f_1 be the function with $\operatorname{dom}(f_1) = \{a_1\}$ and $f(a_1) = b_1$, so that $\operatorname{ran}(f_1) = \{b_1\}$. Then f_1 is (trivially) a finite partial isomorphism between A and B.

Suppose next that f_n is defined with $\operatorname{dom}(f_n) \supseteq \{a_1, \dots, a_n\}$ and $\operatorname{ran}(f_n) \supseteq \{b_1, \dots, b_n\}$. Since B is dense we can apply the first part of the extension lemma to extend f_n to a finite partial isomorphism g between A and B such that $a_{n+1} \in \mathrm{dom}(g)$. Next, since A is dense, we apply the second part of the extension lemma to extend g to a finite partial isomorphism g between A and B such that $b_{n+1} \in \mathrm{ran}(h)$. We then put $f_{n+1} = h$. Then f_{n+1} extends f_n, with $\mathrm{dom}(f_{n+1}) \supseteq \{a_1, \ldots, a_n, a_{n+1}\}$, and $\mathrm{ran}(f_{n+1}) \supseteq \{b_1, \ldots, b_n, b_{n+1}\}$, completing the inductive construction.

Now let $f = \cup_n f_n$. Then f is a well-defined function: If x is in $\mathrm{dom}(f_m) \cap \mathrm{dom}(f_n)$ then $f_n(x) = f_m(x)$ since either f_n extends f_m (for $m \leq n$) or f_m extends f_n (for $n \leq m$). The function f is an extension of every f. Moreover, $\mathrm{dom}(f) = A$ and $\mathrm{ran}(f) = B$ since $a_n \in \mathrm{dom}(f_n) \subseteq \mathrm{dom}(f)$ and $b_n \in \mathrm{ran}(f_n) \subseteq \mathrm{ran}(f)$ for all n. Finally, if $x <_A y$ in A, then $x = a_m$ and $y = a_n$ for some m, n, and with $k = \max(m, n)$ we get $f_k(x) <_B f_k(y)$ (since f_k is a finite partial isomorphism) and so $f(x) <_B f(y)$. Thus f is strictly increasing map from A onto B and hence an order isomorphism between them. □

Corollary 542. *Any countable dense order without endpoints has order type η, and is isomorphic to the set of rational numbers with their usual ordering.*

Corollary 543. *Any nontrivial countable dense order has one of the order types η, $1 + \eta$, $\eta + 1$, or $1 + \eta + 1$.*

Corollary 544. *$\eta + \eta = \eta$. Also, $\eta + 1 + \eta = \eta$, and $\eta\eta = \eta$.*

The last corollary follows from the observation that if α is the order type of a dense order without endpoints, then an order having any other types $\alpha + \alpha$, $\alpha + 1 + \alpha$, or $\alpha\alpha$ will also be a dense order without endpoints.

Both Cantor's theorem and the "back-and-forth" method used to prove it are very powerful and have far-reaching implications. We will illustrate this by using Cantor's theorem to prove (in a later chapter) two classical theorems: Brouwer's Theorem and Sierpinski's Theorem.

A related result which is easier than Cantor's theorem is that any countable order can be order embedded in any dense order.

Theorem 545. *Let A be a countable order and let B be order-dense without endpoints. Then A can be order embedded in B, that is, A is isomorphic to some suborder of B.*

Proof. It suffices to show that there is a strictly increasing $f : A \to B$.

Enumerate $A = \{a_1, a_2, \ldots, a_n, \ldots\}$ and repeatedly apply the extension lemma (first part) to inductively get a sequence $f_1 \subseteq f_2 \subseteq \cdots \subseteq f_n \subseteq \ldots$ of finite partial isomorphisms between A and B with $\mathrm{dom}(f_n) \supseteq \{a_1, \ldots, a_n\}$. Then the function $f = \cup_n f_n$ is a strictly increasing map from A into B. □

Corollary 546. *If X is a nontrivial and order dense, then every countable order is order-isomorphic to some suborder of X. Hence the collection of all order types of suborders of a nontrivial dense order includes all countable order types.*

Corollary 547. *Any order X of type η is a universal countable order, that is, X is a countable order in which every countable order can be embedded.*

Corollary 548. *There are at most $2^{\aleph_0} = \mathfrak{c}$ countable order types.*

Problem 549. *Let E be the set of endpoints of the open intervals removed in the construction of the Cantor set, with the usual inherited order. Express the order type of E as the sum and/or product of some known order types.*

Problem 550 (The Kleene–Brouwer Order). *Let $\mathbf{N}^*$ be the set of all finite sequences (strings) of natural numbers, and say that $u = \langle u_1, u_2, \ldots, u_m \rangle$ precedes $v = \langle v_1, v_2, \ldots, v_n \rangle$ in the Kleene–Brouwer order on $\mathbf{N}^*$ if either $m > n$ and $u_k = v_k$ for all $k \leq n$, or there is $k \leq \min(m, n)$ with $u_k < v_k$ but $u_j = v_j$ for all $j < k$. In other words, u precedes v in the Kleene–Brouwer order on $\mathbf{N}^*$ if either u properly extends v, or none of them is an extension of the other and u lexicographically precedes v. What is the order type of this order?*

Theorem 551. *Every countable order X can be continuously embedded in $\mathbf{Q}$ and in $\mathbf{R}$.*

Problem 552. *Prove Theorem 551.*

[Hint: Between every pair of consecutive points of X, "adjoin" an additional set of points of order type η. The extended order is a countable dense order in which X is continuously embedded.]

However, *there are dense orders without endpoints in which no countable order having limit points can be continuously embedded* (Problem 757).

8.5 $\aleph_0 < \mathfrak{c}$: Another Proof of Uncountability of R

A notable consequence of Cantor's theorem (Theorem 541) is that *every* nontrivial countable dense order has Dedekind gaps. Since a linear continuum is a dense order without gaps, it follows immediately that every continuum, and so $\mathbf{R}$ in particular, must be uncountable.

Proposition 553. *Any nontrivial countable dense order has Dedekind gaps. Consequently, every linear continuum is uncountable.*

Corollary 554 (Uncountability of R). $\mathbf{R}$ *is not countable, i.e.,* $\mathfrak{c} > \aleph_0$.

Proposition 553 is easily derived from Cantor's theorem in various ways. We give three short proofs. Let X be a nontrivial countable dense order. We may assume that X has no endpoints (by removing them if necessary).

Proof (First Proof). By Cantor's theorem X is isomorphic to the ratios (positive rationals), and we had seen that the ratios contain the gap given by $\{\rho \mid \rho^2 < 2\}$ and $\{\rho \mid \rho^2 > 2\}$, hence X must contain a gap too. □

Proof (Second Proof). By Cantor's theorem X must have order type η and another corollary of Cantor's theorem was that $\eta = \eta + \eta$. But any order whose type is expressible as $\alpha + \alpha$, where α is the type of a nonempty order without endpoints, must have a gap. □

Proof (Third Proof). If we remove a point p from X, the resulting order $X \setminus \{p\}$ is still countable, order dense, and without endpoints, and so by Cantor's theorem $X \setminus \{p\}$ is isomorphic with X. But if a point is removed from a dense order without endpoints, the resulting order will have a gap, and so X, being isomorphic to $X \setminus \{p\}$, will have a gap too. □

We thus get a very short proof of uncountability of **R** *by exploiting the power of Cantor's isomorphism theorem on countable dense orders (Theorem 541).*

Remark. In his very first proof of uncountability of **R**, Cantor *directly* showed that every countable dense order has gaps, *without* using the isomorphism theorem on countable dense orders. (Nor did he use the diagonal method which he invented later). Cantor's first proof is given in Appendix A.

The identity $\eta + \eta = \eta$ can also be proved without using Cantor's theorem.

Problem 555. *Let* $\mathbf{Q}^+ := \{r \in \mathbf{Q} \mid r > 0\}$, $\mathbf{Q}^- := \{r \in \mathbf{Q} \mid r < 0\}$, *and* $\mathbf{Q}^* := \mathbf{Q} \setminus \{0\} = \{r \in \mathbf{Q} \mid r \neq 0\}$. *Prove the identity* $\eta + \eta = \eta$ *by finding three explicit order preserving bijections: The first between* **Q** *and* $\mathbf{Q}^+$, *the second between* **Q** *and* $\mathbf{Q}^-$, *and the third between* **Q** *and* $\mathbf{Q}^*$.

8.6 The Order Type of R

We saw that the order property "countable, dense, and without endpoints" characterizes the order type η: An order has type η, or is isomorphic to **Q** with the usual order, if and only if it is countable, dense, and has no endpoints (Cantor's theorem).

There is a similar characterization, also by Cantor, of the order type λ of the real numbers. An important property of the order on the reals is that it is a continuum without endpoints. However, this is not sufficient to characterize the order type λ of the reals, and we will now give some examples of continuums without endpoints which are not isomorphic to **R**. First, we need a definition.

Definition 556 (CCC orders). An order X is said to satisfy the *countable chain condition,* or is called a *CCC* order, if any family of pairwise disjoint nonempty open intervals in X is a countable family.

Clearly, a non-CCC order cannot be embedded in a CCC order.

The real line satisfies the countable chain condition, since given any family of pairwise disjoint nonempty open intervals in **R**, we can pick a rational number in each interval in the family. Distinct rationals will be picked for distinct intervals since the intervals are pairwise disjoint, which gives a one-to-one correspondence between the family and some set of the rational numbers, and so the family must be countable.

The following is an example of a non-CCC continuum.

Problem 557. *Consider the subset* $S = (0, 1) \times [0, 1]$ *of the plane ordered lexicographically. (*S *is the subset obtained by removing the left and right edges of the closed unit square.)*

1. *Verify that the order type of* S *(ordered lexicographically) is* $(1 + \lambda + 1)\lambda$.
2. *Prove that* S *is a continuum without endpoints.*
3. *Show that* S *is not a CCC order.*
4. *Conclude that* S *is not isomorphic to the real numbers with the usual ordering, and so* $(1 + \lambda + 1)\lambda \neq \lambda$.

Thus the non-CCC continuum S cannot be embedded in the CCC continuum **R**, and any order of type $(1 + \lambda + 1)\lambda$ is a continuum without endpoints which is not isomorphic to the real continuum. More examples can be obtained by iterating the above procedure. For example, an order of type $(1+\lambda+1)^2\lambda$ is a continuum without endpoints which cannot be embedded even in S; see Problem 558.

Problem 558. *Let* λ_k *denote the order type* $(1 + \lambda + 1)^k\lambda$ *(with* $\lambda_0 = \lambda$*).*

1. *Show that each order of type* λ_k *is a continuum without endpoints (*$k = 0, 1, 2, \ldots$*).*
2. *Show that if* $m < n$ *then an order of type* λ_n *cannot be embedded in an order of type* λ_m.
3. *Conclude that orders having the distinct types* $\lambda_0, \lambda_1, \ldots$ *must be non-isomorphic continuums without endpoints.*

A property which is stronger than CCC is *separability.*

Definition 559 (Separable Orders). An order X is called *separable* if it contains a countable subset dense in it (i.e., if there is a countable $C \subseteq X$ with $X = C \cup D(C)$).

Recall that a subset A of an order X is dense in X if and only if every nonempty open interval in X contains a point of A. Thus every separable order is CCC (the same proof given above showing **R** is CCC works).

If X is order dense, then a subset A is dense in X if between any two points of X there is a point of A. Thus an order dense order X is separable if and only if X contains a countable subset C such that between any two points of X there is a point of C. For example, the rationals **Q** form a countable subset of **R** with this property, and so **R** is separable. On the other hand, $S = (0, 1) \times [0, 1]$ with lexicographic order is a non-separable continuum.

The property of "being a separable continuum without endpoints" characterizes the ordering of the reals and the order type λ:

Theorem 560 (Cantor's Order Characterization of R). *Every separable continuum without endpoints is isomorphic to the reals with their usual ordering.*

Proof. Let X be a separable continuum without endpoints and let C be a countable dense subset of X. Then the suborder C is a countable dense order without endpoints, and so by Cantor's theorem there is an order isomorphism $f : C \to \mathbf{Q}$.

Regarding $\mathbf{Q}$ as a suborder in $\mathbf{R}$, we see that f extends (uniquely) to an order isomorphism g between X and $\mathbf{R}$, where for $x \in X \smallsetminus C$, we set:

$$g(x) = \sup_{\mathbf{R}}\{f(u) \mid u \in C \text{ and } u < x \text{ in } X\}.$$

Using the density of C in X and of $\mathbf{Q}$ in $\mathbf{R}$, it is readily verified that g is a bijection from X onto $\mathbf{R}$ which preserves order. □

Corollary 561. *An ordering has order type λ (i.e., it is order isomorphic to the set of real numbers with their usual ordering) if and only if it is a separable continuum without endpoints.*

Corollary 562. *Any separable linear continuum has one of the order types λ, $1+\lambda$, $\lambda + 1$, or $1 + \lambda + 1$.*

Problem 563. *For each of the following order types determine if it is separable, dense, and/or Dedekind complete, and identify the ones which are linear continuums. If any of them is identical to a familiar type, indicate so.*

1. λ^2
2. $(1 + \lambda)^2$
3. $(1 + \lambda)\,\lambda$
4. $(1 + \lambda + 1)^2$
5. $\lambda\,\omega$
6. $(1 + \lambda)\,\omega$
7. $(1 + \lambda)\,\omega^3$
8. $(\lambda + 1)\,\omega^2$
9. η^2
10. $(1 + \eta)\,\eta$
11. $\eta\,2$
12. 2η

[Hint: It may be useful to represent each type as a lexicographic product of familiar orders. For example, $(1 + \lambda)\,\lambda$ has the order type of $(0, 1) \times [0, 1)$ ordered lexicographically, and $\lambda\,\omega$ has order type of $\mathbf{N} \times (0, 1)$ ordered lexicographically (note reversal of order of the factors).]

Problem 564. *Let us call a point in an order to be a* removable point *if the suborder obtained by removing this single point is order-isomorphic to the original order. In other words, the point p in the order X is removable if $X \smallsetminus \{p\}$ is order isomorphic to X.*

For each of the following orders, determine which points are removable. All orderings are assumed to be their usual orders, or inherited suborder from the usual order.

1. $\mathbf{N}$.
2. $\mathbf{Z}$.
3. $\mathbf{Q}$.
4. $\mathbf{R}$.

5. *The unit interval* $[0, 1]$.
6. *The set* $\{0\} \cup \{\frac{1}{n} \mid n \in \mathbf{N}\}$.
7. *The set* $\{0\} \cup \left(\bigcup_{n\in\mathbf{N}}\{-\frac{1}{n}, \frac{1}{n}\}\right)$.
8. *The set* $\bigcup_{n\in\mathbf{N}}\{-\frac{1}{n}, \frac{1}{n}\}$.
9. *An order of type* $\omega^2 + \omega$.
10. *An order of type* $\lambda\eta$.
11. *An order of type* $\omega^2 + n$ $(n \in \mathbf{N})$.
12. *An order of type* $\omega\alpha$, α *arbitrary.*

The Suslin Problem

Once it is established that a separable continuum without endpoints must be isomorphic to the real line, the question arises if the result remains true if separability is replaced by the weaker condition of being CCC. This was first asked by Suslin.

The Suslin Problem. Is a CCC continuum without endpoints necessarily order isomorphic to $\mathbf{R}$*?*

The affirmative answer to Suslin's question is known as the *Suslin Hypothesis* (SH). Thus SH is the statement that every CCC continuum without endpoints has order type λ. Like the Continuum Hypothesis, SH is independent of comprehensive set theoretic axiom systems for developing mathematics such as ZFC (Zermelo–Fraenkel Axioms with Choice). This means it has been proved that SH can neither be proved nor be disproved using mathematical principles and methods of proof that are currently accepted as standard (assuming these methods themselves are consistent).

The Suslin Problem has played an important role in the development of axioms and principles of combinatorial set theory (such as constructibility and Jensen's diamond and box axioms) as well as independence proofs (Martins's Axiom and forcing).

8.7 Dedekind Completion

Definition 565 (Dedekind Completion). We say that an order Y is a *Dedekind Completion* of an order X if X is a suborder of Y, Y is Dedekind complete, and every element of $Y \setminus X$ is both an upper limit point of X and a lower limit point of X.

Problem 566. *Prove that if* $Y \supseteq X$ *is a Dedekind Completion of* X*, then* X *is dense in* Y *in the sense that any nonempty open interval in* Y *must contain a point of* X*.*

Theorem 567 (Existence of Dedekind Completion). *Every order* X *has a Dedekind completion.*

Proof. The proof is a straightforward generalization of Dedekind's construction of the real numbers using cuts of rational numbers.

Let X be any order. Without loss of generality, we first replace X by an isomorphic order $\mathsf{E}(X)$:

$$\mathsf{E}(X) := \{\text{Pred}(a) \mid a \in X\},$$

where $\text{Pred}(a) := \{x \in X \mid x < a\}$. Let $\mathsf{E}(X)$ be ordered by the proper set inclusion relation. Then the mapping $a \mapsto \text{Pred}(a)$ is an order isomorphism from X onto $\mathsf{E}(X)$.

Now let

$$\mathsf{H}(X) := \{L \mid (L, X \setminus L) \text{ is a Dedekind gap in } X\},$$

and put $\mathsf{M}(X) := \mathsf{E}(X) \cup \mathsf{H}(X)$, ordered again by proper set inclusion.

Then $\mathsf{M}(X)$ is a Dedekind Completion of $\mathsf{E}(X)$: Here $\mathsf{E}(X)$ plays the role of the rationals and $\mathsf{H}(X)$ plays the role of the irrationals given by Dedekind gaps. □

Problem 568 (Uniqueness of Dedekind Completion). *If the order A is isomorphic to the order B via the order-preserving bijection $f: A \to B$, and if A' and B' are Dedekind completions of A and B, respectively, then there is a unique extension $f' \supseteq f$ which is an order isomorphism between A' and B'.*

[Hint: For $x \in A' \setminus A$, define $f'(x) := sup_{B'}\{f(u) \mid u \in A,\ u <_{A'} x\}$.]

The last result implies that for any order type, there exists a unique order type for its Dedekind Completion.

Definition 569 (Dedekind Completion of Order Types). If τ is an order type, then *the Dedekind Completion of* τ, denoted by $\hat{\tau}$, is the unique order type determined by some (or every) Dedekind completion of an order of type τ.

Problem 570. *Find the Dedekind completions of each of the following types, expressing your answer in terms of known types:*

1. ω
2. ζ
3. η
4. $\omega + {}^*\omega$
5. ${}^*\omega + \omega$
6. ω^2.
7. $\zeta + \zeta$
8. ζ^2
9. $\zeta + \eta$
10. $\lambda + \eta$
11. λ^2
12. $(1 + \lambda)^2$
13. $(1 + \lambda + 1)^2$
14. 2η

Problem 571. *The Dedekind completion of any dense order is also a dense order, and hence, being complete as well, is a continuum.*

Problem 572 (Continuous Embedding in Dedekind Completion). *Every order X is continuously embedded in its Dedekind completion Y, i.e., the inclusion (identity) map from X into Y is a continuous embedding.*

The following shows that the Dedekind completion of X is the "smallest complete order containing X":

Problem 573 (Minimality of Dedekind Completion). *If $X \subseteq Y$ where Y is a complete order, then Y contains suborder (containing X) which is isomorphic to the Dedekind completion of X.*

[Hint: Given a Dedekind completion X' of X, associate each $x \in X' \setminus X$ with the element $\sup_Y\{u \in X \mid u <_{X'} x\}$ of Y.]

Corollary 574. *Every linear continuum contains a suborder isomorphic to the real line (with the usual order). Hence every linear continuum has cardinality $\geq 2^{\aleph_0} = \mathfrak{c}$.*

Proof. Let X be a linear continuum. Since X is order-dense, it contains a suborder of type η, and being complete X contains a suborder of type $\hat{\eta} = \lambda$. □

Problem 575. *Let A be the set consisting of all infinite binary sequences which are eventually constant, except the two sequences "all zeroes" and "all ones," ordered lexicographically. Let B be the set of endpoints of the open intervals removed in the construction of the Cantor set, with the order inherited from the usual order on* **R**.

1. *Show that these two orders are order isomorphic, with each having order type 2η.*
2. *Show that each point of A has an immediate neighbor in A, that is, show that given any $x \in A$ there is a pair of consecutive elements in A with x being one of them.*
3. *Show that A is* dense-in-itself, *that is, every element of A is a limit point of A.*
4. *Let C be the set of points of the Cantor set except 0 or 1. Show that C is a Dedekind completion of B.*

Problem 576. *Prove that the order type of the Cantor set is $1 + \widehat{2\eta} + 1$.*

(In fact, from a result of Brouwer to be proved later (Theorem 1099), it follows that any bounded perfect subset of **R** which does not contain any interval has order type $1 + \widehat{2\eta} + 1$.)

8.8 Properties of Complete Orders and Perfect Sets

Bolzano–Weierstrass and Nested Intervals Properties

We say that an order X has the *Bolzano–Weierstrass property* (or *BW property*) if every bounded infinite subset of X has a limit point in X.

Theorem 577 (Bolzano–Weierstrass). *For an arbitrary order, completeness implies the Bolzano–Weierstrass property.*

Proof. Let E be a bounded infinite subset of the complete order X, say $a \leq E \leq b$ for some $a, b \in X$. Put

$$E_x := \{y \in E \mid y < x\}, \quad \text{and} \quad L := \{x \in X \mid E_x \text{ is finite}\}.$$

Then E_a is finite but E_b is infinite, so $a \in L$ and $L \leq b$, hence $c := \sup_X L$ exists. If now $c \in L$, then c is a lower limit point of E: Since E_c is finite, $c < b$, and for any $z > c$ E_z is infinite and so $E_z \smallsetminus E_c$ is infinite, so there is some $p \in E$ with $c < p < z$.

If $c \notin L$, then c is an upper limit point of E: As $L < c$ and for any $x < c$ there is $y \in L$ with $x < y < c$, so $E_c \smallsetminus E_y$ is infinite hence there is $p \in E$ with $y < p < c$, and so $x < p < c$, □

There are two nested intervals properties ("NIP"s) that we will consider.

1. We say that an order X satisfies the *sequential nested intervals property* if given any nested sequence of nonempty bounded closed intervals in X

$$I_1 \supseteq I_2 \supseteq \cdots \supseteq I_n \supseteq I_{n+1} \supseteq \cdots,$$

   we have $\cap_n I_n \neq \emptyset$.
2. An order is said to satisfy the *strong nested intervals property* if whenever a family F of nonempty bounded closed intervals forms a chain (i.e., for any two intervals $I_1, I_2 \in \mathsf{F}$ either $I_1 \subseteq I_2$ or $I_2 \subseteq I_1$), we have $\cap \mathsf{F} \neq \emptyset$.

Trivially the strong nested intervals property implies the sequential nested intervals property. The following two theorems show how the two nested intervals properties are related to completeness and the Bolzano–Weierstrass property.

Theorem 578 ("BW Implies NIP"). *In an arbitrary order, the Bolzano–Weierstrass property implies the sequential nested intervals property.*

Proof. Let $I_n = [a_n, b_n]$ $(n = 1, 2, \ldots)$ be a nested sequence of nonempty closed intervals in a complete order X, so that

$$a_1 \leq a_2 \leq \ldots a_n \leq a_{n+1} \leq \ldots \quad \cdots \leq b_{n+1} \leq b_n \leq \ldots b_2 \leq b_1.$$

Now either the sequence $\langle a_n \mid n \in \mathbf{N}\rangle$ is eventually constant so that there exist $a \in X$ and $k \in \mathbf{N}$ with $a_n = a$ for all $n \geq k$, in which case $a \in \cap_n I_n$. Or else, the set $L := \{a_n \mid n \in \mathbf{N}\}$ of left endpoints of the intervals I_n is a bounded infinite set and so L has a limit point $c \in X$. Then $c \in \cap_n I_n$, since if $c < a_n$ or $c > b_n$ for some n then c cannot be a limit point of L. □

Theorem 579 (The Strong Nested Intervals Property). *For an arbitrary order, completeness implies the strong nested intervals property.*

Proof. Let A be the set of left endpoints of the intervals in F. Since F is a chain, any right endpoint of any interval in F is an upper bound of A, and so $a := \sup A$ exists.

Then $a \geq p$ for any left endpoint p of any interval in F. Also, $a \leq q$ for any right endpoint q of any interval in F. Hence $a \in \cap$F. □

The above results can be summarized as:

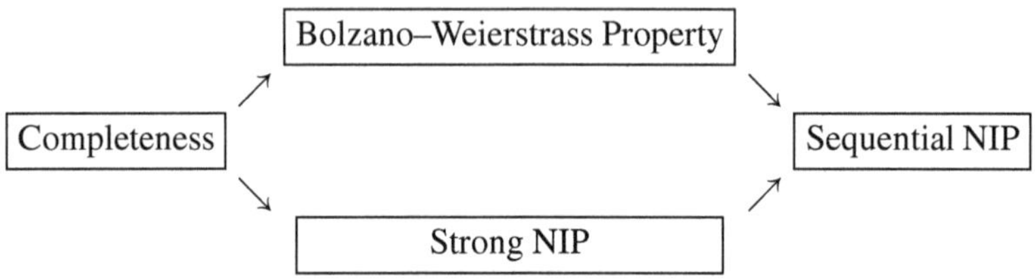

None of the implications above can be reversed and no further implications between the above properties can be obtained. The Bolzano–Weierstrass and Nested Interval properties are strictly weaker than completeness. In fact, there is an order which satisfies both the Bolzano–Weierstrass and the Strong Nested Interval properties but is not complete, and there is an order which satisfies the Sequential NIP but satisfies neither the Bolzano–Weierstrass nor the Strong Nested Intervals properties. Moreover, the Bolzano–Weierstrass and the strong Nested Interval properties are independent of each other. For these counterexamples, see Problem 711 in Chap. 10.

Problem 580. *Show that in a dense order which is separable, all four properties displayed above are equivalent, and therefore completeness is characterized by any of the other three properties.*

Definition 581 (Monotone and Convergent Sequences). Let X be an order and let $\langle x_n \rangle = \langle x_n \mid n \in \mathbf{N} \rangle$ be a sequence of points in X. We say that the sequence $\langle x_n \rangle$ is *monotone increasing* if $x_n \leq x_{n+1}$ for all n. Similarly one defines the notion of *monotone decreasing* sequences. A *monotone sequence* is one which is either monotone increasing or monotone decreasing. We also say that the sequence $\langle x_n \rangle$ *converges to a point* $p \in X$ if for any $a < p$ there exists k such that $x_n > a$ for all $n > k$ and for any $b > p$ there exists k such that $x_n < b$ for all $n > k$.

Problem 582 (The Monotone Convergence Property). *Show that in any order, the Bolzano–Weierstrass property is equivalent to the condition that any bounded monotone sequence converges to some point.*

Dense-in-Itself Orders

Recall that an order X is a dense order if and only if every point of X except the first (if present) is an upper limit point if and only if every point of X except the last (if present) is a lower limit point. In particular, if every element of X is an upper limit point or if every element of X is a lower limit point then X must be order-dense. But if we are given that every element of X is either an upper- or a lower limit point, then, as we will see soon, X may have consecutive points, and may in fact be quite far from being order-dense. We thus get a wider class of orders using the condition "every point of the order is a limit point, upper or lower":

Definition 583 (Density-in-Itself). A subset A of an order X is called *dense-in-itself* if every element of A is a limit point of A (in X), that is if $A \subseteq D(A)$. An order is called *dense-in-itself* if it is dense-in-itself as a subset of itself.

Every dense order is dense-in-itself, but there are dense-in-itself orders which are not order-dense.

Consider the suborder $X = (0, 1] \cup [2, 3)$ of $\mathbf{R}$ as an order by itself. Then X is dense-in-itself but not order-dense as there is no element in X which is between 1 and 2. Another example of an order which is dense-in-itself but not order-dense is the Cantor set $\mathbf{K}$ as an order by itself: There is no element in $\mathbf{K}$ between the two elements $1/3$ and $2/3$.

We can get dense-in-itself orders with lots of consecutive points. The following examples are orders in which every point is a limit point yet every point is one of the two points of a pair of consecutive points.

Problem 584. *Let $X = \{0, 1, 2, 3, 4, 5, 6, 7, 8, 9\}^{\mathbf{N}}$ be the set of "infinite decimal strings" ordered lexicographically, and let Y be the suborder consisting of those members of X which are either eventually 0 or eventually 9 but not all 0 or all 9. Let E be the suborder of $\mathbf{R}$ consisting of the endpoints of the open intervals removed in the construction of the Cantor set. Show that:*

1. *Y and E are isomorphic orders and have order type 2η.*
2. *Any order of type 2η (like Y or E) is dense-in-itself (every point is a limit point).*
3. *In any order of type 2η, every point has either an immediate successor or an immediate predecessor, and consequently no point is a two-sided limit point.*
4. *Any countable dense-in-itself order without endpoints and without any two-sided limit point must have order type 2η.*

Problem 585. *Consider the plane set $T := (0, 1) \times \{0, 1\} = \{\langle x, y\rangle \mid 0 < x < 1, y = 0 \text{ or } 1\}$ be ordered lexicographically, which has order type 2λ. Show that such an order is complete, separable, dense-in-itself, yet every point is one of the two points of a pair of consecutive points.*

The order of the last example cannot be embedded in $\mathbf{R}$, giving us an example of a separable complete order which cannot be embedded in $\mathbf{R}$.

Problem 586. *Show that the set $X = \{0, 1, 2, 3, 4, 5, 6, 7, 8, 9\}^{\mathbf{N}}$ of "infinite decimal strings" ordered lexicographically is order isomorphic to the Cantor set (as a suborder of $\mathbf{R}$). [Hint: Show that X is a Dedekind completion of the suborder consisting of stings which are eventually 0 or eventually 9, which has order type $1 + 2\eta + 1$.]*

Problem 587. *Let A be a subset of an order X.*

1. *Show that if A is a dense-in-itself subset of X, then the suborder A as an order by itself is a dense-in-itself order.*
2. *Give an example to show that the converse of the above fails.*

Problem 588. *Let X be a nontrivial order of order type γ. Show that X is not order-dense if and only if $\gamma = \alpha + 2 + \beta$ for some order types α, β, and X is not dense-in-itself if and only if $\gamma = \alpha + 3 + \beta$ for some order types α, β.*

Problem 589. *Let C be the set of all initial segments of the set $\mathbf{Q}$ of rational numbers. Thus C includes sets of the form $\{x \in \mathbf{Q} \mid x \leq r\}$ (r rational), $\{x \in \mathbf{Q} \mid x < r\}$ (r rational or irrational), as well as $\emptyset$ and $\mathbf{Q}$. Show that C ordered by proper set-inclusion is order-isomorphic to the Cantor set with the usual order.*

Theorem 590. *Let X be an order which is complete and dense-in-itself. Then $\mathbf{R}$ can be order embedded in X and so the cardinality of X is at least $2^{\aleph_0} = \mathfrak{c}$.*

Proof. Let X be an order which is complete and dense-in-itself, and let A be subset of X consisting of all lower limit points of X. Then the suborder A is order-dense, that is between any two points of A there is another point of A: If $x, y \in A$ with $x < y$ then since x is a lower limit point of X there exist $u, v \in X$ with $x < u < v < y$. Now either $u \in A$ or u is not a lower limit point, in which case u will have an immediate successor in X, say u', with $u < u' \leq v < y$. Being an immediate successor, the element u' is not an upper limit point and so must be a lower limit point in X. Thus either $u \in A$ or $u' \in A$, and in either case we have a point of A between x and y.

It follows that A and so X must contain a subset of order type η. But X is complete and so by minimality of Dedekind completion, X will contain a subset of order type $\hat{\eta} = \lambda$. □

Problem 591. *Give an example of a bounded subset A of $\mathbf{R}$ such that no point of A is a limit point of A but the suborder A has order type η.*

Show that for such an A, $D(A)$ must have order type $1 + \widehat{2\eta} + 1$ and $A \cup D(A)$ (called the closure of A) must have order type $1 + \widehat{3\eta} + 1$.

Closed and Perfect Sets

Definition 592. Let X be an order and $A \subseteq X$. We say that *A is closed in X* if A contains all its limit points, that is, if $D(A) \subseteq A$. A subset which is both closed and dense-in-itself is called *perfect.*

Theorem 593. *Let X be a complete order and let $A \subseteq X$. Then A is closed in X if and only if the following two conditions both hold:*

1. *The suborder A is continuously embedded in X.*
2. *A is complete (as an order by itself).*

Problem 594. *Prove Theorem 593.*

Problem 595. *Give an example of closed suborder of a general order which is neither continuously embedded (in the parent) nor complete (by itself).*

Problem 596. *Let X be a complete order. Show that $F \subseteq X$ is closed if and only if for any bounded nonempty subset $A \subseteq F$ we have both* $\sup A \in F$ *and* $\inf A \in F$.

[Hint: Use the fact that if $\sup E$ exists but $\sup E \notin E$ then $\sup E$ must be an upper limit point of E.]

Problem 597. *Let X and Y be orders and let $f: X \to Y$ be continuous. Show that for any $b \in Y$, the set $\{x \in X \mid f(x) \leq b\}$ is a closed subset of X.*

Cardinalities of Perfect Sets in a Complete Order

Since a perfect subset of a complete order, taken as an order by itself, must be dense-in-itself and complete, it follows from Theorem 590 that any nonempty perfect subset of a complete order contains a suborder isomorphic to $\mathbf{R}$, and so must have cardinality $\geq \mathfrak{c}$.

Theorem 598. *Every nonempty perfect subset of a complete order has cardinality $\geq 2^{\aleph_0} = \mathfrak{c}$.*

The notions of closed, dense-in-itself, and perfect sets are due to Cantor.

8.9 Connectedness and the Intermediate Value Theorem

Recall Dedekind's basic definition of the continuum: *An order is a continuum if and only if for any Dedekind partition of the order into nonempty upper and lower segments, at least one of the segments will contain a limit point of the other.* We will now see that for *any* partition of a continuum into two nonempty disjoint sets, at least one of them will contain a limit point of the other. Thus this last stronger condition, known as *topological connectedness,* also characterizes linear continuums. We can state the result equivalently in terms of partitions into a pair of closed sets as follows.

Theorem 599. *Let X be a continuum. If A and B are disjoint closed subsets of X with $A \cup B = X$, then either $A = \emptyset$ and $B = X$ or $A = X$ and $B = \emptyset$.*

Proof. Suppose that A and B are disjoint closed subsets of X with $A \cup B = X$. To get a contradiction, assume that both A and B are nonempty and fix $a \in A$ and $b \in B$. Without loss of generality we assume that $a < b$. Put

$$E := \{x \in B \mid a < x \leq b\}, \quad c := \inf E, \quad \text{and} \quad D := \{x \in X \mid a \leq x < c\}.$$

Note that $c = \inf E$ exists since E is a nonempty bounded set in the complete order X, and thus E, c, and D are all well defined. Also note that $E \subseteq B$ and $D \subseteq A$. Since $E \subseteq B$ and B is closed, we have $c = \inf(E) \in B$. Therefore $a < c$, and so D must be an infinite set with no maximum (since X is a dense order). It follows that $c = \sup D$ must be a limit point of D, and hence of A, so $c \in A$ (as A is closed). But this is a contradiction since $A \cap B = \varnothing$. □

Recall that a subset E of an order is called a segment if whenever $u < z < v$ and $u, v \in E$ then $z \in E$.

Corollary 600 (The Intermediate Value Theorem). *If X is a continuum, Y is any order, and $f: X \to Y$ is continuous, then* $\operatorname{ran}(f) = f[X]$ *is a segment in Y.*

Proof. To get a contradiction assume the conditions of the theorem and suppose that $f[X]$ is not a segment, so that there exist $u < z < v$ in Y such that $u, v \in f[X]$ but $z \notin f[X]$. Put $A := \{x \in X \mid f(x) \leq z\}$ and $B := \{x \in X \mid f(x) \geq z\}$. Then A and B are nonempty disjoint closed sets in X with $A \cup B = X$, which contradicts Theorem 599. □

We summarize these results as characterizations of the continuum.

Problem 601. *Let X be an order. Then the following are equivalent.*

1. X is a continuum*: X is a dense order without Dedekind gaps.*
2. X is topologically connected*: X cannot be partitioned into two disjoint nonempty closed sets.*
3. X satisfies the Intermediate Value Theorem*: For any order Y and any continuous $f: X \to Y$, the image $f[X] = \operatorname{ran}(f)$ is a segment in Y.*

Chapter 9
Well-Orders and Ordinals

Abstract This chapter develops the classical theory of well-orders and ordinals in a naive setting. Ordinals are defined as order types of well-orders, not as von Neumann ordinals. We cover the basic ordinal operations of sum and product, transfinite induction and recursion, uniqueness of isomorphisms and ranks, unique representation of well-orders by initial sets of ordinals, the comparability theorem for well-orders, the division algorithm, remainder ordinals, ordinal exponentiation, and the Cantor Normal Form.

9.1 Well-Orders, Ordinals, Sum, and Product

Cantor invented two remarkable generalizations of the natural numbers extending into the transfinite. One is the notion of cardinal numbers: Two sets have the same cardinal number if their elements can be put in a one-to-one correspondence—without any regard for the ordering between the elements themselves. The other is the notion of an *ordinal number*, which, roughly speaking, represents the "serial position, relative to the beginning, of an object in a que." The number 10 can be used either as a cardinal number (as in "there are 10 students in the room") or as an ordinal number (as in "I am the 10-th person waiting in line"). The distinction becomes sharper if we imagine an ordered infinitely long endless que of people, where each person in the line is the n-th person from the beginning ($n = 0, 1, 2, \dots$). The que has no end, but we can imagine a new person joining in at the end (with infinitely many people ahead), occupying the "ω-th position" in the que. Here the serial or ordinal position of any person is defined as the order type of the part of the que preceding that person. Still another person can be adjoined to the end, who is then called the $\omega + 1$-st person. The serial positions of these last two people, ω and $\omega + 1$, are different ordinal numbers, but in the cardinal sense they both have $\aleph_0$ people ahead. The process of extending such ordered ques by adding new members

A. Dasgupta, *Set Theory: With an Introduction to Real Point Sets*,
DOI 10.1007/978-1-4614-8854-5_9,

at the end while preserving the ordering of the preceding part can be continued indefinitely through the transfinite. The type of orders that can be generated in this way are the *well-orders*.

Criteria for Well-Ordering

Given an order X and $a \in X$, the set of predecessors of a in X will be denoted by $\text{Pred}_X(a)$, or simply by $\text{Pred}(a)$ when the order X is clear from context. For a subset A of X, we say that a is an *immediate successor of A* if $A < a$ and there is no b with $A < b < a$. Note that for $a \in X$ the immediate successor of $\text{Pred}(a)$ is a. In particular, the immediate successor of the empty set is the first element of the order (if the order does not have a first element then the empty set does not have an immediate successor).

Theorem 602 (Equivalent Conditions for Well-Ordering). *Let X be a nonempty linear order. The following conditions are equivalent:*

1. *X is a complete order with a first element, in which every element except the last (if present) has a next element.*
2. *X is a complete order with a first element but without any lower limit point.*
3. *X has a first element, and every Dedekind partition is either a jump or an upper limit cut (i.e., there are no gaps or lower limit cuts).*
4. *Every proper initial segment in X is an initial open interval: If $A \subsetneq X$ is an initial segment then $A = \text{Pred}(a)$ for some $a \in X$.*
5. *Every non-cofinal subset of X has an immediate successor.*
6. *Every nonempty subset of X contains a least element.*
7. *(DC) There is no strictly decreasing infinite sequence in X, i.e., X no suborder of type $^*\omega$.*

Problem 603. *Prove Theorem 602.*

[Hint: The implications $1 \Rightarrow 2 \Rightarrow 3 \Rightarrow 4 \Rightarrow 5 \Rightarrow 6 \Rightarrow 7$ are all routine. For $7 \Rightarrow 6$, note that if $A \subseteq X$ is nonempty but has no minimal element, then the relation R defined on A by $xRy \Leftrightarrow x > y$ satisfies the condition of DC.]

Definition 604 (Well-Orders and Ordinals).

- A *well-order* is an order which is either empty or satisfies any (and so all) of the conditions of the last theorem.
- If X is any order and $A \subseteq X$ we say that *A is well-ordered by the parent order* (or simply that *A is a well-ordered subset of X*) if the suborder on A, inherited from the parent order on X, is a well-order.
- An *ordinal* is the order type of a well-order.

From Part 4 of Theorem 602 we immediately have

Corollary 605. *Let X be a well-order with order type α. The set of all proper initial segments of X, ordered by set inclusion, is isomorphic to X. The set of all initial segments of X has order type $\alpha + 1$.*

Theorem 606. *Every subset (suborder) of a well-order is a well-order. Every finite linear order is a well-order. Any order of type ω, such as the positive integers* **N** *with their usual ordering, is a well-order.*

Corollary 607. *Every finite order type n is an ordinal, and ω is an ordinal.*

Problem 608. *Let X be an order with $X = A \cup B$, where $A < B$. If each of A and B is well-ordered by the parent order on X, then X itself is a well-order.*

Corollary 609. *The sum of two ordinals is an ordinal.*

The above Theorem is a special case of the following more general fact.

Problem 610. *Let X be an order which is expressed as the union of a finite number of subsets, say as $X = \cup_{i=1}^{n} X_i$. If each X_i is a well-ordered subset of X ($i = 1, 2, \ldots, n$) then X is a well-order.*

Using the last corollary, we get more examples of ordinals, such as $\omega + n$ ($n = 1, 2, \ldots$), $\omega 2 = \omega + \omega$, $\omega 2 + n$ ($n = 1, 2, \ldots$), $\omega 3 = \omega 2 + \omega$, etc.

Problem 611. *If A and B are well-orders then so are $B \times A$ and $A \times B$, under lexicographic ordering.*

Corollary 612. *The product of two ordinals is an ordinal.*

Definition 613 (Limit and Successor Ordinals). An ordinal is called a *limit ordinal* if it is the order type of some nonempty well-order without a last element. An ordinal is called a *successor ordinal* if it is the order type of a well-order which has a last element.

Note that limit and successor ordinals must be nonzero, and that α is a successor ordinal if and only if $\alpha = \beta + 1$ for some ordinal β. Examples of successor ordinals are 2, $\omega + 9$, and $\omega 2 + 1$, while ω and $\omega 3$ are limit ordinals.

Problem 614. *Let X be a well-order and let $x \in X$. Prove that exactly one of the following must be true:*

- *x is the first element of X*
- *x is a successor element in X, i.e., x has an immediate predecessor in X*
- *x is an upper limit point in X*

It follows that for every ordinal α, either $\alpha = 0$ or α is a successor ordinal or α is a limit ordinal (the three possibilities are mutually exclusive).

Problem 615. *An order X is a well-order without any (upper) limit point if and only if $\mathrm{Pred}(x)$ is finite for all $x \in X$.*

An order of type ω (such as the set $\mathbf{N}$ of positive integers with the usual ordering) is a nonempty well-order without a last element and without any upper limit point, and is characterized (up to order type) by these properties:

Problem 616. *Let X be a nonempty well-order without a last element and without any upper limit point. Prove that the order type of X must be ω.*

[Hint: Since every element of X must have an immediate successor, there is a function $g: X \to X$ such that $g(x)$ is the immediate successor of x in X. Hence by the principle of recursive definition there is $f: \mathbf{N} \to X$ such that $f(1) =$ the least element of X and $f(n+1) = g(f(n))$ for all $n \in \mathbf{N}$. Show that f maps $\mathbf{N}$ onto X and that f is an order isomorphism of the positive integers (under the usual ordering) with X.]

Problem 617. *Prove that any infinite well-order not containing any limit point must be of type ω.*

Thus ω is the unique limit ordinal which cannot be expressed as $\alpha + \beta$ where α is limit and β is nonzero.

One can also rearrange the elements of $\mathbf{N}$ to get other ordinals. Consider

$$1, 2, 3, 4, \ldots, n, n+1, \ldots \qquad \text{(order type } \omega)$$

$$2, 3, 4, 5, \ldots, n, n+1, \ldots \; 1 \qquad \text{(order type } \omega + 1)$$

$$3, 4, 5, 6, \ldots, n, n+1, \ldots \; 1, 2 \qquad \text{(order type } \omega + 2)$$

The first order has no last element while the other two have last elements, and the last element of the second order, 1, is an upper limit element while the last element of the third order above, 2, is a successor element. Since order isomorphisms preserve all structural properties, so no two of the three orders above are isomorphic and hence the ordinals ω, $\omega + 1$, and $\omega + 2$ are all distinct.

Suppose we put all the odd positive integers before all the even ones but otherwise order them following their usual order. This can be formally defined as an ordering $\langle \mathbf{N}, <_o \rangle$ where $m <_o n$ if and only if m is odd and n is even, or m and n have the same parity and $m < n$. This ordering $\langle \mathbf{N}, <_o \rangle$ is exhibited as

$$1 <_o 3 <_o 5 <_o 7 <_o \cdots \; 2 <_o 4 <_o 6 <_o 8 <_o \cdots,$$

and has order type $\omega + \omega = \omega 2$.

Similarly, we see that the following are all ordinals (where n is any finite ordinal):

$$0,\ 1,\ 2,\ \ldots,\ n,\ \ldots\ \ \omega,\ \omega+1,\ \omega+2,\ \ldots,\ \omega+n,\ \ldots\ \ \omega+\omega=\omega 2,$$

$$\omega 2+1,\ \omega 2+2,\ \ldots\ \ \omega 2+\omega=\omega 3,\ \ \omega 3+1,\ \ \omega 3+2,\ \ \ldots\ \ \omega 3+\omega=\omega 4,\ \ \ldots$$

Problem 618. *Prove that if $\alpha + \beta = \omega$ and $\beta \neq 0$ then $\beta = \omega$. Prove that if $\alpha + \beta = \omega^2$ and $\beta \neq 0$ then $\beta = \omega^2$.*

Problem 619. *Let α be an order type of an ordering with a first element (so that $\alpha = 1 + \beta$ for some order type β). Prove that $(1 + \lambda)\alpha$ is a Dedekind complete order type if and only if α is an ordinal.*

9.2 Limit Points and Transfinite Induction

If P is a property, then we use the predicate notation "$P(a)$" to indicate that "the property P is true of the element a." Recall:

> *The principle of finite induction.* Let A be an ordering which is either finite or is of order type ω. Let a_0 be the first element of A. Suppose that P is any property satisfying (for all $a, b \in A$):
>
> - $P(a_0)$ is true;
> - If $P(a)$ is true and b is an immediate successor of a then $P(b)$ is true.
>
> Then $P(a)$ is true for all $a \in A$.

We will show that a generalized version of the principle of finite induction, called the principle of *transfinite induction*, holds for *all* well-orders. But first let us note that the principle of finite induction, as stated above, does not hold for general well-orders (other than orders which are finite or of type ω).

Example 620. Consider again the ordering $<_o$ on $\mathbf{N}$ of order type $\omega + \omega = \omega 2$, where all the odd natural numbers come before all the even ones:

$$1 <_o 3 <_o 5 <_o 7 <_o \cdots \ 2 <_o 4 <_o 6 <_o 8 <_o \cdots$$

Note that in this ordering 3 is an immediate successor of 1, and 4 is an immediate successor of 2, but 2 is not the immediate successor of any element. In fact for $m, n \in \mathbf{N}$, n is an immediate successor of m in the ordering $<_o$ if $n = m + 2$. Let P be the property of being an odd positive integer. Then $P(a)$ is true for the first element of $\langle \mathbf{N}, <_o \rangle$, namely 1. Also if $P(a)$ is true ("a is odd") and b is an immediate successor of a ("$b = a + 2$") then $P(b)$ is true ("b is odd"). Hence both conditions of the principle of finite induction are true for this ordering $\langle \mathbf{N}, <_o \rangle$. Yet it is false that $P(a)$ holds for all a.

The reason for this failure is easily found. The principle of finite induction holds for only those orderings in which every element can be reached starting from the first element by a "finite number of individual steps of moving to the next immediate successor." And the only orderings in which this last condition is satisfied are the ones which are finite or of type ω. The ordering $\langle \mathbf{N}, <_o \rangle$, which is of type $\omega + \omega$, contains the element 2 which cannot be reached from the first element by a finite number of steps of moving to the immediate successor. In fact 2 is an upper limit point in the ordering $\langle \mathbf{N}, <_o \rangle$.

Recall that for any point x in a well-order X there are three mutually exclusive and exhaustive possibilities:

- x is the first element of X;
- x is a successor element (x is the immediate successor of some element);
- x is a limit element (x is an upper limit point in X; a well-order cannot have a lower limit point).

The principle of finite induction will hold in a well-order so long as the third possibility above (existence of upper limit point) does not arise.

In the above example, the property P of being an odd positive integer was indeed true for all numbers preceding 2 in the ordering $\langle \mathbf{N}, <_o \rangle$, but there was nothing to "induce" the property P to the limit element 2. For that, we will need a condition by which whenever a property holds for certain points, it can be "transferred" or "induced" to hold for any upper limit point of those points. Once we enhance the principle of induction by adding such a clause, it will apply to all well-orderings:

The principle of transfinite induction. Let A be any well-order with first element a_0, and P be any property which satisfies (for all a, b):

- $P(a_0)$ is true;
- If $P(a)$ is true and b is an immediate successor of a then $P(b)$ is true;
- If a is an upper limit point of the set $\{x \in A \mid P(x) \text{ is true}\}$ then $P(a)$ is true.

Then $P(a)$ is true for all $a \in A$.

The proof is straightforward: To get a contradiction, let there be $a \in A$ for which $P(a)$ is not true, and let a be the least such element. Then a can neither be the first element a_0, nor can it be an immediate successor of some other element, and nor can it be an upper limit point, which is a contradiction since these possibilities are exhaustive in a well-order.

One can restate the principle of transfinite induction in terms of sets (instead of "properties") as follows:

The principle of transfinite induction (set version). Let A be a well-order with first element a_0, and let $P \subseteq A$ such that for all a, b:

- $a_0 \in P$;
- If $a \in P$ and b is an immediate successor of a then $b \in P$;
- If a is an upper limit point of P then $a \in P$.

Then $P = A$.

It is possible to combine the three clauses of transfinite induction into a single "strong induction" clause in which we can avoid mentioning "limit point" or "first point." The advantage of such a form is that it covers both finite and transfinite induction via the concise statement "Every strongly inductive subset of a well-order equals the entire order":

The principle of strong induction (finite and transfinite). Let A be any well-order, and let $P \subseteq A$ such that for any $a \in A$:

$$\text{Pred}(a) \subseteq P \Rightarrow a \in P.$$

Then $P = A$.

Strong induction actually characterizes the property of being well-ordered.

Problem 621. *Let X be an order. A subset P of X is called strongly inductive if* $\text{Pred}(a) \subseteq P \Rightarrow a \in P$ *(for all $a \in X$), and the order X is said to satisfy strong induction if every strongly inductive subset of X equals X. Show that X satisfies strong induction if and only if it is a well-order.*

We next give a version of transfinite induction where an assertion can be established for *all* well-orders. In this case, P should be taken to be a *property of orders*, such as being a complete order.

Transfinite induction for all well-orders. Let P be a property of orders such that if every proper initial segment of any well-order X has property P, then X has property P. Then all well-orders have property P.

The proof is again routine: Assume the condition of the theorem but suppose that there is a well-order A which does not have property P. Then some proper initial segment of A will fail to have property P, so we can fix the *least* $a \in A$ such that $\text{Pred}(a)$ does not have property P. But then every proper initial segment of $\text{Pred}(a)$ has property P (by minimality of a) while A does not have property P, contrary to the condition of the theorem.

The following is the corresponding principle in terms of properties of ordinals.

The principle of transfinite induction for all ordinals. Let P be a property of ordinals such that for any ordinal α, if every ordinal less than α has property P, then α has it too. Then all ordinals have property P.

The Principle of Recursive Definition in Sect. 2.10 (as in Theorem 148) also generalizes to well-orders, where it is called the *principle of transfinite recursion.* It is a very useful principle in practice and is often used to define functions on well-orders (or on initial segments of ordinals).

Theorem 622 (Transfinite Recursion). *Let A be any well-order, Y be a nonempty set, $\mathcal{F}$ be the collection of all functions whose domain is an initial segment of A and whose range is contained in Y, and $G\colon A \times \mathcal{F} \to Y$. Then there is a unique function $F\colon A \to Y$ satisfying, for all $a \in A$, the recursion condition:*

$$F(a) = G(a,\ F \restriction \text{Pred}(a)).$$

Proof. The proof is similar to the proof of the Basic Principle of Recursive Definition (Theorem 146 in Sect. 2.10).

Let us say that a function h is *partially G-recursive* if $\mathrm{dom}(h)$ is an initial segment of A, $\mathrm{ran}(h) \subseteq Y$, and $h(a) = G(a, h \restriction \mathrm{Pred}(a))$ for all $a \in \mathrm{dom}(h)$. If h is partially G-recursive then so is $h \restriction I$ if I is any initial segment of $\mathrm{dom}(h)$.

We first have the following uniqueness property: If h, h' are partially G-recursive functions with $a \in \mathrm{dom}(h) \cap \mathrm{dom}(h')$ then $h(a) = h'(a)$. To prove this by transfinite induction, assume that $h(x) = h'(x)$ for all $x \in \mathrm{Pred}(a)$. Then $h(a) = G(a, h \restriction \mathrm{Pred}(a)) = G(a, h' \restriction \mathrm{Pred}(a)) = h'(a)$.

Next, define a relation $F \subseteq A \times Y$ by the condition xFy if and only if there is a partially G-recursive h with $x \in \mathrm{dom}(h)$ and $h(x) = y$. The theorem will be proved if we show that F is a function, $\mathrm{dom}(F) = A$, and F is partially G-recursive. Uniqueness of F follows from the uniqueness property that we just proved.

Assume xFy and xFy'. Then $y = h(x)$ and $y' = h'(x)$ for some partially G-recursive h and h', hence by the uniqueness property that we proved, $h(x) = h'(x)$ and so $y = y'$. Thus F is a function. It is also easy to see that F must be partially G-recursive. Finally, we show that $\mathrm{dom}(F) = A$ by transfinite induction. Suppose that $a \in A$ and $x \in \mathrm{dom}(F)$ for all $x \in \mathrm{Pred}(a)$. Then the function $h := F \restriction \mathrm{Pred}(a)$ is partially G-recursive. Put $b = G(a, h)$, and extend h to $\overline{h} := h \cup \{\langle a, b\rangle\}$. Then $\overline{h}$ is a partially G-recursive function with $a \in \mathrm{dom}(\overline{h})$, hence $a \in \mathrm{dom}(F)$. □

9.3 Well-Orders and Ordinals: Basic Facts

Recall that if A is an initial segment in a well-order X with $A \neq X$ then $A = \mathrm{Pred}(a)$ for some (unique) $a \in X$.

Theorem 623. *Let X be a well-order and $f : X \to X$ be an order preserving (strictly increasing) injection. Then $x \leq f(x)$ for all $x \in X$.*

Proof. Otherwise, there would be a least a such that $f(a) < a$, but then $b = f(a)$ is a still smaller element for which $f(b) < b$, a contradiction. □

A function f such as above need not be onto. For example the mapping $n \mapsto n^2$ is a strictly increasing mapping of **N** into **N**. However, we have:

Corollary 624. *The only order-preserving isomorphism of a well-order onto itself is the identity mapping.*

Proof. If $f : X \to X$ is an order isomorphism of the well-order X onto itself, let $g : X \to X$ be the inverse of f so that $f(g(x)) = x$ for all $x \in X$. Then for any $x \in X$ we have $x \leq f(x)$ and also $x \leq g(x)$, so $f(x) \leq f(g(x)) = x$. □

Corollary 625 (Uniqueness of Isomorphisms). *If A and B are isomorphic well-orders, then there is a unique isomorphism from A onto B.*

Another important immediate corollary of the theorem is:

Corollary 626. *A well-order is never order isomorphic to any of its proper initial segments.*

Problem 627. *Let A be a subset of a well-order X which is strictly bounded above, that is, there is $b \in X$ with $a < b$ for all $a \in A$. Show that the suborder A cannot be isomorphic to X.*

The above facts limit isomorphisms between initial segments of a well-order:

Corollary 628 ("Initial Rigidity of Well-Orders"). *If A and B are initial segments of a well-order X and $f : A \to B$ is an order isomorphism from A onto B, then $A = B$ and f is the identity map on $A = B$.*

In any order X, we define the *rank of an element $a \in X$* to be the order type of $\mathrm{Pred}(a)$. If X is not well-ordered, two different elements may have the same rank. For example, in any order of type ζ, all elements have the same rank $^*\omega$; moreover, in the set $\mathbf{Z}$ of integers with the usual ordering, if m and n are any two integers, then the mapping $x \mapsto x + n - m$ is an order-automorphism of $\mathbf{Z}$ which sends m to n, so that m and n are structurally indistinguishable within the order. The same is true for orderings of type η or λ. The situation for well-orders is strikingly different:

Corollary 629 (Unique Ranks). *In a well-order, distinct initial segments have distinct order types, i.e., distinct elements have distinct ranks. Hence each element in a well-order is uniquely determined by its rank.*

This fact is further exemplified in Theorem 633 below, which exhibits the natural one-to-one correspondence between the elements of a well-order and the set of ordinals representing the ranks of those elements.

Initial rigidity allows a proper definition for comparing ordinals:

Definition 630 (Ordering of Ordinals). Given ordinals α and β with representative well-orders A and B, we define $\alpha < \beta$ if A is order-isomorphic to some proper initial segment of B. We write $\alpha \leq \beta$ for $\alpha < \beta$ or $\alpha = \beta$.

Corollary 631. *The relation $\alpha < \beta$ defined on any set of ordinals is irreflexive and transitive (hence asymmetric).*

(This situation is again specific to well-orders. An attempt to extend this definition to all orderings will fail because asymmetry and irreflexivity will be violated, producing oddities such as $^*\omega < {}^*\omega$ or $\eta < \eta$, and we would even get both $\eta < \eta + 1$ and $\eta + 1 < \eta$ holding at the same time!)

The definition immediately implies that *if α and β are ordinals with corresponding representative well-orders A and B, then $\alpha \leq \beta$ if and only if there is a unique order isomorphism from A onto a unique initial segment of B*. The trichotomy property for $<$ will be established in Theorem 636.

9.4 Unique Representation by Initial Sets of Ordinals

Given an ordinal α and a representative well-order A with order type α, we let $W(\alpha)$ denote the set of order types of proper initial segments of A. Note that this definition of $W(\alpha)$ is independent of the choice of the representative well-order A. Moreover, for every ordinal β, we have $\beta < \alpha$ if and only if β is the order type of some proper initial segment of A. So we can define:

Definition 632. Given any ordinal α, let $W(\alpha) := \{\beta \mid \beta \text{ is an ordinal} < \alpha\}$.

Thus we have:

$$\begin{aligned}
W(0) &= \emptyset, \\
W(1) &= \{0\} \;= W(0) \cup \{0\}, \\
W(2) &= \{0, 1\} \;= W(1) \cup \{1\}, \\
W(3) &= \{0, 1, 2\} \;= W(2) \cup \{2\}, \\
&\dots \\
W(n) &= \{0, 1, 2, \dots, n-1\}, \\
W(n+1) &= \{0, 1, 2, \dots, n\} \;= W(n) \cup \{n\}, \\
&\dots \\
W(\omega) &= \{0, 1, 2, \dots,\ n,\ \dots\}, \\
W(\omega+1) &= \{0, 1, 2, \dots,\ n,\ \dots,\ \omega\} \;= W(\omega) \cup \{\omega\} \\
&\dots \quad \text{etc.}
\end{aligned}$$

Under the relation $<$ on ordinals, the set $W(\alpha)$ of ordinals less than α is itself a well-order of order type α:

Theorem 633 (Representation Theorem for Well-Orders). *Given a well-order A with ordinal α, there is a unique order isomorphism between A and $W(\alpha)$: A strictly increasing bijection $f : A \to W(\alpha)$ defined by $f(a) =$ the rank of a in $A =$ the order type of* Pred(a).

The inverse of this bijection gives a strictly increasing complete enumeration of A indexed by the ordinals less than α:

$$A = \{\, a_0 <_A a_1 <_A a_2 <_A \cdots <_A a_\xi <_A \cdots\} \qquad (\xi < \alpha),$$

with $a_\mu <_A a_\nu$ for all $\mu < \nu < \alpha$, where a_ξ is the unique element in A having rank ξ.

In particular, $W(\alpha)$ is well-ordered by $<$ and its order type (ordinal) is α.

Proof. This is an immediate consequence of the unique rank property (Corollary 629) and the definition of $\alpha < \beta$ for ordinals. □

From the last statement in the theorem it follows that if $W(\alpha) = W(\beta)$ then $\alpha = \beta$. Also, if $\xi < \alpha$, then $W(\xi)$ is a proper initial segment in $W(\alpha)$ with order type ξ so that $W(\xi)$ is the set of predecessors of the element $\xi \in W(\alpha)$ with ξ itself having rank ξ in $W(\alpha)$; and conversely, by uniqueness of ranks, a proper initial segment in $W(\alpha)$ having order type ξ must equal $W(\xi)$.

Corollary 634. *For all ordinals α, β we have:*

1. *$W(\alpha) = W(\beta)$ if and only if $\alpha = \beta$.*
2. *An initial segment of $W(\alpha)$ having order type ξ must equal $W(\xi)$.*
3. *A is an initial segment of $W(\alpha)$ if and only $A = W(\xi)$ for some $\xi \leq \alpha$.*
4. *$W(\alpha) \subsetneq W(\beta)$ if and only if $\alpha < \beta$.*

Example 635. Recall the ordering $<_o$ on $\mathbf{N}$ having order type $\omega + \omega = \omega 2$, where all the odd natural numbers come before all the even ones, as in:

$$1 <_o 3 <_o 5 <_o 7 <_o \cdots \; 2 <_o 4 <_o 6 <_o 8 <_o \cdots$$

The set $W(\omega + \omega) = W(\omega 2)$ is

$$W(\omega 2) := \{0, 1, 2, \ldots, n, \ldots, \omega, \omega + 1, \omega + 2, \ldots, \omega + n, \ldots\}$$

The natural correspondence between $\langle \mathbf{N}, <_o \rangle$ and the ordinals in $W(\omega 2)$ is then seen as:

$$\begin{array}{ccccccccccccccccc}
1 & <_o & 3 & <_o & 5 & <_o & 7 & <_o & \cdots & 2 & <_o & 4 & <_o & 6 & <_o & 8 & <_o \;\cdots \\
\updownarrow & & \updownarrow & & \updownarrow & & \updownarrow & & & \updownarrow & & \updownarrow & & \updownarrow & & \updownarrow & \\
0 & < & 1 & < & 2 & < & 3 & < & \cdots & \omega & < & \omega+1 & < & \omega+2 & < & \omega+3 & < \;\cdots
\end{array}$$

Theorem 636 (Well-Ordering and Ordinal Comparability Theorem). *Given ordinals α and β, exactly one of $\alpha < \beta$, $\beta < \alpha$, and $\alpha = \beta$ holds (trichotomy). Hence if A and B are well-orders, either A is isomorphic to an initial segment of B or B is isomorphic to an initial segment of A.*

Proof. Put $C = W(\alpha) \cap W(\beta)$, so that C is an initial part of $W(\alpha)$ and also of $W(\beta)$. Hence $C = W(\xi)$ (where $\xi =$ order type of C), with both $\xi \leq \alpha$ and $\xi \leq \beta$. But we cannot have $\xi < \alpha$ and $\xi < \beta$, as otherwise we would get $\xi \in C = W(\xi)$ (contradicting irreflexivity of $<$). Hence either $\xi = \alpha$ in which case $\alpha \leq \beta$, or $\xi = \beta$, in which case $\beta \leq \alpha$. □

Thus if A is any set of ordinals, then A must be linearly ordered. In fact, A must be well-ordered, since otherwise we would have a nonempty $B \subseteq A$ without a least element, and then for any $\alpha \in B$ the set $B \cap W(\alpha)$ would be a nonempty subset of $W(\alpha)$ without a least element, a contradiction.

Corollary 637. *Any set of ordinals is linearly ordered and in fact well-ordered.*

Definition 638 (Initial Sets of Ordinals). A set A of ordinals is called an *initial set of ordinals* if $\alpha \in A$ and $\beta < \alpha \Rightarrow \beta \in A$.

For every ordinal α, $W(\alpha)$ is an initial set of ordinals having order type α. In fact, these are the only initial sets of ordinals:

Corollary 639. *Every initial set A of ordinals equals $W(\alpha)$, where α is the order type of A.*

This follows from comparability: If A is an initial set of ordinals with order type α, we cannot have $\alpha \in A$ (since otherwise $W(\alpha)$ would be a proper initial segment of A of order type α), and so $A \subseteq W(\alpha)$ by comparability; hence $A = W(\beta)$ for some β, which implies that $\beta =$ order type of $W(\beta) =$ order type of $A = \alpha$.

Well-Ordered Sum of Ordinals

Recall that we needed the Axiom of Choice to define arbitrary sums of order types (Definition 473) of the form

$$\sum_{i \in I} \alpha_i \qquad (I \text{ an order, } \alpha_i \text{ an order type for each } i \in I).$$

AC was needed twice: First for choosing representative orders of type α_i for each $i \in I$ (existence), and then again for choosing isomorphisms between multiple representatives for each type when showing that the order type of the sum does not depend on the choice of representatives (uniqueness).

A nice consequence of the unique representation property for well-orders is that if each α_i is an ordinal ($i \in I$), then the above sum can be defined in an effective canonical fashion without using AC at all: For the existence part, we can simply let $W(\alpha_i)$ be the canonical representative well-order of type α_i (for each $i \in I$). The uniqueness part follows immediately from the uniqueness of isomorphisms for well-orders.

Theorem 640 (Arbitrary Sum of Ordinals without AC). *If I is any order and α_i is an ordinal for each $i \in I$ then the sum*

$$\sum_{i \in I} \alpha_i$$

is defined and unique even if we do not assume the Axiom of Choice.

Proof. Uniqueness follows from the fact that if X_i and X_i' are representative well-orders of type α_i then there is a unique order isomorphism between X_i and X_i'. For existence, take $W(\alpha_i)$ as the representative well-order for α_i and order $\bigcup_{i \in I} \{i\} \times W(\alpha_i)$ lexicographically (by "first difference"). □

We will be interested in the case where I is a well-order, when the sum itself becomes an ordinal.

Lemma 641. *Suppose that X is an order with $X = \bigcup_{i \in I} S_i$ such that each S_i is well-ordered as a suborder of X (for $i \in I$), $i < j \Rightarrow S_i < S_j$ (for all $i, j \in I$), and I is itself a well-ordered set. Then X is a well-order.*

Proof. Let X, I, and $\langle S_i \mid i \in I \rangle$ be as above, and let E be a nonempty subset of $X = \bigcup_{i \in I} S_i$. Let $J = \{i \in I \mid S_i \cap E \neq \emptyset\}$. Then J is nonempty and since I is a well-order, J contains a smallest member $i_0 \in J$. Then $S_{i_0} \cap E$ is a nonempty subset of S_{i_0}, and since S_{i_0} is a well-ordered subset of X, $S_{i_0} \cap E$ contains a smallest element $p \in S_{i_0} \cap E$. It is then easily verified that p is the least element of E. □

Corollary 642 (Well-ordered sum of ordinals is an ordinal). *If I is well-ordered and α_i is an ordinal for each $i \in I$ then $\sum_{i \in I} \alpha_i$ is an ordinal*

The product $\alpha\beta$ of two ordinals α and β is conveniently viewed as "α repeated β times," or equivalently as "β copies of α." For example, we have:

$$1\omega = 1 + 1 + 1 + \cdots = \sum_{n<\omega} 1 = \omega, \quad 2\omega = 2 + 2 + 2 + \cdots = \sum_{n<\omega} 2 = \omega, \quad \text{etc.,}$$

while

$$\omega^2 = \omega\omega = \sum_{n<\omega} \omega = \omega + \omega + \omega + \cdots$$

We can keep going further using repeated sum. For example, after $\omega^3 = \omega^2\omega = \sum_{n<\omega} \omega^2$, and $\omega^4 = \omega^3\omega = \sum_{n<\omega} \omega^3$, etc., we can get the following larger ordinal which will later be denoted by ω^ω:

$$\sum_{n<\omega} \omega^n = 1 + \omega + \omega^2 + \omega^3 + \cdots + \omega^n + \cdots$$

Problem 643. *1. Simplify the sum $\sum_{n<\omega} n = 1 + 2 + 3 + \cdots + n + \cdots$.*
2. Simplify the sum $\sum_{n<\omega} \omega n = \omega + \omega 2 + \omega 3 + \cdots + \omega n + \cdots$ as a single ordinal.
3. Find a re-ordering of $\mathbf{N}$ having the order type of the previous part.

9.5 Successor, Supremum, and Limit

Given any ordinal α, note that $W(\alpha) \cup \{\alpha\}$ is an initial set of ordinals whose greatest element is α, and we define $S(\alpha)$, *the successor of* α, by

$$\begin{aligned} S(\alpha) := &\text{ the order type of } W(\alpha) \cup \{\alpha\} \\ = &\text{ the unique } \beta \text{ such that } W(\beta) = W(\alpha) \cup \{\alpha\}. \end{aligned}$$

The ordinal $S(\alpha)$ is the least ordinal greater than α and is same as $\alpha + 1$, but the above definition is independent of the notion of sum of ordinals.

If E is any set of ordinals, put:

$$\text{Pred}[E] := \bigcup_{\beta \in E} W(\beta) = \{\alpha \mid \alpha < \beta \text{ for some } \beta \in E\}.$$

Problem 644. *For any set E of ordinals,* Pred$[E]$ *is an initial set of ordinals and therefore equals $W(\gamma)$ for a unique ordinal γ. The ordinal γ is the least upper bound of E, that is we have $\alpha \leq \gamma$ for all $\alpha \in E$ and there is no $\gamma' < \gamma$ such that $\alpha \leq \gamma'$ for all $\alpha \in E$.*

Definition 645. For any set E of ordinals, put

$$\sup E = \sup_{\alpha \in E} \alpha := \text{ the unique ordinal } \gamma \text{ such that } \text{Pred}[E] = W(\gamma).$$

Problem 646. *Show that for any set E of ordinals*

1. *If $E = \emptyset$, then* $\sup E = 0$;
2. *If E has a largest element α, then* $\sup E = \alpha$;
3. *If E is nonempty and has no largest element, then* $\sup E$ *is the unique limit ordinal γ such that $\alpha < \gamma$ for all $\alpha \in E$ and such that for all $\beta < \gamma$ there is $\alpha \in E$ with $\beta < \alpha < \gamma$.*

In the last case above, $\sup E$ is called the *limit of the elements of E*, and denoted by

$$\lim E = \lim_{\alpha \in E} \alpha := \sup E.$$

Problem 647. *For any set E of ordinals, show that*

1. *$E' := E \cup \bigcup_{\alpha \in E} W(\alpha)$ is the smallest initial set of ordinals containing E.*
2. *The order type of E' equals $\sup_{\alpha \in E} S(\alpha) = \sup\{S(\alpha) \mid \alpha \in E\}$.*
3. *$\sup_{\alpha \in E} S(\alpha)$ is the least ordinal greater than all elements of E.*

In particular, *for any set E of ordinals there is an ordinal greater than all elements of E*, with $\sup_{\alpha \in E} S(\alpha)$ being the least such ordinal.

Problem 648. *For any set E of ordinals, show that*

$$\sup_{\alpha \in E} S(\alpha) = \begin{cases} \gamma + 1 & \textit{if } E \textit{ has a largest element } \gamma, \\ \sup E & \textit{otherwise.} \end{cases}$$

Problem 649. *Given a set C of well-orders, effectively construct a well-order whose order type is the supremum of the order types of the well-orders in C.*

Theorem 650 (Transfinite Recursion over all Ordinals). *Let G be a function which assigns an object $G(h)$ to every function h whose domain is an initial set of ordinals (i.e., with* $\mathrm{dom}(h) = W(\alpha)$ *for some ordinal α). Then there exists a unique function F defined for all ordinals such that:*

$$F(\alpha) = G(F \restriction W(\alpha)), \qquad \text{for every ordinal } \alpha.$$

Proof. For each ordinal α, apply Theorem 622, with the well-order A there replaced by $W(\alpha + 1)$, to get a unique function F_α with domain $W(\alpha + 1)$ and satisfying the recursion condition $F_\alpha(\beta) = G(F_\alpha \restriction W(\beta))$ for all $\beta \in W(\alpha + 1)$. Define $F(\alpha) := F_\alpha(\alpha)$. Note that if $\alpha < \beta$ then F_β extends F_α (by uniqueness). Hence for every ordinal α, F extends F_α, and therefore $F(\alpha) = F_\alpha(\alpha) = G(F_\alpha \restriction W(\alpha)) = G(F \restriction W(\alpha))$. □

9.6 Operations Defined by Transfinite Recursion

For given ordinals α and β, one can use transfinite recursion to define the *ordinal sum* $\alpha \dotplus \beta$ as the β-th successor of α, i.e., $\alpha \dotplus \beta$ is obtained starting from α by repeatedly applying the successor operation β times. Here the recursion is done on the second argument β, which means $\alpha \dotplus \beta$ is defined assuming that $\alpha \dotplus \gamma$ has been defined for all $\gamma < \beta$. Breaking up into three cases, we have:

$$\alpha \dotplus \beta = \begin{cases} \alpha & \text{if } \beta = 0; \\ S(\alpha \dotplus \gamma) & \text{if } \beta = S(\gamma) \text{ is the successor of } \gamma; \\ \sup_{\gamma<\beta} \alpha \dotplus \gamma & \text{if } \beta \text{ is a limit ordinal.} \end{cases}$$

Here the first argument α can be regarded as a parameter.

Problem 651. *Show that the above informal definition by transfinite recursion can be cast into the more formal framework of Theorem 650 by first fixing α and taking G_α to be:*

$$G_\alpha(h) := \begin{cases} \alpha & \textit{if } h \textit{ is empty;} \\ \sup\{S(\xi) \mid \xi \in \mathrm{ran}(h) \textit{ and } \xi \textit{ is an ordinal}\} & \textit{otherwise.} \end{cases}$$

That is, for any ordinal α, if we use this G_α in Theorem 650 to obtain F_α with $F_\alpha(\beta) = G(F_\alpha \restriction W(\beta))$ for all β, then $F_\alpha(\beta) = \alpha \dotplus \beta$ for all ordinals β.

From now on, however, we will simply use the informal version of definition by transfinite recursion, without explicitly displaying the function G needed by the formal setup.

Problem 652. *Show that $\alpha \dotplus \beta = \alpha + \beta$ for all ordinals α, β, that is the ordinal sum as defined above by transfinite recursion coincides with the usual sum of order types.*

In a similar way, we can use transfinite recursion to define the *ordinal product* $\alpha \cdot \beta$:

$$\alpha \cdot \beta = \begin{cases} 0 & \text{if } \beta = 0; \\ (\alpha \cdot \gamma) + \alpha & \text{if } \beta = S(\gamma) \text{ is the successor of } \gamma; \\ \sup\limits_{\gamma < \beta} \alpha \cdot \gamma & \text{if } \beta \text{ is a limit ordinal.} \end{cases}$$

Problem 653. *Show that $\alpha \cdot \beta = \alpha\beta$ for all ordinals α, β, that is the ordinal product as defined above by transfinite recursion coincides with the usual product of order types.*

As a result, all rules valid for sums and products of order types will apply to ordinals. In particular, the associative law and the left distributive law hold.

In the definitions above for ordinal sum and product, the limit ordinal clauses say that the each of these operations is "continuous" in the second variable, that is, if β is a limit ordinal so that $\beta = \lim_{\gamma < \beta} \gamma$, then:

$$\lim_{\gamma < \beta} (\alpha + \gamma) = \alpha + \lim_{\gamma < \beta} \gamma, \quad \text{and} \quad \lim_{\gamma < \beta} (\alpha\gamma) = \alpha \lim_{\gamma < \beta} \gamma.$$

Problem 654. *Show that neither the ordinal sum operation nor the ordinal product operation is continuous in the first variable.*

Problem 655. *1. ω is the smallest limit ordinal.*
2. If α is an ordinal then $\alpha + \omega$ is the smallest limit ordinal greater than α.
3. α is a limit ordinal if and only if $\alpha = \omega\beta$ for some ordinal $\beta > 0$.

Problem 656. *If A is a well-order of type α and β is the order type of a suborder of A, then $\beta \leq \alpha$.*

Problem 657. *If $\alpha < \beta$ then $\gamma + \alpha < \gamma + \beta$, and conversely. This gives left-cancellation for addition: If $\gamma + \alpha = \gamma + \beta$ then $\alpha = \beta$. If $\alpha < \beta$ then $\alpha + \gamma \leq \beta + \gamma$.*

For $\gamma > 0$, if $\alpha < \beta$ then $\gamma\alpha < \gamma\beta$ and conversely. This gives left-cancellation for products: If $\gamma\alpha = \gamma\beta$ and $\gamma > 0$ then $\alpha = \beta$. If $\alpha < \beta$ then $\alpha\gamma \leq \beta\gamma$.

Subtraction. If $\alpha \leq \beta$ then there is a unique γ such that $\alpha + \gamma = \beta$. This γ is denoted by $-\alpha + \beta$ and is a form of (one-sided) subtraction for ordinals, so that

$$\beta = \alpha + (-\alpha + \beta), \qquad \text{whenever } \alpha \leq \beta.$$

This follows immediately from the fact that given two well-orders, one of them is isomorphic to a unique initial segment of the other. It is easy to see that we have $-\alpha + (\alpha + \beta) = \beta$ for all ordinals α and β. Note also that if $\alpha \le \beta$, then $-\alpha + \beta$ is the order type of $W(\beta) \setminus W(\alpha)$.

Theorem 658 (Division Algorithm). *If α, β are ordinals with $\alpha > 0$ then there are unique ordinals η and ξ such that*

$$\beta = \alpha\eta + \xi, \qquad \textit{with } \xi < \alpha.$$

Proof. Note that with $\nu = \beta + 1$ we have $\alpha\nu > \beta$. Let μ be the least ordinal such that $\alpha\mu > \beta$, so that $\mu \le \beta + 1$. Then μ must be a successor ordinal since if μ were a limit ordinal, then β would be an upper bound of $E := \{\alpha\nu \mid \nu < \mu\}$, which would imply that $\alpha\mu = \sup E$ (= least upper bound of E) $\le \beta$, which contradicts $\alpha\mu > \beta$. So we can write $\mu = \eta + 1$ for some η. Then $\alpha\eta \le \beta$, and we can put $\xi = -\alpha\eta + \beta$, giving $\beta = \alpha\eta + \xi$. And we must have $\xi < \alpha$, for otherwise we would get $\xi = \alpha + \gamma$ for some γ, giving $\beta = \alpha\eta + \alpha + \gamma \ge \alpha(\eta + 1) = \alpha\mu$, contradicting $\beta < \alpha\mu$.

For uniqueness, suppose that

$$\alpha\eta + \xi = \alpha\eta' + \xi' \qquad \text{with } \xi, \xi' < \alpha.$$

Then $\eta \ge \eta'$, for otherwise $\eta' = \eta + \zeta$ with $\zeta > 0$, so $\alpha\eta + \xi = \alpha\eta' + \xi' = \alpha\eta + \alpha\zeta + \xi'$, and left-cancellation would give $\xi = \alpha\zeta + \xi' \ge \alpha$, contradicting $\xi < \alpha$. Similarly $\eta' \ge \eta$, and so $\eta = \eta'$. Hence $\alpha\eta + \xi = \alpha\eta + \xi'$, and so $\xi = \xi'$ by left-cancellation. □

Problem 659 (Even and Odd Ordinals). *Call an ordinal α* even *if it can be expressed in the form $\alpha = 2\gamma$ and call it* odd *if $\alpha = 2\gamma + 1$ (for some ordinal γ). Show that every ordinal is either even or odd but not both. Show that every limit ordinal is even.*

9.7 Remainder Ordinals and Ordinal Exponentiation

We say that *an ordinal $\beta > 0$ is a remainder of an ordinal γ* if $\gamma = \alpha + \beta$ for some ordinal α. Thus the finite ordinal 3 has as remainders 1, 2, and 3, and 0 has no remainder. In general the finite ordinal $n < \omega$ has exactly n remainders, namely $1, 2, \ldots, n$. The only remainder of the ordinal ω is ω itself.

Problem 660. *An ordinal can have at most finitely many remainders.*

[Hint: Note that the remainders of γ are given by the ordinals of the form $-\alpha + \gamma$ for $\alpha < \gamma$, and the mapping $\alpha \mapsto (-\alpha + \gamma)$ is monotonically decreasing, that is, if $\alpha < \alpha'$ then $-\alpha + \gamma \ge -\alpha' + \gamma$.]

Definition 661. An ordinal $\rho > 0$ is said to be a *remainder ordinal* if the only remainder of ρ is ρ itself, that is, if whenever $\rho = \alpha + \beta$ with $\beta > 0$, we have $\beta = \rho$.

In other words, ρ is a remainder ordinal if it is the type of a nonempty well-order A in which every nonempty terminal segment is isomorphic to the entire order A. It is also easily seen that ρ is a remainder ordinal if and only if $\alpha + \rho = \rho$ for all $\alpha < \rho$.

We have seen that 1 and ω are remainder ordinals. Other than 1, all remainder ordinals must be limit ordinals.

Problem 662. *If α is a remainder ordinal, then so are $\alpha\omega$ and $\omega\alpha$, and $\alpha\omega$ is the smallest remainder ordinal greater than α.*

Thus after 1 and ω, the next remainder ordinal is ω^2, the following one is ω^3, and we get the sequence

$$1,\ \omega,\ \omega^2,\ \omega^3,\ \ldots,\ \omega^n,\ \ldots$$

of the first ω remainder ordinals. Writing $1 = \omega^0$, this sequence consisting of the first ω remainder ordinals can be expressed as $\langle \omega^n \mid n < \omega \rangle$.

Problem 663. *If E is any nonempty set of remainder ordinals without a largest element, then* $\sup E$ *is a remainder ordinal, and so is the least remainder ordinal greater than all the ones in E.*

Thus we have the ordinal

$$\sup_{n<\omega} \omega^n = \sup\{1,\ \omega,\ \omega^2,\ \omega^3,\ \ldots,\ \omega^n,\ \ldots\}$$

as the least remainder ordinal greater than all ω^n, $n < \omega$. This ordinal is denoted by ω^ω.

More generally, for any two ordinals α and β, we can define the *ordinal exponentiation* α^β by transfinite recursion on β:

Definition 664 (Ordinal Exponentiation).

$$\alpha^\beta := \begin{cases} 1 & \text{if } \beta = 0; \\ (\alpha^\gamma)\alpha & \text{if } \beta = S(\gamma) \text{ is a successor ordinal;} \\ \sup\limits_{\gamma<\beta} \alpha^\gamma & \text{if } \beta \text{ is a limit ordinal.} \end{cases}$$

Note that the above definition does not create a conflict with our previous usage of the notation α^n as an abbreviation for $\alpha\alpha\cdots\alpha$ (n factors). From the last clause of the definition we have "continuity in the exponent," i.e., ordinal exponentiation $\alpha, \beta \mapsto \alpha^\beta$ is continuous in the second variable β:

$$\lim_{\gamma<\beta} \alpha^\gamma = \alpha^\beta, \quad \text{where } \beta \text{ is any limit ordinal.}$$

Ordinal exponentiation should be carefully distinguished from cardinal exponentiation. For example, for cardinals we had

$$2^{\aleph_0} > \aleph_0,$$

while with ordinal exponentiation we have:

$$2^{\omega} = \sup_{n<\omega} 2^n = \omega.$$

(For finite ordinals and cardinals, the two notions coincide.)

Using transfinite induction, we can establish the main properties of ordinal exponentiation:

Problem 665 (Properties of Ordinal Exponentiation).

1. $1^{\alpha} = 1$, *and* $0^{\alpha} = 0$ *for* $\alpha > 0$.
2. $\alpha^{\beta}\alpha^{\gamma} = \alpha^{\beta+\gamma}$.
3. $(\alpha^{\beta})^{\gamma} = \alpha^{\beta\gamma}$.
4. *For* $\alpha > 1$, $\alpha^{\beta} \geq \beta$ *and* $\beta < \gamma \Rightarrow \alpha^{\beta} < \alpha^{\gamma}$.

The largest ordinal that we have seen so far is $\omega^{\omega} = \sup_{n<\omega} \omega^n$, but we can proceed further as:

$$\omega^{\omega} + 1 < \omega^{\omega} + \omega < \omega^{\omega} + \omega^2 < \omega^{\omega} + \omega^{\omega} = \omega^{\omega}2 < \omega^{\omega}\omega = \omega^{\omega+1} < \omega^{\omega^2} \ldots$$

on to

$$\omega^{\omega^2} < \omega^{\omega^3} < \cdots, \qquad \text{and in the limit:} \quad \omega^{\omega^{\omega}} = \sup_{n<\omega} \omega^{\omega^n}.$$

In fact, using exponentiation we get:

$$\omega^{\omega} < \omega^{\omega^{\omega}} < \omega^{\omega^{\omega^{\omega}}} < \cdots, \quad \text{and} \quad \varepsilon_0 := \sup\{\omega, \omega^{\omega^{\omega}}, \omega^{\omega^{\omega^{\omega}}}, \ldots\}$$

The ordinal ε_0 has the property $\omega^{\varepsilon_0} = \varepsilon_0$. Ordinals α which satisfy the equation $\omega^{\alpha} = \alpha$ are called *epsilon numbers* (Cantor).

Problem 666. *Show that ε_0 is the smallest epsilon number, and that for every ordinal there is a greater ordinal which is an epsilon number.*

The next epsilon number after ε_0 is called ε_1, the next epsilon number after ε_1 is called ε_2, and so on.

An ordinal is said to be a *countable ordinal* if it is the order type of some countable well-order (i.e., a well-order defined on some countable set). Since ordinal sum and product coincides with the ordinary sum and product of order types, we see that sums and products of countable ordinals are countable ordinals. The countable ordinals are also closed under forming "countable limits":

Problem 667 (CAC). *If E is a countable set of countable ordinals without a largest element, then their limit* $\lim E = \sup E$ *is also a countable ordinal.*

From these facts it follows by transfinite induction that if α and β are countable ordinals then so is α^β.

The countable ordinals thus have quite strong closure properties, and all the ordinals above including ω^ω, ε_0, ε_1, and so on, are countable ordinals.

Definition 668 (Sum-Closed and Product-Closed Ordinals). An ordinal ξ is called *sum-closed* if $\alpha, \beta < \xi \Rightarrow \alpha + \beta < \xi$, and ξ is called *product-closed* if $\alpha, \beta < \xi \Rightarrow \alpha\beta < \xi$.

Problem 669 (Characterization of Remainder Ordinals). *Let ρ be a nonzero ordinal. Then the following conditions are equivalent.*

1. *ρ is a remainder ordinal.*
2. *ρ is a sum-closed ordinal.*
3. *$\rho = \omega^\alpha$ for some ordinal α.*

Problem 670. *An ordinal $\rho > 2$ is product-closed if and only if $\rho = \omega^{\omega^\alpha}$ for some ordinal α.*

Definition 671 (Normal Functions). Suppose that $F(\alpha)$ is an ordinal for each ordinal α. We say that F *is a normal function* if F is increasing, i.e., $\alpha < \beta \Rightarrow F(\alpha) < F(\beta)$, and F is continuous, i.e., $F(\alpha) = \sup_{\beta<\alpha} F(\beta)$ for every limit ordinal α.

Normal functions are frequently encountered in the theory of ordinals. The sum, product, and power functions are normal in the second variable (i.e., when the first argument is held fixed), but not in the first variable.

Problem 672. *Show that a normal function F must have arbitrarily large fixed points, i.e., for each ordinal α there is an ordinal $\beta > \alpha$ with $F(\beta) = \beta$.*

We can generalize the notion of iterated derived sets in orderings using ordinals as follows. Let X be an order and A be a subset. Recall that $D(A)$ denotes the set of limit points of A in X.

Definition 673 (Cantor–Bendixson Derivative). Let X be an order and let $A \subseteq X$. For each ordinal α, define $D^{(\alpha)}(A)$, the α-th iterated derived set of A, by transfinite recursion as follows:

$$
\begin{aligned}
D^{(0)}(A) &:= A, \\
D^{(\alpha+1)}(A) &:= D(D^{(\alpha)}(A)), \\
D^{(\gamma)}(A) &:= \bigcap_{\beta<\gamma} D^{(\beta)}(A) \text{ if } \gamma \text{ is a limit ordinal.}
\end{aligned}
$$

Problem 674. *Let E be any initial set of ordinals, considered as an order by itself. Show that for each ordinal $\alpha > 0$,*

$$D^{(\alpha)}(E) = \{\gamma \in E \mid \gamma = \omega^\alpha \beta \text{ for some } \beta\}.$$

Conclude that if $X := \{\gamma \mid \gamma \leq \omega^\alpha\} = W(\omega^\alpha) \cup \{\omega^\alpha\}$, then in the order X we have $D^{(\alpha)}(X) \neq \emptyset$, but $D^{(\alpha+1)}(X) = \emptyset$.

Problem 675 (Hausdorff's Ordinal Exponentiation). *Let A and B be well-orders with A nonempty and let $a \in A$ be the first element of A. Let E be the set of all functions f from B to A such that $f(x) = a$ for all but finitely many $x \in B$. Thus $E \subseteq A^B$. For $f, g \in E$, define $f <_H g$ if for some $b \in B$ we have $f(b) < g(b)$ with $f(b') = g(b')$ for all $b' > b$.*

1. *Show that the relation $<_H$ linearly orders E.*
2. *Show that E is well-ordered by $<_H$ with order type α^β, where α and β are the order types of A and B, respectively.*

9.8 The Canonical Order on Pairs of Ordinals

Recall the lexicographic order on $W(\alpha) \times W(\alpha)$, of order type α^2. We will now define a different order $\lhd$ on pairs of ordinals called the *canonical order*.

We first partition $W(\alpha) \times W(\alpha)$ into α-many sets P_ξ, $\xi < \alpha$, where

$$P_\xi := \{\langle \mu, \nu \rangle \mid \max(\mu, \nu) = \xi\} \qquad (\xi < \alpha).$$

Note that $P_\xi = \big(W(\xi) \times \{\xi\}\big) \cup \big(\{\xi\} \times W(\xi)\big) \cup \{\langle \xi, \xi \rangle\}$ for each $\xi < \alpha$, and

$$\bigcup_{\xi<\alpha} P_\xi = W(\alpha) \times W(\alpha).$$

Thus the sets P_ξ ($\xi < \alpha$) are pairwise disjoint with union $W(\alpha) \times W(\alpha)$.

Now *each set* P_ξ gets well-ordered by the order it inherits from the lexicographic order on $W(\alpha) \times W(\alpha)$, under which its order type will be $\xi 2 + 1$.

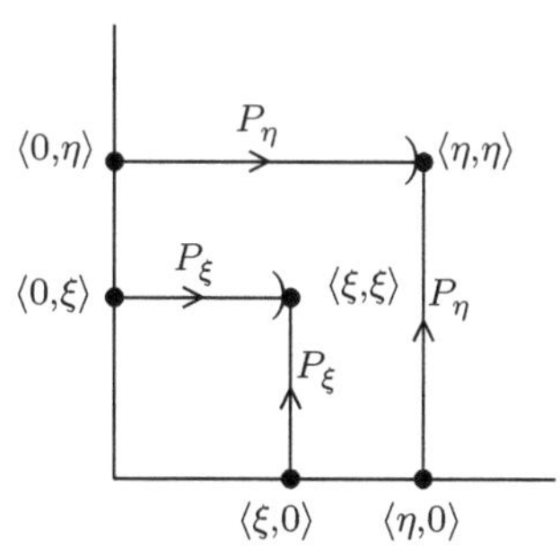

Finally, we define a new order on $W(\alpha)\times W(\alpha)$ by setting $P_\xi < P_\eta$ whenever $\xi < \eta$, and following the inherited lexicographic order within each set P_ξ. Put differently, this order is defined as follows.

Definition 676 (Canonical Order $\lhd$ on Pairs of Ordinals). The canonical order, denoted by $\lhd$, for pairs of ordinals is defined as:

$$\begin{aligned}\langle\mu,\nu\rangle \lhd \langle\gamma,\delta\rangle \Leftrightarrow & \max(\mu,\nu) < \max(\gamma,\delta),\\ & \text{or } \max(\mu,\nu) = \max(\gamma,\delta) \text{ and } \mu < \gamma,\\ & \text{or } \max(\mu,\nu) = \max(\gamma,\delta) \text{ and } \mu = \gamma \text{ and } \nu < \delta.\end{aligned}$$

Since each set P_ξ has order type $\xi 2 + 1$, we immediately get:

Corollary 677. *Under the canonical order $\lhd$, $W(\alpha)\times W(\alpha)$ has order type*

$$\operatorname{OrdTyp}_{\lhd}(W(\alpha)\times W(\alpha)) = \sum_{\xi<\alpha}(\xi 2 + 1),$$

and hence the canonical order is a well-order.

The canonical order on $W(\alpha)\times W(\alpha)$ has nicer properties than the lexicographic order on it.

Problem 678. *1. If $\beta < \alpha$, then $W(\beta)\times W(\beta)$ is an initial segment of $W(\alpha)\times W(\alpha)$ under the canonical order. In fact, we have*

$$W(\beta)\times W(\beta) = \operatorname{Pred}_{\lhd}(\langle 0,\beta\rangle).$$

2. Let $\Phi(\alpha) := \operatorname{OrdTyp}_{\lhd}(W(\alpha)\times W(\alpha)) = \sum_{\beta<\alpha}(\beta 2 + 1)$. Then Φ is a normal function.

Problem 679. *Show that both the properties in Problem 678 fail if the canonical order is replaced by the lexicographic order on $W(\alpha)\times W(\alpha)$. In fact:*

1. *If $1 < \beta < \alpha$ then $W(\beta)\times W(\beta)$ is not an initial segment of $W(\alpha)\times W(\alpha)$ under the lexicographic order.*
2. *If $\Psi(\alpha) := \alpha^2 =$ the order type of $W(\alpha)\times W(\alpha)$ under the lexicographic order, then Ψ is not a normal function.*

Theorem 680. *Let $\alpha > 2$ be a product-closed ordinal. Then the canonical order on $W(\alpha)\times W(\alpha)$ has order type α, i.e., $\operatorname{OrdTyp}_{\lhd}(W(\alpha)\times W(\alpha)) = \alpha$. Hence there is a unique bijection $\psi\colon W(\alpha)\times W(\alpha)\to W(\alpha)$ which preserves order: $\langle\mu,\nu\rangle \lhd \langle\xi,\eta\rangle \Leftrightarrow \psi(\langle\mu,\nu\rangle) < \psi(\langle\xi,\eta\rangle)$.*

Proof. Let $\alpha > 2$ be product-closed, and let $\rho := \operatorname{OrdTyp}_{\lhd}(W(\alpha)\times W(\alpha))$ be the order type of $W(\alpha)\times W(\alpha)$ under the canonical order. The mapping $\xi\mapsto\langle\xi,\xi\rangle$ is a strictly increasing injection (order embedding) of $W(\alpha)$ into $W(\alpha)\times W(\alpha)$, hence $\alpha\le\rho$.

Now let $\beta < \rho$ be given. Then there is $\langle \mu, \nu \rangle \in W(\alpha) \times W(\alpha)$ such that $\text{Pred}_{\lhd}(\langle \mu, \nu \rangle)$ has order type β. Since α is a limit ordinal we can fix δ with $\max(\mu, \nu) < \delta < \alpha$. Then $\text{Pred}_{\lhd}(\langle \mu, \nu \rangle) \subseteq W(\delta) \times W(\delta)$, so

$$\beta \leq \text{OrdTyp}_{\lhd}(W(\delta) \times W(\delta)) = \sum_{\gamma < \delta} (\gamma 2 + 1) \leq (\delta 2 + 1)\delta \leq (\delta 3)\delta < \alpha,$$

since α is product-closed. Thus $\beta < \alpha$ for all $\beta < \rho$, i.e., $\rho \leq \alpha$. So $\alpha = \rho$. □

9.9 The Cantor Normal Form

We generalize the following familiar fact about natural numbers to ordinals: Given a base $b > 1$, every natural number n can be uniquely expressed as

$$n = d_1 b^{p_1} + d_2 b^{p_2} + \cdots + d_k b^{p_k} \quad \begin{cases} p_1 > p_2 > \cdots > p_k \geq 0 \\ 0 < d_1, d_2, \ldots, d_k < b \end{cases}, \ k \in \mathbf{N}.$$

Theorem 681 (Expansion in Powers of a Base). *Let $\beta > 1$ be a fixed "base" ordinal. Then every ordinal $\alpha > 0$ can be expressed as the following "polynomial" in powers of β with nonzero coefficients less than β:*

$$\alpha = \beta^{\zeta_1}\eta_1 + \beta^{\zeta_2}\eta_2 + \cdots + \beta^{\zeta_k}\eta_k \quad \begin{cases} \zeta_1 > \zeta_2 > \cdots > \zeta_k \\ 0 < \eta_1, \eta_2, \ldots, \eta_k < \beta \end{cases}, k \in \mathbf{N}.$$

Proof. Let ν be the least ordinal such that $\beta^{\nu} > \alpha$ (such a ν must exist since $\beta^{\alpha+1} \geq \alpha+1 > \alpha$). Then ν must be a successor ordinal, since otherwise from $\beta^{\gamma} \leq \alpha$ for all $\gamma < \nu$ we would get $\beta^{\nu} \leq \alpha$ (by continuity in the exponent), contradicting $\beta^{\nu} > \alpha$. Hence $\nu = \zeta_1 + 1$ for some ζ_1 satisfying $\beta^{\zeta_1} \leq \alpha$, with ζ_1 being the largest ordinal for which $\beta^{\zeta_1} \leq \alpha$. By division algorithm,

$$\alpha = \beta^{\zeta_1}\eta_1 + \xi_1, \qquad \xi_1 < \beta^{\zeta_1}.$$

Then we get $0 < \eta_1 < \beta$, as $\beta \leq \eta_1$ would imply $\beta^{\zeta_1+1} = \beta^{\zeta_1}\beta \leq \beta^{\zeta_1}\eta_1 \leq \alpha$, contradicting the fact that ζ_1 is the largest ordinal for which $\beta^{\zeta_1} \leq \alpha$.

If now $\xi_1 = 0$, we are done. Otherwise $\xi_1 > 0$ and we repeat the above procedure with α replaced by ξ_1 to get

$$\xi_1 = \beta^{\zeta_2}\eta_2 + \xi_2, \qquad \xi_2 < \beta^{\zeta_2}.$$

Here again we must have $0 < \eta_2 < \beta$, and also $\zeta_2 < \zeta_1$. We thus have:

$$\alpha = \beta^{\zeta_1}\eta_1 + \beta^{\zeta_2}\eta_2 + \xi_2, \qquad \zeta_1 > \zeta_2, \quad 0 < \eta_1, \eta_2 < \beta.$$

Continuing in this fashion we see that the process must stop after a finite number of steps since otherwise we would get an infinite strictly decreasing sequence of ordinals $\zeta_1 > \zeta_2 > \cdots$. □

Note that when α and β are finite ordinals, the above theorem gives the usual representation of ordinary integers with respect to a base, e.g., with $\beta = 10$ we have decimal representation and with $\beta = 2$ we have binary representation.

The case where the base β equals ω is particularly important:

Corollary 682 (Cantor Normal Form). *Every ordinal $\alpha > 0$ can be expressed as a "polynomial in ω with integral coefficients":*

$$\alpha = \omega^{\zeta_1} n_1 + \omega^{\zeta_2} n_2 + \cdots + \omega^{\zeta_k} n_k \quad \begin{cases} \zeta_1 > \zeta_2 > \cdots > \zeta_k \\ 0 < n_1, n_2, \ldots, n_k < \omega \end{cases}, k \in \mathbf{N}.$$

Problem 683. *Every ordinal $\alpha < \omega^\omega$ can be uniquely expressed as the "polynomial in ω with integral exponents and coefficients":*

$$\alpha = \omega^{m_1} n_1 + \omega^{m_2} n_2 + \cdots + \omega^{m_k} n_k, \qquad m_1 > m_2 > \cdots > m_k,$$

with the exponents m_j and coefficients $n_j > 0$ all finite for $j = 1, 2, \ldots, k$.

Remark. The entire basic theory of ordinals and well-orders as presented in this chapter was created by Cantor [6].

Chapter 10
Alephs, Cofinality, and the Axiom of Choice

Abstract This chapter concludes our development of cardinals and ordinals. We introduce the first uncountable ordinal, the alephs and their arithmetic, Hartogs' construction, Zermelo's well-ordering theorem, the comparability theorem for cardinals, cofinality and regular, singular, and inaccessible cardinals, and the Continuum Hypothesis.

10.1 Countable Ordinals, ω_1, and $\aleph_1$

All orders we have encountered so far, including all uncountable orders such as **R** and all ordinals, have been of countable cofinality. (Recall that an order is said to have countable cofinality if it has a countable cofinal subset.) We will now define an ordinal of uncountable cofinality.

Recall that an ordinal is a *countable ordinal* if it is the order type of a well-order defined on a countable set, or equivalently if it is the order type of a well-ordered rearrangement of a subset of **N**. By Cantor's theorem on countable dense orders, every countable order is isomorphic to a suborder of **Q**, hence a countable ordinal could also be defined as an order type of a well-ordered suborder of **Q**. We thus have:

Proposition 684 (Countable Ordinals). *For any ordinal α, the following are equivalent.*

1. *α is a countable ordinal.*
2. *α is the order type of a well-ordered rearrangement of a subset of* **N**.
3. *α is the order type of a well-ordered suborder of* **Q**.
4. *$W(\alpha)$ is countable.*

All ordinals that we have seen so far are countable ordinals, such as:

$$0, 1, 2, \dots, \omega, \omega + 1, \dots, \omega + \omega, \dots, \omega^2, \dots \omega^{\omega}, \dots, \omega^{\omega^{\omega}}, \dots, \varepsilon_0, \dots, \varepsilon_1, \dots$$

A. Dasgupta, *Set Theory: With an Introduction to Real Point Sets*,
DOI 10.1007/978-1-4614-8854-5_10, © Springer Science+Business Media New York 2014

As we will see now, the set of all countable ordinals turns out to be an uncountable well-order. In other words, *the set of ordinals of well-ordered rearrangements of subsets of* **N** *is itself a well-order which is longer than any well-ordered rearrangement of a subset of* **N**.

Let C be the set of all countable ordinals. Then C is an initial set of ordinals (Definition 638), so we have $C = W(\omega_1)$, where ω_1 is the order type of C. Hence *α is a countable ordinal if and only if $\alpha < \omega_1$.*

Definition 685 (Cantor). ω_1 denotes the order type of the set of all countable ordinals, so that $\alpha \in W(\omega_1) \Leftrightarrow W(\alpha)$ is countable. Equivalently, ω_1 is the order type of the set of all ordinals of well-ordered suborders of **Q**.

Since $\omega_1 \notin W(\omega_1)$ and $W(\omega_1)$ contains all countable ordinals, it follows that ω_1 is not a countable ordinal, while every $\alpha < \omega_1$ is countable. Hence ω_1 *is the least uncountable ordinal.* Note also that ω_1 must be a limit ordinal, since the successor of a countable ordinal is a countable ordinal. $W(\omega_1)$ is thus an uncountable well-order without a greatest element, in which every proper initial segment is countable and every nonempty terminal segment is uncountable. These facts are summarized in the following:

Theorem 686. *ω_1 is the smallest uncountable ordinal and is a limit ordinal. The set $W(\omega_1)$ consists of all countable ordinals, and is an uncountable well-order in which every proper initial segment is countable.*

Assuming the countable axiom of choice, the uncountable well-order $W(\omega_1)$ also has *uncountable cofinality,* since limits of countable sequences of countable ordinals are countable ordinals. In other words, if we had a countable cofinal subset E of $W(\omega_1)$, then $W(\omega_1) = \cup_{\alpha\in E} W(\alpha)$ would be countable (being a countable union of countable sets, a fact which uses the countable axiom of choice), contradicting the uncountability of $W(\omega_1)$. This means the limit ordinal ω_1 cannot be expressed as a sequential limit of smaller ordinals:

$$\text{If } \alpha_n < \omega_1 \text{ for all } n < \omega, \text{ then } \lim_{n<\omega} \alpha_n < \omega_1.$$

Hence every cofinal (unbounded) subset of $W(\omega_1)$ is uncountable and so has order type ω_1. Conversely, any uncountable subset of $W(\omega_1)$ is cofinal. Thus:

Theorem 687 (CAC). *The well-order $W(\omega_1)$ has uncountable cofinality. A subset of $W(\omega_1)$ is cofinal (unbounded) if and only if it is uncountable if and only if it has order type ω_1.*

We say that a set E of ordinals is *closed under internal repeated additions* if whenever $\gamma \in E$ and $\beta_\alpha \in E$ for each $\alpha < \gamma$, then $\sum_{\alpha<\gamma} \beta_\alpha \in E$. We say that *$E$ is closed under internal sups* if whenever $\gamma \in E$ and $\beta_\alpha \in E$ for each $\alpha < \gamma$, then $\sup_{\alpha<\gamma} \beta_\alpha \in E$.

Problem 688. *Show that the set* $W(\omega) = \{0, 1, 2, \ldots\} = \{n \mid n < \omega\}$ *of all finite ordinals is closed under addition, multiplication, exponentiation, and internal repeated additions and internal sups.*

Problem 689 (CAC).

1. *Show that* $W(\omega^2) = \{\omega m + n \mid m, n < \omega\}$.
2. *Show that the set* $W(\omega^2) = \{\omega m + n \mid m, n < \omega\}$ *is the smallest set of ordinals containing* 0, 1, *and* ω, *and closed under addition.*
3. *Find the smallest set of ordinals containing* 0, 1, *and* ω, *and closed under both addition and multiplication.*
4. *Find the smallest set of ordinals containing* 0, 1, *and* ω, *and closed under addition, multiplication, and ordinal exponentiation.*
5. *Show that if a set of ordinals containing* 0, 1, *and* ω *is closed under addition and under taking internal sups, then it is closed under internal repeated additions, under multiplication, and under exponentiation. Find the smallest set of ordinals containing* 0, 1, *and* ω, *and closed under addition and taking internal sups.*
6. *Show that if a set of ordinals containing* 0, 1, *and* ω *is closed under internal repeated additions, then it is closed under addition, multiplication, and exponentiation. Find the smallest set of ordinals containing* 0, 1, *and* ω, *and closed under internal repeated additions.*
7. *Show that a set of ordinals which contains* 0 *and is closed under taking successors and countable limits must contain all countable ordinals.*

Problem 690. *Suppose that* $\alpha_\nu > 0$ *is a nonzero countable ordinal for each* $\nu < \omega_1$. *Show that* $\sum_{\nu<\omega_1} \alpha_\nu = \omega_1$. *Are you using the CAC?*

Earlier we saw that orders of type $(1 + \lambda + 1)^k \lambda$ ($k \in \mathbf{N}$) are examples of non-CCC continuums. We now have a different continuum which is not CCC.

Problem 691 (The Long Line). *Let* X *be an order of type* $\lambda + (1 + \lambda)\omega_1$, *and let* Y *be an order of type* $(1 + \lambda + 1)\lambda$. *Show that]*

1. *Both* X *and* Y *are non-CCC linear continuums without endpoints.*
2. *In* X *every nonempty bounded open interval has order type* λ *(and so is isomorphic to* $\mathbf{R}$*), but this is not true in the order* Y.
3. X *has no countable cofinal subset.*
4. *None of the continuums* X *and* Y *can be embedded in the other.*

10.2 The Cardinal $\aleph_1$

The set $W(\omega_1)$ of all countable ordinals has cardinality $> \aleph_0$, and this cardinal number is denoted by $\aleph_1$.

Definition 692 (Cantor). $\aleph_1 := |W(\omega_1)|$.

Problem 693. *A subset of $W(\omega_1)$ is either countable or has cardinality $\aleph_1$.*

So the cardinal $\aleph_1$ is the next larger cardinal after $\aleph_0$ in the following sense:

Problem 694. $\aleph_0 < \aleph_1$ *and there is no cardinal κ such that $\aleph_0 < \kappa < \aleph_1$.*

Problem 695. *Prove that*

1. $n + \aleph_1 = \aleph_1$ *for any finite cardinal n.*
2. $\aleph_0 + \aleph_1 = \aleph_1$.
3. $\aleph_1 + \aleph_1 = \aleph_1$. *[Hint: Use even and odd ordinals.]*
4. $\aleph_1\aleph_0 = \aleph_1$.

Theorem 696. $\aleph_1^2 = \aleph_1$.

Proof. The ordinal ω_1 is product-closed, i.e., $\alpha, \beta < \omega_1 \Rightarrow \alpha\beta < \omega_1$ (since if $W(\alpha), W(\beta)$ are countable then so is $W(\alpha) \times W(\beta)$). Hence by Theorem 680, the canonical order $\lhd$ on $W(\omega_1) \times W(\omega_1)$ has order type ω_1, and so there is a bijection (order isomorphism) from $W(\omega_1) \times W(\omega_1)$ onto $W(\omega_1)$. □

Note that the above is an effective proof of $\aleph_1^2 = \aleph_1$, *without any use of the Axiom of Choice.* A variant proof is obtained using the following problem.

Problem 697. *Show that $f : W(\omega_1) \times W(\omega_1) \to W(\omega_1)$ defined by*

$$f(\alpha, \beta) = \begin{cases} 2\alpha^2 & \text{if } \beta = \alpha, \\ 2(\alpha^2 + \beta) + 1 & \text{if } \beta < \alpha, \\ 2(\beta^2 + \alpha) + 2 & \text{if } \beta > \alpha. \end{cases}$$

is an injection from $W(\omega_1) \times W(\omega_1)$ into $W(\omega_1)$.

One can then use the Cantor–Bernstein theorem to combine the above mapping f with the injection $\alpha \to \langle \alpha, 0 \rangle$ from $W(\omega_1)$ to obtain an effective bijection between $W(\omega_1) \times W(\omega_1)$ and $W(\omega_1)$.

Using the above results, we get many more well-orders of cardinality $\aleph_1$. For example, orders of type $\omega_1 + \omega$, or $\omega_1 2$, or ω_1^2, all have cardinality $\aleph_1$. Note that many of these orders (e.g., any order of type $\omega_1 + \omega$) are uncountable but have countable cofinal subsets.

Closed Unbounded Subsets of $W(\omega_1)$

Recall that a subset A of an order is called closed if A contains all its limit points (Definition 592). In a well-order there are no lower limit points and so A is closed if A contains all its upper limit points, or equivalently if $\sup E \in A$ for all nonempty bounded $E \subseteq A$.

Definition 698 (Club Sets). A subset of $W(\omega_1)$ is called a *club set* if is both closed and unbounded above.

Proposition 699. *The intersection of countably many club sets is a club set.*

Proof. Let $A_1, A_2, \ldots,$ be club sets and let $A = \bigcap_n A_n$. It is easy to see that A is closed, since if x is any (upper) limit point of A then x is an upper limit point of A_n for all n, and so (since A_n is closed) $x \in A_n$ for all n. To see that A is unbounded, fix any countable ordinal μ. Fix also a function $g: \mathbf{N} \to \mathbf{N}$ such that for any $n \in \mathbf{N}$ there are infinitely many m with $g(m) = n$, e.g., g may be taken to be the sequence

$$1,\ 1,\ 2,\ 1,\ 2,\ 3,\ 1,\ 2,\ 3,\ 4,\ 1,\ 2,\ 3,\ 4,\ 5,\ \ldots$$

Since each A_n is unbounded, for any finite set F of countable ordinals there exists $\alpha \in A_n$ such that $\alpha > \xi$ for all $\xi \in F$. Using this we can inductively choose ordinals α_n, for each n, such that each $\alpha_n \in A_{g(n)}$ and is greater than all preceding elements:

$$\mu < \alpha_1 < \alpha_2 < \alpha_3 < \cdots \qquad \text{with } \alpha_n \in A_{g(n)} \text{ for all } n.$$

Let $\alpha := \sup_n \alpha_n$. Then α is a limit point of A_n for all n, hence $\alpha \in A_n$ for all n, and so $\alpha \in \bigcap_n A_n = A$. □

Problem 700. *Prove that if $f: W(\omega_1) \to \mathbf{R}$ is continuous, then f must be eventually constant, i.e., there exists $\alpha < \omega_1$ such that $f(\beta) = f(\alpha)$ for all $\beta > \alpha$.*

[Hint: Put $E[x] := \{\alpha \mid f(\alpha) \geq x\}$ and $F[x] := \{\alpha \mid f(\alpha) \leq x\}$, which are closed subsets of $W(\omega_1)$ for $x \in \mathbf{R}$ (Problem 597). Put $L := \{x \in \mathbf{R} \mid E[x]$ is uncountable$\}$, which is nonempty and bounded above since $\bigcup_{n \in \mathbf{Z}} E[n] = W(\omega_1)$ while $\bigcap_{n \in \mathbf{Z}} E[n] = \emptyset$. Finally, put $p := \sup L$, and show that both $E[p + \frac{1}{n}]$ and $F[p - \frac{1}{n}]$ must be countable for all $n \in \mathbf{N}$.]

10.3 Hartogs' Theorem, Initial Ordinals, and Alephs

The process by which we constructed ω_1 and $\aleph_1$ from ω and $\aleph_0$ can be iterated further as follows. Consider the set of order types of well-orders defined on subsets of $W(\omega_1)$. This is the set of ordinals of well-orders defined on sets of cardinality $\leq \aleph_1$ and is an initial set of ordinals (Definition 638). Hence it equals $W(\omega_2)$ for an ordinal ω_2. Then ω_2 cannot be the ordinal of a well-order of cardinality $\leq \aleph_1$ (since $\omega_2 \notin W(\omega_2)$). Thus $W(\omega_2)$ has cardinality $> \aleph_1$, and ω_2 is the type of a well-order on a set of cardinality $> \aleph_1$ (and is the least such ordinal). We denote the cardinality of the set $W(\omega_2)$ by $\aleph_2$, which gives us a cardinal greater than $\aleph_1$. We can continue this process to get ω_3 and $\aleph_3$, and so on.

The cardinalities of the sets $W(\alpha)$, for various ordinals α, start as:

$$|W(0)| = 0 < |W(1)| = 1 < \cdots < |W(\omega)| = \aleph_0.$$

After this, we have $|W(\alpha)| = \aleph_0$ for all uncountably many α satisfying $\omega \leq \alpha < \omega_1$, since the ordinals in $W(\omega_1) \setminus W(\omega)$ are the types of infinite countable well-orders (this is called the *second number class* by Cantor). The next ordinal α for which we have the $|W(\alpha)| > |W(\beta)|$ for all $\beta < \alpha$ is $\alpha = \omega_1$.

Definition 701 (Initial Ordinals). An ordinal α is an *initial ordinal* if the cardinality of $W(\alpha)$ is greater than that of $W(\beta)$ for all $\beta < \alpha$.

Thus all finite ordinals and the ordinal ω are initial ordinals. The next initial ordinal is ω_1, the following one is ω_2, etc. Note also that if E is a set of initial ordinals, then $\sup_E$ is also an initial ordinal. To generalize, we first define:

Definition 702. If α is an ordinal and A is a set, we write $\alpha \preceq A$ to mean that α is the ordinal of a well-order defined on some subset of A. In other words, $\alpha \preceq A$ if there is an injection from $W(\alpha)$ into A.

For example, $\alpha \preceq \mathbf{N}$ if and only if α is a countable ordinal.

Definition 703 (The Hartogs Set $H(A)$, Ordinal $\omega(A)$, and Cardinal $\aleph(A)$ of a Set). Let A be any set. We define $H(A)$, *the Hartogs set of* A, to be the set of order types of well-orders defined on subsets of A, so that $H(A) = \{\alpha \mid \alpha \preceq A\}$. The order type of $H(A)$, denoted by $\omega(A)$, is called the *Hartogs ordinal of* A, and the cardinality of $H(A)$, denoted by $\aleph(A)$, is called the *Hartogs cardinal of* A.

Theorem 704 (Hartogs' Theorem). *For any set* A*:*

1. *$H(A)$ is an initial set of ordinals, with $H(A) = W(\eta) = \{\alpha \mid \alpha < \eta\}$, where $\eta = \omega(A)$ is the Hartogs ordinal of A.*
2. *$H(A)$ is not equinumerous to any subset of A, and so $\aleph(A) \not\leq |A|$.*
3. *$\omega(A)$ is an initial ordinal with $\omega(A) \not\preceq A$, and so if A is well-ordered with ordinal α then $\omega(A) > \alpha$ and is in fact the least initial ordinal $> \alpha$.*
4. *If A can be well-ordered then $\aleph(A)$ is the next larger cardinal after $|A|$, that is $\aleph(A) > |A|$ and there is no cardinal κ such that $|A| < \kappa < \aleph(A)$.*

Problem 705. *Prove Theorem 704.*

If $|A| = |B|$ then clearly $H(A) = H(B)$. Thus $H(A)$, $\omega(A)$, and $\aleph(A)$ depend only on the cardinality of A. Hence the following definition makes sense.

Definition 706 (κ^+ and $\omega^+(\alpha)$). For a cardinal κ and an ordinal α, define

1. $\kappa^+ := \aleph(A) = |H(A)|$, where A is a set of cardinality κ.
2. $\omega^+(\alpha) := \omega(W(\alpha)) =$ the Hartogs ordinal of $\{\beta \mid \beta < \alpha\}$.

Thus for any cardinal κ, we have $\kappa^+ \not\leq \kappa$, and if κ is the cardinality of a set which can be well-ordered then κ^+ is the least cardinal greater than κ. If α is an ordinal then $\omega^+(\alpha)$ is the least initial ordinal $> \alpha$.

Definition 707 (ω_α **and** $\aleph_\alpha$). For every ordinal α we define the ordinal ω_α by transfinite recursion as:

$$\begin{aligned}\omega_0 &:= \omega,\\ \omega_{\alpha+1} &:= \omega^+(\omega_\alpha) = \text{ least initial ordinal} > \omega_\alpha,\\ \omega_\xi &:= \sup_{\gamma<\xi} \omega_\gamma, \quad \text{if } \xi \text{ is a limit ordinal.}\end{aligned}$$

Finally, define $\aleph_\alpha := |W(\omega_\alpha)| = |\{\nu \mid \nu < \omega_\alpha\}|$.

Theorem 708. *1. Every ω_α is an initial ordinal with $|W(\omega_\alpha)| = \aleph_\alpha$.*
2. $\alpha < \beta \Rightarrow \omega_\alpha < \omega_\beta$ and so $\alpha < \beta \Rightarrow \aleph_\alpha < \aleph_\beta$.
3. $\aleph_{\alpha+1} = \aleph_\alpha^+$, and if ν is a limit ordinal then $\aleph_\nu = \sup_{\alpha<\nu} \aleph_\alpha$.
4. Every infinite initial ordinal equals ω_α for some ordinal α.
5. If an infinite set A can be well-ordered, then $|A| = \aleph_\alpha$ for an ordinal α.

Problem 709. *Prove Theorem 708.*

We thus have the series of *all* initial ordinals as:

$$0 < 1 < 2 < \cdots < \omega < \omega_1 < \omega_2 < \cdots < \omega_\alpha < \cdots,$$

where, after the finite ordinals, $\omega = \omega_0$ is the only countably infinite initial ordinal and ω_α is the α-th uncountable initial ordinal for $\alpha > 0$.

The infinite cardinals in the series

$$\aleph_0 < \aleph_1 < \aleph_2 < \cdots < \aleph_\alpha < \cdots$$

are called *alephs*. By the last part of Theorem 708, the above *series of alephs* gives a well-ordered enumeration, indexed by the ordinals, of all infinite cardinals of well-orderable sets.

We will soon see that every set can be well-ordered using the Axiom of Choice (the well-ordering theorem). Hence, under the Axiom of Choice, *every infinite set has cardinality $\aleph_\alpha$ for some ordinal α* and *every infinite cardinal is an aleph.* Thus, for $\alpha > 0$, *$\aleph_\alpha$ is the α-th uncountable cardinal* (under AC).

Earlier we had proved that $\aleph_1^2 = \aleph_1$. Essentially the same proof yields:

$$\aleph_\alpha^2 = \aleph_\alpha, \qquad \text{and so under AC:} \quad \kappa^2 = \kappa$$

for any infinite cardinal κ. From these relations and the Cantor–Bernstein theorem, the sum and product of any two alephs are determined completely as follows:

Theorem 710. $\aleph_\alpha + \aleph_\beta = \aleph_\alpha\aleph_\beta = \aleph_{\max(\alpha,\beta)} = \max(\aleph_\alpha, \aleph_\beta)$.

We also have, by induction, $\aleph_\alpha^n = \aleph_\alpha$ for any nonzero finite cardinal n.

On the other hand, when the exponent is infinite, it is impossible to compute cardinal powers such as $\aleph_\alpha^{\aleph_\beta}$ or $2^{\aleph_\beta}$ as an aleph even using AC. (Without AC, we cannot prove that such a cardinal is an aleph.)

Problem 711. *Show by examples that for arbitrary orders X:*

1. *The Bolzano–Weierstrass property does not imply the strong Nested Intervals property.*
2. *The strong Nested Intervals property does not imply the Bolzano–Weierstrass property.*
3. *The Bolzano–Weierstrass property together with the strong Nested Intervals property does not imply completeness.*
4. *The sequential NIP does not imply that either the strong Nested Intervals property or the Bolzano–Weierstrass property holds in X.*

[Hint: Consider orders of type $\omega + {}^*\omega_1$, $\omega_1 + {}^*\omega_1$, $\omega_1 + {}^*\omega_2$, etc.]

10.4 Abstract Derivatives and Ranks

Definition 712 (Derivative Operators). Let X be any set. A mapping $\nabla\colon \mathbf{P}(X) \to \mathbf{P}(X)$ will be called a *derivative operator* if $\nabla(E) \subseteq E$ for all $E \subseteq X$. (A derivative may also be referred to as a *contraction* or *reduction*.)

An example of a derivative operator in the context of orders is the following: Put $\nabla(E) := E \cap D(E)$, where $D(E)$ is the set of limit points of E. In other words $\nabla(E)$ is obtained from E by removing the isolated points of E. This is the Cantor–Bendixson derivative, and gives rise to Cantor–Bendixson ranks, which will be studied in detail later. *The notion of derivative as defined here is an abstract version of the Cantor–Bendixson derivative.*

Definition 713 (Strict Derivatives). A derivative operator $\nabla\colon \mathbf{P}(X) \to \mathbf{P}(X)$ will be called *strict* if $\nabla(E) \subsetneq E$ whenever $E \neq \emptyset$.

One can naturally iterate a derivative operator and define by transfinite recursion sets $X^{(\alpha)}$, where α is any ordinal, as follows:

$$\begin{aligned} X^{(0)} &:= X, \\ X^{(\alpha+1)} &:= \nabla\big(X^{(\alpha)}\big), \quad \text{and,} \\ X^{(\xi)} &:= \bigcap_{\alpha<\xi} X^{(\alpha)}, \quad \text{if } \xi \text{ is a limit ordinal.} \end{aligned}$$

The sets $X^{(\alpha)}$ decrease with α so that we have

$$X = X^{(0)} \supseteq X^{(1)} \supseteq \cdots \supseteq X^{(\alpha)} \supseteq X^{(\alpha+1)} \supseteq \cdots,$$

but the following theorem shows that the process must "stabilize" at an ordinal, i.e., there is μ such that $X^{(\mu+1)} = X^{(\mu)}$ (and so $X^{(\alpha)} = X^{(\mu)}$ for all $\alpha > \mu$). It is a theorem which, in an abstract setting, assigns ordinal ranks to certain elements of the set X without using the Axiom of Choice. We will have many occasions to use the general framework of this theorem.

Theorem 714 (Rank Decomposition for Derivative Operators). *Let X be a set, $\nabla\colon \mathsf{P}(X) \to \mathsf{P}(X)$ be a derivative operator, and $\eta = \omega(\mathsf{P}(X))$ be the Hartogs ordinal for $\mathsf{P}(X)$. For each $\alpha < \eta$, define the set $X^{(\alpha)}$, called the α-th iterated derivative of X, by transfinite recursion as follows.*

$$
\begin{aligned}
X^{(0)} &:= X, \\
X^{(\alpha+1)} &:= \nabla\big(X^{(\alpha)}\big), \qquad \textit{and,} \\
X^{(\xi)} &:= \bigcap_{\alpha<\xi} X^{(\alpha)}, \qquad \textit{if } \xi \textit{ is a limit ordinal.}
\end{aligned}
$$

Then

1. *The sets $X^{(\alpha)}$ decrease with α, and there exists a unique least ordinal $\mu < \eta$ for which $X^{(\mu)} = X^{(\mu+1)}$, so that $X^{(\alpha)} = X^{(\mu)}$ for all $\alpha > \mu$, but $X^{(\alpha)} \subsetneq X^{(\mu)}$ for $\alpha < \mu$:*

$$X = X^{(0)} \supsetneq X^{(1)} \supsetneq \cdots X^{(\alpha)} \supsetneq X^{(\alpha+1)} \supsetneq \cdots X^{(\mu)} = X^{(\mu+1)} = X^{(\infty)},$$

where $X^{(\infty)}$ denotes the "stabilized smallest set $X^{(\mu)}$" among the $X^{(\alpha)}$.

2. *The set $X \setminus X^{(\infty)} = X \setminus X^{(\mu)}$ is partitioned as:*

$$X \setminus X^{(\infty)} = \bigcup_{\alpha<\mu} X^{(\alpha)} \setminus X^{(\alpha+1)}, \qquad \textit{with } X^{(\alpha)} \setminus X^{(\alpha+1)} \neq \emptyset \textit{ for all } \alpha < \mu.$$

3. *Consequently, if for each $x \in X \setminus X^{(\infty)}$ we put $\rho(x) = \rho_\nabla(x) :=$ the least $\alpha < \mu$ such that $x \in X^{(\alpha)} \setminus X^{(\alpha+1)}$, then $\rho = \rho_\nabla\colon X \setminus X^{(\infty)} \to W(\mu)$ is the unique "ordinal rank function" such that for any $x \in X \setminus X^{(\infty)}$ and any ordinal α:*

$$\rho(x) = \alpha \;\Leftrightarrow\; x \in X^{(\alpha)} \setminus X^{(\alpha+1)}.$$

If $\rho(x) = \alpha$, we say that the element x has rank α, and thus $X^{(\alpha)} \setminus X^{(\alpha+1)}$ consists precisely of the elements of rank α. (Put $\rho(x) = \infty$ if $x \in X^{(\infty)}$.)

4. *If ∇ is a strict derivative then $X^{(\mu)} = X^{(\infty)} = \emptyset$, and so $\mathrm{dom}(\rho) = X$, i.e., $\rho\colon X \to W(\mu)$, and every element in X has an ordinal rank.*

Proof. 1. The fact that the sets $X^{(\alpha)}$ decrease is immediate from their definition via a routine transfinite induction. If we had $X^{(\alpha)} \neq X^{(\alpha+1)}$ for all $\alpha < \eta$, then the mapping $\alpha \mapsto X^{(\alpha)}$ is easily seen to be a one-to-one mapping from $W(\eta)$

into $\mathbf{P}(X)$, which is impossible since η is the Hartogs ordinal for $\mathbf{P}(X)$. Hence $X^{(\alpha)} = X^{(\alpha+1)}$ for some ordinal $\alpha < \eta$, and we can therefore take μ to be the least such ordinal.
2. This is immediate from the definition of the sets $X^{(\alpha)}$.
3. The function $\rho = \rho_\nabla$ is readily defined as the relation:

$$\rho_\nabla := \{\langle x, \alpha\rangle \in X \times W(\mu) \mid x \in X^{(\alpha)} \smallsetminus X^{(\alpha+1)}\}.$$

4. If ∇ is strict, then $X^{(\infty)} \neq \emptyset$ would contradict $\nabla(X^{(\infty)}) = X^{(\infty)}$. □

Remarks. (1) Note that the theorem does not use the Axiom of Choice. (2) The relation on $X \smallsetminus X^{(\infty)}$ defined by $\rho(x) \leq \rho(y)$ is called a *pre-well-ordering.*

Problem 715 (Monotone Operators). *A mapping* $\nabla: \mathbf{P}(X) \to \mathbf{P}(X)$ *is called* monotone *if* $A \subseteq B \Rightarrow \nabla(A) \subseteq \nabla(B)$. *Show that if* ∇ *is a derivative on* X *which is also monotone, then the set* $X^{(\infty)}$ *is the largest fixed point of* ∇, *that is, we have* $\nabla(X^{(\infty)}) = X^{(\infty)}$, *and if* $\nabla(E) = E$ *then* $E \subseteq X^{(\infty)}$.

Problem 716. *Show that if a derivative* $\nabla: \mathbf{P}(X) \to \mathbf{P}(X)$ *has the property that* $|E \smallsetminus \nabla(E)| \leq 1$ *for all* E, *then the rank function* $\rho = \rho_\nabla$ *must be one-to-one on* $X \smallsetminus X^{(\infty)}$. *If, in addition,* ∇ *is strict, then* $\rho: X \to W(\mu)$ *is a bijection from* X *onto the set* $W(\mu)$ *of ordinals* $< \mu$.

10.5 AC, Well-Ordering Theorem, Cardinal Comparability

Recall the two versions of Axiom of Choice that we have seen earlier:

- *Axiom of Choice, Partition Version.* Every partition has a choice set. If $\mathcal{P}$ is a family of pairwise disjoint nonempty sets then there is a "choice set" C for the partition $\mathcal{P}$ satisfying $|C \cap E| = 1$ for every $E \in \mathcal{P}$.
- *Axiom of Choice, Choice Function Version.* Every family of nonempty sets has a choice function. Equivalently, for any set A there is choice function $\varphi: \mathbf{P}^*(A) \to A$ with $\varphi(E) \in E$ for all $E \in \mathbf{P}^*(A)$, where $\mathbf{P}^*(A) := \mathbf{P}(A) \smallsetminus \{\emptyset\}$ is the collection of all nonempty subsets of a set A.

We had seen that the two forms are equivalent. In the theorem below, we use the Choice Function version to show that the Axiom of Choice implies the *well-ordering theorem,* which says that every set can be well-ordered.

Theorem 717 (Zermelo's Well-Ordering Theorem). *The Axiom of Choice implies the well-ordering theorem (that every set can be well-ordered).*

Proof. Let A be an arbitrary set with a choice function $\varphi: \mathbf{P}^*(A) \to A$. The idea of the proof is to use the choice function φ to well-order A as follows: Let

$$a_0 = \varphi(A), \quad a_1 = \varphi(A \smallsetminus \{a_0\}), \quad a_2 = \varphi(A \smallsetminus \{a_0, a_1\}), \text{ etc.,}$$

and in general we keep putting

$$a_\alpha = \varphi(A \smallsetminus \{a_\beta \mid \beta < \alpha\}),$$

until we exhaust A. We now formalize this using the framework of abstract iterated derivatives and ranks (Sect. 10.4, Theorem 714).

Define a strict derivative operator $\nabla \colon \mathbf{P}(A) \to \mathbf{P}(A)$ by

$$\nabla(E) := \begin{cases} E \smallsetminus \{\varphi(E)\} & \text{if } E \neq \emptyset, \\ E & \text{if } E = \emptyset. \end{cases}$$

Let $A^{(\alpha)}$ denote the α-th iterated derivative of A, and μ be the least ordinal with $A^{(\mu)} = A^{(\mu+1)} = A^{(\infty)}$. Then $A^{(\infty)} = \emptyset$ since ∇ is strict, and so there is a rank function $\rho = \rho_\nabla \colon A \to W(\mu)$ such that for all $x \in A$, $\rho(x) = \alpha$ (i.e., x has rank α) if and only if $x \in A^{(\alpha)} \smallsetminus \nabla(A^{(\alpha)})$. Since $A^{(\alpha)} \smallsetminus \nabla(A^{(\alpha)})$ has at most one member $\varphi(A^{(\alpha)})$, no two distinct elements can have the same rank, i.e., the rank function ρ is injective (in fact a bijection from A onto $W(\mu)$). So the ordering defined on A by the relation

$$x < y \;\Leftrightarrow\; \rho(x) < \rho(y) \qquad (x, y \in A)$$

is a well-ordering of A (of order type μ). □

Since the well-ordering theorem clearly implies the Axiom of Choice, we have:

Corollary 718. *The Axiom of Choice is equivalent to the well-ordering theorem.*

If A and B are any two sets then using the well-ordering theorem we can well-order each of them, and consequently by comparability of well-orders one of them must be isomorphic to an initial segment of the other; in particular, there is an injection from one of the sets into the other. Thus we have:

Theorem 719 (Cardinal Comparability). *The well-ordering theorem implies that for any two sets one of them is equinumerous to a subset of the other, and therefore that cardinal comparability holds: For any two cardinals κ and μ either $\kappa \leq \mu$ or $\mu \leq \kappa$, and thus exactly one of $\kappa < \mu$, $\kappa = \mu$, $\kappa > \mu$ is true.*

Conversely, cardinal comparability implies the well-ordering theorem. To see this, let A be any set and let $H(A)$ be the Hartogs set of all ordinals of well-orderings defined on subsets of A. Then by Hartogs' theorem $H(A)$ is not equinumerous with any subset of A and so by cardinal comparability A is equinumerous with some subset of $H(A)$, which is well-ordered. Hence A itself can be well-ordered. It follows that:

Theorem 720. *The well-ordering theorem is equivalent to cardinal comparability.*

The above results can be summarized as:

Theorem 721 (Equivalents of AC). *Without using the Axiom of Choice, we can prove that any of the following assertions implies the others:*

1. *The Axiom of Choice (either the partition or the choice function version).*
2. *The Well-Ordering Theorem: Any set can be well-ordered.*
3. *Cardinal Comparability: If κ, μ are cardinals then either $\kappa \le \mu$ or $\mu \le \kappa$.*

Note that by the well-ordering theorem, any set is equinumerous with $W(\alpha)$ for some α, and hence to $W(\beta)$ for some initial ordinal $W(\beta)$, and so the cardinal number of any infinite set is an aleph. Thus we have a *well-ordered enumeration of all cardinals* as

$$0 < 1 < 2 < \cdots < \aleph_0 < \aleph_1 < \cdots < \aleph_\omega < \cdots < \aleph_\alpha < \cdots$$

Since every infinite cardinal is an aleph and the alephs are well-ordered under the relation $<$ for comparing cardinals, it follows that any set of cardinals is well-ordered. This allows us to use phrases like "the least cardinal with such and such property." Moreover, any set of cardinals has a unique cardinal as the least upper bound (supremum) of the given set.

All the facts of the last paragraph assume the Axiom of Choice, under which the infinite cardinals (as alephs) correspond naturally to the infinite initial ordinals ω_α in a one-to-one fashion, allowing us to informally identify the infinite cardinals with the infinite initial ordinals ω_α.

Corollary 722 (AC). *Under the Axiom of Choice, every infinite cardinal is an aleph, and therefore for any two infinite cardinals κ and μ, we have $\kappa + \mu = \kappa\mu = \max(\kappa, \mu)$. In particular, $\kappa^2 = \kappa$ for all infinite cardinals κ.*

Problem 723. *AC holds if and only if $\kappa < \kappa^+$ for every infinite cardinal κ.*

Problem 724 (Tarski). *The Axiom of Choice follows from the assumption that $\kappa^2 = \kappa$ for all infinite cardinals κ.*

[Hint: Given any infinite set A, fix a well-ordered B with $|B| = \aleph(A) = |A|^+$ and $A \cap B = \emptyset$. If $\kappa^2 = \kappa$ for all κ, then $|A||B| \leqslant |A| + |B|$, so there is an injection $f: A \times B \to A \cup B$. Now $f[\{a\}\times B] \cap B \neq \emptyset$ for all $a \in A$, so we get an injective $g: A \to B$, where $g(a) :=$ the least element in $f[\{a\}\times B] \cap B$.]

10.6 Cofinality: Regular and Inaccessible Cardinals

Proposition 725. *Let X be a nonempty order without a greatest element and suppose that $|X| = \aleph_\alpha$. Then X has a well-ordered cofinal subset whose order type is $\le \omega_\alpha$. In particular, any countable order without a greatest element has a cofinal subset of type ω.*

Proof. Since $|X| = \aleph_\alpha$, we can enumerate the elements of X indexed by ordinals $< \omega_\alpha$ as

$$X = \{a_\nu \mid \nu < \omega_\alpha\}.$$

Define the subset I of $W(\omega_\alpha)$ by

$$I := \{\nu \mid \nu < \omega_\alpha \text{ and } a_\nu > a_\xi \text{ in } X \text{ for all } \xi < \nu\}.$$

I is nonempty since $0 \in I$, and if $\xi < \nu$ are in I then $a_\xi < a_\nu$ in X. Hence the suborder C of X defined by

$$C := \{a_\nu \mid \nu \in I\}$$

has the same order type as I, and so is well-ordered with type $\leq \omega_\alpha$.

Finally, C is cofinal in X. For otherwise, we could get the least $\beta < \omega_\alpha$ with $a_\beta > x$ in X for all $x \in C$. Then for any $\xi < \beta$ if $\xi \in I$ then $a_\xi \in C$ so we would have $a_\xi < a_\beta$, and if $\xi \notin I$ then there would be the least $\mu < \xi$ such that $a_\mu > a_\xi$ and for this μ we must have $a_\mu \in C$ which implies $a_\xi < a_\mu < a_\beta$. In either case, we have $a_\xi < a_\beta$ in X for any $\xi < \beta$. Hence $\beta \in I$, so $a_\beta \in C$, contrary to our assumption. □

Corollary 726. *Let X be an order without a last element. If X has countable cofinality, then X has a cofinal subset of order type ω, that is, there exists a strictly increasing sequence $x_1 < x_2 < \cdots < x_n < \cdots$ in X such that $\{x_n \mid n \in \mathbf{N}\}$ is cofinal in X.*

Corollary 727. *Let X be a nonempty well-order without a largest element. Then the least ordinal μ such that X has a cofinal subset of type μ is an infinite initial ordinal $\mu = \omega_\alpha$.*

Proof. Let C be a cofinal subset of X of order type μ and let $|C| = \aleph_\alpha$ so that $\omega_\alpha \leq \mu$. If μ were not an initial ordinal, we would have $\omega_\alpha < \mu$. By the proposition, there is $E \subseteq C$ such that E is cofinal in C and the order type of E is $\leq \omega_\alpha$. Since E is cofinal in C and C is cofinal in X, therefore E is cofinal in X. Hence X has a cofinal subset of type $\leq \omega_\alpha < \mu$, a contradiction. □

From the corollary it follows that for every well-order X there is a unique smallest cardinal μ such that X has a cofinal subset of cardinality μ.

Definition 728 (Cofinality of Well-Orders and Ordinals). The *cofinality of a well-order* X is the least cardinal μ such that X has a cofinal subset of cardinality μ. The *cofinality of an ordinal* α is the cofinality of the well-order $W(\alpha) = \{\beta \mid \beta < \alpha\}$.

From the definition and the previous corollary it follows that if α is the ordinal of a nonempty well-order X without a largest element, then the cofinality of α equals $\aleph_\mu$ if and only if X has a cofinal subset of type ω_μ but of no smaller type.

The cofinality of any successor ordinal is 1. For a limit ordinal α, the cofinality of α equals $\aleph_\mu$ if and only if the least order type of subsets cofinal in $W(\alpha)$ (which must be an initial ordinal) equals ω_μ. In particular, the cofinality of any countable limit ordinal is $\aleph_0$, while the cofinality of ω_1 is $\aleph_1$.

For the rest of this section we will assume the Axiom of Choice so that every cardinal κ equals an aleph $\kappa = \aleph_\alpha$.

Definition 729 (Cofinality of Cardinals (AC)). The *cofinality of a cardinal* $\kappa = \aleph_\alpha$, denoted by $\mathrm{cf}(\kappa)$, is the cofinality of the ordinal ω_α, i.e., it is the least cardinal μ such that $W(\omega_\alpha)$ has a cofinal subset of cardinality μ.

Note that the definition of cofinality for cardinals *requires* the Axiom of Choice.

Theorem 730 (AC). *For any cardinal* $\kappa > 0$,

1. $\mathrm{cf}(\kappa) \leq \kappa$.
2. $\mathrm{cf}(\mathrm{cf}(\kappa)) = \mathrm{cf}(\kappa)$.

Proof. The first part is immediate. For the second part, suppose that $\kappa = \aleph_\alpha$, $\mathrm{cf}(\kappa) = \aleph_\mu$, and $\mathrm{cf}(\aleph_\mu) = \aleph_\nu$, so that $\kappa \geq \mu \geq \nu$. Then $W(\omega_\alpha)$ has a cofinal subset C of order type ω_μ, but of no smaller type. Since $\mathrm{cf}(\aleph_\mu) = \aleph_\nu$, so $W(\omega_\mu)$, and hence the isomorphic order C, has a cofinal subset of order type ω_ν. But if E is cofinal in C of order type ω_ν, then E will be cofinal also in $W(\omega_\alpha)$, and so ω_μ will be $\leq$ the order type of E which is ω_ν. Hence $\omega_\mu = \omega_\nu$, and so $\aleph_\mu = \aleph_\nu$. □

Problem 731. *If* α *is a successor ordinal, then* $\mathrm{cf}(\aleph_\alpha) = \aleph_\alpha$. *If* α *is a limit ordinal, then* $\mathrm{cf}(\aleph_\alpha)$ *equals the cofinality of* α.

[Hint: For the first part use the fact that $\aleph_\xi^2 = \aleph_\xi$.]

Thus $\mathrm{cf}(\aleph_0) = \aleph_0$, $\mathrm{cf}(\aleph_1) = \aleph_1$, while $\mathrm{cf}(\aleph_\omega) = \aleph_0$.

The following definition is based on cofinalities of cardinals and therefore assumes the Axiom of Choice.

Definition 732 (Successor, Limit, Regular and Singular Cardinals). An infinite cardinal κ is a *successor cardinal* if $\kappa = \mu^+$ for some cardinal μ; otherwise κ is a *limit cardinal.* κ is *regular cardinal* if $\mathrm{cf}(\kappa) = \kappa$; otherwise κ is a *singular cardinal.*

Thus $\aleph_\alpha$ is a successor cardinal if and only if α is a successor ordinal, and κ is a singular cardinal if $\mathrm{cf}(\kappa) < \kappa$. For every $\alpha < \omega$, the cardinal $\aleph_\alpha$ is regular, and $\aleph_\omega$ is the smallest singular infinite cardinal (the next singular cardinal being $\aleph_{\omega+\omega}$). Since every successor cardinal is regular, singular cardinals must be limit cardinals. Also, $\mathrm{cf}(\kappa)$ must be regular for any cardinal κ, since $\mathrm{cf}(\mathrm{cf}(\kappa)) = \mathrm{cf}(\kappa)$. We record these facts in the following proposition.

Proposition 733. cf(κ) *is a regular cardinal for every infinite cardinal* κ. *Every successor cardinal is regular. Hence singular cardinals must be limit cardinals.*

Let us say that an infinite ordinal ν is a *regular ordinal* if there is no cofinal subset of $W(\nu)$ having order type less than ν. By the following problem, the regular ordinals can be identified with the regular cardinals since they correspond to each other in a natural one-to-one fashion.

Problem 734. *Show that if* ν *is an regular ordinal then* ν *is an initial infinite ordinal and so* $\nu = \omega_\alpha$ *for some* α. *Moreover, the ordinal* ω_α *is regular if and only if the cardinal* $\aleph_\alpha$ *is regular.*

Problem 735 (AC). *Let* κ *be an infinite cardinal and* A *a set with* $|A| = \kappa$.

1. cf(κ) *is the least cardinal* μ *such that* A *can be expressed as the union of* μ *pairwise disjoint sets each of cardinality* $< \kappa$. *This assertion remains true even if we do drop the qualifier "pairwise disjoint."*
2. cf(κ) *is the least cardinal* μ *such that* κ *can be expressed as*

$$\kappa = \sum_{i \in I} \kappa_i, \quad \text{where } |I| = \mu \text{ and } \kappa_i < \kappa \text{ for } i \in I.$$

3. cf(κ) *is the least cardinal* μ *such that* κ *can be expressed as*

$$\kappa = \sup_{i \in I} \kappa_i, \quad \text{where } |I| = \mu \text{ and } \kappa_i < \kappa \text{ for } i \in I.$$

As mentioned before, very little can be said about cardinal exponentiation when the exponent is infinite, but we do have:

Theorem 736 (König). *If* $\kappa \geq 2$ *and* $\mu \geq \aleph_0$ *are cardinals, then* $\text{cf}(\kappa^\mu) > \mu$.

Proof. It suffices to show that if $\kappa_i < \kappa^\mu$ for all $i \in I$ with $|I| = \mu$, then $\sum_{i \in I} \kappa_i < \kappa^\mu$. So assume that $\kappa_i < \kappa^\mu$ for all $i \in I$ where $|I| = \mu$.

Using the original König's Inequality and the fact that $\mu^2 = \mu$, we get:

$$\sum_{i \in I} \kappa_i < \prod_{i \in I} \kappa^\mu = (\kappa^\mu)^\mu = \kappa^{\mu^2} = \kappa^\mu. \qquad \square$$

It follows that the cardinality of the continuum $\mathfrak{c} = 2^{\aleph_0}$ has cofinality $> \aleph_0$, and so $\mathbf{R}$ cannot be expressed as the union of countably many sets each of cardinality less than $\mathfrak{c}$.

Problem 737. $2^{\aleph_0} = \aleph_n$ *for some* $n \in \mathbf{N}$ *if and only if* $\aleph_\omega^{\aleph_0} > 2^{\aleph_0}$, *and* $2^{\aleph_0} > \aleph_n$ *for all* $n \in \mathbf{N}$ *if and only if* $\aleph_\omega^{\aleph_0} = 2^{\aleph_0}$.

Problem 738 (Hausdorff's Formula). $\aleph_{\mu+1}^{\aleph_\nu} = \aleph_{\mu+1}\aleph_\mu^{\aleph_\nu}$.

$(\omega_\alpha, {}^*\omega_\beta)$ Gaps in Orders

We illustrate the use of cofinalities by giving characterizations for each of the weak completeness properties of orders introduced in Sect. 8.8, namely, the Bolzano–Weierstrass property, the strong nested intervals property, and the sequential nested intervals property. This is done by analyzing the cofinality and coinitiality of Dedekind gaps in orders. By the *cofinality of an order* X we will mean the least ordinal μ such that X has a cofinal subset of order type μ. The *coinitiality of an order* X is the cofinality of the reverse order ${}^*X$. By Corollary 727, if X has no last element, then the cofinality of X is an initial ordinal ω_α.

Definition 739 ($(\omega_\alpha, {}^*\omega_\beta)$ gaps in orderings). An $(\omega_\alpha, {}^*\omega_\beta)$ *gap* in an order X is a Dedekind partition L, U of X such that L has cofinality ω_α and U has coinitiality ω_β, that is, ω_α is the least ordinal μ such that L has a cofinal subset of type μ and ω_β is the least ordinal ν such that U has a coinitial subset of type ${}^*\nu$.

Thus for any $(\omega_\alpha, {}^*\omega_\beta)$ gap, ω_α and ω_β must be regular ordinals (i.e., $\aleph_\alpha$ and $\aleph_\beta$ must be regular cardinals). In fact, we have the following.

Problem 740. *A Dedekind partition L, U of an order X is an $(\omega_\alpha, {}^*\omega_\beta)$ gap if and only if $\aleph_\alpha$ and $\aleph_\beta$ are regular cardinals and there exist $L' \subseteq L$ and $U' \subseteq U$ such that L' has order type ω_α, U' has order type ${}^*\omega_\beta$, for all $x \in L$ there is $y \in L'$ with $y > x$, and for all $x \in U$ there is $y \in U'$ with $y < x$.*

Thus an ordinary Dedekind gap is simply an $(\omega_\alpha, {}^*\omega_\beta)$ gap for some α, β. In a countable order, a Dedekind gap must be an $(\omega_0, {}^*\omega_0)$ (or $(\omega, {}^*\omega)$) gap.

Problem 741 (Characterizing the Bolzano–Weierstrass Property). *Show that an order X satisfies the Bolzano–Weierstrass property if and only if it has no $(\omega_\alpha, {}^*\omega_\beta)$ gap with $\alpha = 0$ or $\beta = 0$ (i.e., it has no $(\omega_\alpha, {}^*\omega)$ or $(\omega, {}^*\omega_\beta)$ gaps, or in other words, in every Dedekind gap, both the cofinality and coinitiality are uncountable).*

Problem 742 (Characterizing the Strong NIP). *Show that an order X satisfies the strong nested intervals property if and only if it has no $(\omega_\alpha, {}^*\omega_\alpha)$ gap (i.e., has no "symmetric" gap with identical cofinality and coinitiality).*

Problem 743 (Characterizing the Sequential NIP). *Show that an order X satisfies the sequential nested intervals property if and only if it has no $(\omega, {}^*\omega)$ gaps.*

Problem 744. *Show that if X is an order which (a) has the Bolzano–Weierstrass property, (b) has the strong nested intervals property, and (c) has cardinality $\leq \aleph_1$, then X must be complete. Show that none of the three conditions in the hypothesis of this result can be dropped.*

Inaccessible Cardinals

The uncountable initial ordinals we have seen so far, $\omega_1, \omega_2, \ldots, \omega_\omega$ etc., all have the property that the α-th uncountable initial ordinal ω_α is much larger than the ordinal α. For example $1 < \omega_1$, $2 < \omega_2$, $\omega < \omega_\omega$, and even $\omega_1 < \omega_{\omega_1}$. Equivalently, $\aleph_\alpha$ has much larger cardinality than the set $W(\alpha) = \{\beta \mid \beta < \alpha\}$ for any $\alpha < \omega_1$ or even for any $\alpha < \omega_{\omega_1}$. Hence any ordinal α such that $\omega_\alpha = \alpha$ (the α-the uncountable initial ordinal equals α itself), or equivalently any ordinal such that the cardinality of $W(\alpha) = \{\beta \mid \beta < \alpha\}$ equals $\aleph_\alpha$, must be larger than all the above ordinals.

Problem 745. *Show that if*

$$\alpha = \sup\{\omega,\ \omega_\omega,\ \omega_{\omega_\omega},\ \omega_{\omega_{\omega_\omega}},\ \ldots\},$$

then $\omega_\alpha = \alpha$ *and so* $|\{\beta \mid \beta < \alpha\}| = \aleph_\alpha$.

All uncountable limit cardinals we have seen so far, such as $\aleph_\omega$, $\aleph_{\omega+\omega}$, $\aleph_{\omega_1}$, etc., are singular. For the ordinal α of the last problem satisfying $\omega_\alpha = \alpha$, $\aleph_\alpha$ is a much larger limit cardinal but is still singular, since $\mathrm{cf}(\aleph_\alpha) = \aleph_0$.

Problem 746. *Show that* $\alpha \mapsto \omega_\alpha$ *is a normal function, and so for every ordinal* α *there is* $\beta > \alpha$ *with* $\omega_\beta = \beta$.

Definition 747. An uncountable cardinal is *weakly inaccessible* if it is a regular limit cardinal.

Problem 748. *Show that if* $\aleph_\alpha$ *is a weakly inaccessible cardinal then* $\omega_\alpha = \alpha$ *and so* $|\{\beta \mid \beta < \alpha\}| = \aleph_\alpha$.

A weakly inaccessible must be quite large, but we also define:

Definition 749. A cardinal κ is a *strong limit* if $2^\mu < \kappa$ for all $\mu < \kappa$, and κ is *strongly inaccessible* if it is uncountable, regular, and a strong limit.

A strong limit cardinal is clearly a limit cardinal, and hence a strongly inaccessible cardinal is weakly inaccessible. It is not possible to show that strongly inaccessible cardinals exist,[1] nor that weakly inaccessible cardinals exist.

Cardinals which cannot be shown to exist using the standard axioms of set theory are called *large cardinals*. Inaccessible cardinals are the simplest examples of large cardinals. The subject area of large cardinals studies much larger cardinals and the consequences of adding their existence as axioms. Such axioms have resolved some classical mathematical problems which could not be decided under the usual axioms, although one famous problem has been stubborn in resisting resolution via large cardinals. It is the Continuum Hypothesis, which we discuss next.

---

[1] From the standard axioms for set theory (such as ZFC), assuming they are consistent; this follows from a result of Gödel known as the second incompleteness theorem.

10.7 The Continuum Hypothesis

Both $2^{\aleph_0}$ and $\aleph_1$ are uncountable cardinals, and Cantor conjectured that

$$2^{\aleph_0} = \aleph_1,$$

an assertion known as the *Continuum Hypothesis* (CH). Cantor implicitly assumed the Axiom of Choice, by which $2^{\aleph_0}$ is an uncountable aleph, i.e.,

$$2^{\aleph_0} = \aleph_\alpha \text{ for some } \alpha \geq 1, \quad \text{and so:} \quad 2^{\aleph_0} \geq \aleph_1.$$

Hence under AC, CH is equivalent to the statement that every set of reals is either countable or equinumerous to **R**. (Without AC, we cannot prove that **R** can be well-ordered, or even that **R** has a subset of cardinality $\aleph_1$.)

The CH is perhaps the most famous problem of set theory, and the problem of settling it is known as *the continuum problem*.[2] All attempts to settle the CH by Cantor and by other early twentieth-century mathematicians failed, even though for most effectively defined sets of reals, they could prove them to be either countable or equinumerous to **R**. Cantor and Bendixson proved the important result that every *closed* set of reals must either be countable or be equinumerous to **R**, which can be informally expressed by saying that "the closed sets satisfy the CH." Other mathematicians extended the result to show that a larger class of sets known as analytic sets satisfy the CH. We will prove both these results in the next part (Corollary 1081, Theorem 1160). However, Lusin introduced other effectively defined sets of reals which could not be proved to satisfy the CH. The magnitude of the cardinal $2^{\aleph_0}$ turned out to be very difficult to estimate, and much of research in set theory was driven by investigations into the continuum problem.

By König's theorem, we have

$$\operatorname{cf}\left(2^{\aleph_0}\right) > \aleph_0,$$

so $2^{\aleph_0}$ cannot equal any cardinal of countable cofinality, e.g.,

$$2^{\aleph_0} \neq \aleph_0, \quad 2^{\aleph_0} \neq \aleph_\omega, \quad 2^{\aleph_0} \neq \aleph_{\omega^2}, \quad 2^{\aleph_0} \neq \aleph_{\omega_1+\omega}, \text{ etc.}$$

That is about as much as we can say about the magnitude of $2^{\aleph_0}$ as an aleph using the current standard axioms of set theory. In later research, first Gödel introduced the notion of *constructible sets* to show that CH cannot be disproved (assuming that the standard axioms of set theory are consistent). Then Cohen invented the powerful technique of *forcing* to show that CH cannot be proved either. Extending Cohen's result, Solovay showed that one can consistently assume that $2^{\aleph_0}$ equals $\aleph_\alpha$ for any α so long as $\aleph_\alpha$ has uncountable cofinality.

---

[2]It was the first in Hilbert's celebrated list of problems presented in 1900.

Problem 750. *Show without using the Axiom of Choice that*

$$\aleph_1 \leq 2^{2^{\aleph_0}}.$$

[Hint: For each $\alpha < \omega_1$, consider the subcollection of $\mathbf{P(Q)}$ consisting of those subsets of $\mathbf{Q}$ which are well-ordered and have order type α.]

Problem 751. *Define the cardinals* $\beth_n$, $n = 0, 1, 2, \ldots,$ *as*

$$\beth_0 := \aleph_0, \qquad \text{and} \qquad \beth_{n+1} := 2^{\beth_n}.$$

Thus $\beth_1 = \mathfrak{c}$, $\beth_2 = \mathfrak{f}$, *etc. Show without using the Axiom of Choice that*

$$\aleph_n \leq \beth_{2n}, \qquad \text{for all } n = 0, 1, 2, \ldots.$$

Under the Axiom of Choice, one can define $\beth_\alpha$ for all ordinals α and it readily follows that $\aleph_\alpha \leq \beth_\alpha$ for all α. On the other hand, by the Cohen–Solovay result mentioned above, no inequality of the form

$$2^{\aleph_0} \leq \aleph_\alpha,$$

however large α may be, can be obtained using the usual axioms of set theory.

Problem 752. *Show that CH holds if and only if* $\aleph_1^{\aleph_0} < \aleph_2^{\aleph_0}$.

Problem 753. *Show that CH implies* $\aleph_n^{\aleph_0} = \aleph_n$ *for all finite cardinals* $n > 0$.

[Hint: Use induction, and the fact that $\mathrm{cf}(\aleph_n) > \aleph_0$ for $n > 0$.]

The Generalized Continuum Hypothesis

After CH, Hausdorff introduced the much stronger statement:

$$2^{\aleph_\alpha} = \aleph_{\alpha+1} \quad \text{for every ordinal } \alpha,$$

which is known as the *Generalized Continuum Hypothesis* (GCH). It is a much stronger assumption and the cardinal power $\aleph_\alpha^{\aleph_\beta}$ can be completely determined using GCH. Using the notation $\beth_\alpha$ as above, GCH holds if and only if $\aleph_\alpha = \beth_\alpha$ for α. Under an additional assumption called the Axiom of Foundation, GCH becomes equivalent to the statement

$$2^\kappa = \kappa^+ \quad \text{for every cardinal } \kappa,$$

which readily implies the Axiom of Choice.

Problem 754. *Show (without using AC) that if* $2^\kappa = \kappa^+$ *for every cardinal* κ, *then the Axiom of Choice holds.*

Problem 755. *Under GCH, show that if* $\aleph_\mu \leq \aleph_{\nu+1}$ *then* $\aleph_\mu^{\aleph_\nu} = \aleph_{\nu+1}$.

η_1-Orderings and the CH

The η_1 orderings were introduced and studied by Hausdorff as a generalization of the order type η to the next higher cardinal $\aleph_1$. The behavior of η_1 orderings of cardinality $\aleph_1$ is similar to that of countable dense orderings without endpoints (Problem 766, Problem 767, Corollary 768). However, as we will see, the problem is inextricably linked to the CH, since the existence of η_1 orderings of cardinality $\aleph_1$ is equivalent to the CH (Problem 764).

Definition 756 (η_1 orderings). An order X is an η_1 *ordering* if for any countable $A, B \subseteq X$, if $A < B$ then there exists $p \in X$ such that $A < p < B$.

Problem 757. *Every η_1 ordering is a nontrivial dense order without endpoints in which every countable subset is bounded (both above and below).*

In particular, every sequence is bounded, but no strictly increasing sequence is convergent (i.e., a strictly increasing sequence does not have a supremum).

[Hint: The sets A and B in the definition are allowed to be empty.]

Problem 758. *An order is an η_1 order if and only if it is a nontrivial dense linear ordering without endpoints in which every sequence is bounded, no strictly monotone sequence is convergent, and there are no $(\omega, {}^*\omega)$ gaps.*

Problem 759. *In an η_1 ordering a nonempty open interval $\{x \mid a < x < b\}$ is an η_1 ordering, but not all open segments are η_1 orderings.*

Problem 760. *An η_1 order X contains suborders isomorphic to Y for every order Y with $|Y| \leqslant \aleph_1$ (and so X has suborders of type α for every $\alpha < \omega_2$).*

Definition 761 (Lexicographic Powers). If X is an order and α is any ordinal, we define an order on $X^{W(\alpha)}$ by defining, for $a, b \in X^{W(\alpha)}$, $a < b$ if and only if there is $\xi < \alpha$ such that $a(\xi) < b(\xi)$ and $a(\eta) = b(\eta)$ for all $\eta < \xi$.

Problem 762. *Let H_1 be the suborder of the lexicographic power $\{0,1\}^{W(\omega_1)}$ consisting of all binary ω_1-sequences with a last 1, that is:*

$$H_1 := \left\{ \langle a_\nu \rangle_{\nu < \omega_1} \in \{0,1\}^{W(\omega_1)} \mid \exists \beta < \omega_1 (a_\beta = 1 \text{ and } a_\nu = 0\ \forall \nu > \beta) \right\}.$$

1. *H_1 is an η_1 ordering of cardinality $\mathfrak{c}$.*
2. *Let $c = \langle c_\nu \mid \nu < \omega_1 \rangle \in \{0,1\}^{W(\omega_1)}$ be the binary ω_1-sequence defined by setting $c_\nu = 1$ if ν is even, and $c_\nu = 0$ if ν is odd. Let $L = \{a \in H_1 \mid a < c\}$ and $U = \{a \in H_1 \mid c < a\}$. Show that L, U form a $(\omega_1, {}^*\omega_1)$ gap in H_1.*
3. *Show that H_1 has exactly $2^{\aleph_1}$ many $(\omega_1, {}^*\omega_1)$ gaps.*

Problem 763. *Show that every η_1 ordering contains a subset isomorphic to* $\mathbf{R}$ *and so has cardinality at least* $\mathfrak{c}$.

Problem 764. *Show that the Continuum Hypothesis is equivalent to the statement that there exists an η_1 ordering of cardinality $\aleph_1$.*

The Dedekind completion of an η_1 ordering cannot be an η_1 ordering. However, we have the following result.

Problem 765 (($\omega_1, {}^*\omega_1$) Completion). *Suppose that X is a suborder of an order Y such that for any $p \in Y \setminus X$, p is a two-sided limit point of X and the sets $L := \{x \in X \mid x < p\}$ and $U := \{x \in X \mid p < x\}$ form a $(\omega_1, {}^*\omega_1)$ gap in X. Show that if X is an η_1 ordering then so is Y.*

Conclude that for any η_1 ordering X, the "$(\omega_1, {}^*\omega_1)$ completion of X," i.e., the ordering obtained by "filling in" all the $(\omega_1, {}^*\omega_1)$ gaps in X, is an η_1 ordering which has no $(\omega_1, {}^*\omega_1)$ gap.

The following problems form the "η_1 analogues" of Cantor's theorem on countable dense linear orders without endpoints (characterizing the order type η) and the proof that **R** is uncountable that follows from it. Note, however, that the results are vacuous unless we assume CH since without CH there are no η_1 orderings of cardinality $\aleph_1$ (Problem 764).

Problem 766. *Any two η_1 orderings of cardinality $\aleph_1$ are order isomorphic.*

[Hint: Mimic Cantor's "back-and-forth" proof.]

Problem 767. *Any η_1 ordering of cardinality $\aleph_1$ must have $(\omega_1, {}^*\omega_1)$ gaps.*

[Hint: Removing a point from any η_1 ordering produces an η_1 ordering with a $(\omega_1, {}^*\omega_1)$ gap.]

Corollary 768. *Any η_1 ordering without $(\omega_1, {}^*\omega_1)$ gaps has cardinality $> \aleph_1$.*

If X is an η_1 ordering of cardinality $\aleph_1$, then by Problem 765 the "$(\omega_1, {}^*\omega_1)$ completion" of X will be an η_1 ordering without $(\omega_1, {}^*\omega_1)$ gaps, and so will have cardinality $> \aleph_1$. Thus just as a countable dense linear order has uncountably many irrational Dedekind gaps, similarly every η_1 ordering of cardinality $\aleph_1$ has more than $\aleph_1$ $(\omega_1, {}^*\omega_1)$ gaps.

Chapter 11
Posets, Zorn's Lemma, Ranks, and Trees

Abstract This chapter covers the very basics of the following topics: Partial orders, Zorn's Lemma and some of its applications, well-founded relations and ranks on them, trees, König's Infinity Lemma, well-founded trees, and Ramsey's theorem.

11.1 Partial Orders

A linear order $<$ is an irreflexive transitive relation which is also connected, i.e., if $x \neq y$ then either $x < y$ or $y < x$ (any two distinct elements are comparable). By dropping this last condition of comparability, we get the more general notion of a *partially ordered set* or simply a *poset.*

Definition 769. A *strict poset* is a pair $\langle P, \prec \rangle$ where P is a set and $\prec$ is a binary relation on P which is irreflexive ($x \nprec x$ for all x) and transitive (if $x \prec y$ and $y \prec z$ then $x \prec z$) on the set P. (Note that "$a \nprec b$" means that "$a \prec b$ is false.")

It is easy to verify that if $\langle P, \prec \rangle$ is a strict poset, then the relation $\prec$ is asymmetric on P (for all $x, y \in P$, if $x \prec y$ then $y \nprec x$).

Posets also come in an essentially equivalent "reflexive" variety:

Definition 770. A *reflexive poset* is a pair $\langle P, \preceq \rangle$ where P is a set and $\preceq$ is a binary relation on P which is reflexive ($x \preceq x$ for all x), antisymmetric (if $x \preceq y$ and $y \preceq x$ then $x = y$), and transitive (if $x \preceq y$ and $y \preceq z$ then $x \preceq z$) on the set P.

It is easy to verify that if $\langle P, \prec \rangle$ is a strict poset, then the relation $\preceq$ on P defined by $x \preceq y \Leftrightarrow x \prec y$ or $x = y$ makes $\langle P, \preceq \rangle$ a reflexive poset, from which the original strict poset can be recovered by defining $x \prec y \Leftrightarrow x \preceq y$ and $x \neq y$. Similarly, one can start from a reflexive poset $\langle P, \preceq \rangle$, then get a strict poset by setting $x \prec y \Leftrightarrow x \preceq y$ and $x \neq y$, and recover back the original reflexive poset as before. This gives, for each set P, a natural one-to-one correspondence between the strict posets and reflexive posets over P.

A. Dasgupta, *Set Theory: With an Introduction to Real Point Sets*,
DOI 10.1007/978-1-4614-8854-5_11,

Thus the notions of reflexive and strict posets are essentially variants of the same concept, analogous to that of inclusive set inclusion $\subseteq$ and proper set inclusion $\subsetneq$. From now on we will use the term *poset* to denote either a reflexive or a strict poset, as determined by context or notation.

Problem 771. *Every linear order is a poset. Every subset (restriction) of a poset is a poset.*

Definition 772. Let $\langle P, \preceq \rangle$ be poset, $A \subseteq P$, and $a \in P$. We say that

1. a is *lower bound of* A if $a \preceq x$ for all $x \in A$. *Upper bounds* are similarly defined.
2. a is a *least element of* A, written as $a = \min(A)$, if a is lower bound of A which is also a member of A. *Greatest elements* are similarly defined.
3. a is a *minimal element of* A if $a \in A$ and there is no $x \in A$ distinct from a with $x \preceq a$. *Maximal elements* are similarly defined.
4. a is the *least upper bound* or *supremum* of A, written as $a = \vee A$ or $a = \sup A$, if a is an upper bound of A and $a \preceq x$ for every upper bound x of A. *Greatest lower bounds* or *infimums* are similarly defined.

Note that a set A can have at most one least element, at most one greatest element, at most one supremum (least upper bound), and at most one infimum (greatest lower bound).

Definition 773. Let x and y be elements of a poset $\langle P, \preceq \rangle$. We say that:

1. x and y are *comparable* if either $x \preceq y$ or $y \preceq x$; otherwise, x and y are *incomparable*.
2. x and y are *compatible* if there is z such that $z \preceq x$ and $z \preceq y$; otherwise, x and y are *incompatible*.

In a linear order, every pair of elements are comparable and therefore compatible. In a poset with a least element, every pair of elements are compatible.

Definition 774. Let A be a subset of poset $\langle P, \preceq \rangle$.

1. A is called an *initial part of the poset* $\langle P, \preceq \rangle$ or a *downward closed subset of* P if for all $x, y \in P$, $x \in A$ and $y \preceq x \Rightarrow y \in A$.
2. A is *bounded above* (in P) if there is an element of P which is an upper bound of A.
3. A is called a *chain* if A is linearly ordered by $\preceq$, i.e., if any two elements of A are comparable.
4. A is called an *antichain* if any two elements of A are incompatible.

A most important example of a poset is obtained by taking any family of sets with set inclusion $\subseteq$ as the ordering relation.

Problem 775. *Let X be a set, and $P = \mathrm{P}(X)$. Then both $\langle P, \subseteq \rangle$ and $\langle P, \supseteq \rangle$ are posets. What are the least and greatest elements of $\langle P, \subseteq \rangle$? If A is the collection of all nonempty proper subsets of X, then what are the minimal and maximal elements* *of A in the poset $\langle P, \subseteq \rangle$? Does A have least or greatest elements? If B is an initial part of $\langle P, \subseteq \rangle$ which is also a chain, what can you say about B?*

Problem 776. *Let P be the set of nonnegative integers and for $x, y \in P$ define $x \preceq y \Leftrightarrow x$ divides y. Then $\langle P, \preceq \rangle$ is a poset. Does P have least or greatest elements? Let $A = \{n \in P \mid n \geq 2\}$. What are the minimal elements of A? Give an example of an infinite initial part of P which is a chain.*

Theorem 777 (A representation theorem for posets). *Let $\langle P, \preceq \rangle$ be a reflexive poset. Then there is a set X, a subset $\mathcal{S} \subseteq \mathbf{P}(X)$ such that $\langle P, \preceq \rangle$ is isomorphic to $\langle \mathcal{S}, \subseteq \rangle$; that is, there is a bijection $F : P \to \mathcal{S}$ such that $\forall x, y \in P$, $x \preceq y \Leftrightarrow F(x) \subseteq F(y)$.*

Proof. Define, for $x \in P$, $F(x) := \{y \in P \mid y \preceq x\}$. Now put $X := P$, and $\mathcal{S} := \{F(x) \mid x \in P\}$. □

Problem 778. *Find a chain C in the poset $\langle \mathbf{P}(\mathbf{N}), \subseteq \rangle$ such that C is order isomorphic to $\mathbf{R}$ under its usual ordering.*

[Hint: Try using $\mathbf{P}(\mathbf{Q})$ instead of $\mathbf{P}(\mathbf{N})$.]

Problem 779. *Consider the poset $P = \mathbf{N} \setminus \{1\}$ with divisibility as the ordering relation. Find a necessary and sufficient condition for a subset to be an antichain.*

Definition 780 (Increasing Maps, Embeddings, and Isomorphisms). Suppose that $\langle P, \prec_P \rangle$ and $\langle Q, \prec_Q \rangle$ are strict posets, and let $f : P \to Q$. We say that

1. f is *strictly increasing* if $x \prec_P y \Rightarrow f(x) \prec_Q f(y)$ (for all $x, y \in P$).
2. f is an *embedding* if $x \prec_P y \Leftrightarrow f(x) \prec_Q f(y)$ (for all $x, y \in P$).
3. f is an *isomorphism* if f is a bijective embedding of P onto Q.

Problem 781. *Suppose that $\langle S, < \rangle$ is a linear order, $\langle P, \prec \rangle$ is a strict poset, and $f : S \to P$. Show that if f is strictly increasing, then it must be an embedding.*

11.2 Zorn's Lemma

An extremely useful consequence of AC is known as *Zorn's Lemma,* which asserts that if in a poset every chain is bounded above then the poset has a maximal element.

Theorem 782 (Zorn's Lemma). *The Axiom of Choice implies that if every chain in a poset is bounded above then the poset has a maximal element.*

Proof. Let $\langle P, \prec \rangle$ be a poset in which every chain is bounded above, and let $\varphi : \mathbf{P}^*(P) \to P$ be a choice function (so that $\varphi(E) \in E$ whenever E is a nonempty subset of P). For each $E \subseteq P$, let $U(E)$ be the set of those elements of E which are upper bounds of its complement $P \setminus E$:

$$U(E) := \{x \mid x \in E \text{ and } y \prec x \text{ for all } y \in P \setminus E\}.$$

Define a derivative operator $\nabla\colon \mathsf{P}(P) \to \mathsf{P}(P)$ by

$$\nabla(E) := \begin{cases} E \smallsetminus \{\varphi(U(E))\} & \text{if } U(E) \text{ is non empty,} \\ E & \text{otherwise.} \end{cases}$$

As usual, let $P^{(\alpha)}$ denote the α-th iterated ∇-derivative of P. We then have a least ordinal μ such that $P^{(\mu+1)} = P^{(\mu)}$, which means the set $C := P \smallsetminus P^{(\mu)}$ contains all its upper bounds. Now C is partitioned as $C = \cup_{\alpha<\mu} P^{(\alpha)} \smallsetminus P^{(\alpha+1)}$, and for each $\alpha < \mu$, the set $P^{(\alpha)} \smallsetminus P^{(\alpha+1)}$ is a singleton whose member is an upper bound of $P \smallsetminus P^{(\alpha)}$. Thus C is a chain. Let p be an upper bound of C. Since C contains all its upper bounds, $p \in C$ is the greatest element of C, and so p must be a maximal element of P. □

The converse of the above implication is also true.

Proposition 783. *Zorn's Lemma implies the Axiom of Choice: If it is true that any poset in which every chain is bounded above must contain a maximal element, then the Axiom of Choice holds.*

Proof. Let $\mathcal{P}$ be any family of pairwise disjoint nonempty sets, and consider the collection $\mathcal{C}$ of those subsets A of $\cup\mathcal{P}$ such that $|A \cap E| \leq 1$ for every $E \in \mathcal{P}$. Then $\mathcal{C}$ forms a poset under set inclusion in which every chain is easily verified to have an upper bound. Hence $\mathcal{C}$ has a maximal element M, for which we will have $|M \cap E| = 1$ for every $E \in \mathcal{P}$. Thus Zorn's Lemma is another equivalent of AC. □

In many applications of AC, Zorn's Lemma facilitates and simplifies proofs. For example, using Zorn's Lemma, one can readily establish the following standard mathematical result:

Problem 784. *In any vector space, every linearly independent subset is contained in a maximal linearly independent subset (called a* basis*).*[1]

The next problem gives another equivalent of AC known as the *Hausdorff Maximal Principle.*

Problem 785. *Show without using the Axiom of Choice that Zorn's Lemma is equivalent to the statement that every chain in a poset is contained in some maximal chain.*

Combining all the equivalents of AC that we have obtained, we get:

Theorem 786. *The following conditions are equivalent to each other:*

1. The Axiom of Choice, Partition Version. *Every partition has a choice set.*

[1]Blass has shown that this statement is actually equivalent to the Axiom of Choice.

2. The Axiom of Choice, Choice Function Version. *Every family of nonempty sets has a choice function.*
3. The Well-Ordering Theorem. *Every set can be well-ordered.*
4. Cardinal Comparability. *If κ, μ are cardinals then either $\kappa \leq \mu$ or $\mu \leq \kappa$.*
5. Zorn's Lemma. *A poset in which every chain is bounded above has at least one maximal element.*
6. Hausdorff Maximal Principle. *Every chain in a poset is contained in a maximal chain.*

11.3 Some Applications and Examples

We will now see some examples of applications of Zorn's Lemma and the Hausdorff Maximal Principle. Throughout this section, we will assume the Axiom of Choice.

Almost Disjoint Families

The following is a direct generalization of Definition 392.

Definition 787 (Almost Disjoint Family). If X is an infinite set with $\kappa = |X|$, we say that $\mathsf{C} \subseteq \mathbf{P}(X)$ is an *almost disjoint family* of subsets of X if

1. $|E| = \kappa$ for all $E \in \mathsf{C}$.
2. If $E_1, E_2 \in \mathsf{C}$ and $E_1 \neq E_2$ then $|E_1 \cap E_2| < \kappa$.

Let X be an infinite set, and define a relation $\subset^*$ on $\mathbf{P}(X)$ by $A \subset^* B \Leftrightarrow |A \smallsetminus B| < \kappa$ and $|B \smallsetminus A| = \kappa$.

Problem 788. *$\mathbf{P}(X)$ is a poset under $\subset^*$ in which the minimal elements are precisely the subsets of X of cardinality $< \kappa = |X|$.*

Let us remove the minimal elements of $\langle \mathbf{P}(X), \subset^* \rangle$ to obtain the "subposet" $\mathbf{P}_*(X) := \{E \mid E \subseteq X \text{ and } |E| = \kappa\}$, which does not have any minimal element. Now note that C *is an almost disjoint family of subsets of X if and only if* C *is an antichain in the poset* $\langle \mathbf{P}_*(X), \subset^* \rangle$.

Also, since $\kappa^2 = \kappa$, $\mathbf{P}_*(X)$ has an antichain of size κ. We will show that if κ is a regular cardinal, then antichains of size $> \kappa$ can be obtained.

Lemma 789. *If $|X| = \kappa$ is a regular cardinal and C is an almost disjoint family of subsets of X with $|\mathsf{C}| = \kappa$, then there is $E \subseteq X$ such that $E \notin \mathsf{C}$ and $\mathsf{C} \cup \{E\}$ is still almost disjoint.*

Proof. Assume $\kappa = |X|$ is regular with $\kappa = \aleph_\alpha$ (say). Since $|\mathsf{C}| = \kappa$, we can enumerate C as $\mathsf{C} = \{E_\beta \mid \beta < \omega_\alpha\}$, where $|E_\xi \cap E_\eta| < \kappa$ for $\xi \neq \eta$. Now for each $\beta < \omega_\alpha$, we must have $E_\beta \smallsetminus \bigcup_{\xi<\beta} E_\xi = E_\beta \smallsetminus \bigcup_{\xi<\beta}(E_\beta \cap E_\xi)$ nonempty, since

$|E_\beta| = \kappa$ while $| \bigcup_{\xi<\beta} (E_\beta \cap E_\xi)| < \kappa$ by regularity of κ. Hence by the Axiom of Choice we can pick $x_\beta \in E_\beta$ for each $\beta < \omega_\alpha$. Let $E := \{x_\beta \mid \beta < \omega_\alpha$. Then $|E| = |W(\omega_\alpha)| = \kappa$ since $x_\xi \neq x_\eta$ for $\xi \neq \eta$. Moreover, for any $\beta < \omega_\alpha$, we have $E_\beta \cap E \subseteq \{x_\xi \mid \xi \leq \beta\}$, so $|E_\beta \cap E| \leqslant |W(\beta)| < \kappa$. □

Now, since $\kappa^2 = \kappa$, we may fix a pairwise disjoint family C_0 of subsets of X all having size κ. Notice that the union of any chain of almost disjoint families is itself an almost disjoint family. Hence the poset consisting of all the almost disjoint families containing C_0 and ordered by inclusion (of families) has the property that every chain in this poset has an upper bound. By Zorn's Lemma we may fix a maximal almost disjoint family C containing C_0. Then $|\mathrm{C}| \geqslant \kappa$, but we cannot have $|\mathrm{C}| = \kappa$ since then C would not be maximal by the above Lemma. Hence we have:

Theorem 790. *Let X be a set of regular infinite cardinality $\kappa = |X|$. Then there exists an almost disjoint family C of subsets of X with $|\mathrm{C}| > \kappa$.*

For the case where $\kappa = \aleph_0$, we saw in Problem 393 that one can obtain an almost disjoint family of size $2^{\aleph_0} = \mathfrak{c}$ in a highly effective fashion. The above theorem generalizes the result to larger cardinalities in a weaker fashion and is not effective.

Problem 791. *Assuming the Continuum Hypothesis show that a set of cardinality $\aleph_1$ has an almost disjoint family of size $2^{\aleph_1}$.*

[Hint: Problem 393 can help.]

Short Linear Orders

Definition 792 (Short Orders). A linear order is called *short* if it does not contain any suborder of type ω_1 or $^*\omega_1$.

In other words, X is short if X does not contain a strictly increasing or strictly decreasing ω_1-sequence.

Problem 793. *A suborder of a short order is short. An order which is a countable union short suborders is itself short. If X and Y are short orders, then $X \times Y$ with the lexicographic ordering is short.*

Since the usual ordering on the separable continuums $\mathbf{R}$ and $[0, 1]$ is short, we can get examples of many short orders, such as the (lexicographically ordered) non-CCC continuums $[0, 1]^k$, for $k = 2, 3, \ldots$.

Problem 794. *Let X be an order and Y is a short suborder such that for all $x, y \in X$ if there is $z \in X$ with $x < z < y$ then there is a $w \in Y$ with $x < w < y$. Then X is short.*

We can manufacture more examples of short orders by taking "countable lexicographic powers" as follows.

Problem 795. *If X is short and $\alpha < \omega_1$, then the lexicographic power $X^{W(\alpha)}$ is short.*

[Hint: Use transfinite induction on α. Note that $X^{W(\alpha+1)}$ is order isomorphic to $X^{W(\alpha)} \times X$ with lexicographic order. If α is a limit ordinal, fix $a \in X$ and consider the suborder Y of X consisting of those elements of X which take the eventually constant value of a. Then by induction hypothesis, Y is a countable union of short suborders, and an application of Problem 794 shows that X must be short.]

Problem 796. *For any ordinal α, the lexicographic power $\{0,1\}^{W(\omega_\alpha)}$ does not contain any suborder of type $\omega_{\alpha+1}$ or ${}^*\omega_{\alpha+1}$.*

The following main result on short orders is proved using Zorn's Lemma.

Theorem 797 (Hausdorff). *If an order X is a union of $\aleph_1$-many short suborders, then X can be embedded in any η_1 order. In particular, every short order can be embedded in any η_1 order.*

Proof. The proof is based on the following extension lemma.

Lemma 798. *If A is a short linear order, B is an η_1 order, $S \subseteq A$, and $f: S \to B$ is strictly increasing, then f can be extended to a strictly increasing map from all of A into B.*

Proof (Lemma). Let F be the family of all strictly increasing functions which extend f and map some subset of A into B. Partially order F under extension. Then by Zorn's Lemma, there is a maximal member $g \in \mathsf{F}$. We claim that $\operatorname{dom}(g) = A$. Otherwise there would exist $a \in A \setminus \operatorname{dom}(g)$. Let $L := \{x \in \operatorname{dom}(g) \mid x < a\}$ and $R := \{x \in \operatorname{dom}(g) \mid a < x\}$. Since A is short, there exist countable sets P and Q with P cofinal in L and Q coinitial in R. Put $C := g[P]$ and $D := g[Q]$. Then C and D are countable subsets of B with $C < D$ (in B). Since B is an η_1 order, there is $b \in B$ with $C < \{b\} < D$. Define an extension h of g with $\operatorname{dom}(h) = \operatorname{dom}(g) \cup \{a\}$ by setting $h(a) = b$ and $h(x) = g(x)$ for $x \in \operatorname{dom}(g)$. Then h is a strictly increasing proper extension of g, contradicting the maximality of g. □

To finish the proof of the theorem, let A be an order such that $A = \bigcup_{\alpha<\omega_1} A_\alpha$ where each suborder A_α is short. We can assume that the sets A_α increase with α (since otherwise we could replace A_α by $\bigcup_{\beta\leq\alpha} A_\beta$). Now let B be any η_1 order. Using the lemma and the Axiom of Choice, we can build by transfinite induction strictly increasing functions $f_\alpha: A_\alpha \to B, \alpha < \omega_1$ such that if $\alpha < \beta < \omega_1$, then f_β extends f_α. Then the common extension of all the functions f_α is a strictly increasing map from A to B. □

Problem 799. *Any order X with $|X| > \mathfrak{c}$ has a suborder of type ω_1 or ${}^*\omega_1$.*

Recall that in Problem 762, we defined H_1 as the suborder of the lexicographic power $\{0,1\}^{W(\omega_1)}$ consisting of all binary ω_1-sequences with a last 1. H_1 was an η_1 order of cardinality $\mathfrak{c}$ containing $2^{\aleph_1}$ many $(\omega_1, {}^*\omega_1)$ gaps.

Problem 800. *H_1 can be expressed as the union of $\aleph_1$ short suborders. Hence, every η_1 order contains a suborder isomorphic to H_1.*

Some Posets Containing η_1 Chains

The four related posets below all contain chains which are η_1 orders. Consequently they contain chains of order type α and $^*\alpha$ for each ordinal $\alpha < \omega_2$.

Problem 801 (Orders of Magnitude for Positive Sequences). *Let S denote the set $(\mathbf{R}^+)^{\mathbf{N}}$ of all sequences of positive real numbers, and for any $x = \langle x_n \mid n \in \mathbf{N}\rangle \in \mathrm{S}$ and $y = \langle y_n \mid n \in \mathbf{N}\rangle \in \mathrm{S}$ define*

$$x \prec y \text{ if and only if } \lim_{n\to\infty} \frac{x_n}{y_n} = 0.$$

($x \prec y$ is often written in the "little-oh notation" as $x_n = o(y_n)$.) Show that

1. *Under the relation $\prec$, S is a poset of size $\mathfrak{c}$.*
2. *For any countable subset C of S, there exist $x, y \in \mathrm{S}$ such that $x \prec C \prec y$.*
3. *If A and B are countable chains in S with $A \prec B$ (i.e., $x \prec y$ for all $x \in A$ and $y \in B$), then there is $p \in \mathrm{S}$ such that $A \prec p \prec B$.*
4. *Every maximal chain in S is an η_1 ordering.*

Problem 802 (Orders of Infinity for Sequences with Limit ∞). *Let M be the set of all sequences of natural numbers $f \in \mathbf{N}^{\mathbf{N}}$ which approach $+\infty$, i.e., with $\lim_n f(n) = +\infty$. For $f, g \in \mathrm{M}$ define $f \prec g$ if and only if $\lim_n (g(n) - f(n)) = +\infty$. Show that*

1. *Under the relation $\prec$, M is a poset of size $\mathfrak{c}$.*
2. *For any countable $C \subseteq \mathrm{M}$, there exist $f, g \in \mathrm{M}$ such that $f \prec C \prec g$.*
3. *If A and B are countable chains in M with $A \prec B$ (i.e., $f \prec g$ for all $f \in A$ and $g \in B$), then there is $h \in \mathrm{M}$ such that $A \prec h \prec B$.*
4. *Every maximal chain in M is an η_1 ordering.*

Problem 803 (Ordering on $\mathrm{P}(\mathrm{N})$ modulo finite sets). *Let P be the collection of all subsets A of $\mathbf{N}$ such that both A and its complement are infinite. For $A, B \in \mathrm{P}$ define $A \prec B$ if and only if $A \setminus B$ is finite and $B \setminus A$ is infinite. Assuming the Axiom of Choice, show that:*

1. *Under the relation $\prec$, P is a poset of size $\mathfrak{c}$.*
2. *A and B are incompatible if and only if $A \cap B$ is finite.*
3. *Every antichain of size $\aleph_0$ is properly contained in another antichain.*
4. *There is an antichain of cardinality $\mathfrak{c}$. [Hint: See Problem 393.]*
5. *If $\mathrm{X}, \mathrm{Y} \subseteq \mathrm{P}$ are countable chains such that $A \prec B$ for all $A \in \mathrm{X}$ and $B \in \mathrm{Y}$, then there is M such that $A \prec M \prec B$ for all $A \in \mathrm{X}$ and $B \in \mathrm{Y}$.*
6. *Any maximal chain in P must be an η_1 ordering.*

Problem 804 (The Strict Dominating Order). *For $f, g \in \mathbf{N}^{\mathbf{N}}$, say that g dominates f and write $f \prec^* g$ if and only if there is m such that $f(n) < g(n)$ for all $n \geq m$. Show that in the poset $\mathsf{H} := \langle \mathbf{N}^{\mathbf{N}}, \prec^* \rangle$:*

1. *Every $f \in \mathbf{N}^{\mathbf{N}}$ has an immediate successor, that is, there is $g \in \mathbf{N}^{\mathbf{N}}$ such that $f \prec^* g$ and there is no h with $f \prec^* h \prec^* g$.*
2. *Every countable subset is bounded above.*
3. *If $f_1 \prec^* f_2 \prec^* \cdots \prec^* f_n \prec^* f_{n+1} \prec^* \cdots \prec^* f$, then there is g with $f_n \prec^* g \prec^* f$ for all n. Thus no strictly increasing sequence has a supremum.*
4. *Let A and B be countable chains in this poset with $A \prec^* B$ (i.e., $f \prec^* g$ for all $f \in A$ and $g \in B$). If either A has no maximum or B has no minimum, then there is h such that $A \prec^* h \prec^* B$.*
5. *The poset H contains a chain which is an η_1 ordering.*

[Hint: Note that if $f \prec g$ in M then $f \prec^* g$ in H (but not conversely), and so any chain in M is a chain in H, and by Problem 802 M contains η_1 chains.]

It is a celebrated result of Hausdorff [60] that the poset H as well as the posets S, P, and M, all contain $(\omega_1, {}^*\omega_1)$ gaps (i.e., they contain maximal chains with $(\omega_1, {}^*\omega_1)$ gaps). We will not prove the result in full generality, but it is easy to derive it from the Continuum Hypothesis.

Proposition 805 (CH). *All four posets above have $(\omega_1, {}^*\omega_1)$ gaps.*

Proof. Assume the CH. In the posets S, P, and M, any maximal chain is an η_1 chain of size $\aleph_1$, and so has $(\omega_1, {}^*\omega_1)$ gaps by Problem 767.

To get an $(\omega_1, {}^*\omega_1)$ gap in H, start with a maximal chain C in M having a Dedekind partition L, U, where L has a cofinal subset of type ω_1 and U has a coinitial subset of type ${}^*\omega_1$. Now note that one cannot have an element f with $L \prec^* f \prec^* U$ in H, since that would imply $L \prec f \prec U$ in M. Extending C to a maximal chain in H thus retains the $(\omega_1, {}^*\omega_1)$ gap of C. □

For a proof of this result without assuming the Continuum Hypothesis, see [35] or [34].

11.4 Well-Founded Relations and Rank Functions

Well-founded relations can be viewed as a generalization of well-orders.

Definition 806. Let R be a relation on a set A and let $B \subseteq A$. Given $x \in B$, we say that *x is an R-minimal element of B* if there is no $y \in B$ with yRx. We write $\min_R[B]$ for the set of R-minimal elements of B.

Definition 807. We say that the relation *R is well-founded on the set A* if every nonempty subset of A has at least one R-minimal element.

We say that $\langle A, R\rangle$ *is a well-founded structure* if R is a well-founded relation on A.

Note that a well-founded relation must be asymmetric and hence irreflexive.

Problem 808 (DC). *A relation R on a set A is not well-founded if and only if X contains an "infinite sequence of R-descending elements," that is, there is a sequence of elements $x_1, x_2, \dots, x_n, \dots \in X$ such that $x_{n+1} R x_n$ for all $n \in \mathbf{N}$.*

Clearly, every well-order is a well-founded relation.

Problem 809. *Show that the strict divisibility relation on $\mathbf{N}$, defined by $xRy \Leftrightarrow x$ divides y and $x \neq y$, is well-founded.*

Problem 810 (Transfinite Induction on Well-Founded Structures). *Let R be a well-founded relation on the set A, and $B \subseteq A$. Suppose that for any $a \in A$, if $\{x \in A \mid xRa\} \subseteq B$ then $a \in B$. Then $B = A$.*

Given a well-founded relation R on a set A, the elements of A can be classified into distinct *ordinal ranks* as follows: The R-minimal elements of A are said to have rank 0. We then remove the elements of rank 0 from the set A to get the set A', and the minimal elements of A' are said to have rank 1. In general, using transfinite recursion on the ordinal α, we can define the elements of A of rank α to be the minimal elements of the subset obtained by removing from A all elements having rank $< \alpha$. We can continue this process through the ordinals until the set A is exhausted. This procedure is readily formalized using the framework of abstract iterated derivatives and ranks (Theorem 714, Sect. 10.4), when we define the derivative operator ∇ by $\nabla(E) := E \smallsetminus \min_R[E]$.

Theorem 811 (Canonical Decomposition of Well-Founded Relation). *Let R be a well-founded relation on the set A. Then there is a unique ordinal μ and a unique partition $\langle A_\alpha \mid \alpha < \mu\rangle$ of A into pairwise disjoint nonempty sets such that for every $\alpha < \mu$, A_α consists of the set of R-minimal elements of $A \smallsetminus \bigcup_{\beta<\alpha} A_\beta$, that is,*

$$A_\alpha = \min_R \left[A \smallsetminus \bigcup_{\beta<\alpha} A_\beta \right].$$

Proof. The result follows directly from Theorem 714 when we define a derivative operator $\nabla \colon \mathbf{P}(A) \to \mathbf{P}(A)$ by

$$\nabla(B) := B \smallsetminus \min_R[B],$$

and define the sets A_α as

$$A_\alpha := A^{(\alpha)} \smallsetminus A^{(\alpha+1)},$$

where $A^{(\alpha)}$ denotes the α-th iterated derivative of A. In particular, $A^{(\alpha+1)} = \nabla(A^{(\alpha)})$ and $A \smallsetminus A^{(\alpha)} = \cup_{\beta<\alpha} A_\beta$. Theorem 714 guarantees the existence of a unique least ordinal μ such that $A^{(\mu+1)} = A^{(\mu)}$. Since R is well-founded on A, the derivative ∇ is strict, and so we have $A^{(\mu)} = \emptyset$. It follows that $\langle A_\alpha \mid \alpha < \mu \rangle = \langle A^{(\alpha)} \smallsetminus A^{(\alpha+1)} \mid \alpha < \mu \rangle$ is a partition of A satisfying the condition of the theorem.

Uniqueness of the partition follows by routine transfinite induction. □

Note that in the framework of Theorem 714, the set A_α above consists precisely of the elements of rank α. Thus rank can also be defined in terms of the sets A_α as follows.

Definition 812 (Ranks on Well-Founded Relations). Let R be a well-founded relation on A, and let $\langle A_\alpha \mid \alpha < \mu \rangle$ be the canonical decomposition of $\langle A, R \rangle$ as stated in the theorem.

1. For each $x \in A$, the *rank of the element* x, denoted by $\rho_R(x)$, is defined to be the unique ordinal ν such that $x \in A_\nu$.
2. The ordinal μ will be called the *rank of the well-founded structure* $\langle A, R \rangle$ and will be denoted by $\text{rank}_R(A)$.
3. The mapping $x \mapsto \rho_R(x)$ from A to the set $W(\mu)$ of ordinals is called the *canonical rank function* for the well-founded structure $\langle A, R \rangle$.

Remark. With ∇ as the derivative operator defined by

$$\nabla(B) := B \smallsetminus \min_R[B],$$

the rank function ρ_R in the above definition is same as the rank function $\rho = \rho_\nabla$ of Theorem 714 for the abstract derivative ∇.

Problem 813. *Show that if R is a well-founded relation on a set A having rank μ, then the canonical rank function $\rho_R \colon A \to W(\mu)$ is surjective, and hence we have:*

$$\mu = \text{rank}_R(A) = \sup_{x \in A}(\rho_R(x) + 1).$$

Problem 814. *Show that for a well-founded structure $\langle A, R \rangle$, the canonical rank function ρ_R satisfies:*

$$xRy \Rightarrow \rho_R(x) < \rho_R(y) \qquad \textit{(for all } x, y \in A\textit{).}$$

Definition 815. Say that ρ *is a rank function for a relation* R *on a set* A (or ρ *is rank function for* $\langle A, R \rangle$) if ρ maps A to a set of ordinals and ρ is strictly increasing, i.e., if for any $x, y \in A$, $\rho(x)$ and $\rho(y)$ are ordinals and

$$xRy \Rightarrow \rho(x) < \rho(y).$$

We say that a relation R on a set A *admits a rank function* if there is some rank function ρ for $\langle A, R\rangle$.

The following is an important characterization of well-founded relations.

Problem 816. *A relation is well-founded if and only if it admits a rank function.*

By Theorem 811, for every well-founded relation one can effectively determine a rank function for it, namely the canonical rank function. The canonical rank function can itself be characterized as follows.

Problem 817. *Let R be a well-founded relation on a set A, and let ρ be any rank function for $\langle A, R\rangle$. Show that ρ equals the canonical rank function ρ_R if and only if*

$$\text{For every } x \in A\text{:} \quad \rho(x) = \sup\{\rho(y) + 1 \mid y \in A,\ yRx\},$$

where we take $\sup \emptyset = 0$.

If ρ, σ are rank function for $\langle A, R\rangle$, we write $\rho \preceq \sigma$ to denote $\rho(x) \leq \sigma(x)$ for all $x \in A$. The following characterizes the canonical rank function as the unique "least one."

Problem 818. *Let R be a well-founded relation on a set A, and let ρ_R be the canonical rank function for $\langle A, R\rangle$. Show that $\rho_R \preceq \rho$ for any rank function ρ for $\langle A, R\rangle$. Conversely, if ρ^* is a rank function for $\langle A, R\rangle$ such that $\rho^* \preceq \rho$ for every rank function ρ for $\langle A, R\rangle$, then $\rho^* = \rho_R$.*

Problem 819. *Let R be a relation on A, S be a well-founded relation on B, and let $f\colon A \to B$ be strictly increasing: $xRy \Rightarrow f(x)Sf(y)$. Then R is well-founded on A, and the rank of $\langle A, R\rangle$ is at most the rank of $\langle B, S\rangle$.*

Definition 820. Let R be a relation on a set A, and $x, y \in A$.

1. x *is an R-predecessor of* y if xRy.
2. $B \subseteq A$ is *downward R-closed* if $v \in B, uRv \Rightarrow u \in B$.
3. We say that x *is an R-ancestor of* y, and write xR_*y, if every downward R-closed subset of A containing all R-predecessors of y also contains x.
4. $\overleftarrow{R}[y] := \{x \mid xR_*y\}$ denotes the set of all R-ancestors of y.

Problem 821. *Let R be a relation on a set A. Then xR_*y if and only if there exist $n \geq 2$ and $u_1, u_2, \ldots, u_n$ such that $u_1 = x$, $u_n = y$, and $u_k R u_{k+1}$ for $1 \leq k < n$, and so R_* is the smallest transitive relation containing R, i.e., R_* is the* transitive closure *of R.*

Problem 822. *Let R be a relation on a set A. Show that R_* is well-founded on A if and only if R is well-founded on A.*

As a result we have the following strengthening of Problem 810.

Problem 823 (Strong Induction on Well-Founded Relations). *Let R be a well-founded relation on the set A, and $B \subseteq A$. Suppose that for any $a \in A$, if every R-ancestor of a is in B then $a \in B$. Then $B = A$.*

Note that if R is well-founded on A, then for any $a \in A$, R is well-founded on $\overleftarrow{R}[a]$, and so $\overleftarrow{R}[a]$ by itself becomes a well-founded structure under the relation R, which we may call the *well-founded substructure* consisting of the R-ancestors of a. The following useful proposition shows how the distinct rank functions on the parent structure and on the substructures are related.

Proposition 824. *Let $\langle A, R\rangle$ be a well-founded structure and let $a \in A$. Then*

1. *The canonical rank function $\rho_{\overleftarrow{R}[a]}$ for the well-founded substructure $\overleftarrow{R}[a]$ is the restriction of the canonical rank function ρ_R for $\langle A, R\rangle$.*
2. $\rho_R(a) = \operatorname{rank}_R(\overleftarrow{R}[a])$, *that is the canonical rank of a in $\langle A, R\rangle$ equals the rank of the substructure $\overleftarrow{R}[a]$.*
3. $\operatorname{rank}_R(A) = \sup_{b\in A}\big(\operatorname{rank}_R(\overleftarrow{R}[b]) + 1\big)$.

Proof. The first part follows from Problem 817 by transfinite induction, using the fact that $\overleftarrow{R}[a]$ is downward R-closed.

Since $\rho_R(a) = \sup\{\rho_R(x) \mid xRa\} = \sup\{\rho_{\overleftarrow{R}[a]}(x) \mid xRa\} = \operatorname{rank}_R(\overleftarrow{R}[a])$, the second part follows.

The last part follow from the second part. □

Problem 825 (Transfinite Induction for Well-Founded Structures). *Let P be a property which satisfies the following condition: For any well-founded relation R on any set A, if every substructure $\overleftarrow{R}[a]$ $(a \in A)$ has property P, then $\langle A, R\rangle$ itself has property P. Then every well-founded structure has property P.*

The following problem gives an example of a well-founded relation whose inverse relation is also a nontrivial well-founded relation.

Problem 826. *Let $X = \mathrm{P}^*(\mathbf{N})$ be the set of all finite subsets of $\mathbf{N}$ and define a relation P on X by the condition aPb if and only $a \subsetneq b$ and either $a = \emptyset$ or* $\min a > \max(b \smallsetminus a)$. *Let $R = P^{-1}$ be the inverse relation of P. Show that*

1. *aPb holds if and only if b can be partitioned as $b = c \cup a$ $(c \cap a = \emptyset)$, where $c \neq \emptyset$ and $x < y$ for all $x \in c$ and $y \in a$.*
2. *Both $\langle X, P\rangle$ and $\langle X, R\rangle$ are well-founded strict posets.*
3. *In $\langle X, P\rangle$, the element $\emptyset$ is the least element (hence the unique minimal element), and every singleton is an immediate P-successor of $\emptyset$.*
4. *If $a \in X$ has n elements ($|a| = n$), what is the P-rank of a?*
5. *In $\langle X, P\rangle$, every element has finite P-rank, and for every $n \in \mathbf{N}$ there is an element having P-rank n.*
6. *There is no strictly P-increasing infinite sequence.*

7. *R is a well-founded relation on X, in which $\emptyset$ is the R-greatest element and every singleton is an immediate R-predecessor of $\emptyset$.*
8. *$\{1\}$ is an R-minimal element in X.*
9. *What are the R-predecessors of $\{2\}$? Of $\{3\}$?*
10. *Draw a diagram showing the R-predecessors of $\{4\}$ and how they are related by the relation R.*
11. *What is the R-rank of $\{2\}$? Of $\{3\}$? Of $\{4\}$?*
12. *What are the R-minimal elements in X?*
13. *For each $n \in \mathbf{N}$ find the R-rank of $\{n\}$.*
14. *What is the R-rank of $\emptyset$?*
15. *What is the rank of the well-founded relation $\langle X, R \rangle$? Of $\langle X, P \rangle$?*

Problem 827. *A positive integer is called* square free *if it is a product of distinct primes. We regard* 1 *as square free. If $a \in \mathbf{N}$ is square free with*

$$a = q_1 q_2 \cdots q_n, \qquad \text{where } q_1 < q_2 < \cdots q_n \text{ are increasing primes,}$$

then we say that q_k is the k-th prime factor of a *($k = 1, 2, \ldots, n$). Let A be the set of those square free positive integers a such that, with q as the smallest prime factor of a, the total number of prime factors of a either does not exceed q or does not exceed the value of the q-th prime factor of a. In particular, $1 \in A$. Define a relation R on A by the condition*

$$aRb \Leftrightarrow a, b \in A \text{ and } b \text{ is a proper divisor of } a.$$

1. *Show that $\langle A, R \rangle$ is a well-founded structure, that is, the relation R is well-founded on A.*
2. *Find the ranks of the elements* 6, 10, *and* 21
3. *Characterize the R-minimal elements of A.*
4. *Find the rank of the element* 1 *and of the structure $\langle A, R \rangle$.*

11.5 Trees

Definition 828. A poset $\langle T, \prec \rangle$ is called a *tree* if either $T = \emptyset$, or T has a least element root(T) (called the *root of T*) and the set of predecessors of any element is well-ordered.

The elements of a tree will often be referred to as *nodes*.

If $\langle T, \prec \rangle$ is a tree, we will often refer to the underlying set T as the tree so long as the relation $\prec$ can be understood from context.

Evidently, every well-order is a tree, and every tree is well-founded. The poset $\langle X, P\rangle$ from Problem 826 is tree.

Definition 829. Let T be a tree under the relation $\prec$.

1. We say that $T' \subseteq T$ is a *subtree* of T if T' downward closed, i.e., if $x \in T'$ and $y \prec x$ imply $y \in T'$.
2. The *height of an element* $x \in T$, denoted by $\mathrm{ht}_T(x)$ or $\mathrm{ht}(x)$, is the order type (ordinal) of the set $\{y \in T \mid y \prec x\}$ of predecessors of x.
3. For any ordinal α, the *α-th level of* T, denoted by $\mathrm{Lev}_\alpha(T)$, is defined as the set of all elements of T with height α. (So $x \in \mathrm{Lev}_\alpha(T) \Leftrightarrow \mathrm{ht}(x) = \alpha$.)
4. A node $v \in T$ is a *child* of a node $u \in T$ if v is an immediate successor of u, i.e., if $u \prec v$ and $\mathrm{ht}(v) = \mathrm{ht}(u) + 1$.
5. B is a *branch* of T if B is a chain and is downward closed, i.e., if B is linearly ordered subtree of T.

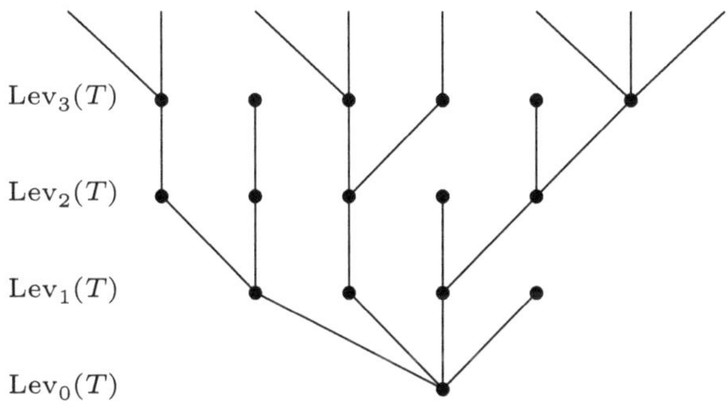

A Tree T Drawn Growing Upward

The following facts are immediate.

Problem 830. *Any subtree of a tree is a tree. If a tree has an element of height α, then it has elements of every height $\beta < \alpha$. The levels of a tree are pairwise disjoint, and so form a partition of the tree.*

Problem 831. *Show that for every tree T there is an ordinal α such that no element of T has height α.*

[Hint: If η is the Hartogs ordinal for $\mathbf{P}(T)$, then the levels $\mathrm{Lev}_\alpha(T)$, $\alpha < \eta$, are pairwise disjoint sets so all of them cannot be non empty.]

Definition 832. The *height of a tree* T, denoted by $\mathrm{ht}(T)$, is the least ordinal α such that no element of T has height α.

Definition 833. A tree T is said to be *finitely branching* if every node has at most finitely many children (immediate successors), i.e., if for any $x \in T$, the set $\{y \in T \mid x \prec y \text{ and } \mathrm{ht}(y) = \mathrm{ht}(x) + 1\}$ is finite.

Problem 834. *The poset $\langle X, P\rangle$ from Problem 826 is tree. What is its height? Is it finitely branching?*

Problem 835. *Let $\langle T, \prec\rangle$ be a tree. Regarding it as a well-founded structure, show that the height of an element x, $\mathrm{ht}(x)$, is same as $\rho_{\prec}(x)$, the canonical $\prec$-rank of x, and that the height of the tree, $\mathrm{ht}(T)$, equals $\mathrm{rank}_{\prec}(T)$, the rank of the structure $\langle T, \prec\rangle$.*

Problem 836. *A tree in which every level is finite must be finitely branching. Is the converse true?*

The Tree A^* of Strings over a Set A

The following example gives an especially important type of tree that will concern us.

Example 837. Let A be a nonempty set and let A^* denote the set of all strings (finite sequences) consisting of elements of A. Then A^* is a tree under the relation $\subset$, where $u \subset v$ stands for "u is a (proper) initial prefix of v."

Two important special cases of this example are obtained by taking $A = \{0, 1\}$, giving us the *full binary tree* $\{0, 1\}^*$ where every node has exactly two immediate successors, and by taking $A = \mathbf{N}$ which gives a tree $\mathbf{N}^*$ in which every node has infinitely many immediate successors.

Problem 838. *Consider the tree A^* of all finite strings over A ($A \neq \emptyset$). Let $\kappa = |A|$ be the cardinality of A. Show that*

1. *A^* is a tree under the relation $\subset$ with root ε and height ω.*
2. *The height of any element is its length (as a string).*
3. *Every node in A^* has κ-many immediate successors, so A^* is finitely branching if and only if A is finite.*
4. *$|\mathrm{Lev}_n(A^*)| = \kappa^n$ for any finite $n = 0, 1, 2, \ldots$, so if A is finite then every level of A^* is finite.*
5. *A branch in the tree A^* is infinite if and only if it is maximal.*

Definition 839 (Trees over A). We say that T is a *tree over A* if T is a subtree of the tree A^* of strings from A (under the string prefix relation $\subset$).

We now obtain a "representation theorem" for trees of height at most ω.

Problem 840. *Let $\langle T, \prec\rangle$ be a tree of height at most ω. Then T is isomorphic to a tree over A for some A. That is, there is a set A, a subtree $T' \subseteq A^*$, and a bijection $f : T \to T'$ such that for all $x, y \in T$, $x \prec y \Leftrightarrow f(x) \subset f(y)$ (x precedes y in T if and only if $f(x)$ is an initial prefix of $f(y)$).*

Moreover, if T is also countable, then one can take A to be the set $\mathbf{N}$.

Thus *every countable tree of height $\leq \omega$ is isomorphic to some tree over* $\mathbf{N}$.

Problem 841. *Let $\langle \mathbf{N}, | \rangle$ denote the poset of natural numbers under the (strict) divisibility relation $|$. Show that every countable tree of height $\leq \omega$ can be embedded (as a poset) into $\langle \mathbf{N}, | \rangle$.*

[Hint: By the previous problem, it suffices to embed $\mathbf{N}^*$ into $\langle \mathbf{N}, | \rangle$.]

Although we are primarily interested in countable trees of height ω, the representation theorem can be generalized to arbitrary trees as well.

Problem 842. *Given a set A and an ordinal α, let $A^{<\alpha}$ denote the set of all functions whose domain is a proper initial segment of $W(\alpha)$ and whose range is contained in A. For $u, v \in A^{<\alpha}$, let $u \subset v$ if and only if v extends u, that is, if there exist ordinals $\alpha < \beta$ such that* $\mathrm{dom}(u) = W(\alpha)$, $\mathrm{dom}(v) = W(\beta)$, *and for all $\gamma < \alpha$ we have $u(\gamma) = v(\gamma)$. Show that*

1. *For any set A and any ordinal α, $\langle A^{<\alpha}, \subset \rangle$ is a tree of height α.*
2. *Every tree is isomorphic to a subtree of $\langle A^{<\alpha}, \subset \rangle$, for some set A and some ordinal α.*

Remark. With the notation of the last problem, the tree A^* of all finite sequences from A can be denoted by $A^{<\omega}$.

11.6 König's Lemma and Well-Founded Trees

König's Infinity Lemma

Problem 843. *Show that if T is a tree of height $\leq \omega$, then T is finitely branching if and only if every level of T is finite.*

Give an example of a finitely-branching tree with some infinite levels.

Theorem 844 (The König Infinity Lemma). *Let T be a tree of height ω in which every level is finite. Then T has an infinite branch.*

The result is often expressed by saying "every finitely branching infinite tree has an infinite branch."

Proof. Let $\langle T, \prec \rangle$ be a tree of height ω in which every level is finite.

Since T has height ω so it has elements of height n for every $n < \omega$, and so T is infinite. For each $x \in T$, let $\mathrm{Succ}(x) := \{y \mid x \prec y \text{ or } x = y\}$. Note that for any $x \in T$ and any $n < \omega$, we have

$$x \in \mathrm{Lev}_n(T) \Rightarrow \mathrm{Succ}(x) = \{x\} \cup \bigcup \{\mathrm{Succ}(y) \mid x \prec y, y \in \mathrm{Lev}_{n+1}(T)\}.$$

Note that since every level of T is finite, so the big union above is actually a finite union.

Let x_0 be the least element of T. Then $\text{Succ}(x_0) = T$, so $\text{Succ}(x_0)$ is infinite, with

$$\text{Succ}(x_0) = \{x_0\} \cup \bigcup \{\text{Succ}(y) \mid x_0 \prec y, y \in \text{Lev}_1(T)\}.$$

Since the big union above is finite while the left side is infinite, there must be at least one $x_1 \in \text{Lev}_1(T)$ such that $\text{Succ}(x_1)$ is infinite. Fix such an x_1 (with $\text{Succ}(x_1)$ infinite). Then

$$\text{Succ}(x_1) = \{x_1\} \cup \bigcup \{\text{Succ}(y) \mid x_1 \prec y, y \in \text{Lev}_2(T)\}.$$

Again, since the big union above is finite but the left side is infinite, we can fix $x_2 \in \text{Lev}_2(T)$ such that $\text{Succ}(x_2)$ is infinite. Continuing in this fashion, we get a sequence

$$x_0 \prec x_1 \prec x_2 \prec \cdots \prec x_n \prec x_{n+1} \prec \cdots \qquad \big(x_n \in \text{Lev}_n(T)\big).$$

Then $B := \{x_n \mid n = 0, 1, 2, \dots\}$ is an infinite branch through T. □

Well-Founded Trees

Since every tree is well-founded, the term "well-founded tree" seems redundant. However, it has the following special meaning in the context of trees.

Definition 845. A *well-founded tree* is a tree $\langle T, P\rangle$ such that the inverse relation $R = P^{-1}$ is well-founded on T.

We saw an example of a well-founded tree in Problem 826.

Problem 846 (DC). *Show that a tree is well-founded if and only if it has no infinite branch.*

Definition 847 (Ranks in Well-Founded Trees). Let $\langle T, P\rangle$ be a well-founded tree, so that the inverse relation $R = P^{-1}$ is well-founded on T, and let ρ_R be the canonical rank function for the well-founded structure $\langle T, R\rangle$.

The *rank of an element* $x \in T$ is defined as $\rho_R(x)$.

The *rank of the tree* T, denoted by $\text{rank}(T)$, is defined as $\rho_R(r)$ where $r = \text{root}(T)$ is the P-least element of T. We put $\text{rank}(T) = 0$ if $T = \emptyset$.

Note on terminology. If $\langle T, P\rangle$ is a well-founded tree with the inverse relation $R = P^{-1}$ well-founded on T, then the term "height" applies to the relation P and the term "rank" applies to the relation R. E.g., the "rank of an element x" is the R-rank of x in the well-founded structure $\langle T, R\rangle$, while the "height of x" is the P-height of x in the tree $\langle T, P\rangle$.

Problem 848. *A well-founded tree has height at most ω.*

By Problems 840 and 848, every well-founded tree is isomorphic to a tree over A for some set A, so the study of well-founded trees can be limited to trees over (some set) A, i.e., to subtrees of A^*.

Problem 849. *Define a subtree T of $\mathbf{N}^*$ by*

$$T := \{u \in \mathbf{N}^* \mid u = \varepsilon \text{ or } u = \langle u_1, u_2, \ldots, u_n \rangle,\ n \in \mathbf{N}, \text{ and } u_1 \geq n\}.$$

1. *Show that $\langle X, P \rangle$ is a well-founded tree of rank $\omega + 1$.*
2. *What is the rank of ε? What is the rank of the element $\langle 7, 9, 2 \rangle$?*

Problem 850. *Give an example of a well-founded tree over $\mathbf{N}$ which has rank ω^2 but height ω.*

Problem 851. *A nonempty well-founded tree T has finite rank if and only if it has finite height, and in this case* $\operatorname{ht}(T) = \operatorname{rank}(T) + 1$.

Problem 852. *If A is a finite set then any well-founded tree over A must be finite and so must have rank $< \omega$.*

Problem 853. *If $T \subseteq A^*$ is a well-founded tree over a set A, then* $\operatorname{rank}(T)$ *equals the rank of the well-founded structure $\langle T \smallsetminus \{\varepsilon\}, \supset \rangle$.*

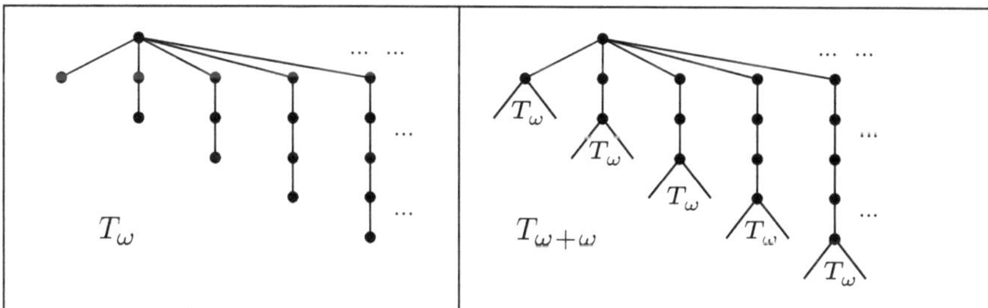

T_ω: A well founded tree of rank ω | $T_{\omega+\omega}$: A well founded tree of rank $\omega+\omega$

Problem 854 (Truncated Ranks). *For a tree T over $\mathbf{N}$ and $u \in \mathbf{N}^*$, put*

$$T^{(u)} := \{v \in \mathbf{N}^* \mid u * v \in T\} \qquad (\textit{i.e., } T^{(u)} \textit{ is } T \textit{ truncated at } u.)$$

Then:

1. *$T^{(u)}$ is a tree, and $(T^{(u)})^{(v)} = T^{(u*v)}$. If T is well-founded then so is $T^{(u)}$ with* $\operatorname{rank}(T^{(u)}) \leq \operatorname{rank}(T)$.
2. *The rank function on well-founded trees satisfies, and is the unique function satisfying, the recursion equation*

$$\text{rank}(T) = \sup\{\text{rank}(T^{(\langle n\rangle)}) + 1 \mid \langle n\rangle \in T,\ n \in \mathbf{N}\}, \quad \textit{with } \sup(\emptyset) := 0.$$

3. *If $T \not\subseteq \{\varepsilon\}$, then T is well-founded of rank α if and only if $T^{(\langle n\rangle)}$ is well-founded of rank $< \alpha$ for all $n \in \mathbf{N}$ and for each $\xi < \alpha$ there is $v \in \mathbf{N}^*$ with* $\text{len}(v) > 0$ *such that $T^{(v)}$ has rank ξ.*

Existence of Well-Founded Trees of Every Rank. Our remaining task is to show that for every countable ordinal $\alpha < \omega_1$ there is countable well-founded tree (over $\mathbf{N}$) having rank α. (Using the full Axiom of Choice, one can also show that for every ordinal α there is a well-founded tree of rank α.)

Definition 855. If $T \subseteq A^*$ is a tree over A and $a \in A$, define:

$$\begin{aligned} a * T &:= \{\varepsilon\} \cup \{\langle a\rangle * u \mid u \in T\} \\ &= \{\varepsilon\} \cup \{\langle a, u_1, u_2, \ldots, u_n\rangle \mid \langle u_1, u_2, \ldots, u_n\rangle \in T,\ n \geq 0\} \end{aligned}$$

Problem 856. *Show that*

1. *If $T \subseteq \mathbf{N}^*$ is a nonempty well-founded tree over $\mathbf{N}$ and $n \in \mathbf{N}$, then $n * T$ is well-founded tree over $\mathbf{N}$ with*

$$\text{rank}(n * T) = \text{rank}(T) + 1.$$

In particular, for every well-founded tree $T \subseteq \mathbf{N}^*$ over $\mathbf{N}$ one can effectively find a well-founded tree $T' \subseteq \mathbf{N}^*$ such that $\text{rank}(T') > \text{rank}(T)$.

2. *If $\langle T_n \mid n \in \mathbf{N}\rangle$ is a sequence of well-founded trees over $\mathbf{N}$, then*

$$T := \{\varepsilon\} \cup \bigcup_{n \in \mathbf{N}} n * T_n$$

is a well-founded tree over $\mathbf{N}$, and if $T_n \neq \emptyset$ for some n, then

$$\text{rank}(T) = \sup_{n \in \mathbf{N}} \big(\text{rank}(T_n) + 1\big).$$

Notice that every countable well-founded tree must have rank $< \omega_1$. Using the last problem (and use of Choice), one obtains the following converse result.

Problem 857 (CAC). *Show that for every ordinal $\alpha < \omega_1$, there is a well-founded tree over $\mathbf{N}$ having rank α.*

Finally, under the full Axiom of Choice one gets the existence of well-founded trees of every possible rank.

Problem 858 (AC). *Show that for every ordinal α, there is a well-founded tree of rank α.*

11.7 Ramsey's Theorem

A popular puzzle says that in any group of six or more people there are three people who are mutual acquaintances or mutual strangers. Ramsey's Theorem, which we prove below, says that in any infinite group of people there are infinitely many people who are mutual strangers or mutual acquaintances.

Definition 859. For any set X and $n \in \mathbf{N}$, we let $[X]^n$ denote the family of n-element subsets of X:

$$[X]^n := \{E \mid E \subseteq X \text{ and } |E| = n\}.$$

If $[X]^n$ is partitioned into k sets as $[X]^n = \bigcup_{i=1}^k X_i$, then a subset $H \subseteq X$ is said to be *homogeneous* for the partition if $[H]^n \subseteq X_i$ for some $i = 1, 2, \ldots, k$.

Similarly, if $f\colon [X]^n \to \{1, 2, \ldots, k\}$ then $H \subseteq X$ is said to be *homogeneous* for f if f is constant on $[H]^n$.

The puzzle above can now be stated as follows: If $|X| \geqslant 6$ and if $[X]^2$ is partitioned as $[X]^2 = X_1 \cup X_2$ (with $X_1 \cap X_2 = \emptyset$) then there is $H \subseteq X$ with $|H| = 3$ and $[H]^2 \subseteq X_i$ for some $i \in \{1, 2\}$ (i.e., H is homogeneous).

We could state the result equivalently using functions: If $|X| \geqslant 6$ and $f\colon [X]^2 \to \{1, 2\}$, then there is $H \subseteq X$ such that $|H| = 3$ and f is constant on $[H]^2$, i.e., $f(\{x, y\}) = f(\{u, v\})$ for all $x, y, u, v \in H$, with $x \neq y$ and $u \neq v$.

Theorem 860 (Ramsey's Theorem). *If X is an infinite set and $f\colon [X]^2 \to \{1, 2\}$, then some infinite $H \subseteq X$ is homogeneous for f.*

Proof. We will use König's Infinity Lemma to prove the theorem.

Without loss of generality we assume $X = \mathbf{N}$, so that $f\colon [\mathbf{N}]^2 \to \{1, 2\}$. Also for each 2-element set $\{m, n\}$ with $m < n$, we will write $f(m, n)$ for $f(\{m, n\})$. For each nonempty $E \subseteq \mathbf{N}$, we put

$$E^{(1)} := \{n \in E \mid n > \min(E) \text{ and } f(\min(E), n) = 1\}, \quad \text{and}$$

$$E^{(2)} := \{n \in E \mid n > \min(E) \text{ and } f(\min(E), n) = 2\}.$$

Then $E^{(1)}$ and $E^{(2)}$ are disjoint, and $E = \{\min(E)\} \cup E^{(1)} \cup E^{(2)}$, so $E^{(1)} \cup E^{(2)}$ contains all but one member of E.

We now define a finitely branching tree T of height at most ω consisting of *nonempty* subsets of $\mathbf{N}$ in which every node has at most two children, so that $|\operatorname{Lev}_n(T)| \leq 2^n$ for all $n = 0, 1, 2, \ldots$.

Let $\mathbf{N}$ be the root node of T, and for each node $E \in T$, we take each of $E^{(1)}$ and $E^{(2)}$, provided that it is nonempty, to be a child of E. Since the union of the children of a node contains all but one member of the node and since T is finitely branching, it follows by induction that the union of the nodes of level n contains all but finitely many natural numbers. Hence $\mathrm{Lev}_n(T) \neq \emptyset$ for all $n = 0, 1, 2, \ldots$, and so T has height ω. By König's Infinity Lemma, T has an infinite branch, say

$$E_0 \supsetneq E_1 \supsetneq E_2 \supsetneq \cdots \supsetneq E_n \supsetneq E_{n+1} \supsetneq \cdots$$

where $E_0 = \mathbf{N}$, and E_{n+1} equals $E_n^{(1)}$ or $E_n^{(2)}$ for all n. Hence there is $b \in \{1, 2\}$ such that $E_{n+1} = E_n^{(b)}$ for infinitely many n, and so there are natural numbers $n_1 < n_2 < \cdots < n_k < \cdots$ with $E_{n_k+1} = E_{n_k}^{(b)}$ for all k. Put $a_k := \min(E_{n_k})$. Then for $k < m$ we have $a_m \in E_{n_m} \subseteq E_{n_k+1} = E_{n_k}^{(b)}$, hence $f(a_k, a_m) = b$. Therefore the set $H = \{a_1, a_2, \ldots\}$ is homogeneous for f. □

Definition 861 (Arrow Notation). If κ, μ are cardinals and $n, k \in \mathbf{N}$, we write

$$\kappa \to (\mu)^n_k$$

to denote the statement "For any sets X with $|X| = \kappa$ and any function $f\colon [X]^n \to \{1, 2, \ldots, k\}$, there is a homogeneous $H \subseteq X$ with $|H| = \mu$."

Thus, Ramsey's Theorem says that $\aleph_0 \to (\aleph_0)^2_2$. This is a special case of the following more general result.

Theorem 862 (General Ramsey Theorem). *For all $n, k \in \mathbf{N}$, we have:*

$$\aleph_0 \to (\aleph_0)^n_k.$$

We will not prove this result, but the reader may try it as a challenging exercise (it can be proved using induction on n).

A sequence $\langle x_n \rangle$ in an order or a partial order is *monotone increasing* if $x_m \leq x_n$ for all $m < n$, and it is *strictly increasing* if $x_m < x_n$ for all $m < n$. Similar definitions are given for decreasing sequences. A sequence is *monotone* if it is either monotone increasing or monotone decreasing.

Problem 863. *Use the General Ramsey Theorem to show that every infinite sequence in a linear order has a subsequence which is either strictly decreasing or strictly increasing or constant. Conclude that every infinite sequence in a linear order has a monotone subsequence.*

[Hint: For a sequence $\langle x_n \rangle$ in an order define $f\colon [\mathbf{N}]^2 \to \{1, 2, 3\}$ by setting, for $m < n$, $f(m, n) := 1$ if $x_m < x_n$, $:= 2$ if $x_m > x_n$, and $:= 3$ if $x_m = x_n$.]

Problem 864. *Every infinite sequence in a partial order has a subsequence which is either monotone or consists of pairwise incomparable elements.*

Problem 865. *Show that $2^{\aleph_0} \not\to (\aleph_1)^2_2$. Hence $\aleph_1 \not\to (\aleph_1)^2_2$.*

[Hint: If $2^{\aleph_0} \to (\aleph_1)^2_2$, argue as in Problem 863 to show that every linear order of size $2^{\aleph_0}$ has a suborder of type ω_1 or of type $^*\omega_1$, a contradiction.]

Problem 866. *Show that* $2^{\aleph_\alpha} \not\to (\aleph_{\alpha+1})^2_2$.

[Hint: Use Problem 796.]

Remark. The General Ramsey Theorem, for each $n \geq 3$, is closely related to König's Infinity Lemma. In a sense, each can be "easily derived" from the other without using any other "strong" theorems. This vague statement is made precise in an area of mathematical logic known as *reverse mathematics,* where strengths of mathematical statements are studied relative to weaker base subsystems. See [75] for more details.

Chapter 12
Postscript II: Infinitary Combinatorics

Abstract The topics of the last chapter (Chap. 11) naturally lead to the area of *Infinitary Combinatorics,* which is beyond the scope of this text. This postscript to Part II is intended to be a link for the reader to begin further study in the area. We indicate how the obvious generalizations of three separate topics of the last chapter, namely short orders, König's Infinity Lemma, and Ramsey's Theorem, converge naturally to the notion of a *weakly compact cardinal,* an example of a large cardinal. In addition, it is shown how Suslin's Problem is equivalent to the existence of Suslin trees. Finally, we briefly mention Martin's Axiom and Jensen's Diamond principle $\diamondsuit$, and their implications for the Suslin Hypothesis.

Note: *Throughout this postscript we will assume the Axiom of Choice without explicitly mentioning it.*

12.1 Weakly Compact Cardinals

An interesting property of (linear) orders is that any sequence in an order has a monotone subsequence. This can be proved in various ways, e.g., Problem 863 derived it from Ramsey's Theorem. The property can be stated equivalently as: *Any order of size $\aleph_0$ has a suborder of order type ω or $^*\omega$.*

On the other hand **R** is a short linear order (i.e., **R** has no subset of type ω_1 or $^*\omega_1$) and $\aleph_1 \leqslant |\mathbf{R}|$, so we have the contrasting fact: *There are orders of size $\aleph_1$ which do not have any suborder of order type ω_1 or $^*\omega_1$.*

Definition 867. We will say that an infinite cardinal $\kappa = \aleph_\alpha$ has the *monotone order property*[1] if every linear order of cardinality $\aleph_\alpha$ has a suborder of type ω_α or $^*\omega_\alpha$.

[1]This terminology is not a standard one.

A. Dasgupta, *Set Theory: With an Introduction to Real Point Sets*,
DOI 10.1007/978-1-4614-8854-5_12,

Thus $\aleph_0$ has the monotone order property but $\aleph_1$ does not. More generally:

Proposition 868. *An infinite cardinal with the monotone order property is a strong limit. Hence, no successor cardinal has the monotone order property.*

Proof. Suppose that $\kappa = \aleph_\alpha$ has the monotone order property. If we had $2^{\aleph_\beta} \geqslant \kappa = \aleph_\alpha$ for some $\beta < \alpha$, then the lexicographic power $\{0,1\}^{W(\omega_\beta)}$ would have a suborder of type ω_α or $^*\omega_\alpha$ and hence also a suborder of type $\omega_{\beta+1}$ or $^*\omega_{\beta+1}$ (as $\beta + 1 \leq \alpha$), contradicting Problem 796. □

We now have:

Theorem 869. *Any uncountable cardinal $\kappa = \aleph_\alpha$ having the monotone order property is regular, and therefore strongly inaccessible.*

Proof. Otherwise, we would get $\aleph_\alpha = \sum_{\xi<\beta} \aleph_{\alpha_\xi}$ where $\beta < \alpha$ and $\alpha_\xi < \alpha$ for all $\xi < \beta$. For each $\xi < \beta$ fix an order X_ξ of order type $^*\omega_{\alpha_\xi}$, and let $X := \bigcup_{\xi<\beta}(\{\xi\} \times X_\xi)$ be equipped with the lexicographic order. Then X does not contain any subset of order type ω_α or $^*\omega_\alpha$, but $|X| = \aleph_\alpha$. □

A similar property of cardinals can be generalized from Ramsey's Theorem. We saw that $\aleph_0 \to (\aleph_0)^2_2$ (Ramsey's Theorem), while $\aleph_1 \not\to (\aleph_1)^2_2$. (Problem 865). As in the case for the monotone order property, this immediately implies that if $\kappa \to (\kappa)^2_2$ then κ must be a strong limit:

Proposition 870. *Let κ be an infinite cardinal satisfying $\kappa \to (\kappa)^2_2$. Then κ is a strong limit. In particular, κ cannot be a successor cardinal.*

Proof. Essentially same as that for Proposition 868, but use Problem 866. □

Theorem 871. *If κ is uncountable and $\kappa \to (\kappa)^2_2$, then κ is regular and therefore strongly inaccessible.*

Proof. If κ were singular, we would get a partition of the form $X = \bigcup_{i\in I} X_i$ where $|X| = \kappa$, $|I| < \kappa$, and $|X_i| < \kappa$ for all $i \in I$. Define $f\colon [X]^2 \to \{1,2\}$ by setting $f(\{x,y\}) := 1$ if $x, y \in X_i$ for some $i \in I$, and $:= 2$ otherwise. A homogeneous set for the partition would then yield a contradiction. □

The third and last property of cardinals that we will consider is generalized from König's Infinity Lemma.

Definition 872. A cardinal $\kappa = \aleph_\alpha$ is said to have the *tree property* if any tree T of height ω_α in which each level has cardinality $< \kappa$ has a branch of height ω_α (a branch of height ω_α can be equivalently described as a chain $C \subseteq T$ such that $\mathrm{Lev}_\xi(T) \cap C \neq$ for all $\xi < \omega_\alpha$).

König's Infinity Lemma is the assertion that $\aleph_0$ has the tree property. A result of Aronszajn says that $\aleph_1$ does not have the tree property. This means there is a tree T of height ω_1 in which all levels are countable yet T has no branch of height ω_1. Such trees are known as *Aronszajn trees.*

However, we cannot prove that a cardinal having the tree property must be inaccessible. Still, the following important result shows how closely the three properties just considered are related.

Theorem 873. *For any uncountable cardinal κ, the following are equivalent:*

1. *κ has the monotone order property.*
2. *$\kappa \to (\kappa)_2^2$.*
3. *κ is strongly inaccessible and has the tree property.*

For a proof, see [14] or [41].

Problem 874. *Prove that 2 implies 1 in the above theorem.*

Cardinals which satisfy any (and so all) of the conditions of Theorem 873 can also be characterized by a metamathematical property of infinitary languages known as "weak compactness," and so are called *weakly compact cardinals.*

Definition 875 (Weakly Compact Cardinals). A cardinal is said to be *weakly compact* if it satisfies any of the conditions of Theorem 873.

A lot more can be said about weakly compact cardinals. For example, they are not only strongly inaccessible, but also preceded by an equal number of strong inaccessibles. In fact, if κ is weakly compact, then there are arbitrarily large cardinals $\aleph_\alpha < \kappa$ such that $\aleph_\alpha$ is the ω_α-th strongly inaccessible cardinal. For more on weakly compact cardinals, see [14, 37].

The notion of a weakly compact cardinal is thus a good example of a *large cardinal.* As mentioned before, such cardinals, being at least inaccessible, cannot be shown to exist using the standard axioms of set theory (assuming these axioms are consistent) due to a result known as Gödel's second incompleteness theorem. Asserting the existence of a large cardinal is therefore known as a *large cardinal hypothesis* or an *axiom of strong infinity.* As we will see later, certain large cardinal hypotheses, such as that of the existence of a *measurable cardinal,* can have significant implications for ordinary mathematics.

12.2 Suslin's Problem, Martin's Axiom, and $\diamondsuit$

Recall *Suslin's Problem,* which asks:

The Suslin Problem. Is every CCC continuum without endpoints necessarily order isomorphic to **R***?*

The mechanism of trees can be used to put Suslin's Problem in a more useful combinatorial form. Note that a negative answer to Suslin's Problem amounts to the existence of a linear continuum which is CCC but not separable. Such a continuum is called a *Suslin line.*

Recall that an Aronszajn tree is a tree of height ω_1 in which all levels and all chains are countable. Aronszajn proved that such trees exist. We now consider a stronger property: A tree is called a *Suslin tree* if it is an Aronszajn tree in which there are no uncountable antichains.

Definition 876 (Suslin Lines and Trees). A *Suslin line* is a linear continuum which is CCC but not separable. A *Suslin tree* is a tree of height ω_1 in which all chains and all antichains are countable.

The following result reduces Suslin's Problem to a question about trees.

Theorem 877. *There is a Suslin line if and only if there is a Suslin tree.*

To outline a proof of Theorem 877, we need some definitions and lemmas. If u is a node in a tree T, we will use the notation $\mathrm{Succ}_\alpha^T(u)$ to denote the extensions of u of height α, i.e., $\mathrm{Succ}_\alpha^T(u) := \{v \in \mathrm{Lev}_\alpha(T) \mid u \preceq v\}$.

Definition 878. A Suslin tree T is *normal* if every node has extensions in all higher levels below ω_1, i.e., $\mathrm{Succ}_\alpha^T(u) \neq \emptyset$ for every α with $\mathrm{ht}(u) < \alpha < \omega_1$.

Lemma 879. *Let T be a Suslin tree and let*

$$T^+ := \{u \in T \mid \mathrm{Succ}_\alpha^T(u) \neq \emptyset \textit{ for every } \alpha \textit{ with } \mathrm{ht}(u) < \alpha < \omega_1\}.$$

Then T^+ is a normal Suslin tree.

Proof. T^+ is a nonempty subtree of T, and it suffices to show that for each $u \in T^+$, $T^+ \cap \mathrm{Succ}_\alpha^T(u) \neq \emptyset$ for all α with $\mathrm{ht}(u) < \alpha < \omega_1$.

Suppose $T^+ \cap \mathrm{Succ}_\alpha^T(u) = \emptyset$ with $u \in T^+$ and $\mathrm{ht}(u) < \alpha < \omega_1$. Then for each $v \in \mathrm{Succ}_\alpha^T(u)$, there is α_v with $\alpha < \alpha_v < \omega_1$ such that $\mathrm{Succ}_{\alpha_v}^T(v) = \emptyset$. As $\mathrm{Succ}_\alpha^T(u)$ is countable, we may fix $\beta < \omega_1$ with $\beta > \alpha_v$ for all $v \in \mathrm{Succ}_\alpha^T(u)$. But then $\mathrm{Succ}_\beta^T(u) = \bigcup_{v \in \mathrm{Succ}_\alpha^T(u)} \mathrm{Succ}_\beta^T(v) = \emptyset$, a contradiction. □

For $u, v \in T$, we write $u \sim v \Leftrightarrow u$ and v have the same set of predecessors. Then $\sim$ is an equivalence relation on T such that each equivalence class $[u]_\sim$ is contained in some single level of T.

Lemma 880. *If there is a Suslin tree then there is a normal Suslin tree in which $[u]_\sim$ is infinite for all u with* $\mathrm{ht}(u) > 0$.

Proof. Let T be a Suslin tree, which we assume to be normal by the last lemma. Note that if $\mathrm{ht}(u) < \alpha < \beta < \omega_1$ then $1 \leq |\mathrm{Succ}_\alpha^T(u)| \leq |\mathrm{Succ}_\beta^T(u)|$. Also, if $u \in T$ then $\{v \in T \mid u \preceq v\}$ cannot be a chain, and in fact $|\mathrm{Succ}_\alpha^T(u)| \geq 2$ for some $\alpha > \mathrm{ht}(u)$. Repeating the process infinitely many times, we see that there is $\beta > \mathrm{ht}(u)$ such that $\mathrm{Succ}_\beta^T(u)$ is infinite. More generally, for any countable subset $C \subseteq T$, we can get β such that $\mathrm{Succ}_\beta^T(u)$ is infinite for all $u \in C$. Now define increasing ordinals $\langle \alpha(\xi) \mid \xi < \omega_1 \rangle$ as follows: Let $\alpha(0) = 0$, and for $\xi > 0$ let $\delta := \sup\{\alpha(\eta) \mid \eta < \xi\}$ and put $\alpha(\xi) :=$ the least $\beta > \delta$ such that $\mathrm{Succ}_\beta^T(u)$ is infinite for all $u \in \mathrm{Lev}_\delta(T)$. Finally, $T' := \bigcup_{\xi<\omega_1} \mathrm{Lev}_{\alpha(\xi)}(T)$ with the inherited order is a tree with the required property. □

Proof (of Theorem 877, Outline). Suppose first that there is a Suslin line, i.e., a linear continuum L without endpoints which is CCC but not separable. We define a ω_1-sequence of nonempty open intervals $\langle (a_\alpha, b_\alpha) \mid \alpha < \omega_1 \rangle$ (with $a_\alpha < b_\alpha$) using transfinite recursion: For each $\alpha < \omega_1$, the countable set $E_\alpha := \{a_\beta, b_\beta \mid \beta < \alpha\}$ is not dense, so there are $c < d$ with $(c,d) \cap E_\alpha \neq \emptyset$, and therefore by density, we can choose and fix a_α, b_α such that $c < a_\alpha < b_\alpha < d$. Let $T := \{L\} \bigcup \{(a_\alpha, b_\alpha) \mid \alpha < \omega_1\}$, and order T by reverse inclusion. Then T is a Suslin tree.

Conversely, suppose that T is a Suslin tree. By Lemma 880, we may assume that T is a normal Suslin tree in which $[u]_\sim$ is infinite for all u with $\mathrm{ht}(u) > 0$. For each equivalence class $[u]_\sim = [u]$, fix an order $<_{[u]}$ on $[u]$ of order type η. Now define a linear order $<_*$ on all of T by setting $x <_* y$ if and only if either $x \prec y$ in T, or there exist $u \preceq x$ and $v \preceq y$ with $u \sim v$ and $u <_{[u]} v$. Finally, take the Dedekind completion of the order $<_*$ to get a Suslin line. □

Martin's Axiom

Recall that the affirmative answer to Suslin's Problem is called the *Suslin Hypothesis* or SH. Equivalently, SH is the statement that there is no Suslin line. SH cannot be settled one way or the other using the standard axioms of set theory. The Continuum Hypothesis (CH) cannot decide SH either. However, an important combinatorial principle called *Martin's Axiom* (MA) implies that if CH fails then SH must be true, i.e., MA + not-CH $\Rightarrow$ SH.

Definition 881. Let $\langle P, \preceq \rangle$ be a poset (partial order).

1. $\langle P, \preceq \rangle$ is said to satisfy the *countable chain condition* (CCC) if every antichain in P is countable.
2. A subset $D \subseteq P$ is called *dense* if for all $u \in P$ there is $v \in D$ with $v \preceq u$.
3. A nonempty subset $G \subseteq P$ is called *filter* if $u \in G$ and $u \preceq v \Rightarrow v \in G$ (G is upward closed), and for all $u, v \in G$ there is $w \in G$ with $w \preceq u$ and $w \preceq v$ (G is downward directed).

Martin's Axiom (MA). Martin's Axiom (MA) says: *"If $\langle P, \preceq \rangle$ is a CCC partial order and* D *is a family of dense subsets of P with* $|\mathrm{D}| < 2^{\aleph_0}$ *then there is a filter G such that $G \cap D \neq \emptyset$ for all $D \in$* D."

Martin's Axiom is an immediate consequence of CH, and so is of interest only when CH fails. (If CH fails, then MA implies that cardinals strictly between $\aleph_0$ and $2^{\aleph_0}$ have many of the properties of $\aleph_0$.)

Proposition 882. CH $\Rightarrow$ MA.

[Hint: Let $\langle P, \preceq\rangle$ be a CCC partial order and let D be a family of dense subsets of P with $|\mathrm{D}| < 2^{\aleph_0}$. By CH, D is countable and can be enumerated, say as $\mathrm{D} = \{D_1, D_2, \dots\}$. Fix $d_1 \in D_1$, and use density to inductively pick $d_{n+1} \in D_{n+1}$ such that $d_{n+1} \preceq d_n$. Then put $G = \{u \mid d_n \preceq u \text{ for some } n\}$.]

Theorem 883. MA + not-CH $\Rightarrow$ SH.

Proof. Assume MA + not-CH. To get a contradiction, suppose that there is a Suslin tree T. By Lemma 879 we may assume that T is a normal Suslin tree.

Consider T as a poset with the *reverse order* of T, with the root as the largest element. Since T is a Suslin tree, T has CCC. Now for each $\alpha < \omega_1$, put $D_\alpha := \bigcup_{\beta \geq \alpha} \mathrm{Lev}_\beta(T)$. Since T is normal, each D_α is dense. Since we are assuming that CH is false, $|\{D_\alpha \mid \alpha < \omega_1\}| < 2^{\aleph_0}$. Hence by MA, there is a filter G with $G \cap D_\alpha \neq \emptyset$ for all $\alpha < \omega_1$. But then G must be a branch of height ω_1 through T, contradicting that T is a Suslin tree. □

The condition MA + not-CH has been shown to be consistent with the standard axioms of set theory. Therefore, by the above theorem, we can consistently assume that there are no Suslin lines.

Jensen's Diamond Principle ◊

Another important combinatorial principle is the *Diamond Principle* due to Jensen, which is denoted by the symbol ◊.

The Diamond Principle. ◊ says: *"There are sets A_α, $\alpha < \omega_1$, such that for all $A, C \subseteq W(\omega_1)$ if C is a club set then $A \cap W(\alpha) = A_\alpha$ for some $\alpha \in C$."*

CH is an immediate consequence of ◊. Also, ◊ has been shown to be consistent with the standard axioms of set theory (since it follows from the axiom of constructibility devised by Gödel). One important application of ◊ is the following result, which we state without proof.

Theorem 884. ◊ $\Rightarrow$ not-SH.

It follows that the negation of the Suslin Hypothesis (not-SH) is also consistent with the standard axioms of set theory. Combined with the consistency of SH, this means that SH cannot be settled using the usual axioms. In other words, neither SH nor its negation can be derived from the standard axioms: *The Suslin Hypothesis is independent of the standard axioms of set theory,* assuming that these axioms are themselves consistent.

MA and ◊ have many interesting properties and applications, but we conclude our brief discussion of infinitary combinatorics and refer the reader to some texts for further study.

Good introductions to infinitary combinatorics can be found in [14, 35, 41, 48]. For more advanced treatments see [34, 37, 44].

Part III
Real Point Sets

Introduction to Part III

This part focuses exclusively on the real line **R**. Cantor's work not only gave birth to the theory of transfinite, but was also instrumental in the development of *point set topology*, which, roughly speaking, is the study of limits and continuity in a general setting. Topological notions such as closed sets, dense-in-itself sets, and perfect sets were first introduced by Cantor.

The opening chapter, much of which is very elementary, introduces base representation via interval trees, *Cantor systems*, and *generalized Cantor sets*. The next chapter deals with basic topology of the real line.

The material of the chapter on Heine–Borel and Baire-Category Theorems is often called "measure and category." It is shown that G_δ sets satisfy the Continuum Hypothesis, and that perfect sets have cardinality $\mathfrak{c}$.

The chapters on Cantor–Bendixson analysis and on Brouwer's and Sierpinski's Theorems are somewhat more special. An application of the ordinals is illustrated by the method of Cantor–Bendixson analysis, giving a complete enumeration of the $\aleph_1$ distinct "homeomorphism types" of countable compact sets. The proofs of Brouwer's and Sierpinski's Theorems given here illustrate how the Cantor–Dedekind theory of order can be used to give somewhat elementary proofs of some relatively advanced topological results.

The chapter on Borel and analytic sets touches on the rudiments of descriptive set theory, and proves that the analytic sets have the perfect set property—the best possible result that can be proved using the usual axioms of set theory. They are also shown to be Lebesgue measurable (and having the Baire property) using the Ulam matrix decomposition for coanalytic sets. To obtain a non-Borel analytic set, a direct effective proof of the boundedness theorem for the set of codes of well-founded trees is given (since with no access to product spaces, the standard method of diagonalizing universal sets cannot be used).

The postscript chapter for this part gives a detailed account of Ulam's analysis of the measure problem leading to the notion of measurable cardinals, and a brief discussion of Lusin's problem for the projective sets.

Chapter 13
Interval Trees and Generalized Cantor Sets

Abstract This elementary chapter applies the nested intervals theorem to obtain base expansion of real numbers via trees of uniformly subdivided nested closed intervals, with detailed illustrations for ternary expansions. The construction of the Cantor set is then generalized to *Cantor systems* (systems of nested intervals indexed by binary trees), to formally introduce generalized Cantor sets.

13.1 Intervals, Sup, and Inf

Definition 885. An *interval* is a set having one of the forms

$$(a,b),\ [a,b],\ (a,b],\ [a,b),\ (a,\infty),\ [a,\infty),\ (-\infty,b),\ (-\infty,b],\ \mathbf{R},\ \emptyset.$$

An *open interval* is an interval having one of the forms

$$(a,b),\ (a,\infty),\ (-\infty,b),\ \mathbf{R},\ \emptyset.$$

A *closed interval* is an interval having one of the forms

$$[a,b],\ [a,\infty),\ (-\infty,b],\ \mathbf{R},\ \emptyset.$$

An interval is *proper* if it contains at least two points.

Note that each of $\emptyset$ and $\mathbf{R}$ is both an open interval and a closed interval. Moreover every singleton set $\{a\} = [a,a]$ is a closed interval (improper).

Definition 886 (Bounds, Sup, and Inf). Let $A \subseteq \mathbf{R}$.

1. $u \in \mathbf{R}$ is an *upper bound* of A, written as $u \geq A$ or $A \leq u$, if $u \geq x$ for all $x \in A$.
2. $u \in \mathbf{R}$ is a *strict upper bound* of A, written as $u > A$ or $A < u$, if $u > x$ for all $x \in A$.

A. Dasgupta, *Set Theory: With an Introduction to Real Point Sets*,
DOI 10.1007/978-1-4614-8854-5_13, © Springer Science+Business Media New York 2014

3. A is *bounded* if there exist $a, b \in \mathbf{R}$ such that $a \leq A \leq b$.
4. $p \in \mathbf{R}$ is a *least upper bound* or *supremum* of A if p is an upper bound of A and no point $q < p$ is an upper bound of A. If a least upper bound of A exists, it will be unique and will be denoted by $\sup A$. We will also use the notation

$$\sup_{x \in A} f(x)$$

to denote $\sup f[A]$.
5. We write $u = \max A$ if $A \leq u$ and $u \in A$, that is when u is the greatest element of A.

Lower bounds (greatest), infimum, $\inf A$, $\min A$, etc, are similarly defined.

Recall (from our definitions of these notions for orders) that $a = \sup A$ if and only if either $a = \max A$ or $a > A$ and a is an upper limit point of A. Moreover, since $\mathbf{R}$ is a complete order, for every nonempty set A if A is bounded above then $\sup A$ exists, and if A is bounded below then $\inf A$ exists.

The Nested Intervals Theorem in R

Recall that the completeness of $\mathbf{R}$ implies the sequential nested interval property. The following variant (and immediate consequence) of that result will be used frequently in this part of the book.

Theorem 887 (Nested Intervals Theorem in R). *Suppose that*

$$I_1 \supseteq I_2 \supseteq \cdots \supseteq I_n \supseteq I_{n+1} \supseteq \cdots$$

is a sequence of nested intervals satisfying the following conditions.

1. *Each interval is nonempty and closed, where we allow the closed intervals to be unbounded, having the form* $[a, \infty)$, *or* $(-\infty, a]$, *or* $(-\infty, \infty) = \mathbf{R}$.
2. $\lim_{n \to \infty} \operatorname{len}(I_n) = 0$, *that is for any* $\epsilon > 0$ *there is* k *such that* $\operatorname{len}(I_n) < \epsilon$ *for* $n \geq k$.

Then the intersection of the intervals is a singleton,

$$\bigcap_n I_n = \{a\} \qquad \textit{for some } a \in \mathbf{R}.$$

Note that while we are allowing unbounded closed intervals in the nested sequence, the second condition implies that at most finitely many of those intervals can be unbounded.

Problem 888. *Show that the second condition in the theorem above cannot be dropped.*

Problem 889. *Show that we cannot replace "closed" by "open" in the above theorem.*

Problem 890. *Show that the above theorem remains valid if we replace "closed" by "open" and assume that* $\inf I_n < \inf I_{n+1}$ *and* $\sup I_{n+1} < \sup I_n$.

13.2 Interval Subdivision Trees

In this section we explain how the familiar method of decimal expansions of numbers in the interval $[0, 1]$ naturally leads to iterated subdivisions of intervals forming a tree structure. Instead of the base 10 (decimal) system, one can use any fixed base.

Using a Fixed Base b

Given a positive integer $b \geq 2$ (the base) and a closed interval I, we subdivide I into b equal closed subintervals each of length $\frac{1}{b}\,\mathrm{len}(I)$ and write these subintervals as:

$$I[0], \quad I[1], \quad \ldots, \quad I[b-1] \qquad \text{(base } b\text{)}.$$

Thus $I[d]$ is the d-th subinterval in this subdivision of I into b equal subintervals, for $d = 0, 1, \ldots, b-1$.[1]

The process is then further iterated as follows. If d_1 and d_2 are two b-ary digits (i.e., $d1, d2 \in \{0, 1, \ldots, b-1\}$), then we let $I[d_1 d_2]$ denote *the* d_2*-th subinterval of* $I[d_1]$,. Thus in the first stage I is subdivided into b equal subinterval $I[0]$, $I[1]$, $\ldots$, $I[b-1]$, and then each of these b subintervals $I[d_1]$ is further subdivided into b more smaller sub-subintervals $I[d_1 0], I[d_1 1], \ldots I[d_1(b-1)]$, giving a total of b^2 sub-subintervals at the second stage.

We can continue iterating the process, giving b^n intervals at stage n.

[1] Our notation is ambiguous since $I[1]$ could denote the second sub interval in two, or three, or ten (or any other numbers) equal subdivisions of I. It would be more correct, but more clumsy, to write $I_b[0], I_b[1], \ldots, I_b[b-1]$ in place of $I[0], I[1], \ldots, I[b-1]$. Since the base b is generally fixed throughout a situation, it is understood from context, and dropping the subscript b does not cause any confusion.

The Ternary Subdivision Tree

We illustrate the system and notation for the specific case where $b = 3$ (the ternary system) and $I = [0, 1]$, the unit interval. The general case of any $b \geq 2$ is so similar that we will not discuss it separately.

The initial three ternary subdivisions of $I = [0, 1]$ are:

$$I[0] = [0, \tfrac{1}{3}], \quad I[1] = [\tfrac{1}{3}, \tfrac{2}{3}], \quad \text{and} \quad I[2] = [\tfrac{2}{3}, 1] \qquad (\text{base } b = 3).$$

These three intervals each have length $\frac{1}{3}$. Then each of these is further subdivided into three equal sub-subintervals, giving a total of nine sub-subintervals, each of length $\frac{1}{9}$:

$$I[00],\ I[01],\ I[02];\ \ I[10],\ I[11],\ I[12];\ \ I[20],\ I[21],\ I[22],$$

where $I[00] = [0, \frac{1}{9}]$, $I[01] = [\frac{1}{9}, \frac{2}{9}]$, etc, with $I[22] = [\frac{8}{9}, 1]$.

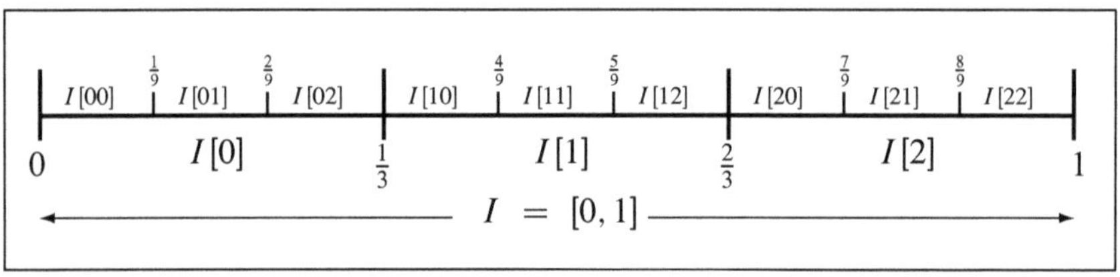

This process is further continued to obtain $9 \times 3 = 27$ sub-sub-subintervals each of length $1/27$, denoted by $I[000]$, $I[001]$, $I[002]$, $I[010]$, …, $I[222]$.

Regarding subintervals as descendents of intervals containing them, the entire systems can be arranged in the form of a tree:

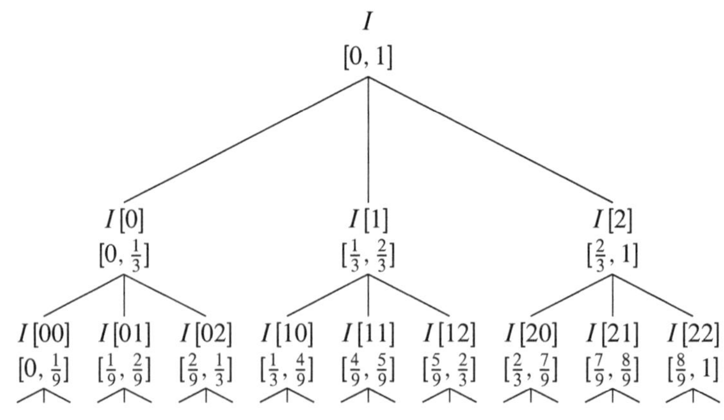

In general, at stage n, there will be 3^n subintervals of the form $I[d_1 d_2 \ldots d_n]$.

Ternary Strings

Notice how at stage n, the 3^n intervals of the form $I[d_1 d_2 \ldots d_n]$ are indexed by ternary strings $d_1 d_2 \ldots d_n$ formed out of the three "letters" from the set $\{0, 1, 2\}$, which is called the *ternary alphabet*. The *finite ternary strings* themselves are of various lengths:

$$\varepsilon,\ 0,\ 1,\ 2,\ 00,\ 01,\ 02,\ 10,\ 11,\ 12,\ 20,\ 21,\ 22,\ 000,\ 001,\ 002,\ 100,\ \ldots$$

Here ε is the empty string (of length zero), and there are 3^n ternary strings of length n.

The finite ternary strings themselves are naturally arranged in an infinite tree by regarding string prefixes (i.e., initial segments) as ancestors:

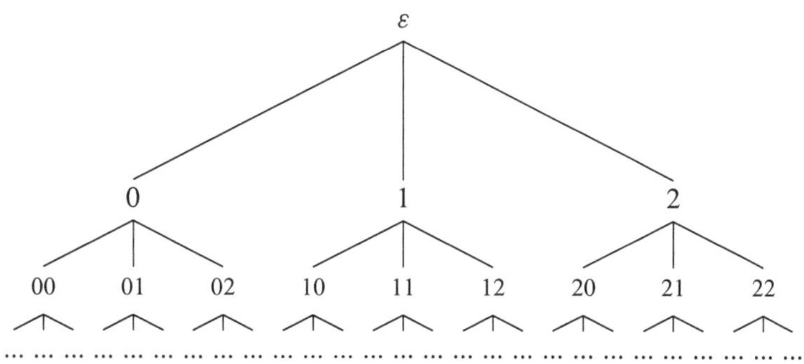

Thus the system of ternary subdivision of intervals, when arranged by the relation of containment of intervals, is naturally mapped by the tree of ternary words arranged by the relation of prefixes of words. Clearly, this mapping is a one-to-one correspondence between the nodes which transforms string prefix relations into interval containment relations. We thus have a natural representation of the ternary interval tree by the tree of finite ternary strings.

13.3 Infinite Branches Through Trees

An *infinite branch* through the above ternary tree is an infinite set of nodes (i.e., strings)

$$\{\,\varepsilon,\ d_1,\ d_1 d_2,\ \ d_1 d_2 d_3,\ \ldots,\ d_1 d_2 \ldots d_n,\ \ldots\}$$

containing exactly one node of each length $n = 0, 1, 2, \ldots$ in which the shortest node ε of length 0 is the root ("the branch starts at the root"), and each node $d_1 d_2 \ldots d_n$ of length n is obtained from the previous node $d_1 d_2 \ldots d_{n-1}$ length

$n - 1$ by appending a single digit d_n. Note that we usually draw trees "growing downward," so the branch "grows downward" in the tree as well.

Infinite Branches as Infinite Digit Strings

The infinite branch above, namely $\varepsilon, d_1, d_1d_2, d_1d_2d_3, \ldots$, can be identified with the infinite ternary sequence $d_1d_2 \cdots d_n \cdots \in \{0, 1, 2\}^{\mathbf{N}}$, since by taking finite initial prefixes of an infinite string, *each branch through the tree is represented uniquely by a single infinite ternary string.*

For example, the initial prefixes of the constant infinite ternary string $000000\cdots$ are ε, 0, 00, 000, etc, so the infinite ternary string $000000\cdots$ represents the leftmost branch of the ternary tree, while the infinite string $111111\cdots$ represents the center-most string going straight down right through the middle of the tree. As another example, the bold segments in the following figure illustrates a "zigzag" infinite branch represented by the infinite string $0202020202\cdots$.

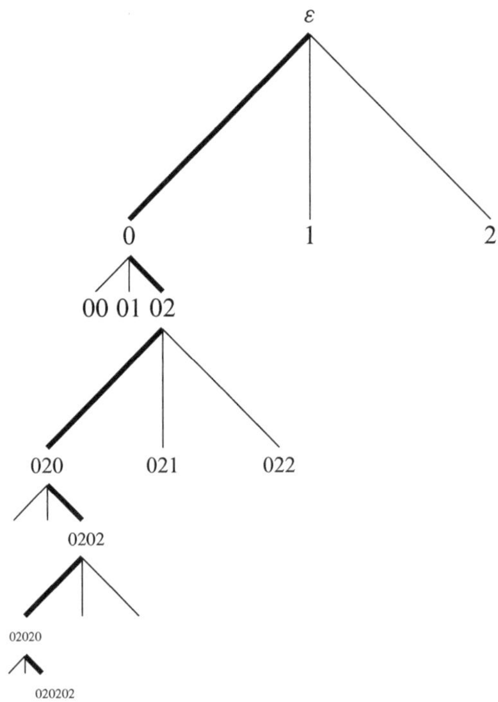

So by taking the finite initial prefixes of any given infinite ternary string, we get a set of nodes growing down through the ternary tree forming an infinite branch. And conversely, any infinite branch through the ternary tree produces an infinite ternary string as its "limit."

Thus, we have *a natural one-to-one correspondence between infinite ternary strings and the infinite branches through the ternary tree.*

Infinite Branches as Nested Intervals

Say that a sequence of bounded closed intervals is a *nested ternary sequence* if the first interval in the sequence is the unit interval, and each succeeding interval equals either the left-third, the middle-third, or the right-third closed subinterval of the preceding interval in the sequence. Note that an infinite branch through the ternary tree determines a nested ternary sequence of intervals.

Specifically, given an infinite ternary string $d_1 d_2 d_3 \ldots$ (with $d_n \in \{0, 1, 2\}$ for all n), the intervals indexed by initial prefixes of $d_1 d_2 d_3 \ldots$ form the nested ternary sequence of intervals

$$[0, 1] = I[\varepsilon] \supseteq I[d_1] \supseteq I[d_1 d_2] \supseteq \cdots \supseteq I[d_1 d_2 \cdots d_n] \supseteq \cdots.$$

We thus also have a *natural one-to-one correspondence of infinite ternary strings with nested ternary sequences of intervals, via infinite branches through the ternary tree.* This is the basis of ternary expansions.

Ternary Expansions

In the nested ternary sequence of intervals displayed above, we have $\text{len}(I[\varepsilon]) = 1$, $\text{len}(I[d_1]) = \frac{1}{3}$, $\text{len}(I[d_1 d_2]) = \frac{1}{9}$, and so on, so that $\text{len}(I[d_1 d_2 \cdots d_n]) = \frac{1}{3^n} \to 0$ as $n \to \infty$. Hence by the Nested Interval Property the above sequence of nested intervals must contain a unique real number in their intersection, i.e.,

$$\bigcap_{n=1}^{\infty} I[d_1 d_2 \cdots d_n] = \{x\}, \qquad \text{for a unique } x \in \mathbf{R}.$$

In this case, we say that *the infinite ternary string* $d_1 d_2 d_3 \cdots$ *represents the real number* x, or that x *has a ternary expansion* $d_1 d_2 d_3 \cdots$. This is also expressed as:

$$x = 0 \cdot d_1 d_2 d_3 \cdots d_n \cdots \quad \text{in ternary expansion.}$$

Thus *every infinite ternary string determines a unique real number in* $[0, 1]$ *via the nested intervals associated with the initial prefixes of the infinite string.* Conversely, we have:

Theorem 891. *Any real number in* $[0, 1]$ *has a ternary expansion.*

The proofs of the theorems above and below are left as exercises.

Theorem 892 (Ternary Expansion in $[0, 1]$**).** *For each finite ternary string* $u \in \{0, 1, 2\}^*$ *let* $I[u]$ *denote the corresponding iterated subinterval in the system of ternary subdivision over* $[0, 1]$. *Let* $(d_n)_{n=1}^{\infty} = d_1 d_2 \cdots d_n \cdots$ *be an infinite ternary string, and put* $[u_n, b_n] := I[d_1 d_2 \cdots d_n]$.

Then for any $x \in [0, 1]$, *the conditions below are equivalent to each other:*

1. $x = 0 \cdot d_1 d_2 \cdots d_n \cdots$ *in ternary expansion.*
2. $\bigcap_{n=1}^{\infty} I[d_1 d_2 \cdots d_n] = \{x\}$, *i.e., x is the unique member of the intersection of the nested intervals* $I[\varepsilon] \supseteq I[d_1] \supseteq I[d_1 d_2] \supseteq \cdots \supseteq I[d_1 d_2 \cdots d_n] \supseteq \cdots$.
3. $x = \sum_{n=1}^{\infty} \frac{d_n}{3^n}$ *(infinite series expansion).*
4. $x = \lim_{n\to\infty} a_n$ *(limit of the left endpoint of the nested intervals).*
5. $x = \lim_{n\to\infty} b_n$ *(limit of the right endpoint of the nested intervals).*

A similar result is true for any fixed base b.

Problem 893. *Find the values of the following infinite series expansions:*

1. $0 \cdot 111111 \cdots$ *(ternary)*
2. $0 \cdot 111111 \cdots$ *(binary)*
3. $0 \cdot 111111 \cdots$ *(decimal)*
4. $0 \cdot 1022222222 \cdots$ *(ternary)*
5. $0 \cdot 1100000000 \cdots$ *(ternary)*
6. $0 \cdot 02020202 \cdots$ *(ternary)*

[Hint: Use the fact that the sum of a convergent geometric series $a + ar + ar^2 + \cdots$ is $a/(1-r)$, where $|r| < 1$.]

Problem 894. *Let* $b \in \{2, 3, 4, \ldots\}$ *be a fixed base. A rational number x is said to be b-*adic *if* $x = m/b^n$ *for some* $m \in \mathbf{Z}$ *and* $n \in \mathbf{N}$. *We use the terms* dyadic *for 2-adic and* triadic *for 3-adic.*

1. *Prove that a real number in* $[0, 1]$ *has a ternary expansion which is eventually constant to a digit value of* 0 *or* 2 *if and only if it is a triadic rational in* $[0, 1]$.
2. *Formulate and prove similar results for the binary (base* $b = 2$*) and the decimal (base* $b = 10$*) systems.*
3. *Prove that* $0 < x < 1$ *has multiple ternary expansions if and only if x is a triadic rational. More specifically, show that every triadic rational x,* $0 < x < 1$, *has exactly two ternary expansions, and every other real in* $[0, 1]$ *has a unique ternary expansion,*

Problem 895. *An infinite digit sequence* $d_1 d_2 \cdots d_n \cdots$ *is said to be* repeating *if there is finite block of digits which eventually keeps repeating (formally: if there is* $r \in \mathbf{N}$ *and* $k \in \mathbf{N}$ *such that* $d_{n+r} = d_n$ *for all* $n > k$*).*

1. *Prove that* $0 < x < 1$ *has a repeating ternary expansion if and only if x is a rational number in* $[0, 1]$
2. *Prove that the same result is true for any fixed base* $b \in \{2, 3, 4, \ldots\}$.

Problem 896. *Show that the Cantor set consists of those reals in the unit interval which admit a ternary expansion in which the digit* 1 *does not occur.*

Problem 897. *Show that* $\frac{1}{4}$ *is a member of the Cantor set, but* $8e/7\pi$ *is not. Give an example of an irrational member of the Cantor set.*

Problem 898. *Show that every real number in* $[0, 2]$ *can be expressed as a sum of two members of the Cantor set.*

[Hint: First show that every real in $[0, 1]$ can be expressed as the sum of two reals in $[0, 1]$ each of which has a ternary expansion not containing the digit 2.]

13.4 Cantor Systems and Generalized Cantor Sets

The following definition is a direct generalization of the binary tree of intervals that was used in the construction of the Cantor set.

Definition 899 (Cantor Systems). A family $J = \langle J_u \mid u \in \{0,1\}^* \rangle$ of sets indexed by the binary tree $\{0,1\}^*$ is called a *Cantor System* if for each binary string $u \in \{0,1\}^*$:

1. J_u is a bounded proper closed interval, i.e., $J_u = [a, b]$ for some $a < b$;
2. $J_{u^\frown 0}, J_{u^\frown 1} \subseteq J_u$;
3. $J_{u^\frown 0} \cap J_{u^\frown 1} = \emptyset$;
4. For any infinite binary sequence $b = b_1 b_2 \cdots b_n \cdots \in \{0,1\}^{\mathbf{N}}$,

$$\lim_{n\to\infty} \operatorname{len}(J_{b|n}) = \lim_{n\to\infty} \operatorname{len}(J_{b_1 b_2 \cdots b_n}) = 0.$$

Note that the notation $b|n$ denotes the finite initial prefix of the infinite sequence b consisting of its first n entries, i.e., $b|n := b_1 b_2 \cdots b_n$. Also, recall that the notation $u^\frown d$ denotes the string which is obtained by appending the string u with the digit d, so that $\operatorname{len}(u^\frown d) = \operatorname{len}(u) + 1$.

Let $\langle J_u \mid u \in \{0,1\}^* \rangle$ be a Cantor system. If $b = b_1 b_2 \cdots b_n \cdots \in \{0,1\}^{\mathbf{N}}$ is an infinite binary sequence, then $J_{b_1} \supseteq J_{b_1 b_2} \supseteq J_{b_1 b_2 b_3} \supseteq \cdots$ forms a nested sequence of nonempty closed intervals whose lengths approach zero, and so their intersection must be a singleton. Hence each infinite binary sequence $b = b_1 b_2 \cdots b_n \cdots \in \{0,1\}^{\mathbf{N}}$ determines a unique real number x_b such that

$$\{x_b\} = \bigcap_n J_{b|n}.$$

Moreover, note that distinct infinite binary sequences determine distinct real numbers: If $b = b_1 b_2 \cdots b_n \cdots$ and $c = c_1 c_2 \cdots c_n \cdots$ are distinct infinite binary sequences, then there exists a least k such that $b_k \neq c_k$. Then $J_{b_1 b_2 \cdots b_k}$ and $J_{c_1 c_2 \cdots c_k}$ are disjoint subintervals of $J_{b_1 b_2 \cdots b_{k-1}} = J_{c_1 c_2 \cdots c_{k-1}}$, so $x_b \neq x_c$. By setting $\varphi(b) := x_b$, we get an injective mapping $\varphi\colon \{0,1\}^{\mathbf{N}} \to \mathbf{R}$. Thus *every Cantor system* $J = \langle J_u \mid u \in \{0,1\}^* \rangle$ *effectively determines a unique one-to-one mapping* φ *from* $\{0,1\}^{\mathbf{N}}$ *into the reals such that*

$$\text{for all } b \in \{0,1\}^{\mathbf{N}}: \qquad \{\varphi(b)\} = \bigcap_n J_{b|n}.$$

The set of all real numbers $x_b = \varphi(b)$ as b ranges over all possible infinite binary sequences, that is the range of the function φ, is called *the set generated by the Cantor system* $\langle J_u \mid u \in \{0,1\}^*\rangle$.

Definition 900 (Set Generated by a Cantor System). The *set generated by the Cantor system* $J = \langle J_u \mid u \in \{0,1\}^*\rangle$ is the set P of real numbers defined by the condition:

$$x \in P \quad \text{if and only if} \quad \text{there exists } b \in \{0,1\}^{\mathbf{N}} \text{ such that } x \in \bigcap_n J_{b|n}.$$

Definition 901 (Generalized Cantor Sets). A set is called a *generalized Cantor set* or a *Cantor-like set* if it is generated by some Cantor system.

We summarize the above discussion in:

Proposition 902. *If P is the generalized Cantor set generated by a Cantor system $J = \langle J_u \mid u \in \{0,1\}^*\rangle$, then the function φ above maps $\{0,1\}^{\mathbf{N}}$ bijectively onto P.*

Hence every generalized Cantor set is effectively bijective with $\{0,1\}^{\mathbf{N}}$ and so has cardinality $\mathfrak{c} = 2^{\aleph_0}$.

The bijection in the above proposition can be viewed as a correspondence between the infinite branches of the binary tree and the points of the generalized Cantor set P being generated: Each infinite branch through the binary tree determines a sequence of nested intervals, which in turn determines a point of the set P.

Problem 903. *Let $\langle J_u \mid u \in \{0,1\}^*\rangle$ be a Cantor system. If $J_u \cap J_v \neq \emptyset$, then show that one of the binary strings u and v must be an extension of the other (i.e., either u is an initial prefix of v or v is an initial prefix of u).*

Problem 904. *Let $\langle J_u \mid u \in \{0,1\}^*\rangle$ be a Cantor system which generates the generalized Cantor set P. Show that*

1. *For any $x \in P$ and $\delta > 0$ there is $u \in \{0,1\}^*$ such that $x \in J_u$ and $\operatorname{len}(J_u) < \delta$.*
2. *Every J_u contains some point of P.*

Problem 905. *Let P be the generalized Cantor set generated by a Cantor system $\langle J_u \mid u \in \{0,1\}^*\rangle$, and for each $n = 0, 1, 2, \ldots$ let F_n be the union of the 2^n disjoint intervals J_u where u is a binary string of length n, that is,*

$$F_n := \bigcup \{J_u \mid \operatorname{len}(u) = n\}.$$

Show that

$$P = \bigcap_n F_n.$$

Chapter 14
Real Sets and Functions

Abstract This chapter covers the basic topology of the real line. Many of the notions of this chapters, such as derived sets, closed sets, dense-in-itself sets, and perfect sets, were first introduced by Cantor during his study of the real continuum.

14.1 Open Sets

Definition 906 (Open Sets). A set $G \subseteq \mathbf{R}$ is called *open* if every point of G belongs to some open interval contained in G, that is if for every $p \in G$, there is an open interval I such that $p \in I \subseteq G$.

Note that every nonempty open set contains a nonempty open interval and hence must be uncountable. Thus no nonempty countable set can be open. Also, no nonempty bounded closed interval is open.

Problem 907. *Show that in the definition of open sets we can replace "open interval" with "bounded open interval."*

Problem 908. *Show that a set is open if and only if it can be expressed as the union of some family of open intervals.*

Problem 909. *1. The empty set $\emptyset$ and* **R** *are open sets.*
2. The union of any collection of open sets is open.
3. The intersection of finitely many open sets is open.

Problem 910. *Show that the intersection of infinitely many open sets may not be open.*

Problem 911 (Countable Base). *Let* $\mathcal{B} := \{(p,q) \mid p,q \in \mathbf{Q}, p < q\}$ *be the collection of all nonempty bounded open intervals with rational endpoints. Then* $\mathcal{B}$ *is a countable collection of open intervals, and every open set is a (countable) union of members of* $\mathcal{B}$.

Problem 912. *Show that there are exactly* $2^{\aleph_0} = \mathfrak{c}$ *many open sets.*

A. Dasgupta, *Set Theory: With an Introduction to Real Point Sets*, DOI 10.1007/978-1-4614-8854-5_14, © Springer Science+Business Media New York 2014

Problem 913 (The Countable Chain Condition). *Show that every family of pairwise disjoint nonempty open sets is countable, and hence every family of pairwise disjoint nonempty open intervals is countable.*

[Hint: Every nonempty open interval contains a rational number.]

Problem 914. *Let $A \subseteq \mathbf{R}$ and suppose that $\forall x, y \in A$, $x < z < y \Rightarrow z \in A$. Show that A must be an interval. If in addition A is open show that A must be an open interval.*

[Hint: If A is nonempty bounded, put $a = \inf A$, $b = \sup A$, and show that A must be one of (a, b), $[a, b]$, $(a, b]$ or $[a, b)$. If A is neither bounded above nor bounded below, A must be $\mathbf{R}$. If A is bounded below but not above, A must be one of (a, ∞) or $[a, \infty)$ where $a = \inf A$. Etc.]

Problem 915. *Let C be a collection of open intervals having nonempty intersection (so there is p such that $p \in I$ for every $I \in \mathsf{C}$). Show that the union $\bigcup\mathsf{C}$ of all the intervals in C is itself an open interval.*

In particular, the union of two open intervals having nonempty intersection is an open interval.

Problem 916. *Let C be a family of pairwise disjoint nonempty open intervals, and let $G = \bigcup\mathsf{C}$. Show that if I is a nonempty open interval contained in G, then I is contained in a unique member of C.*

Problem 917. *Let G be a nonempty open set and for $x, y \in G$ write $x \sim y$ if there is an open interval containing both x and y. Show that $\sim$ is an equivalence relation on G which partitions G into nonempty open intervals.*

From the last few problems we get a *canonical decomposition* of each open set into a unique countable family of disjoint open intervals.

Theorem 918 (Canonical Decomposition of Open Sets into Disjoint Open Intervals). *Every open set can be expressed as the union of a unique (countable) family of pairwise-disjoint nonempty open intervals.*

14.2 Limit Points, Isolated Points, and Derived Sets

For general orders, we had defined the notions of upper and lower limit points, derived sets, bounds, supremum, etc, and the same definitions apply for $\mathbf{R}$:

Definition 919 (Limit Points and Derived Sets). Let $A \subseteq \mathbf{R}$.

1. A point $p \in \mathbf{R}$ is an *upper limit point* of A if for all $x < p$ there is y such that $x < y < p$. Lower limit points are defined similarly.

2. A point is a *limit point* of A if it is either a lower or an upper limit point of A. The set of all limit points of A is called the *derivative* or *derived set* of A and will be denoted by $D(A)$.
3. A point $p \in A$ is an *isolated point* of A if p is not a limit point of A, that is if $p \in A \setminus D(A)$.
4. A limit point of A which is both an upper and a lower limit point of A will be called a *two-sided limit point* of A; otherwise it is a one-sided limit point of A.

Problem 920. *For each of the following sets, find $D(A)$.*

1. $\mathbf{Z}$.
2. $\{1/n \mid n \in \mathbf{N}\}$.
3. $\mathbf{Q} \cap (0, 1)$.
4. $\{1/2^n + 1/2^{m+n} \mid m, n \in \mathbf{N}\}$.

Problem 921. *Give an example of an infinite set A such that A has arbitrarily close points (for any $p > 0$ there are $x, y \in A$ with $0 < |x - y| < p$) but A has no limit points ($D(A) = \emptyset$).*

Problem 922. *Let E be the set all points $x \in [0, 1]$ having a ternary expansion $x = \sum_n x_n/3^n$ for which there is k such that $x_n = 0$ or 2 for $n < k$ and $x_n = 1$ for all $n \geq k$ (i.e., any point $x \in E$ has ternary expansion of the form $x = 0 \cdot x_1 x_2 \cdots x_{k-1} 111111 \cdots$ with $x_1, x_2, \ldots, x_{k-1} \in \{0, 2\}$).*

1. *Which points of A are limit points of A?*
2. *Find $D(A)$.*

Problem 923. *Show that $p \in \mathbf{R}$ is a limit point of A if and only if every open interval containing p contains a point of A other than p if and only if every open interval containing p contains infinitely many points of A.*

Show that p is an isolated point of A if and only if $I \cap A = \{p\}$ for some open interval I.

Show that p is not a limit point of A if and only if $I \cap A \subseteq \{p\}$ for some open interval I containing p.

Problem 924 (Properties of $D(A)$). *For any sets A and B we have:*

1. $A \subseteq B \Rightarrow D(A) \subseteq D(B)$.
2. $D(A \cup B) = D(A) \cup D(B)$.
3. $D(D(A)) \subseteq D(A)$.

Problem 925. *Give an example of a set A such that*

$$\emptyset \subsetneq D(D(A)) \subsetneq D(A) \subsetneq A$$

(all inclusions being proper).

14.3 Closed, Dense-in-Itself, and Perfect Sets

Definition 926 (Cantor). A set $A \subseteq \mathbf{R}$ is called

1. *Closed* if every limit point of A is in A, i.e. if $D(A) \subseteq A$.
2. *Dense-in-itself* if every point of A is a limit point of A, i.e. if $A \subseteq D(A)$.
3. *Perfect* if it is both closed and dense-in-itself, i.e., if $D(A) = A$.

Some examples:

- Any finite set is closed. The set $D(A)$ of limit points of any set A is closed (recall that $D(D(A)) \subseteq D(A)$ in orders, and so in $\mathbf{R}$ too).
- The set $\mathbf{Z}$ of integers is closed but not dense-in-itself, while the set $\mathbf{Q}$ of rational numbers is dense-in-itself but not closed.
- Every proper closed interval is perfect. The Cantor set is perfect.
- The set $A := \{1, \frac{1}{2}, \frac{1}{3}, \ldots, \frac{1}{n}, \ldots\}$ is not closed since A has a (unique) limit point 0 which is not in A. But adjoining this limit point 0 to the set A gives a closed set $A \cup \{0\}$. This method is fully general, and leads to the notion of *closure.*

Definition 927 (Closure). The *closure* $\overline{A}$ of A is the set $\overline{A} := A \cup D(A)$.

Theorem 928. *For any set A, its closure $\overline{A} = A \cup D(A)$ is closed. In fact, $\overline{A}$ is the smallest closed set containing A.*

Proof. We have:

$$D(\overline{A}) = D(A \cup D(A)) = D(A) \cup D(D(A)) \subseteq D(A) \cup D(A) = D(A) \subseteq \overline{A},$$

and so $\overline{A}$ is closed.

Now if B is any closed set containing A, then since $D(B) \subseteq B$,

$$\overline{A} = A \cup D(A) \subseteq A \cup D(B) \subseteq A \cup B = B.$$

Thus the closed set $\overline{A}$ is contained in every closed set containing A, and hence $\overline{A}$ is the smallest closed set containing A. □

It follows immediately from the definitions of closure and of closed sets that:

$$A \text{ is closed if and only if } A = \overline{A}.$$

Problem 929. *$p \in \overline{A}$ if and only if every open interval containing p has nonempty intersection with A. Hence $p \notin \overline{A}$ if and only if there is an open interval I containing p such that $I \cap A = \emptyset$.*

Problem 930. *Prove that $\overline{\overline{A}} = \overline{A}$ and $\overline{A \cup B} = \overline{A} \cup \overline{B}$.*

Problem 931. *Let A be a closed set. If B is a nonempty bounded subset of A then* $\inf B \in A$ *and* $\sup B \in A$. *In particular, if A is nonempty, closed, and bounded, then* $\inf A \in A$ *and* $\sup A \in A$.

Proposition 932. *A set is closed if and only if its complement is open.*

Proof. Let A and B be complements of each other so that $A \cap B = \emptyset$ and $A \cup B = \mathbf{R}$. We show that A is closed if and only if B is open.

If A is closed so that $\overline{A} = A$, then for any $x \in B$ we have $x \notin \overline{A}$ hence there is an open interval I such that $x \in I$ and $I \cap A = \emptyset$, which means $x \in I \subseteq B$. Thus B is open.

If B is open then for any $x \in B$ there is an open interval I with $x \in I \subseteq B$, hence $x \in I$ and $I \cap A = \emptyset$, and so $x \notin \overline{A}$. Thus no point of B is in $\overline{A}$ which means $\overline{A} \subseteq A$, and so A is closed. □

Corollary 933. *1. The empty set $\emptyset$ and $\mathbf{R}$ are closed.*
2. The intersection of any collection of closed sets is closed.
3. The union of finitely many closed sets is closed.

Problem 934. *If A is dense-in-itself, G is open, and $A \cap G \neq \emptyset$, then $A \cap G$ is a nonempty dense-in-itself set.*

Proposition 935. *The Cantor set is perfect.*

Proof. Let $\mathbf{K}$ be the Cantor set. Then $\mathbf{K} = \cap_n \mathbf{K}_n$, where $\mathbf{K}_n$ is the union of 2^n closed intervals obtained at stage n of the construction of the Cantor set. Now each $\mathbf{K}_n$ is closed, being a finite union of closed intervals. Hence $\mathbf{K}$ is closed, being the intersection of the sequence of closed sets $\mathbf{K}_n$.

To see tat $\mathbf{K}$ is dense-in-itself: Given any $x \in \mathbf{K}$ and any open interval (a, b) with $a < x < b$, pick n large enough so that $1/3^n < \min(x - a, b - x)$. Since $x \in \mathbf{K}_n$, so x is in one of the 2^n closed intervals of length $1/3^n$ making up $\mathbf{K}_n$, say in $[c, d]$. Then $[c, d] \subseteq (a, b)$ since $d - c = 1/3^n$. Now both c and d are in $\mathbf{K}$, but either $c \neq x$ or $d \neq x$, and so (a, b) contains a point of $\mathbf{K}$ other than x. Thus x is a limit point of $\mathbf{K}$. Hence $\mathbf{K}$ is dense-in-itself.

Thus $\mathbf{K}$ is perfect. □

Example 936. The complement of $\mathbf{K}$ is open, and therefore can be decomposed into a unique family of disjoint open intervals. Note that this decomposition of $\mathbf{R} \setminus \mathbf{K}$ consists of the two unbounded open intervals $(-\infty, 0)$ and $(1, \infty)$ as well as infinitely many bounded open intervals

$$(\tfrac{1}{3}, \tfrac{2}{3}),\ (\tfrac{1}{9}, \tfrac{2}{9}),\ (\tfrac{7}{9}, \tfrac{8}{9}),\ (\tfrac{1}{27}, \tfrac{2}{27}),\ (\tfrac{4}{27}, \tfrac{5}{27}),\ (\tfrac{19}{27}, \tfrac{20}{27}),\ (\tfrac{25}{27}, \tfrac{26}{27}),\ \ldots$$

Problem 937. *Show that there are exactly $\mathfrak{c}$ many closed and perfect sets. How many dense-in-itself sets are there?*

Problem 938. *If A and B are closed sets with $A \cap B = \emptyset$ then there exist open sets U and V with $A \subseteq U$, $B \subseteq V$, and $U \cap V = \emptyset$ (any two disjoint closed sets can be separated by disjoint open sets).*

[Hint: Let U be the union of all open intervals $I = (a,b)$ such that $A \cap I \neq \emptyset$ but $B \cap (a - \text{len}(I), b + \text{len}(I)) = \emptyset$. Similarly define V.]

Definition 939 (Eventual Containment). A sequence $\langle x_n \rangle_{n\in\mathbf{N}}$ is *eventually in a set* A if there is $m \in \mathbf{N}$ such that $x_n \in A$ for all $n \geq m$.

Definition 940 (Convergence and Limit). A sequence $\langle x_n \rangle_{n\in\mathbf{N}}$ of real numbers *converges to a real number* x, written as $\langle x_n \rangle \to x$ or as $x_n \to x$ *as* $n \to \infty$, if for any open interval I containing x, the sequence is eventually in I. If $\langle x_n \rangle \to x$, we also say that x is a *limit of the sequence* $\langle x_n \rangle_{n\in\mathbf{N}}$.

Problem 941 (Uniqueness of Limits of Sequences). *If* $\langle x_n \rangle \to x$ *and* $\langle x_n \rangle \to x'$ *then* $x = x'$. *(A limit of a sequence, if it exists, is unique.)*

Thus the limit of a convergent sequence is also written as:

$$\lim_{n\to\infty} x_n \quad \text{or} \quad \lim_n x_n.$$

Definition 942 (Cauchy Sequences). A sequence $\langle x_n \rangle_{n\in\mathbf{N}}$ of real numbers is a *Cauchy Sequence* if for any $\epsilon > 0$ there is $k \in \mathbf{N}$ such that $|x_m - x_n| < \epsilon$ for all $m, n \geq k$.

Proposition 943 (The Cauchy Criterion for Convergence). *A sequence* $\langle x_n \rangle_{n\in\mathbf{N}}$ *is convergent if and only if it is Cauchy.*

Proof. If $\langle x_n \rangle \to x$, then, given any $\epsilon > 0$ we can fix k such that $|x_n - x| < \frac{\epsilon}{2}$ for all $n \geq k$, so $|x_m - x_n| \leq |x_m - x| + |x - x_n| < \frac{\epsilon}{2} + \frac{\epsilon}{2} = \epsilon$ for all $m, n \geq k$.

Conversely, suppose that $\langle x_n \rangle_{n\in\mathbf{N}}$ is a Cauchy sequence. Then we can fix k such that $|x_m - x_n| < 1$ for all $m.n \geq k$, and so $x_n \in [x_k - 1, x_k + 1]$ for all $n \geq k$. Thus $\langle x_n \rangle_{n\in\mathbf{N}}$ is eventually in the interval $I_1 := [x_k - 1, x_k + 1]$. Now note that if a sequence is eventually in an interval $[a,b]$ of length $\ell = b - a$, then it is either eventually in $[a, b - \frac{\ell}{3}]$ or in $[a + \frac{\ell}{3}, b]$. Hence, starting with I_1, we can recursively define a nested sequence of intervals $I_1 \supseteq I_2 \supseteq \cdots$ such that $\text{len}(I_{m+1}) = \frac{2}{3}\,\text{len}(I_m)$ and $\langle x_n \rangle_{n\in\mathbf{N}}$ is eventually in I_m for all m. Let x be in $\bigcap_m I_m$. Since $\text{len}(I_m) \to 0$, any open interval I containing x must also contain some I_m, and so $\langle x_n \rangle_{n\in\mathbf{N}}$ is eventually in I. Thus $\langle x_n \rangle \to x$. □

Problem 944 (CAC). $x \in \overline{A}$ *if and only if there is a sequence* $\langle x_n \rangle_{n\in\mathbf{N}}$ *converging to* x *with* $x_n \in A$ *for all* $n \in \mathbf{N}$.

14.4 Dense, Discrete, and Nowhere Dense Sets

Definition 945. 1. A set $A \subseteq \mathbf{R}$ is called *everywhere dense* or simply *dense* (in $\mathbf{R}$) if every point of $\mathbf{R} \smallsetminus A$ is a limit point of A, that is if $\overline{A} = \mathbf{R}$.

2. More generally, we say that *A is dense on B* if every point of $B \setminus A$ is a limit point of A, that is if $B \subseteq \overline{A}$. If in addition $A \subseteq B$, we say that *A is a dense subset of B* or that *the subset A is dense in B*.

For example, $\mathbf{Q}$ is (everywhere) dense in $\mathbf{R}$.

Problem 946. *Let E be the set of end points of the open intervals removed in the construction of the Cantor set. Show that E is a dense subset of the Cantor set.*

Proposition 947 (CAC). *Every set has a countable dense subset.*

Proof. Let $E \subseteq \mathbf{R}$ be nonempty, and consider the collection

$$\mathsf{C} := \{E \cap (p,q) \mid p, q \in \mathbf{Q}, E \cap (p,q) \neq \emptyset\}$$

of nonempty sets which are intersections of E with open intervals with rationals endpoints. Then C is a countable collection of nonempty subsets of E, and so by the Countable Axiom of Choice there is a function $\varphi\colon \mathsf{C} \to \bigcup\mathsf{C}$ such that $\varphi(A) \in A$ for all $A \in \mathsf{C}$. Put:

$$D := \{\varphi(A) \mid A \in \mathsf{C}\}.$$

Then D is a countable subset of E. We claim that D is dense in E, that is $E \subseteq \overline{D}$. It suffices to show that for every $x \in E$ and any open interval I with $x \in I$ we have $I \cap D \neq \emptyset$.

Let $x \in E$, and let $I = (a,b)$ be an open interval with $x \in I$. We can fix rational numbers p and q such that $a < p < x < q < b$, so that $E \cap (p,q) \neq \emptyset$ and so $E \cap (p,q) \in \mathsf{C}$. Put $y = \varphi(E \cap (p,q))$. Then $y \in E \cap (p,q) \subseteq I$ and $y \in D$, so $I \cap D \neq \emptyset$. □

Problem 948. *A set A is everywhere dense if and only if every nonempty open interval (or nonempty open set) has nonempty intersection with A.*

A set A is dense on a set B if and only if every nonempty open interval which has nonempty intersection with B also has nonempty intersection with A.

Problem 949. *If A is dense on B and B is dense on C then A is dense on C.*

Problem 950. *Find two disjoint sets each of which is dense.*

Problem 951. *A dense subset of a dense-in-itself set is dense-in-itself.*

Proposition 952. *In any set, all but countably many points are limit points. That is, the set $A \setminus D(A)$ of isolated points of a set A is countable.*

Proof. For each $x \in A \setminus D(A)$, since x is an isolated point of A, we can choose an open interval (a,b) with $(a,b) \cap A = \{x\}$, and then fix rational numbers p_x, q_x such that $a < p_x < x < q_x < b$. For each $x \in A \setminus D(A)$ we have

$$(p_x, q_x) \cap A = \{x\},$$

and so if $x \neq y$ are in $A \setminus D(A)$ then $(p_x, q_x) \neq (p_y, q_y)$. Hence the mapping

$$x \mapsto (p_x, q_x)$$

is a one-to-one mapping from $A \setminus D(A)$ into the countable family of intervals with rational endpoints, and so $A \setminus D(A)$ must be countable. □

Definition 953. A set $A \subseteq \mathbf{R}$ is called *discrete* if each point of A is an isolated point of A, that is if $A \cap D(A) = \emptyset$.

By the previous proposition, every discrete set is countable. Some examples of discrete subsets of **R** are **N**, **Z**, and the set $\{\frac{1}{n} \mid n \in \mathbf{N}\}$.

Problem 954. *Show that the union of two* closed *discrete sets is discrete.*

Problem 955. *Show that the set* $A := \mathbf{N} \cup \{n\sqrt{2} \mid n \in \mathbf{N}\}$ *is discrete, but that* A *has arbitrarily close points, that is for any* $\epsilon > 0$ *there exist* $p, q \in A$ *with* $0 < |p - q| < \epsilon$.

Definition 956 (Nowhere Dense Sets). If a set is dense on some nonempty open interval, we call it *somewhere dense*; otherwise, we call it *nowhere dense.*

Clearly any subset of a nowhere dense is nowhere dense.

Problem 957. *A set is nowhere dense if and only if its closure does not contain any nonempty open interval. Hence a closed set is nowhere dense if and only if it does not contain any nonempty open interval.*

Since the Cantor set is closed and does not contain any nonempty open interval, we have:

Proposition 958. *The Cantor set is nowhere dense.*

Problem 959. *A set* A *is nowhere dense if and only if every nonempty open interval contains a nonempty open subinterval which is disjoint from* A.

Problem 960. *Prove that a set is nowhere dense if its complement contains a dense open set.*

Problem 961. *The intersection of two dense open sets is a dense open set.*

Problem 962. *The union of two nowhere dense sets is nowhere dense.*

Problem 963. *Consider the collection* C *of the open intervals removed in the construction of the Cantor set. Since the open intervals in* C *are nonempty and pairwise disjoint,* C *can be naturally ordered by the usual order on* **R**: *For* $I, J \in \mathrm{C}$ *we have* $I < J$ *if and only if* $x < y$ *for all* $x \in I$ *and all* $y \in J$. *Show that under this natural ordering,* C *becomes a dense order of type* η.

Problem 964. *Give an example of a countable set disjoint from the Cantor set which is dense on the Cantor set.*

Problem 965. *Give an example of an infinite discrete set $A \subseteq \mathbf{R}$ such that the suborder A is a dense order, that is, for any $x < y$ in A there is z in A with $x < z < y$.*

Problem 966. *Give an example of a discrete subset $A \subseteq \mathbf{R}$ such that $D(A)$ is an uncountable perfect set.*

Problem 967. *Show that the closure of a discrete set is nowhere dense, and so if A is discrete then $D(A)$ is nowhere dense closed.*

Problem 968.* *Show that if E is nowhere dense closed then $E = D(A)$ for some discrete set A.*

Condensation Points

Definition 969 (Condensation Points). A point p is called *condensation point* of a set A if every open interval containing p contains uncountably many points of A.

Clearly, every condensation point of a set is a limit point of the set.

The following is an important generalization of Proposition 952.

Theorem 970. *All but countably many points of a set are condensation points.*

Proof. Let C be the set of condensation points of a set A. We show that $A \setminus C$ is countable. Put

$$\mathsf{H} := \{A \cap (p,q) \mid p,q \in \mathbf{Q},\ |A \cap (p,q)| \leq \aleph_0\}.$$

Then H is a countable collection of countable subsets of A, and so $E := \cup\mathsf{H}$ is a countable subset of A. We claim that $A \setminus C \subseteq E$: Let $x \in A \setminus C$, so that x is not a condensation point of A, and hence there is an open interval (a,b) with $x \in (a,b)$ and $|A \cap (a,b)| \leq \aleph_0$. Fix $p,q \in \mathbf{Q}$ such that $a < p < x < q < b$, so that $A \cap (p,q) \in \mathsf{H}$, therefore $A \cap (p,q) \subseteq E$, and so $x \in E$. Hence $A \setminus C \subseteq E$, and so $A \setminus C$ is countable. □

Clearly, if x is a condensation point of A, and B is any countable subset of A not containing x, then $A \setminus B$ will still have x as a condensation point. Hence by the theorem, if C is the set of condensation points of an uncountable set A, then C is an uncountable set in which every point is a condensation point. Therefore we have:

Corollary 971. *The set of condensation points of any uncountable set forms a nonempty subset which is dense-in-itself.*

Using a routine argument, we have the following result.

Problem 972. *The set of condensation points of a closed set is closed.*

The important *Cantor–Bendixson Theorem* is now an immediate corollary.

Corollary 973 (The Cantor–Bendixson Theorem). *Any uncountable closed set is the union of a nonempty perfect set and a countable set.*

The Cantor–Bendixson Theorem will be proved again more effectively in Chap. 16 (Theorem 1079 and Corollary 1080 in Sect. 16.2).

We had seen in Theorem 598 that every nonempty perfect subset of a complete order has cardinality $\mathfrak{c}$. Hence it follows from the Cantor–Bendixson Theorem that "the closed sets satisfy the Continuum Hypothesis" in the following sense.

Corollary 974. *Every closed set is either countable or has cardinality* $\mathfrak{c}$.

A different proof of the last result (a proof without using Theorem 598) will be given in Chap. 15 (Corollary 1051).

Problem 975. *Show that all but countably many points of a set are two-sided limit points of the set.*

Problem 976. *Find an example of a nonempty dense-in-itself set in which no point is an upper limit point of the set.*

Generalized Cantor Sets Are Perfect Nowhere Dense

Theorem 977. *Every generalized Cantor set is a bounded perfect nowhere dense set which is bijective with* $\{0,1\}^{\mathbb{N}}$ *(and so has cardinality* $\mathfrak{c} = 2^{\aleph_0}$*).*

Proof. Let A be a generalized Cantor set generated by the Cantor system $\langle J_u \mid u \in \{0,1\}^* \rangle$. We had already seen that A is bijective with $\{0,1\}^{\mathbb{N}}$ and hence has cardinality $\mathfrak{c} = 2^{\aleph_0}$.

First note that A is bounded, as $A \subseteq J_\varepsilon = [a,b]$ for some $a < b$. Also, A is closed, as (by Problem 905) it is an intersection of a collection of closed sets. To see that A is dense-in-itself, fix $p \in A$ and an infinite binary string $b_1 b_2 \cdots b_n \cdots$ such that $p \in \bigcap_n J_{b_1 b_2 \cdots b_n}$. Given an open interval (c,d) containing p, choose a sufficiently large n for which $\text{len}(J_{b_1 b_2 \cdots b_n}) < \min(p-c, d-p)$, so that $J_{b_1 b_2 \cdots b_n} \subsetneq (c,d)$. Let u be the finite binary string $b_1 b_2 \cdots b_n$. Then at most one of the disjoint intervals $J_{u^\frown 0}$ and $J_{u^\frown 1}$ can contain p, so by taking v to be either $u^\frown 0$ or $u^\frown 1$, we can assume that $p \notin J_v$. Now fix any infinite binary string extending v and let q be the unique point in the intersection of the corresponding nested sequence of intervals. Then $q \in J_v$ and so $p \neq q$ while $p \in (c,d) \cap A$. Thus every open interval containing p contains a point of A distinct from p, which means $p \in D(A)$. Since p was chosen arbitrarily from A, it follows that every point of A is a limit point, so A is dense-in-itself. Hence A is perfect.

To show that A is nowhere dense, it will suffice to show that A does not contain any proper interval. Let (c,d) be a nonempty open interval. We show that A does not contain all points of (c,d). Fix $p \in (c,d)$. If $p \notin A$ we are done, so assume that $p \in A$ and let $b_1 b_2 \cdots b_n \cdots$ be an infinite binary string such that $p \in \bigcap_n J_{b_1 b_2 \cdots b_n}$.

Since $\lim_n \operatorname{len}(J_{b_1 b_2 \cdots b_n}) = 0$, we can choose a sufficiently large n for which $\operatorname{len}(J_{b_1 b_2 \cdots b_n}) < \min(p - c, d - p)$ so that $J_{b_1 b_2 \cdots b_n} \subsetneq (c, d)$. Let u be the finite binary string $b_1 b_2 \cdots b_n$. Since $J_{u^\frown 0}$ and $J_{u^\frown 1}$ are disjoint closed subintervals of J_u, there must exist some $q \in J_u$ which does not belong to any of the intervals $J_{u^\frown 0}$ or $J_{u^\frown 1}$ (since by Theorem 599 an interval cannot be the disjoint union of two nonempty closed intervals). We now claim that $q \notin A$. To see this, assume that $q \in A$ (to get a contradiction), and fix a finite binary string v with $\operatorname{len}(v) > \operatorname{len}(u)$ such that $q \in J_v$. Then $q \in J_u \cap J_v$, so $J_u \cap J_v \neq \emptyset$, and so v must be an extension of u. But since $\operatorname{len}(v) > \operatorname{len}(u)$, v must either be an extension of $J_{u^\frown 0}$ or be an extension of $J_{u^\frown 1}$, which implies that q is either in $J_{u^\frown 0}$ or in $J_{u^\frown 1}$, contradicting the fact that q is not in any of these two intervals. □

We will later prove (in the chapter on Brouwer's theorem) that a converse of the result also holds: *Every bounded perfect nowhere dense set is a generalized Cantor set, that is A can be generated by some Cantor system.*

14.5 Continuous Functions and Homeomorphisms

We had already defined continuous functions for orders, but we now want to define continuity for functions which are only partially defined on $\mathbf{R}$, i.e., for functions whose domain may be a proper subset of $\mathbf{R}$. It is assumed that the reader is familiar with this notion of continuity through elementary calculus.

Definition 978 (Continuous Functions). Let $A \subseteq \mathbf{R}$ and let $f : A \to \mathbf{R}$.

1. We say that f *is continuous at a point* $p \in A$ if for any open interval J containing $f(p)$ there is an open interval I containing p such that for all x, $x \in I \cap A \Rightarrow f(x) \in J$.
2. We say that f *is continuous (on A)* if for all $p \in A$ f is continuous at p.

Problem 979. *Enumerate the rational numbers as* $\mathbf{Q} = \{r_n \mid n \in \mathbf{N}\}$ *where* $r_m \neq r_n$ *for* $m \neq n$, *and define* $f : \mathbf{R} \to \mathbf{R}$ *by:*

$$f(x) := \begin{cases} 1/n & \text{if } x = r_n \text{ and } n \in \mathbf{N}, \\ 0 & \text{otherwise.} \end{cases}$$

Show that f is continuous at each irrational point, but is discontinuous (not continuous) at each rational point.

Problem 980. *Show that if A is a discrete set then any function defined on A is continuous.*

Problem 981. *Let* $f : \mathbf{R} \to \mathbf{R}$. *Show that each of the following conditions is necessary and sufficient for* f *to be continuous on* $\mathbf{R}$.

1. *For any $x \in \mathbf{R}$ and any open interval J containing $f(x)$ there is an open interval I containing x such that $f[I] \subseteq J$.*
2. *For any open interval J the inverse image $f^{-1}[J]$ is an open set.*
3. *For any open set G the inverse image $f^{-1}[G]$ is an open set.*
4. *For any closed set F the inverse image $f^{-1}[F]$ is a closed set.*

Problem 982. *Show that $f: A \to \mathbf{R}$ is continuous if and only if for any $x \in A$ and any sequence $\langle x_n \rangle_{n \in \mathbf{N}}$ with $x_n \in A$ for all n, if $\langle x_n \rangle \to x$ then $\langle f(x_n) \rangle \to f(x)$.*

Problem 983. *Let $f: \mathbf{R} \to \mathbf{R}$ be continuous. Show that for any sequence of nested closed intervals $I_1 \supseteq I_2 \supseteq \cdots \supseteq I_n \supseteq I_{n+1} \supseteq \cdots$, if $\operatorname{len}(I_n) \to 0$ as $n \to \infty$ then $f[\bigcap_n I_n] = \bigcap_n f[I_n]$.*

We had earlier proved the Intermediate Value Theorem for general linear continuums (Corollary 600). Since an interval in $\mathbf{R}$ is a linear continuum, the IVT remains true for this case.

Theorem 984 (IVT). *Let I be an interval and $f: I \to \mathbf{R}$ be continuous. Then $f[I]$ is an interval. In other words, if $a < b$ are in I and if $f(a) < y < f(b)$ or if $f(a) > y > f(b)$ then there is $x \in (a, b)$ such that $f(x) = y$.*

Let $f: \mathbf{R} \to \mathbf{R}$. For $a, b \in \mathbf{R}$, let f_a^b denote the function obtained from f by redefining its value at a to b, i.e., $f_a^b(x) := b$ if $x = a$ and $f_a^b(x) := f(x)$ otherwise. We say that f has a *removable discontinuity at the point a* if f is discontinuous at a but f_a^b is continuous at a for some $b \in \mathbf{R}$. (In terms of limits, this means that $\lim_{x \to a} f(x)$ exists but does not equal $f(a)$.)

Problem 985. *Let $f: \mathbf{R} \to \mathbf{R}$ and let E be the set of points at which f has a removable discontinuity. Then E is countable.*

[Hint: To each $a \in E$, assign rational numbers p, q, r, s such that $a \in (p, q)$ and for all $x \in (p, q)$ with $x \neq a$ we have $f(x) \in (r, s)$ but $f(a) \notin (r, s)$.]

Homeomorphisms and Homeomorphic Sets

Definition 986. Let $A, B \subseteq \mathbf{R}$. A *homeomorphism from A to B* is a bijection f from A onto B such that both f and f^{-1} are continuous. The sets A and B are *homeomorphic* if there is a homeomorphism from A onto B.

For example, the mapping $x \mapsto x/(1 + |x|)$ is a homeomorphism from $\mathbf{R}$ onto the open interval $(-1, 1)$.

Problem 987. *Let $f: \mathbf{R} \to \mathbf{R}$ be continuous and strictly increasing. Show that $f[\mathbf{R}]$ is an open interval (which may be unbounded) and that f is a homeomorphism from $\mathbf{R}$ onto $f[\mathbf{R}]$.*

Problem 988. *Any two infinite discrete sets are homeomorphic.*

Problem 989. *The set* $\{0\} \cup \{1/n \mid n \in \mathbf{N}\}$ *is homeomorphic to the set* $\{0\} \cup \{1/n \mid n \in \mathbf{N}\} \cup \{-1/n \mid n \in \mathbf{N}\}$*, but not to* $\{0\} \cup \{1/n \mid n \in \mathbf{N}\} \cup \mathbf{N}$.

Problem 990. *No two of* $[0, 1]$, $[0, 1)$, *and* $(0, 1)$ *are homeomorphic.*

Problem 991*. *The sets* $[0, 1] \cap \mathbf{Q}$ *and* $(0, 1) \cap \mathbf{Q}$ *are homeomorphic.*

Problem 992. *Let I and J be nonempty open intervals (none, one, or both of which may be unbounded), A be a countable dense subset of I, and B be a countable dense subset of J. Show that there is a homeomorphism of I onto J such that $f[A] = B$, and conclude that A and B must be homeomorphic.*

[Hint: Use Cantor's theorem on countable dense orders.]

If A and B are homeomorphic via f, then f will preserve all internal properties of points and subsets of A involving limit points. For example, if A has exactly three limit points, then B the same will be true for B. If $p \in A$ and $E \subseteq A$, then p is a limit point of E if and only if $f(p)$ is a limit point of $f[E]$, E is a dense subset of A if and only if $f[E]$ is dense subset of B, and so on.

Problem 993. *Let A be homeomorphic to B. Show that*

1. *A is discrete if and only if B is discrete.*
2. *A contains a proper interval if and only if B contains a proper interval.*
3. *A is dense-in-itself if and only if B is dense-in-itself.*

Internal properties of sets which are preserved by homeomorphisms are called *topological properties.* Thus the properties listed in the last problem are topological. If two sets are homeomorphic then they will share all topological properties and are said to have identical (internal) topological structures.

Properties of a subset A of $\mathbf{R}$ which express how A, its points or other subsets of A relate to the parent $\mathbf{R}$ (i.e., properties which express how A is situated within $\mathbf{R}$) are in general not topological. This is illustrated by the following problem.

Problem 994. *None of the following properties of subsets of* $\mathbf{R}$ *is a topological property, i.e., it is not preserved by homeomorphisms in general.*

1. *Being bounded.*
2. *Being closed.*
3. *Being everywhere dense.*
4* *Being nowhere dense.*

Problem 995. *None of the properties of being closed and being nowhere dense is a topological property, but if A and B are homeomorphic* closed *sets then A is nowhere dense if and only if B is nowhere dense.*

While none of the individual properties of being closed and being bounded is a topological property, the combined property of being both closed and bounded becomes a topological property and is known as *compactness,* but the proof of this fact requires the Heine–Borel theorem which is covered in the next chapter.

Space Filling Peano Curves

By a (parametric) *continuous curve* $\langle x(t), y(t)\rangle$, $0 \le t \le 1$, *in the plane* we mean a pair of continuous functions $t \mapsto x(t)$ and $t \mapsto y(t)$, $0 \le t \le 1$. The classic example is that of the unit circle $x(t) = \cos 2\pi t$, $y(t) = \sin 2\pi t$. Peano showed the surprising result that there is a continuous curve in the plane which fills up the entire unit square $[0, 1] \times [0, 1]$! We can readily derive Peano's result using the "identification" of the Cantor set **K** with the set $\{0, 1\}^{\mathbf{N}}$ of infinite binary sequences (review Sects. 6.6 and 6.7).

For each $\alpha: \mathbf{N} \to \mathbf{N}$, define $\mathbf{h}_\alpha: \mathbf{K} \to [0, 1]$ by:

$$\mathbf{h}_\alpha(x) := \sum_{n=1}^{\infty} \frac{x_{\alpha(n)}}{2^n} \qquad \left(x = \textstyle\sum_{n=1}^{\infty} \frac{2x_n}{3^n} \in \mathbf{K},\ \langle x_n\rangle \in \{0, 1\}^{\mathbf{N}}\right),$$

Problem 996. *If* $\alpha: \mathbf{N} \to \mathbf{N}$ *is injective then* $\mathbf{h}_\alpha$ *is a continuous surjection from* **K** *onto* $[0, 1]$. *If* $\alpha, \beta: \mathbf{N} \to \mathbf{N}$ *are injective with disjoint ranges, then for all* $\langle a, b\rangle \in [0, 1] \times [0, 1]$ *there is* $x \in \mathbf{K}$ *with* $a = \mathbf{h}_\alpha(x)$, $b = \mathbf{h}_\beta(x)$.

Let $\mathbf{h} := \mathbf{h}_\iota$ where $\iota: \mathbf{N} \to \mathbf{N}$ is the identity map $\iota(n) = n$.

Problem 997. *For the function* $\mathbf{h} = \mathbf{h}_\iota$, *if* (a, b) *is a component open interval of the complement of the Cantor set, then* $\mathbf{h}(a) = \mathbf{h}(b)$.

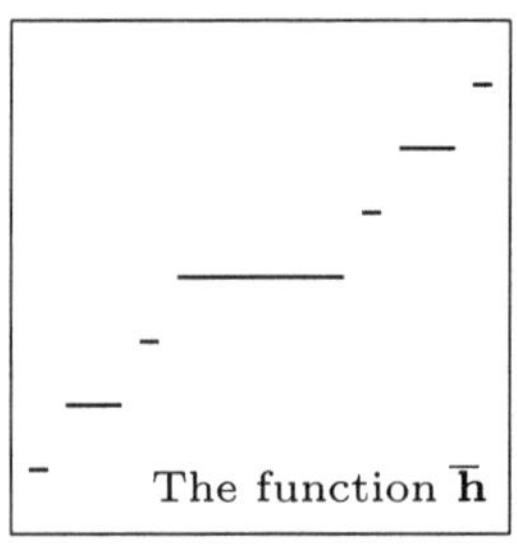

By the last problem, one can extend **h** to a function $\overline{\mathbf{h}}: [0, 1] \to [0, 1]$ by "joining $\mathbf{h}(a)$ and $\mathbf{h}(b)$ with a horizontal line segment" over each component open interval of $[0, 1] \smallsetminus \mathbf{K}$. The resulting function $\overline{\mathbf{h}}$ will be a continuous function (which maps **K** onto $[0, 1]$). More generally, we have:

Problem 998 (Cantor Ternary Functions). *Let* $g: \mathbf{K} \to [a, b]$ *be continuous. Then there is a unique continuous extension* $\overline{g}: [0, 1] \to [a, b]$ *of* g *such that* $\overline{g}$ *is linear on each component open interval of* $[0, 1] \smallsetminus \mathbf{K}$.

Now fix injective $\alpha, \beta : \mathbf{N} \to \mathbf{N}$ with disjoint ranges, e.g., $\alpha(n) = 2n - 1$ and $\beta(n) = 2n$, and put $\boldsymbol{\lambda} := \overline{\mathbf{h}_\alpha}$, $\boldsymbol{\rho} := \overline{\mathbf{h}_\beta}$. By Problem 998, $\langle \boldsymbol{\lambda}(t), \boldsymbol{\rho}(t) \rangle$, $0 \le t \le 1$, is a continuous curve in the plane, and by Problem 996 it fills up the unit square. We thus have:

Corollary 999 (Peano Curve). *Let $\alpha(n) := 2n - 1$, $\beta(n) := 2n$, $\boldsymbol{\lambda} := \overline{\mathbf{h}_\alpha}$, and $\boldsymbol{\rho} := \overline{\mathbf{h}_\beta}$. Then $\langle \boldsymbol{\lambda}(t), \boldsymbol{\rho}(t) \rangle$, $0 \le t \le 1$, is a continuous curve in the plane which fills up the entire unit square.*

Chapter 15
The Heine–Borel and Baire Category Theorems

Abstract This chapter starts with the Heine–Borel theorem and its characterization of complete orders, and then uses Borel's theorem to give a measure-theoretic proof that **R** is uncountable. Other topics focus on measure and category: Lebesgue measurable sets, Baire category, the perfect set property for G_δ sets, the Banach–Mazur game and Baire property, and the Vitali and Bernstein constructions.

15.1 The Heine–Borel Theorem

Earlier, we had encountered the Bolzano–Weierstrass and Nested Intervals properties for complete orders and saw that none of those properties characterize complete orders. On the other hand, the Heine–Borel theorem, which has very wide applicability, gives a stronger condition which actually characterizes complete orders.

Definition 1000. Let A be a set and C be a collection of sets. We say that *A is covered by* C or C *covers* A if every element of A is a member of some set in C, that is if

$$A \subseteq \bigcup \mathrm{C}.$$

We also say that *A is covered by the sets $A_1, A_2, \ldots, A_n$* if A is covered by the collection $\{A_1, A_2, \ldots, A_n\}$.

Theorem 1001. *Let $[a,b]$ be a bounded closed interval and let* C *be a collection of open sets which covers $[a,b]$. Then $[a,b]$ can be covered by finitely many sets from* C.

Proof. Let $a < b$ and let C be a collection of open sets with

$$[a,b] \subseteq \bigcup \mathrm{C}.$$

A. Dasgupta, *Set Theory: With an Introduction to Real Point Sets*,
DOI 10.1007/978-1-4614-8854-5_15,

Let A be the set consisting of those real numbers $x \in [a,b]$ such that the interval $[a,x]$ can be covered by finitely many sets from C. In other words, we have $x \in A$ if and only if $a \leq x \leq b$ and there are sets $G_1, G_2, \ldots, G_n \in$ C (for some natural number n) such that $[a,x] \subseteq \cup_{k=1}^{n} G_k$. Then $A \subseteq [a,b]$ is bounded and nonempty (since $a \in A$), and so with

$$c := \sup A$$

we have $c \in [a,b]$. Since $c \in [a,b]$ there is an open set $G \in$ C such that $c \in G$, and so we can fix an open interval (p,q) such that $c \in (p,q) \subseteq G$.

Since $p < c$ and $c = \sup A$, there is $r \in A$ such that $p < r < c$. Then $r \in [a,b]$ and $[a,r]$ can be covered by finitely many sets from C, say

$$[a.r] \subseteq \bigcup_{k=1}^{n} G_k, \quad G_k \in \text{C for } k = 1,2,\ldots,n \quad (n \in \mathbf{N}).$$

Putting $G_{n+1} := G$, we see that $[r,c] \subseteq G_{n+1}$, and so

$$[a,c] = [a,r] \cup [r,c] \subseteq \bigcup_{k=1}^{n+1} G_k,$$

hence $[a,c]$ can be covered by finitely many sets from C and so $c \in A$. Moreover, if we had $c < b$, we could choose s with $c < s < \min(b,q)$ and so we would get $[r,s] \subseteq G = G_{n+1}$. Then $[a,s] = [a,r] \cup [r,s]$ would again be covered by the sets $G_1, G_2, \ldots, G_n, G_{n+1}$, which would imply $s \in A$, which is a contradiction since $s > c = \sup A$. Therefore $c = b$, hence $b \in A$, and so $[a,b]$ can be covered by finitely many sets from C. □

The following generalization of Theorem 1001 is usually called *the Heine–Borel theorem.* It is sometimes paraphrased as "every open cover of a bounded closed set has a finite subcover."

Corollary 1002 (Heine–Borel). *Let E be a bounded closed set and let* C *be a collection of open sets which covers E. Then E can be covered by finitely many sets from* C.

Proof. Since E is bounded there exist a, b with $E \subseteq [a,b]$. Let $G = \mathbf{R} \setminus E$, so that G is open, and put C$'$ = C $\cup \{G\}$. Then C$'$ covers $\mathbf{R}$, since for any x either $x \in E$ in which case $x \in \cup$C or else $x \in G$. Hence C$'$ covers $[a,b]$, so $[a,b]$, and hence E, can be covered by finitely many sets from C$'$, say $E \subseteq \cup_{k=1}^{n} G_k$ with $G_k \in$ C$'$ for $k = 1,2,\ldots,n$. Now from the sets G_k, $k = 1,2,\ldots,n$, we can remove any G_k which equals G, and the remaining sets will still cover E (since $E \cap G = \emptyset$), giving us finitely many sets from C which covers E. □

Corollary 1003. *If $F_1, F_2, \ldots, F_n, \ldots$ are bounded closed sets, G is an open set, and $\bigcap_{n=1}^{\infty} F_n \subseteq G$, then $\bigcap_{k=1}^{m} F_k \subseteq G$ for some $m \in \mathbf{N}$.*

Proof. Put $G_n := \mathbf{R} \setminus F_n$. Then $\bigcap_n F_n \subseteq G$ means that $F_1 \subseteq G \cup \left(\bigcup_{n=2}^{\infty} G_n \right)$, and so by the Heine–Borel theorem there is m such that $F_1 \subseteq G \cup \left(\bigcup_{k=2}^{m} G_k \right)$, that is, $\bigcap_{k=1}^{m} F_k \subseteq G$. □

Corollary 1004. *The intersection of a nested sequence of nonempty bounded closed sets is nonempty.*

Problem 1005. *Let A be a subset of $\mathbf{R}$. Show that if every covering of A by a collection of open sets has a finite subcollection which also covers A, then A is closed and bounded.*

The following is also known as the Heine–Borel Theorem (in a "necessary and sufficient" form).

Corollary 1006 (The Heine–Borel Theorem). *A subset A of $\mathbf{R}$ is closed and bounded if and only if every covering of A by a collection of open sets has a finite subcollection which also covers A.*

This last and final version of the Heine–Borel theorem is a characterization of the property of being a closed and bounded subset of $\mathbf{R}$ using a property involving covering by open sets. It can be used to show that the property of being a closed and bounded subset of $\mathbf{R}$ is preserved by homeomorphisms, that is, it is a topological property. Thus even though the individual properties of being closed or being bounded are not topological, the combined property becomes a topological property, and is termed as *compactness.*

Problem 1007. *Show that if A and B are homeomorphic subsets of $\mathbf{R}$ and one of them is closed and bounded then so must be the other. In other words, compactness is a topological property.*

It should be noted that the proof of Theorem 1001 uses only the completeness property of $\mathbf{R}$, and hence can be generalized for orders. We will say that *an order X satisfies the Heine–Borel condition* if for any bounded closed interval I covered by a collection of open intervals there are finitely many of those open intervals which covers I.

Problem 1008. *Let X be an order without endpoints, Show that X is complete if and only if X satisfies the Heine–Borel condition.*

Definition 1009. The *length* of an interval I is defined as the nonnegative quantity $\operatorname{len}(I) := \sup I - \inf I$, and the *total length* of a sequence $\langle I_n \rangle$ of intervals is defined as the nonnegative sum $\sum_n \operatorname{len}(I_n)$.

We will now apply the Heine–Borel theorem to obtain the following result known as *Borel's Theorem* which is useful in theory of Lebesgue-measure: *An interval $[a, b]$ cannot be covered by a sequence of open intervals having total length $< b - a$.*

First, we need the following proposition.

Proposition 1010. *Let $[a,b]$ be a closed interval and suppose that (a_k,b_k) is an open interval with $a_k < b_k$ for each $k = 1,2,\ldots,n$. If*

$$[a,b] \subseteq \bigcup_{k=1}^{n}(a_k,b_k),$$

then we have

$$b - a < \sum_{k=1}^{n}(b_k - a_k).$$

Proof. The proof is done by induction on n for the statement of the proposition. We assume that $a \leq b$.

If $n = 1$, then $[a,b] \subseteq (a_1,b_1)$, so $a_1 < a \leq b < b_1$, hence $b - a < a_1 - b_1 = \sum_{k=1}^{n}(b_k - a_k)$.

For the induction step, suppose that the proposition is true for $n = m$, and let (a_k,b_k) be open intervals with $a_k < b_k$ for each $k = 1,2,\ldots,m+1$ such that

$$[a,b] \subseteq \bigcup_{k=1}^{m+1}(a_k,b_k).$$

Then b is a member of one of the intervals covering $[a,b]$, which we may assume to be (a_{m+1},b_{m+1}) without loss of generality. Then $a_{m+1} < b < b_{m+1}$. If now $a_{m+1} < a$, then $b - a < b_{m+1} - a_{m+1}$ and we are done, so assume $a \leq a_{m+1}$. Then since $[a,a_{m+1}] \cap (a_{m+1},b_{m+1}) = \emptyset$, we must have

$$[a,a_{m+1}] \subseteq \bigcup_{k=1}^{m}(a_k,b_k),$$

and hence by induction hypothesis we have:

$$a_{m+1} - a < \sum_{k=1}^{m}(b_k - a_k),$$

and so

$$\begin{aligned} b - a < (a_{m+1} - a) + (b_{m+1} - a_{m+1}) &< \sum_{k=1}^{m}(b_k - a_k) + (b_{m+1} - a_{m+1}) \\ &= \sum_{k=1}^{m+1}(b_k - a_k). \end{aligned}$$

Thus the result holds for $n = m + 1$. □

An application of the Heine–Borel theorem now immediately yields Borel's Theorem that no interval I can be covered by a sequence of open intervals having total length less than the length of I.

Theorem 1011 (Borel's Theorem). *Let $[a,b]$ be a closed interval and suppose that (a_k, b_k) is an open interval with $a_k < b_k$ for each $k \in \mathbf{N}$. If*

$$[a,b] \subseteq \bigcup_{k=1}^{\infty}(a_k, b_k),$$

then we have

$$b - a < \sum_{k=1}^{\infty}(b_k - a_k).$$

Problem 1012. *If A and B are disjoint bounded closed sets then there exists $p > 0$ such that no interval of length $> p$ intersects both A and B.*

[Hint: All intervals $I = (a,b)$ with $(2a-b, 2b-a) \cap B = \emptyset$ form an open cover of A, and so has a finite subcover $I_1, I_2, \ldots, I_n$. Take $p = \min_k \operatorname{len}(I_k)$.]

15.2 Sets of Lebesgue Measure Zero

Definition 1013 (Lebesgue Measure Zero). $E \subseteq \mathbf{R}$ is said to be a *set of Lebesgue measure zero* or simply a *measure zero set* if E can be covered by sequences of intervals of arbitrarily small total length, i.e., if for any positive number $\epsilon > 0$ there is a sequence of open intervals $\langle (a_n, b_n) \mid n \in \mathbf{N} \rangle$ with

$$E \subseteq \bigcup_{n=1}^{\infty}(a_n, b_n) \quad \text{and} \quad \sum_{n=1}^{\infty}(b_n - a_n) < \epsilon.$$

Clearly a subset of a set of measure zero is of measure zero. We also have:

Proposition 1014. *A countable union of measure zero sets has measure zero.*

Proof. Let A_n have measure zero for each $n \in \mathbf{N}$, and let $\epsilon > 0$. Then for each m, since A_m has measure zero, there is a sequence $\langle (a_{m,n}, b_{m,n}) \mid n \in \mathbf{N} \rangle$ such that

$$A_m \subseteq \bigcup_{n=1}^{\infty}(a_{m,n}, b_{m,n}) \quad \text{and} \quad \sum_{n=1}^{\infty}(b_{m,n} - a_{m,n}) < \frac{\epsilon}{2^m}.$$

Then we have

$$\bigcup_{m=1}^{\infty} A_m \subseteq \bigcup_{m=1}^{\infty}\bigcup_{n=1}^{\infty}(a_{m,n}, b_{m,n})$$

with

$$\sum_{m=1}^{\infty}\sum_{n=1}^{\infty}(b_{m,n} - a_{m,n}) < \sum_{m=1}^{\infty}\frac{\epsilon}{2^m} = \epsilon.$$

As $\epsilon > 0$ was arbitrary, it follows that $\cup_m A_m$ has measure zero, □

Since every singleton set is of measure zero, we get:

Corollary 1015. *Every countable set has measure zero.*

But not all measure zero sets are countable:

Problem 1016. *The Cantor set has measure zero.*

[Hint: The closed set $\mathbf{K}_n$ found at stage n of the construction of the Cantor set consists of 2^n disjoint closed intervals each of length $1/3^n$, and so $\mathbf{K}_n$ can be covered by 2^n open intervals of total length $2(2^n/3^n)$.]

On the other hand, Borel's Theorem (Theorem 1011) immediately gives examples of sets not having measure zero:

Corollary 1017. *If $a < b$ then the interval $[a,b]$ does not have measure zero. Hence a measure zero set cannot contain a nonempty open interval.*

Corollary 1018. *If $a < b$ then the interval $[a,b]$ is uncountable.*

Thus we have another proof that $\mathbf{R}$ is uncountable.

Small Sets, Ideals, and σ-Ideals

The above results indicate that in a sense the sets of measure zero are "small subsets" of $\mathbf{R}$. Some other examples of collections of sets which may be regarded as "small" are:

- The finite subsets of $\mathbf{R}$.
- The countable subsets of $\mathbf{R}$.
- The nowhere dense subsets of $\mathbf{R}$.
- The closed discrete subsets of $\mathbf{R}$.

In general, the notion of "small subsets of **R**" can be axiomatized as:

1. Ø is small but **R** is not small.
2. A subset of a small set is small.
3. The union of two (or finitely many) small sets is small.

All the above examples satisfy these properties. (In addition, they satisfy the stronger condition that no small set contains a nonempty open interval.)

Any collection of sets which satisfies the three conditions listed above is called an *ideal* of sets.

The collection of countable sets and the collection of measure zero sets satisfy an additional fourth property:

4. The union of countably many small sets is small.

Ideals which satisfy this additional property are called σ*-ideals*. The sets of measure zero form an important σ-ideal.

The Borel Conjecture

Definition 1019 (Strong Measure Zero). A set A is said to have *strong measure zero* if for any sequence $\langle \epsilon_n \mid n \in \mathbf{N}\rangle$ of positive numbers $\epsilon_n > 0$, there exists a sequence $\langle (a_n, b_n) \mid n \in \mathbf{N}\rangle$ of open intervals such that

$$b_n - a_n < \epsilon_n \text{ for all } n \in \mathbf{N} \quad \text{and} \quad A \subseteq \bigcup_{n=1}^{\infty} (a_n, b_n).$$

Problem 1020. *Show that the collection of sets of strong measure zero is a σ-ideal containing all countable sets.*

Problem 1021. *Show that the Cantor set does not have strong measure zero.*

The assertion that every set of strong measure zero is countable is known as *the Borel conjecture*. It can neither be proved nor be disproved using the usual axioms of set theory (provided that these axioms are consistent).

15.3 Lebesgue Measurable Sets

Definition 1022 (Lebesgue Measurable Sets). We say that $A \subseteq \mathbf{R}$ is *Lebesgue Measurable* (or simply *measurable*) if for any $\epsilon > 0$ there is an open set G and a closed set F with $F \subseteq A \subseteq G$, such that $G \setminus F$ can be covered by a sequence of open intervals of total length less than ϵ, i.e., there is a sequence $\langle I_n \mid n \in \mathbf{N}\rangle$ of open intervals such that $G \setminus F \subseteq \cup_n I_n$ and $\sum_n \operatorname{len}(I_n) < \epsilon$.

We will let $\mathbf{L}$ denote the collection of all Lebesgue measurable sets.

The results in the following problem are immediate from the definition.

Problem 1023. *Show that*

1. *The complement of a measurable set is measurable.*
2. *Any set of measure zero is measurable.*
3. *Every interval is measurable.*

Proposition 1024. *Let A be a Lebesgue measurable set. If A is not of measure zero then A contains an uncountable closed set and hence a perfect set.*

Proof. Suppose that A does not contain any uncountable closed set. We show that then A must have measure zero.

Given $\epsilon > 0$, fix closed F and open G with $F \subseteq A \subseteq G$ and a sequence $\langle I_n \rangle$ of open intervals covering $G \setminus F$ with $\sum_n \text{len}(I_n) < \frac{\epsilon}{2}$. By assumption, F must be countable, so there is a sequence $\langle J_n \rangle$ of open intervals covering F with $\sum_n \text{len}(J_n) < \frac{\epsilon}{2}$. The combined sequence of intervals $I_1, J_1, I_2, J_2, \dots$ covers A and has total length $< \epsilon$. Since ϵ is arbitrary, it follows that A has measure zero. $\square$

Problem 1025. *A is measurable if and only if $A \cap (a,b)$ is measurable for all $a < b$.*

[Hint: If $F_n \subseteq (n, n+2)$ is closed for all $n \in \mathbf{Z}$, then $\bigcup_{n \in \mathbf{Z}} F_n$ is closed.]

Proposition 1026. *Every open set is measurable.*

Proof. By the last problem, it suffices to show that every bounded open set is measurable. Let G be an open set contained in (a,b), and suppose that $\epsilon > 0$. Since G is open, we have $G = \bigcup_n (a_n, b_n)$ for some sequence of pairwise disjoint open intervals (a_n, b_n), $n = 1, 2, \dots$.

Note that if $I_1, I_2, \dots, I_n$ are finitely many pairwise disjoint open subintervals of (a,b), then by rearranging them we can assume $I_k = (c_k, d_k)$, $k = 1, 2, \dots, n$ with $a \le c_1 < d_1 \le c_2 < d_2 \le \dots \le c_n < d_n \le b$, and so the total length of the intervals $\sum_{k=1}^n \text{len}(I_k)$ cannot exceed $b - a$. In particular, we have $\sum_{k=1}^n (b_k - a_k) \le b - a$ for any n, and so $\sum_{n=1}^\infty (b_n - a_n) \le b - a$. Hence there exists $m \in \mathbf{N}$ such that $\sum_{n=m+1}^\infty (b_n - a_n) < \frac{\epsilon}{2}$.

Now put $F = \bigcup_{n=1}^m [a'_n, b'_n]$, where $a'_n = a_n + \frac{\epsilon}{4m}$ and $b'_n = b_n - \frac{\epsilon}{4m}$. Then F is a closed subset of G and moreover $G \setminus F$ is covered by the open intervals (a_n, a'_n), $n = 1, 2, \dots, m$, (b'_n, b_n), $n = 1, 2, \dots, m$, and (a_n, b_n), $n = m+1, m+2, \dots$, whose total length is less than $\frac{\epsilon}{4} + \frac{\epsilon}{4} + \frac{\epsilon}{2} = \epsilon$. $\square$

Theorem 1027. *Let $\langle A_n \mid n \in \mathbf{N} \rangle$ be a sequence of measurable sets. Then $A := \bigcup_n A_n$ is measurable.*

Proof. Let $a < b$ be arbitrary reals. It suffices to show that the set $A^* := A \cap (a,b)$ is measurable. Let $\epsilon > 0$ be given.

Put $A_n^* := A_n \cap (a,b)$. Then A_n^* is measurable for each n, and so we can find closed F_n and open G_n such that $F_n \subseteq A_n^* \subseteq G_n$ and $G_n \setminus F_n$ is covered by a sequence of intervals of total length $< \frac{\epsilon}{2^{n+1}}$, where we assume that $G_n \subseteq (a,b)$ for all n (by replacing G_n by $G_n \cap (a,b)$ if necessary). Thus the open set $V := \cup_n(G_n \setminus F_n)$ can be covered by a sequence of open intervals of total length $< \sum_{n=1}^{\infty} \frac{\epsilon}{2^{n+1}} = \frac{\epsilon}{2}$.

Let $G := \cup_n G_n$, so that G is an open set contained in (a,b), and let $U_n := G \setminus F_n$. Then U_n is open, hence measurable, and so there is a closed $H_n \subseteq U_n$ such that $U_n \setminus H_n$ is covered by a sequence of intervals of total length $< \frac{\epsilon}{2^{n+1}}$, for each n. Therefore the open set $U := \cup_n(U_n \setminus H_n)$ can be covered by a sequence of open intervals of total length $< \sum_{n=1}^{\infty} \frac{\epsilon}{2^{n+1}} = \frac{\epsilon}{2}$.

Note that we have $\cap_n H_n \subseteq \cap_n U_n \subseteq V$, and hence by the Heine–Borel theorem there is m such that $\cap_{k=1}^{m} H_k \subseteq V$.

Now put $F = \cup_{k=1}^{m} F_k$. Then F is closed with $F \subseteq A^* \subseteq G$, and

$$G \setminus F = \bigcap_{k=1}^{m} U_k \subseteq \bigcap_{k=1}^{m} H_k \cup \bigcup_{k=1}^{m}(U_k \setminus H_k) \subseteq V \cup U.$$

Since each of V and U can be covered by a sequence of open intervals of total length less than $\frac{\epsilon}{2}$, it follows that $G \setminus F$ can be covered by a sequence of intervals of total length less than ϵ. □

By the above results, the collection $\mathbf{L}$ of all Lebesgue measurable sets contains all measure zero sets, intervals, open and closed sets, and is closed under taking countable unions and complements of sets, and hence under forming countable intersections as well. This makes $\mathbf{L}$ a very comprehensive collection of sets, and in fact a *sigma-algebra* of sets, discussed in Sect. 18.1.

The basic result in Lebesgue measure theory is Lebesgue's landmark theorem that the length function for intervals can be uniquely extended to a nonnegative *countably additive* function on $\mathbf{L}$:

Theorem 1028 (Lebesgue). *There is* $m\colon \mathbf{L} \to [0,\infty]$ *such that*

1. m *is* countably additive*: If* $A_1, A_2, \ldots$ *are pairwise disjoint measurable sets, then* $m(\bigcup_n A_n) = \sum_n m(A_n)$.
2. $m(I) = \operatorname{len}(I)$ *for any interval* I *(thus* $m(\emptyset) = 0$*).*

Proof. The proof is given in Appendix B. □

Such a function m must be *uniquely defined on* $\mathbf{L}$. This and several other important immediate consequences are derived in the following corollary:

Corollary 1029. *Suppose that* m *is as in Theorem 1028,* $r \in \mathbf{R}$*, and* A*,* B*,* $A_1, A_2, \ldots, A_n, \ldots$ *are measurable sets. The following properties hold:*

1. Monotonicity: $A \subseteq B \Rightarrow m(A) \leq m(B)$.
2. Countable Subadditivity: $m(\bigcup_n A_n) \leq \sum_n m(A_n)$.
3. Uniqueness: *If* $m'\colon \mathbf{L} \to [0,\infty]$ *also satisfies Theorem 1028 then* $m' = m$.
4. Outer Regularity: *For any* $\epsilon > 0$ *there is open* $G \supseteq A$ *with* $m(G \setminus A) < \epsilon$.
5. Translation Invariance: $m(A + r) = m(A)$*, where* $A + r := \{x + r \mid x \in A\}$.

6. CCC Property: *If $\langle E_i \mid i \in I\rangle$ is an arbitrary family of pairwise disjoint measurable sets, then $m(E_i) = 0$ for all but countably many $i \in I$.*

Proof. 1. $A \cap (B \smallsetminus A) = \emptyset$, so $m(B) = m(A) + m(B \smallsetminus A) \geq m(A)$.
2. The sets $B_n := A_n \smallsetminus \bigcup_{k<n} A_k$ are pairwise disjoint with $\bigcup_n B_n = \bigcup_n A_n$.
3. Suppose m' also satisfies the two conditions of Theorem 1028. Let E be measurable and $\epsilon > 0$. Fix closed F and open G with $F \subseteq E \subseteq G$ and a sequence of intervals $\langle I_n \rangle$ covering $G \smallsetminus F$ with $\sum_n \text{len}(I_n) < \epsilon$. Since G is open, it is a disjoint union of intervals and so $m(G) = m'(G)$. Hence $m(E) + m(G \smallsetminus E) = m(G) = m'(G) = m'(E) + m'(G \smallsetminus E)$. Now $m(G \smallsetminus E) \leq m(\bigcup_n I_n) \leq \sum_n \text{len}(I_n) < \epsilon$, and similarly, $m'(G \smallsetminus E) < \epsilon$. Hence $m(E)$ and $m'(E)$ cannot differ by more than ϵ.
4. Immediate from definition of measurability and monotonicity.
5. Follows from (4), as intervals and so open sets are translation invariant.
6. For any m, n, the set $\{i \in I \mid m([n, n+1] \cap E_i) \geq \frac{1}{m}\}$ is finite, and $m(E_i) > 0$ if and only if $m([n, n+1] \cap E_i) > 0$ for some n. □

Definition 1030 (Lebesgue Measure). *Lebesgue Measure* is the unique function $m: \mathbf{L} \to [0, \infty]$ which is countably additive and satisfies $m([a, b]) = b - a$ for all $a \leq b$.

15.4 F_σ and G_δ Sets

Definition 1031. A is called an F_σ set if it can be expressed as a countable union of closed sets, that is if $A = \cup_n A_n$ for some sequence $\langle A_n \mid n \in \mathbf{N}\rangle$ of closed sets.

B is called a G_δ set if it can be expressed as a countable intersection of open sets, that is if $B = \cap_n B_n$ for some sequence $\langle B_n \mid n \in \mathbf{N}\rangle$ of open sets.

Problem 1032. *A set is F_σ if and only if its complement is G_δ, and so a set is G_δ if and only if its complement is F_σ.*

Problem 1033. *1. Every countable set is an F_σ set.*
2. *Every open interval is an F_σ set.*
3. *Every open set is an F_σ set, and so every closed set is a G_δ set.*
4. *Every open set and every closed set is both an F_σ set and a G_δ set.*
5. *The union of countably many F_σ sets is an F_σ set, and the intersection of countably many G_δ sets is a G_δ set.*
6. *The intersection of two F_σ sets is an F_σ set, and the union of two G_δ sets is a G_δ set.*

While all open sets and closed sets are both F_σ and G_δ, the set $\mathbf{Q}$ of rational numbers is an F_σ set which is neither open nor closed. Thus the collection of F_σ sets (as well as the collection of G_δ) is strictly larger than the collection of open sets or the closed sets. We will see later that the set of rational numbers is not a G_δ.

Problem 1034. *Give an example of a set which is both F_σ and G_δ, but neither open nor closed.*

Problem 1035. *All* F_σ *sets and* G_δ *sets are measurable.*

Problem 1036. *The following conditions are equivalent for any* $A \subseteq \mathbf{R}$.

1. *A is Lebesgue measurable.*
2. $G \smallsetminus F$ *has measure zero for some* F_σ F *and* G_δ G *with* $F \subseteq A \subseteq G$.
3. $A = F \cup E$ *for some* F_σ *set* F *and measure zero set* E.
4. $A = G \smallsetminus E$ *for some* G_δ *set* G *and measure zero set* E.

Many important sets in the theory of real functions are F_σ or G_δ sets.

Problem 1037. *Let* $f : \mathbf{R} \to \mathbf{R}$ *and* C *be the set of points at which* f *is continuous. Show that* C *is a* G_δ *set.*

[Hint: Let $G_n := \bigcup \{(a,b) \mid \text{For all } x, y \in (a,b), |f(x) - f(y)| < \frac{1}{n}\}$. Then $C = \bigcap_n G_n$.]

15.5 The Baire Category Theorem

Theorem 1038 (Baire Category, Baire 1899). *The intersection of countably many dense open sets is dense.*

Proof. Let $G = \cap_n G_n$, where each G_n is a dense open set.

The proof will be similar to the proof of uncountability of **R** in that we will build a sequence nested closed intervals $I_1 \supseteq I_2 \supseteq \cdots \supseteq I_n \supseteq \cdots$, but here at each stage n, we will make sure that I_n is contained in G_n.

We will make use of the fact that every nonempty open set must contain some closed interval $I = [a,b]$ with $a < b$.

To show that $G = \cap_n G_n$ is dense, let (a,b) be a nonempty open interval so that $a < b$. It will show that $G \cap (a,b)$ is nonempty.

Since G_1 is a dense open set, so $G_1 \cap (a,b)$ will be nonempty open, and so $G_1 \cap (a,b)$ will contain a closed interval $I_1 = [a_1, b_1]$ with $a_1 < b_1$. Since G_2 is a dense open set, so $G_2 \cap (a_1, b_1)$ will be nonempty open, and so $G_2 \cap (a_1, b_1)$ will contain a closed interval $I_2 = [a_2, b_2]$ with $a_2 < b_2$. Continuing in this way, we get a sequence of closed intervals $I_n = [a_n, b_n]$, $n = 1, 2, \ldots$, with $a_n < b_n$ such that

$$I_1 \supseteq I_2 \supseteq \cdots \supseteq I_n \supseteq \cdots \quad \text{and} \quad I_n \subseteq G_n \text{ for all } n.$$

By the nested intervals property there is $p \in \cap_n I_n$. But since $I_n \subseteq G_n$ for all n, we have $p \in \cap_n G_n = G$. Thus $p \in G \cap (a,b)$ and thus $G \cap (a,b) \neq \emptyset$. □

Corollary 1039 (Baire Category). *The union of countably many nowhere dense sets cannot contain a nonempty open interval.*

Proof. Let $A_1, A_2, \ldots, A_n, \ldots$ be nowhere dense sets. Then for each n there is a dense open set G_n disjoint from A_n. If (a,b) is any nonempty open interval, then since $\cap_n G_n$ is dense (by the theorem), (a,b) contains a point $p \in \cap_n G_n$ Then

$p \notin \cup_n A_n$, and so $\cup_n A_n$ does not contain the interval (a,b). Since (a,b) was an arbitrarily chosen interval, it follows that no nonempty open interval is contained in $\cup_n A_n$. □

Definition 1040. A set is called *meager* or *of first category* if it can be expressed as the union of countably many nowhere dense sets. A set is called *comeager* or *residual* if its complement is meager.

Thus the Baire category theorem says that a meager set cannot contain a nonempty open interval, or equivalently that a comeager set must be dense.

Problem 1041. *The collection of meager sets form a σ-ideal. In particular, we have*

1. *The subset of a meager set is meager.*
2. *The union of countably many meager sets is meager.*
3. *Every countable set is meager.*
4. *No interval is meager. In particular,* **R** *is not meager, and so the complement of a meager set cannot be meager.*

Recall that all the conditions of the last problem are satisfied if "meager" is replaced by "measure zero." But the following result shows that these two collections are very different.

Proposition 1042. **R** *(or more generally any interval) can be partitioned into two disjoint sets one of which is meager and the other has measure zero.*

Proof. Fix an enumeration of the set $\mathbf{Q}$ of rational numbers, say $\mathbf{Q} = \{r_n \mid n \in \mathbf{N}\}$. For each m, n, let $I_{m,n}$ be any open interval of length $1/2^{m+n}$ containing r_n (e.g., say $I_{m,n} = (a_{m,n}, b_{m,n})$, where $a_{m,n} = r_n - 1/2^{m+n+1}$ and $b_{m,n} = r_n + 1/2^{m+n+1}$). Now put

$$G_m := \bigcup_n I_{m,n}, \quad G := \bigcap_m G_m, \quad \text{and} \quad F := \mathbf{R} \smallsetminus G.$$

Then $G_m \supseteq \mathbf{Q}$, so G_m is a dense open set, so by the Baire category theorem G is a comeager dense G_δ, and hence its complement F is a meager F_σ. Also for each m, $\mathbf{Q} \subseteq G \subseteq G_m \subseteq \cup_n I_{m,n}$ with

$$\sum_{n=1}^{\infty} \operatorname{len}(I_{m,n}) = \sum_{n=1}^{\infty} \frac{1}{2^{m+n}} = \frac{1}{2^m} \sum_{n=1}^{\infty} \frac{1}{2^m} = \frac{1}{2^m}.$$

Thus for each m, G can be covered by a sequence of open intervals of total length $1/2^m$. Hence G has measure zero. Thus $\mathbf{R} = G \cup F$, with G having measure zero, and F being meager. □

Problem 1043. *Every F_σ set whose complement is dense must be meager.*

A meager set can be dense. For example, the countable dense set $\mathbf{Q}$ of rational numbers is both meager and of measure zero. The following is a consequence of the Baire category theorem.

Proposition 1044. *The set* $\mathbf{R} \setminus \mathbf{Q}$ *of irrational numbers is a* G_δ *set which is not* F_σ. *Hence, the set* $\mathbf{Q}$ *of rational numbers is an* F_σ *set which is not* G_δ.

Proof. If $\mathbf{R} \setminus \mathbf{Q}$ were an F_σ set, it would be meager since any F_σ set with dense complement must be meager. But that would imply $\mathbf{R} = (\mathbf{R} \setminus \mathbf{Q}) \cup \mathbf{Q}$ is the union of two meager sets and hence itself meager, a contradiction. □

Problem 1045. *Show (in contrast to the example of Problem 979) that there cannot be any function* $f : \mathbf{R} \to \mathbf{R}$ *such that* f *is continuous at each rational point and discontinuous at each irrational point.*

Problem 1046. *Show that* $([0, 1] \cap \mathbf{Q}) \cup ([2, 3] \setminus \mathbf{Q})$ *is neither* G_δ *nor* F_σ.

Problem 1047. *Show that the set of irrational numbers contains a translated copy of the Cantor set.*

[Hint: Consider all translates of the Cantor set by rational numbers.]

15.6 The Continuum Hypothesis for G_δ Sets

In this section we will show that G_δ sets satisfy continuum hypothesis in the sense that every G_δ set is either countable or has cardinality $\mathfrak{c}$.

Theorem 1048. *Every nonempty dense-in-itself* G_δ *set* E *contains a generalized Cantor set and so there is an injective* $\varphi : \{0, 1\}^{\mathbf{N}} \to E$ *with* $\operatorname{ran}(\varphi)$ *a perfect set.*

In particular, every nonempty dense-in-itself G_δ *set has cardinality* $\mathfrak{c}$.

Proof. Let E be a nonempty dense-in-itself set with

$$E = \bigcap_n G_n,$$

where G_n is an open set for each $n \in \mathbf{N}$. We will say that *the interior of a closed interval* $[a, b]$ *meets a set* A if $(a, b) \cap A \neq \emptyset$. Then we have:

Lemma. For every closed interval I whose interior meets E and every $n \in \mathbf{N}$, there exist disjoint closed subintervals J and K of I such that $J, K \subseteq G_n$, $0 < \operatorname{len}(J) < 1/n$, $0 < \operatorname{len}(K) < 1/n$, and each of the interiors of J and K meets E.

Proof. Suppose that the interior of $I = [a, b]$ meets E and $n \in \mathbf{N}$. Since every nonempty open set having nonempty intersection with E contains infinitely many points of E and $(a, b) \cap E$ is nonempty, we can fix $u, v \in E$ with $a < u < v < b$. Fix also t, w such that $a < u < t < w < v < b$. Then $u \in G_n \cap (a, t)$ and since $G_n \cap (a, t)$ is open, we can choose p, q such that $u \in (p, q) \subseteq G_n \cap (a, t)$. Similarly, we can choose r, s such that $v \in (r, s) \subseteq G_n \cap (w, b)$. Finally choose p', q', r', s' such that $p < p' < u < q' < q$, $r < r' < v < s' < s$, and $q' - p'$, $s' - r' < 1/n$. Now putting $J = [p', q']$ and $K = [r', s']$ we get the conclusion of the lemma. □

Now fix $c \in E$, a, b with $a < c < b$, and put $I_\varepsilon = [a, b]$. Then I_ε satisfies the condition of the lemma, so there exist disjoint closed subintervals I_0 and I_1 of I_ε such that $I_0, I_1 \subseteq G_1$, $0 < \text{len}(J) < 1$, $0 < \text{len}(K) < 1$, and each of the interiors of I_0 and I_1 meets E. Continuing this process and using the lemma repeatedly, we can build a family of closed intervals $\langle I_u \mid u \in \{0,1\}^*\rangle$ such that for every binary string u of length n we have:

1. $I_u \subseteq G_n$ and $0 < \text{len}(I_u) < 1/n$.
2. The interior of I_u meets E.
3. $I_{u^\frown 0}, I_{u^\frown 1} \subseteq I_u$.
4. $I_{u^\frown 0} \cap I_{u^\frown 1} = \emptyset$.

The family $\langle I_u \mid u \in \{0,1\}^*\rangle$ is thus a Cantor system, and hence determines an injective $\varphi \colon \{0,1\}^{\mathbf{N}} \to \mathbf{R}$ such that for each infinite binary sequence $z = z_1 z_2 \cdots z_n \cdots$ we have

$$\bigcap_n I_{z_1 z_2 \cdots z_n} = \{\varphi(z)\}.$$

Then $\text{ran}(\varphi)$ is a generalized Cantor set (hence perfect). But since $I_u \subseteq G_n$ for any binary string of length n, it follows that for any infinite binary sequence $z = z_1 z_2 \cdots z_n \cdots$,

$$\{\varphi(z)\} = \bigcap_n I_{z_1 z_2 \cdots z_n} \subseteq \bigcap_n G_n = E,$$

and so $\varphi(z) \in E$. Hence $\text{ran}(\varphi) \subseteq E$, so E contains the generalized Cantor set $\text{ran}(\varphi)$. □

Since every closed set is G_δ, so every perfect set is a dense-in-itself G_δ, and we have another proof of Theorem 598 for the case of $\mathbf{R}$:

Corollary 1049 (Cantor). *A nonempty perfect set in* $\mathbf{R}$ *has cardinality* $\mathfrak{c}$.

Since the set of rational numbers is a countable dense-in-itself set, Theorem 1048 gives another proof of the following:

Corollary 1050. *The set* $\mathbf{Q}$ *of rational numbers is not a* G_δ *set, and hence the set of irrational numbers is not an* F_σ *set.*

We now have the result that the G_δ sets, and therefore the closed sets, satisfy the continuum hypothesis.

Corollary 1051. *Every uncountable* G_δ *set contains a generalized Cantor set and hence has cardinality* $\mathfrak{c}$.

Proof. Let E be an uncountable G_δ set and let C be its set of condensation points. By Corollary 971, C is a nonempty dense-in-itself set. By Theorem 970, $E \smallsetminus C$ is countable and so is an F_σ set, and hence $C = E \smallsetminus (E \smallsetminus C)$ is G_δ. By Theorem 1048, C contains a generalized Cantor set (which is perfect and of cardinality $\mathfrak{c}$). □

Corollary 1052 (Cantor). *Any uncountable closed subset of* $\mathbf{R}$ *contains a generalized Cantor set and hence has cardinality* $\mathfrak{c}$.

We will later show that the continuum hypothesis is satisfied by a larger class of sets called analytic sets.

Note that a set contains a generalized Cantor set if and only if it contains a nonempty perfect set. Hence we make the following definition.

Definition 1053 (The Perfect Set Property). A set is said to have the *perfect set property* if it is either countable or contains a perfect set (or equivalently, contains a generalized Cantor set).

A collection of sets is said to have the *perfect set property* if every set in the family has the perfect set property.

Thus Corollaries 1052 and 1051 are simply stating that the closed sets and the G_δ sets, respectively, have the perfect set property.

15.7 The Banach–Mazur Game and Baire Property

For each $A \subseteq \mathbf{R}$, a two person infinite game of perfect information called the *Banach–Mazur game* G_A^{**} is defined as follows:

Two players I and II alternately play an infinite sequence of bounded closed intervals of positive length with player I going first:

$$\text{The game } G_A^{**}: \quad \begin{array}{ccccccccc} \text{I} & I_1 & & & & I_3 & & & \\ & & \searrow & & \nearrow & & \searrow & & \nearrow \quad \cdots \\ \text{II} & & & I_2 & & & & I_4 & \end{array}$$

Rules:

1. Each I_n is a closed interval of finite positive length.
2. Each player's move must be contained in the opponent player's previous move, so that we have a nested sequence of intervals:

$$I_1 \supseteq I_2 \supseteq I_3 \supseteq I_4 \supseteq \cdots \supseteq I_{2n-1} \supseteq I_{2n} \supseteq \cdots$$

3. $0 < \operatorname{len}(I_1) < \infty$ and $0 < \operatorname{len}(I_{n+1}) \le \frac{1}{2} \operatorname{len}(I_n)$.

Any play of the game as above therefore defines a unique real number p in the intersection of the sequence $\langle I_n \rangle$ of nested intervals: $p \in \bigcap_n I_n$.

Winning conditions: Player I wins the above play of the game G_A^{**} if $p \in A$; otherwise, Player II wins.

Problem 1054 (Mazur). *Let $A \subseteq \mathbf{R}$. Show that in the game G_A^{**} above:*

1. *If A is meager then Player II has a "winning strategy," i.e., Player II can force a win no matter how Player I plays.*
2. *If there is a nonempty open set U such that $U \setminus A$ is meager then Player I has a winning strategy.*

[Hint: Mimic the proof of the Baire category theorem.]

Note. Mazur invented the game G_A^{**} and proved the above results. He then asked if the converses are true. Banach showed that the answer is yes and won a bottle of wine as a prize from Mazur.

Definition 1055 (Baire Property). A set $E \subseteq \mathbf{R}$ is said to have the *Baire property* if there is an open set U such that $A \triangle U = (A \setminus U) \cup (U \setminus A)$ is meager. The class of all sets with Baire property will be denoted by $\mathbf{Y}$.

Corollary 1056. *If $A \subseteq \mathbf{R}$ has Baire property, then either A is meager or $U \setminus A$ is meager for some nonempty open set U.*

Problem 1057. *If $A \subseteq \mathbf{R}$ has Baire property, then the game G_A^{**} is "determined," i.e., at least one of the players has a winning strategy.*

Problem 1058. *Every open set and every meager set has Baire property.*

Proposition 1059. *Let A be a set with Baire property. If A is non-meager, then A contains a perfect set.*

Proof. By Corollary 1056, fix a nonempty open U such that $U \setminus A$ is meager, and let F be a meager F_σ set with $(U \setminus A) \subseteq F$. Then $U \setminus F$ is a $\mathbf{G}_\delta$ set which must be uncountable (if $U \setminus F$ were countable then $U \subseteq (U \setminus F) \cup F$ would be meager, but no nonempty open set is meager). Hence by the perfect set property for $\mathbf{G}_\delta$ sets, $U \setminus F$ contains a perfect set. But then, A contains a perfect set, since $(U \setminus F) \subseteq A$. □

Proposition 1060. *If A has Baire property then so does $\mathbf{R} \setminus A$. If a sequence of sets $A_1, A_2, \ldots$ all have Baire property, then so does their union $\bigcup_n A_n$.*

Proof. Let A have Baire property. Fix open U with $A \triangle U$ meager. Put $V = \mathbf{R} \setminus \overline{U}$, so that V is open and the boundary set $\overline{U} \setminus U$ is nowhere dense (and hence meager). Now notice that $(\mathbf{R} \setminus A) \triangle V$ is contained in $(A \triangle U) \cup (\overline{U} \setminus U)$ which is meager, hence $\mathbf{R} \setminus A$ has Baire property.

Next assume that A_n has Baire property for each $n \in \mathbf{N}$, and fix open U_n such that $A_n \triangle U_n$ is meager. Then the countable union $\bigcup_n (A_n \triangle U_n)$ is also meager. Now let $U := \bigcup_n U_n$. Then U is open and $(\bigcup_n A_n) \triangle U \subseteq \bigcup_n (A_n \triangle U_n)$, and so $\bigcup_n A_n$ has Baire property. □

By the above results, the collection **Y** of all sets having Baire property contains all meager sets, intervals, open and closed sets, and is closed under taking countable unions and complements of sets, and hence under forming countable intersections as well. Hence **Y** also contains all $\mathbf{F}_\sigma$ and all $\mathbf{G}_\delta$ sets as well. Thus, like the collection **L** of all Lebesgue measurable sets, the collection **Y** of sets with Baire property is also a large collection of sets forming a *sigma-algebra* (Sect. 18.1).

Problem 1061. *Suppose that A and B are sets with Baire property, and U and V are open sets with $A \triangle U$ and $B \triangle V$ both meager. If $A \cap B$ is meager, then $U \cap V = \emptyset$.*

[Hint: By the Baire category theorem, no nonempty open set is meager. Now the open set $U \cap V$ is contained in the meager set $(A \cap B) \cup (U \setminus A) \cup (V \setminus B)$.]

Corollary 1062 (CCC Property). *If $\langle A_i \mid i \in I \rangle$ are pairwise disjoint sets with Baire property, then A_i is meager for all but countably many $i \in I$.*

Problem 1063 (Translation Invariance). *If E has Baire property then so does $E + p$, where $E + p := \{x + p \mid x \in E\}$.*

An infinite binary sequence $\langle x_n \rangle \in \{0, 1\}^{\mathbf{N}}$ is called 1-*normal* if the relative frequency of 1s among the first n bits approaches the limiting value $\frac{1}{2}$, i.e., if $\lim_{n\to\infty} \frac{1}{n}(\sum_{k=1}^{n} x_k) = \frac{1}{2}$. It is a celebrated result of Borel that the set N_1 of all $x \in [0, 1]$ which admit a 1-normal binary representation has measure 1.

Problem 1064. *Show that N_1 is meager.*

[Hint: "A dyadic interval $\left(\frac{k-1}{2^n}, \frac{k}{2^n}\right)$ fixes the first n bits."]

15.8 Vitali and Bernstein Sets

Vitali Sets

Definition 1065 (Vitali). A set $V \subseteq \mathbf{R}$ is said to be a *Vitali set* if the following conditions hold:

1. $(V + r) \cap (V + s) = \emptyset$ for all distinct rational numbers $r \neq s$ (in $\mathbf{Q}$).
2. $\bigcup_{r \in \mathbf{Q}} (V + r) = \mathbf{R}$.

Theorem 1066 (AC). *Vitali sets exist.*

Proof. Recall the equivalence relation on $\mathbf{R}$ defined by $x \sim_{\mathbf{Q}} y \Leftrightarrow x - y \in \mathbf{Q}$. The corresponding partition $\mathbf{R}/\mathbf{Q}$ of $\mathbf{R}$ consists of sets of the form $a + \mathbf{Q} := \{a + r \mid r \in \mathbf{Q}\}$, any two of which are either identical or disjoint. Now note that V is Vitali set if and only if it is a choice set for the partition $\mathbf{R}/\mathbf{Q}$. Hence we obtain a Vitali set using AC. □

Theorem 1067. *A Vitali set cannot be Lebesgue measurable and cannot have Baire property. Hence there are subsets of* $\mathbf{R}$ *which are neither Lebesgue measurable nor have the Baire property.*

Proof. Let V be a Vitali set.

Suppose that, if possible, V is measurable. By translation invariance we have $m(V+r) = m(V)$ for all r, and $\mathbf{R}$ is a countable union of sets of the form $V + r$, hence V cannot have measure zero.

Hence there exist $a < b$ such that $m([a,b] \cap V) > 0$. Put $W := [a,b] \cap V$. Then $\langle W + r \mid r \in \mathbf{Q} \cap (0,1) \rangle$ is a family of pairwise disjoint measurable sets all having the same positive measure $m(W)$, hence by countably additivity of m the union $\bigcup\{W + r \mid r \in \mathbf{Q} \cap (0,1)\}$ has infinite measure. On the other hand this union is contained in $(a, b+1)$ and so has finite measure $\le b + 1 - a$. We thus get a contradiction.

We next show that V cannot have Baire property. Suppose, if possible, there is an open set U such that $V \triangle U$ is meager. Then U is nonempty since if U were empty, then V and so $V + r$ would be meager for all r, and $\mathbf{R}$ would be a countable union of meager sets, which is impossible.

Hence U is nonempty open, and we can fix $a < b$ with $(a,b) \subseteq U$. Put $W := (a,b) \cap V$. Then $(a,b) \smallsetminus W$ is meager. Now fix any rational $0 < r < b - a$. Then $W \cap (W + r) = \emptyset$, so by Problem 1061, $(a,b) \cap (a+r, b+r) = \emptyset$, which is a contradiction since $\frac{1}{2}(a + r + b) \in (a,b) \cap (a+r, b+r)$. □

Feferman showed that the existence of a Vitali set cannot be proved without using the Axiom of Choice, and even if the full use of AC is allowed, no effectively defined set can be proved (without additional axioms) to be a Vitali set.

Note that the main property of Lebesgue measurable sets and sets having Baire property that was used in the above proof is *translation invariance.* We next show a very different method for obtaining non-measurable sets.

Bernstein Sets

Definition 1068 (Bernstein). A set B is said to be a *Bernstein set* if neither B nor its complement $\mathbf{R} \smallsetminus B$ contains any nonempty perfect set.

Theorem 1069 (AC). *Bernstein sets exist.*

Proof. By the Axiom of Choice every infinite cardinal is an aleph, and so we can fix α such that $\mathfrak{c} = \aleph_\alpha$. Since there are exactly $\mathfrak{c}$ nonempty perfect sets, we can enumerate them as $\langle P_\nu \mid \nu < \omega_\alpha \rangle$. Fix a choice function φ for the collection of all nonempty subsets of $\mathbf{R}$. Now define, by transfinite recursion, two sequences of reals $\langle a_\nu \mid \nu < \omega_\alpha \rangle$ and $\langle b_\nu \mid \nu < \omega_\alpha \rangle$ as follows.

$$a_\nu := \varphi\big(P_\nu \smallsetminus \{a_\xi, b_\xi \mid \xi < \nu\}\big), \qquad b_\nu := \varphi\big(P_\nu \smallsetminus (\{a_\nu\} \cup \{a_\xi, b_\xi \mid \xi < \nu\})\big).$$

The elements a_ν and b_ν are well defined since for each ν, the perfect set P_ν has cardinality $\mathfrak{c}$ while the sets being removed from P_ν have cardinality $< \mathfrak{c}$. Also we have $a_\mu \neq b_\nu$ for all $\mu, \nu < \omega_\alpha$.

Finally, put $A = \{a_\nu \mid \nu < \omega_\alpha\}$ and $B = \{b_\nu \mid \nu < \omega_\alpha\}$. Then $A \cap B = \emptyset$, with $a_\nu \in P_\nu \cap A$ and $b_\nu \in P_\nu \cap B$ for all $\nu < \omega_\alpha$. Thus every nonempty perfect set has nonempty intersection with each of the disjoint sets A and B, and so both A and B must be Bernstein sets. □

Let B be a Bernstein set. By Propositions 1024 and 1059, both $A \cap B$ and $A \setminus B$ are non-measurable for any measurable set A of nonzero measure, and both $A \cap B$ and $A \setminus B$ fail to have Baire property for any non-meager set A with Baire property. This gives a stronger result than Theorem 1067:

Corollary 1070 (AC). *Every set not of measure-zero has non-measurable subsets. Every non-meager set has subsets without the Baire property.*

Consequently, there are measurable sets without Baire property and there are sets with Baire property which are non-measurable.

Notice the highly non-effective way of obtaining Vitali and Bernstein sets. In both cases, an uncountable number of choices were essential.

It turns out that this is unavoidable: No non-measurable set can be proved to exist using only countably many choices. By a famous result of Solovay, one can consistently assume that all sets are Lebesgue measurable and that the Axiom Dependent Choices holds (assuming that the usual axioms are consistent with the existence of an inaccessible cardinal). So any proof that non-measurable sets exist will need uncountably many choices.

Vitali and Bernstein sets are relevant to the question if Lebesgue measure can be extended to a measure which is defined for *all* sets of reals:

The Measure Extension Problem (Lebesgue). *Does there exist a countably additive function $\mu \colon \mathbf{P}(\mathbf{R}) \to [0, \infty]$ which extends Lebesgue measure?*

This question was fully analyzed by Ulam, which opened up the field of *large cardinal numbers*. But this is a topic for Postscript III (Chapter 19), where we will present a detailed account of Ulam's work.

Chapter 16
Cantor–Bendixson Analysis of Countable Closed Sets

Abstract We devote this chapter to the Cantor–Bendixson analysis of countable closed sets. We first prove the effective Cantor–Bendixson theorem which decomposes a closed set into an effectively countable set and a perfect set. We then obtain a full topological classification for the class of countable closed bounded subsets of **R**: The Cantor–Bendixson rank is shown to be a complete invariant for the relation of homeomorphism between these sets, and the countable ordinals $\omega^{\nu}n+1$ ($\nu < \omega_1$, $n \in \mathbf{N}$) are shown to form an exhaustive enumeration, up to homeomorphism, of the countable closed bounded sets into $\aleph_1$ many pairwise non-homeomorphic representative sets.

Countable closed sets arose in Cantor's study of trigonometric series. The analysis requires iterations of derived sets (i.e., orders of limit points) into the transfinite, and it naturally leads to the notion of transfinite ordinals. Roughly speaking, this is how Cantor was led to his creation of ordinal numbers, and went on to eventually create set theory. We will briefly discuss the background in Sect. 16.4.

16.1 Homeomorphisms of Orders and Sets

Homeomorphic Order Types

A bijection f from an order X onto an order Y is a *homeomorphism* if both f and its inverse f^{-1} are continuous. Two orders are *homeomorphic* if there is a homeomorphism between them. Clearly, isomorphic orders are homeomorphic. If X and X' are isomorphic orders and Y and Y' are isomorphic orders, then X is homeomorphic to Y if and only if X' is homeomorphic to Y'. Hence we can speak of *homeomorphic order types*.

A. Dasgupta, *Set Theory: With an Introduction to Real Point Sets*,
DOI 10.1007/978-1-4614-8854-5_16, © Springer Science+Business Media New York 2014

Problem 1071. *If X and Y are orders, then $f: X \to Y$ is a homeomorphism if and only if f is a bijection and for every $A \subseteq X$ and $p \in X$, $p \in D(A)$ in X if and only if $f(p) \in D(f[A])$ in Y.*

Problem 1072. *Show that*

1. *ω is homeomorphic to ζ.*
2. *$\omega + n$ is homeomorphic to $\omega + 1$ for any $n \in \mathbf{N}$.*
3. *$\omega + 1 + \omega^*$ is homeomorphic to $\omega + 1$.*
4. *$\zeta + 1$ not homeomorphic to $\omega + 1$.*
5. *$\eta + 1$ is homeomorphic to η.*
6. *$1 + \lambda$ is homeomorphic to $\lambda + 1$.*
7. *$\eta + \eta$ is homeomorphic to η, but $\lambda + \lambda$ is not homeomorphic to λ.*

Problem 1073. *If α, β, etc are order types, then:*

1. *α is homeomorphic to α^*.*
2. *If α and β both are order types of orders with endpoints, then $\alpha + \beta$ is homeomorphic to $\beta + \alpha$.*
3. *If I is an ordered set and if for each $i \in I$ α_i and β_i are order types with endpoints with α_i homeomorphic to β_i, then $\sum_{i \in I} \alpha_i$ is homeomorphic to $\sum_{i \in I} \beta_i$.*

Homeomorphisms Between Subsets of R and Orders

A subset $Y \subseteq \mathbf{R}$ is said to be *homeomorphic to an order* X if there is a bijection f from X onto Y such that for every $A \subseteq X$ and $p \in X$, $p \in D(A)$ in X if and only if $f(p) \in D(f[A])$ in $\mathbf{R}$. If $Y \subseteq \mathbf{R}$ and X and X' are isomorphic orders, then Y is homeomorphic to X if and only if Y is homeomorphic to X'. Hence we can talk about a subset Y of $\mathbf{R}$ being homeomorphic to an order type α.

Problem 1074. *Let $A = \{0, \frac{1}{2}, \frac{2}{3}, \dots, \frac{n-1}{n}, \dots\} \cup \{2\}$, so that the order type of A is $\omega + 1$. Show that A is not homeomorphic to $\omega + 1$, and in fact that A is homeomorphic to ω.*

The above problem shows that *if $A \subseteq \mathbf{R}$ has order type α, then A may fail to be homeomorphic to α.* The reason for the failure in the above is easily found: While 2 is a limit point of A when the suborder A is considered as an order by itself, 2 is not a limit point of A as a subset of $\mathbf{R}$.

Problem 1075. *Show that there is a subset of $\mathbf{R}$ having order type η which is homeomorphic to ω. Conclude that for any infinite countable order type α there is a subset A of $\mathbf{R}$ having order type α which is homeomorphic to ω.*

Recall that $A \subseteq \mathbf{R}$ is said to be *continuously order-embedded* if whenever $p \in A$ is a limit point of $E \subseteq A$ when A is considered as an order by itself, then p is a limit point of A in $\mathbf{R}$. Then we have

Proposition 1076. *If $A \subseteq \mathbf{R}$ is continuously order-embedded in $\mathbf{R}$ and the order type of A is α, then A is homeomorphic to α.*

Recall also that there are two important cases when $A \subseteq \mathbf{R}$ can be guaranteed to be continuously order-embedded in $\mathbf{R}$:

1. If A is closed, then A is continuously order-embedded in $\mathbf{R}$. (Theorem 593)
2. If every point of A is a two-sided limit point of A, then A is continuously order-embedded in $\mathbf{R}$. (Theorem 532)

Problem 1077. *Give an example of a subset A of $\mathbf{R}$ which is dense-in-itself (every point of A is a limit point of A) but which is not continuously order-embedded in $\mathbf{R}$.*

Problem 1078. *Let X be an order and let $f: X \to \mathbf{R}$ be a strictly increasing function which continuously order-embeds X in $\mathbf{R}$. Then the image $f[X]$ is closed and bounded in $\mathbf{R}$ if and only if X is a complete order with endpoints.*

[Hint: Use Theorem 593.]

16.2 The Cantor–Bendixson Theorem and Perfect Sets

Theorem 1079 (Cantor–Bendixson). *Let A be any nonempty closed subset of $\mathbf{R}$ and for each $\alpha < \omega_1$ define*

$$F_\alpha := D^{(\alpha)}(A) = \text{ the } \alpha\text{-th iterated derived set of } A.$$

Then

1. *$\langle F_\alpha \mid \alpha < \omega_1 \rangle$ are decreasing closed sets so that*

$$F_0 \supseteq F_1 \supseteq \cdots \supseteq F_n \supseteq F_{n+1} \cdots F_\omega \supseteq F_{\omega+1} \supseteq \cdots F_\alpha \supseteq F_{\alpha+1} \supseteq \cdots$$

2. *The set*

$$H := \bigcup_{\alpha<\omega_1} (F_\alpha \setminus F_{\alpha+1}) = F_0 \setminus \bigcap_{\alpha<\omega_1} F_\alpha$$

   is countable, in an effective fashion. In particular, for each $\alpha < \omega_1$, the set $F_\alpha \setminus F_{\alpha+1}$ is countable.
3. *There exists a least $\mu < \omega_1$ such that $F_{\mu+1} = F_\mu$, and so $F_\alpha = F_\mu$ for all $\alpha \geq \mu$.*
4. *For the ordinal μ above, F_μ is either empty or nonempty perfect (hence uncountable), and so if A is countable then $F_\mu = \emptyset$.*
5. *If A is countable and bounded then the ordinal μ above is a successor ordinal.*

Proof. 1. This is obtained by transfinite induction as follows. $F_0 = A$ is given to be closed. For any set E, $D(E)$ is always closed and if E is closed then $E \supseteq D(E)$. Thus if F_α is closed, then $F_{\alpha+1} = D(F_\alpha)$ is a closed subset of F_α. Finally, since the intersection of any family of closed sets is closed, therefore if α is a limit ordinal then $F_\alpha = \bigcap_{\beta<\alpha} F_\beta$ is a closed subset of F_β for each $\beta < \alpha$.

2. Let B be the family of all open intervals with rational endpoints. Since B is countable, we can enumerate it as

$$\mathrm{B} = \{V_n \mid n \in \mathbf{N}\}.$$

For each $x \in H$, fix the unique $\alpha = \alpha_x$ such that $x \in F_{\alpha_x} \setminus F_{\alpha_x+1}$, and then (noting that $x \in F_{\alpha_x} \setminus D(F_{\alpha_x})$ is an isolated point of F_{α_x}) choose the least $n = n_x \in \mathbf{N}$ such that $V_n \cap F_{\alpha_x} = \{x\}$. This defines a mapping $x \mapsto n_x$ from H into $\mathbf{N}$. We claim that this mapping is injective. To see this, suppose that $x \neq y$ are in H. Now if $\alpha_x < \alpha_y$ then $V_{n_x} \cap F_{\alpha_x+1} = \emptyset$ but $V_{n_y} \cap F_{\alpha_x+1} \supseteq V_{n_y} \cap F_{\alpha_y} \neq \emptyset$, and so $n_x \neq n_y$. If $\alpha_x > \alpha_y$ then similarly $n_x \neq n_y$. Finally, if $\alpha_x = \alpha_y = \alpha$ (say), then $V_{n_x} \cap F_\alpha = \{x\} \neq \{y\} = V_{n_y} \cap F_\alpha$, so again we have $n_x \neq n_y$. Thus the mapping $x \mapsto n_x$ from H into $\mathbf{N}$ is injective and so H is countable.

3. If we had $F_\alpha \setminus F_{\alpha+1} \neq \emptyset$ for all $\alpha < \omega_1$, the set H would be uncountable (because it would then be the union of ω_1-many pairwise disjoint nonempty sets), a contradiction. Hence $F_\alpha \setminus F_{\alpha+1} = \emptyset$ or $F_\alpha = F_{\alpha+1}$ for some α, and we can fix μ to be the least such α.

4. This is immediate since $F_{\mu+1} = D(F_\mu)$ and a nonempty set E is perfect if and only if $D(E) = E$.

5. If $A = F_0$ is nonempty, countable, and bounded, note first that μ is the least ordinal such that $F_\mu = \emptyset$. Since F_0 is nonempty, so $\mu > 0$. Finally μ cannot be a limit ordinal since if F_α is nonempty, closed, and bounded for each $\alpha < \mu$, then by the Heine–Borel Theorem $\bigcap_{\alpha<\mu} F_\alpha$ would be nonempty as well. Hence μ is a successor ordinal.

□

A consequence of the proof of the theorem is that a countable closed set is *effectively countable*, that is, each countable closed set A determines a unique effectively defined injection from A into $\mathbf{N}$. More generally, we have:

Corollary 1080 (The Cantor–Bendixson Theorem). *Every closed set is the union of an effectively countable set and a set E with $D(E) = E$.*

Corollary 1081. *Every uncountable closed set contains a nonempty perfect set. Hence closed sets "satisfy the continuum hypothesis," that is, if A is closed then either $|A| \leq \aleph_0$ or $|A| = 2^{\aleph_0}$.*

Corollary 1082. *If A is a nonempty countable closed bounded set, then there is a unique ν such that $D^{(\nu)}(A) \neq \emptyset$ but $D^{(\nu+1)}(A) = \emptyset$. In this case, $D^{(\nu)}(A)$ is a nonempty finite set.*

Proof. The first statement follows from the last two parts of the theorem. For the second statement, note that if $D^{(\nu)}$ were infinite. then being bounded it would have a limit point by the Bolzano–Weierstrass theorem, which would imply $D^{(\nu+1)}(A) \neq \emptyset$, a contradiction. □

Recall that a point p is said to be *condensation point of a set A* if every open interval containing p contains uncountably many points of A.

Problem 1083. *Let A be a nonempty closed set, and put $F_\alpha = D^{(\alpha)}(A)$ and $H = \bigcup_{\alpha<\omega_1}(F_\alpha \smallsetminus F_{\alpha+1})$ as in Theorem 1079. Also put $P = \bigcap_{\alpha<\omega_1} F_\alpha$. Show that*

1. *H consists precisely of the non-condensation points of A and P consists precisely of the condensation points of A.*
2. *Assuming $P \neq \emptyset$, show that P is the largest perfect set contained in A.*

From the above problem it follows that in a closed set A, all except countably many points of A are condensation points of A.

16.3 Ordinal Analysis of Countable Closed Bounded Sets

Using Corollary 1082 we can make the following definition.

Definition 1084 (CB-rank, Cantor–Bendixson rank). If A is a nonempty countable closed bounded set, we define its *CB-rank* (Cantor–Bendixson rank) to be the pair ν, n where ν is the unique ordinal such that $D^{(\nu)}(A) \neq \emptyset$ but $D^{(\nu+1)}(A) = \emptyset$, and $n = |D^{(\nu)}(A)|$.

Thus the CB-rank of a nonempty countable closed bounded set A equals ν, n if and only if $D^{(\nu)}(A)$ is a nonempty finite set with n elements.

Problem 1085. *Let $A \subseteq \mathbf{R}$ be a closed bounded set with exactly one limit point. Show that A is countable and must be homeomorphic to $\omega+1$. What are the possible order types for A?*

Proposition 1086. *Let A and B be homeomorphic countable closed bounded subsets of* $\mathbf{R}$. *Then A and B have identical CB-ranks.*

Proof. Let $f: A \to B$ be a homeomorphism of A onto B. Since A and B are closed, we have $D(A) \subseteq A$, $D(B) \subseteq B$. Since f is a homeomorphism, we have $f[D(A)] = D(B)$. More generally, by a routine transfinite induction we have $f[D^{(\alpha)}(A)] = D^{(\alpha)}(B)$ for all ordinals α, and the result follows. □

Proposition 1087. *If α is an infinite successor ordinal, then α is homeomorphic to $\omega^\zeta n + 1$ for some ordinal $\zeta > 0$ and some $n \in \mathbf{N}$.*

Proof. By Cantor Normal Form, we have

$$\alpha = \omega^{\zeta_1} n_1 + \omega^{\zeta_2} n_2 + \cdots + \omega^{\zeta_k} n_k, \quad k \in \mathbf{N}, \begin{cases} \zeta_1 > \zeta_2 > \cdots > \zeta_k, \\ n_1, n_2, \ldots, n_k \in \mathbf{N}. \end{cases}$$

Since α is an infinite successor ordinal, we must have $\zeta_k = 0$ (so that $\omega^{\zeta_k} = 1$) and $k \geq 2$, so we can write

$$\begin{aligned} \alpha &= \omega^{\zeta_1} n_1 + \omega^{\zeta_2} n_2 + \omega^{\zeta_3} n_3 + \cdots + \omega^{\zeta_{k-1}} n_{k-1} + n_k \\ &= \omega^{\zeta_1} n_1 + (1 + \beta) \\ &= (\omega^{\zeta_1} n_1 + 1) + \beta, \end{aligned}$$

for some ordinal $\beta < \omega^{\zeta_1}$ which must be either zero or a successor ordinal (since α is a successor ordinal).

If $\beta = 0$, then $\alpha = \omega^{\zeta_1} n_1 + 1$ and we are done. Otherwise, β is a successor ordinal, and since $\mu + \nu$ is homeomorphic to $\nu + \mu$ for successor ordinals μ and ν, it follows that α is homeomorphic to

$$\beta + (\omega^{\zeta_1} n_1 + 1) = \omega^{\zeta_1} n_1 + 1,$$

where the last equality holds since $\beta < \omega^{\zeta_1}$, $n_1 > 0$, and ω^{ζ_1} is a remainder ordinal and so "absorbs any smaller ordinal as a summand from the left." □

Corollary 1088. *Let X be a well-order with a last element. Then X is homeomorphic to the order $W(\omega^{\zeta} n + 1) = \{\alpha \mid \alpha \leq \omega^{\zeta} n\}$ for some ordinal ζ and some $n \in \mathbf{N}$.*

Definition 1089. Let $\langle I_n \mid n \in \mathbf{N} \rangle$ with $I_n = [a_n, b_n]$ be a sequence of closed intervals, with $a_n < b_n$ for each $n \in \mathbf{N}$. We say that *the sequence of intervals $\langle I_n \mid n \in \mathbf{N} \rangle$ converges to a point $p \in \mathbf{R}$* and write $\langle I_n \mid n \in \mathbf{N} \rangle \to p$ if for any a, b with $a < p < b$ there is $m \in \mathbf{N}$ such that $I_n \subseteq (a, b)$ for all $n \geq m$.

Proposition 1090. *Let $I_n = [a_n, b_n]$ and $J_n = [c_n, d_n]$, $n \in \mathbf{N}$, be closed intervals such that $a_n < b_n$ and $c_n < d_n$ for each $n \in \mathbf{N}$. Assume that the sequence Let $\langle I_n \mid n \in \mathbf{N} \rangle$ and $\langle J_n \mid n \in \mathbf{N} \rangle$ each be a pairwise disjoint sequence of closed intervals, and let $A_n \subseteq (a_n, b_n)$ and $B_n \subseteq (c_n, d_n)$ for each $n \in \mathbf{N}$ with A_n homeomorphic to B_n for each $n \in \mathbf{N}$. Suppose that $p, q \in \mathbf{R}$ with $p \notin I_n$ and $q \notin J_n$ for any n, and that $\langle I_n \mid n \in \mathbf{N} \rangle \to p$ and $\langle J_n \mid n \in \mathbf{N} \rangle \to q$. Then the sets*

$$A := \bigcup_{n \in \mathbf{N}} A_n \cup \{p\} \quad \text{and} \quad B := \bigcup_{n \in \mathbf{N}} B_n \cup \{q\}$$

are homeomorphic.

Proof. For each n, fix a homeomorphism $\langle f_n \rangle : A_n \to B_n$ (using AC), and define $f: A \to B$ by setting $f(p) = q$ and $f(x) = f_n(x)$ if $x \in A_n$. Clearly f is a bijection from A onto B.

To show that f is continuous, suppose that $\langle x_n \rangle \to x$ in A.

Case 1: $x \neq p$. Then $x \in A_m$ for some m. Since $A_m \subseteq (a_m, b_m)$, so $a_m < x < b_m$ and hence there is k such that $x_n \in (a_m, b_m)$ for all $n \geq k$. Then $x_n \in A_m$ for all $n \geq k$, and thus $\langle x_n \mid n \geq k \rangle$ is a sequence in A_m converging to $x \in A_m$. Since $f_m: A_m \to B_m$ is continuous, the sequence $\langle f_m(x_n) \mid n \geq k \rangle$ converges to $f_m(x)$ in B_m.

Case 2: $x = p$. In this case we show that $\langle f(x_n) \mid n \in \mathbf{N} \rangle$ converges to $f(p) = q$. Suppose that $c < q < d$. Since the sequence of intervals $\langle J_n \rangle$ converges to q, there is m such that $J_n \subseteq (c, d)$ for all $n \geq m$. Since $p \notin \bigcup_{n<m} I_n$, we can choose a, b with $a < p < b$ such that $I_j \cap (a, b) = \emptyset$ for $j < m$. Since $\langle x_n \rangle$ converges to p, there is k such that $x_n \in (a, b)$ for $n \geq k$. Then for any $n \geq k$, $x_n \notin I_j$ for any $j < m$, so for all $n \geq k$ either $x_n = p$ or $x_n \in A_j$ for some $j \geq m$. Hence for any $n \geq k$ we have $f(x_n) = q$ or $f(x_n) \in B_j$ for some $j \geq m$, which implies $f(x_n) \in (c, d)$. This shows that $\langle f(x_n) \rangle$ converges to $f(p) = q$.

Thus f is continuous. Similarly f^{-1} is continuous. □

Proposition 1091. *Let $\langle [a_n, b_n] \mid n \in \mathbf{N} \rangle$ be a sequence of pairwise disjoint closed intervals converging to p, where $p \notin \bigcup_n [a_n, b_n]$, and suppose that for each n A_n is a countable closed set contained in (a_n, b_n) with the CB-rank of A_n being α_n, k_n. Let $\alpha = \sup_n \alpha_n$, and $A = \bigcup_n A_n \cup \{p\}$.*

1. *If $\alpha_n = \alpha$ for infinitely many n, then the CB-rank of A is $\alpha + 1, 1$.*
2. *If $\alpha_n < \alpha$ for all n so that $\alpha = \sup_n \alpha_n$ is a limit ordinal, then the CB-rank of A is $\alpha, 1$.*

Proof. For the first part, note that we have $D^{(\alpha)}(A_n)$ is finite for all n and is nonempty for infinitely many n, hence $p \in D^{(\alpha)}(A)$ and so $D^{(\alpha)}(A)$ is a closed set of order type $\omega + 1$ with greatest element p. Therefore $D^{(\alpha+1)}(A) = \{p\}$.

For the second part, note that $D^{(\alpha)}(A_n) = \emptyset$ for all n, but for every $\beta < \alpha$ we have $D^{(\beta)}(A_n) \neq \emptyset$ for infinitely many n, and so $p \in D^{(\beta)}(A)$ for all $\beta < \alpha$. Hence $p \in \bigcap_{\beta<\alpha} D^{(\beta)}(A) = D^{(\alpha)}(A)$. It follows that $D^{(\alpha)}(A) = \{p\}$. □

Corollary 1092. *Let $0 < \alpha < \omega_1$, $n \in \mathbf{N}$, and let E be a closed and bounded subset of $\mathbf{R}$ having order type $\omega^\alpha n + 1$. Then the CB-rank of E is α, n.*

Proof. First note that since $\omega^\alpha n + 1 = (\omega^\alpha + 1) + (\omega^\alpha + 1) + \cdots + (\omega^\alpha + 1)$ (with n summands), so E can be partitioned into n closed sets $E_1 < E_2 < \cdots < E_n$ with the order type of each E_k being $\omega^\alpha + 1$. Since for disjoint closed sets A and B we have $D^{(\beta)}(A \cup B) = D^{(\beta)}(A) \cup D^{(\beta)}(B)$, it suffices to show that the CB-rank of $\omega^\alpha + 1$ is $\alpha, 1$. But this follows from the previous proposition by a routine transfinite induction argument, since if α is a successor ordinal with $\alpha = \beta + 1$ then $\omega^\alpha + 1 = \sum_n (\omega^\beta + 1) + 1$, and if α is a limit ordinal then $\omega^\alpha + 1 = \sum_n (\omega^{\alpha_n} + 1) + 1$ where $\alpha_n < \alpha$ for all n with $\sup_n \alpha_n = \alpha$. □

Proposition 1093. *For all* $0 < \alpha < \omega_1$ *and* $n \in \mathbf{N}$, *there is a closed and bounded subset of* $\mathbf{R}$ *having order type* $\omega^\alpha n + 1$ *and hence having CB-rank* α, n.

Proof. Recall that every countable order X can be continuously order-embedded in $\mathbf{R}$ (Theorem 551), and that the embedded image is closed and bounded in $\mathbf{R}$ if X is complete and with endpoints (Problem 1078). Since $\omega^\alpha n + 1$ is a countable complete order with endpoints, the result follows.

(Alternatively, one can inductively build a continuously order-embedded set $A \subseteq \mathbf{R}$ of order type $\omega^\alpha + 1$ ($\alpha < \omega_1$) by taking subsets $A_1 < A_2 < \cdots < A_n < \cdots$ and ordinals $\alpha_n, n \in \mathbf{N}$, such that A_n is a closed set of order type $\omega^{\alpha_n}+1$, $\sup_n(\alpha_n+1) = \alpha$, and $A = \cup_n A_n \cup \{p\}$ where $p = \sup \cup_n A_n$.) □

Theorem 1094. *Every nonempty countable closed bounded subset* A *of* $\mathbf{R}$ *is homeomorphic to a countable successor ordinal.*

Proof. The result is obvious if A is finite, so assume that A is infinite.

It suffices to show that every infinite countable closed bounded set is homeomorphic to a well-ordered countable closed set with a greatest element.

The proof will be by induction on the Cantor–Bendixson rank of A.

Let ν, n ($\nu > 0, n \in \mathbf{N}$) be the CB-rank of A and suppose that the result is true for all sets having CB-rank μ, m with $\mu < \nu$. We first do the proof for the case $n = 1$. Then $D^{(\nu)} = \{p\}$ for some $p \in \mathbf{R}$. Since the set A is nowhere dense, for each $x < y$ we can choose a, b with $x < a < b < y$ and $[a, b] \cap A = \emptyset$. Hence we can choose sequences $\langle a_n\rangle$, $\langle b_n\rangle$, $\langle c_n\rangle$, $\langle d_n\rangle$ and sequences of sets $\langle A_n\rangle$, $\langle C_n\rangle$ such that

1. $a_1 < b_1 < \cdots a_n < b_n < \cdots < p < \cdots < c_n < d_n < \cdots < c_1 < d_1$;
2. $\sup_n a_n = \sup_n b_n = p$ and $\inf_n c_n = \inf_n d_n = p$;
3. $A_n \subseteq (a_n, b_n)$ and $C_n \subseteq (c_n, d_n)$ are closed sets;
4. $A = (\cup_n A_n) \cup \{p\} \cup (\cup_n B_n)$.

To do this, start with $a_1 < \inf A$ and $d_1 > \sup A$, and choose b_1, a_2 such that

$$\max(a_1, p-1) < b_1 < a_2 < p \quad \text{and} \quad [b_1, a_2] \cap A = \emptyset.$$

Next choose b_2, a_3 such that

$$\max(a_2, p-1/2) < b_2 < a_3 < p \quad \text{and} \quad [b_2, a_3] \cap A = \emptyset,$$

and so on. Similarly complete the sequences $\langle c_n\rangle$ and $\langle d_n\rangle$. Then put $A_n = (a_n, b_n) \cap A = [a_n, b_n] \cap A$, and $C_n = (c_n, d_n) \cap A = [c_n, d_n] \cap A$, so that A_n and C_n are closed subsets A, with $A = (\cup_n A_n) \cup \{p\} \cup (\cup_n B_n)$.

Now put, for each $n \in \mathbf{N}$,

$$I_{2n-1} = [a_n, b_n],\ I_{2n} = [c_n, d_n] \quad \text{and} \quad E_{2n-1} = A_n,\ E_{2n} = C_n.$$

Then note that the sequence $\langle I_n \rangle$ is a pairwise disjoint sequence of closed interval converging to p with $p \notin I_n$ for any n, and that $A = \cup_n E_n \cup \{p\}$.

Now for each n, the closed countable set E_n has CB-rank μ, m for some μ, m with $\mu < \nu$, since otherwise $D^{(\nu)}(E_n)$ would be nonempty, and so $D^{(\nu)}(A)$ will contain a point distinct from p, a contradiction. Hence by induction hypothesis each E_n, if nonempty, will be homeomorphic to a countable successor ordinal. Since each countable successor ordinal can be continuously order-embedded in any interval as closed set, we can choose a well-ordered countable closed subset F_1 of $[0, \frac{1}{2}]$ such that E_1 is homeomorphic to F_1. Similarly, choose a well-ordered countable closed subset F_2 of $[\frac{2}{3}, \frac{3}{4}]$ such that E_2 is homeomorphic to F_2. In general, choose a well-ordered countable closed subset F_n of $[\frac{2n-2}{2n-1}, \frac{2n-1}{2n}]$ such that E_n is homeomorphic to F_n. Being bounded and closed, each set F_n, if nonempty, must have a greatest element. Finally put:

$$F := \bigcup_{n \in \mathbf{N}} F_n \cup \{1\},$$

which is a well-ordered countable closed set with the greatest element 1. Then by Proposition 1090, The set A is homeomorphic to the set F, and therefore to a countable successor ordinal.

If $D^{(\nu)}(A)$ has n elements with $n > 1$, then we can order the elements of $D^{(\nu)}(A)$ as $D^{(\nu)}(A) = \{p_1 < p_2 < \cdots < p_n\}$, and then choose elements $a_1, a_2, \ldots, a_n$ and $b_1, b_2, \ldots, b_n$ such that

$$a_1 < p_1 < b_1 < a_2 < p_2 < b_2 < \cdots < a_n < p_n < b_n,$$

$a_1 < \inf A$, $b_n > \sup A$, and

$$[b_1, a_2] \cap A = \emptyset, \quad [b_2, a_3] \cap A = \emptyset, \quad \ldots, \quad [b_{n-1}, a_n] \cap A = \emptyset.$$

Putting $H_k = A \cap [a_k, b_k] = A \cap (a_k, b_k)$ for each $k = 1, 2, \ldots, n$, we note that each H_k is a countable closed set with $D^{(\nu)}(H_k) = \{p_k\}$, hence H_k is homeomorphic to a countable successor ordinal α_k. Thus $A = \cup_{k=1}^{n} H_k$ is homeomorphic to the countable successor ordinal $\alpha_1 + \alpha_2 + \cdots + \alpha_n$. □

Corollary 1095. *Let A be a countably infinite closed bounded set with CB-rank ν, n, with $0 < \nu < \omega_1$ and $n \in \mathbf{N}$. Then A is homeomorphic to the ordinal $\omega^\nu n + 1$, and hence to any closed subset of $\mathbf{R}$ having order type $\omega^\nu n + 1$.*

Proof. By the last theorem, A is homeomorphic to a countably infinite successor ordinal α. By Proposition 1087, α is homeomorphic to $\omega^\mu m + 1$ for some μ, m. Hence A and $\omega^\mu m + 1$ are homeomorphic, and so they must have identical CB-ranks. Therefore $\mu = \nu$ and $m = n$. □

We thus arrive at our main result:

Corollary 1096 (Classification of countable closed bounded sets). *Consider the series of countably infinite successor ordinals of the form:*

$$\omega^{\nu} n + 1, \quad 0 < \nu < \omega_1,\ n \in \mathbf{N}.$$

Every countably infinite closed bounded set is homeomorphic to one and only one of the ordinals above. Conversely, for each $0 < \nu < \omega_1$ *and* $n \in \mathbf{N}$ *there exists a countably infinite closed bounded set having order type* $\omega^{\nu} n + 1$.

Hence the above series gives a complete enumeration, up to homeomorphism, of all countably infinite closed bounded sets into $\aleph_1$ *many pairwise non-homeomorphic sets.*

Remark. Although we are dealing exclusively with sets of reals, the above result can be stated in the more general context of *topological spaces.* We will not define topological spaces or the relevant related notions, but the general statement is that *every countable compact Hausdorff space* X *with* $|X| \geq 2$ *is homeomorphic to* $\omega^{\nu} n + 1$ *for some unique* $\alpha < \omega_1$ *and* $n \in \mathbf{N}$. This follows immediately from the above result since every countable compact Hausdorff space is homeomorphic to a subset of $\mathbf{R}$. (An alternative proof that every countable compact Hausdorff space is homeomorphic to a countable ordinal can be given using the Sorgenfrey topology on the reals.)

16.4 Cantor and Uniqueness of Trigonometric Series

A *trigonometric series* is a series of the form

$$a_0 + \sum_{n=1}^{\infty} a_n \cos nx + b_n \sin nx,$$

which may or may not converge for a given value of $x \in [0, 2\pi]$. If the above series converges for all x, it defines a periodic function. A special type of trigonometric series are the familiar *Fourier series*, where the coefficients a_n and b_n are given as $a_n = \frac{1}{\pi}\int_0^{2\pi} f(t) \cos nt\, dt$ and $b_n = \frac{1}{\pi}\int_0^{2\pi} f(t) \sin nt\, dt$ $(n = 1, 2, \ldots)$ for some function f integrable on $[0, 2\pi]$. However, there are everywhere convergent trigonometric series which are not Fourier series.

Cantor's work began with the *uniqueness problem for trigonometric series,* which asks this: If two trigonometric series converge and agree everywhere then will they necessarily have identical coefficients? More precisely, if for all $x \in [0, 2\pi]$ we have

$$a_0 + \sum_{n=1}^{\infty} a_n \cos nx + b_n \sin nx = c_0 + \sum_{n=1}^{\infty} c_n \cos nx + d_n \sin nx,$$

then does it follow that $a_n = c_n$ and $b_n = d_n$ for all n?

Cantor's first important result on the uniqueness problem was that the above is indeed true.

Theorem (Cantor 1870). If for all $x \in [0, 2\pi]$ the series on both sides of the equation displayed above converge and are equal, then $a_n = c_n$ and $b_n = d_n$ for all n.

Cantor then continued to work on extending the result to the case where the hypothesis is weakened to allow for an "exception set" $E \subseteq [0, 2\pi]$ on which the two series may not agree (or fail to converge). In other words, given $E \subseteq [0, 2\pi]$, *the uniqueness problem for an exception set* E asks:

If two trigonometric series agree outside E *(that is, if the above equality holds for all* $x \in [0, 2\pi] \setminus E$*), then do they necessarily have identical coefficients?*

If the answer to this question is yes, we say that the exception set E is a *set of uniqueness.* Thus Cantor's first result above says that the empty set $E = \emptyset$ is a set of uniqueness.

Cantor next established that E is a set of uniqueness if E is an arbitrary finite set. He went on to point out that E is a set of uniqueness if E is any closed set with a finite number of limit points (that is, $D(E)$ is finite), or if E is a closed set whose set of limit points in turn have only finitely many limit points (that is, $D^{(2)}(E)$ is finite), and so on for any finite number of iterations of orders of limit points. In our terminology, Cantor essentially established the following.

Cantor Uniqueness Theorem (Cantor 1872). *If* $E \subseteq [0, 2\pi]$ *is closed and of "finite Cantor–Bendixson rank," that is if* $D^{(n)}(E)$ *is finite for some* $n \in \mathbf{N}$*, then* E *is a set of uniqueness.*

The next natural extension would be to consider limit points of order ω, and ask if E is a set of uniqueness when $D^{(\omega)}(E)$ is finite; and one can proceed further through the transfinite ordinals. This is indeed true, and in fact any countable closed set is a set of uniqueness, but that was proved much later. When Cantor was investigating the uniqueness problem, notions such as "transfinite ordinal number" and "countable set" did not exist, and after establishing his theorem stated above, Cantor created and developed such foundational concepts almost single-handedly, giving birth to set theory. Thus Cantor's quest for generalizing his uniqueness results led him to consider transfinite iterations of the operation of forming the derived set (the set of limit points of a set), and then on to far-reaching abstractions such as countable and uncountable sets, the topology of real point sets, the theory of orders, order types, well-ordered sets, transfinite ordinals, and cardinals.

Busy in his creation and development of the theory of the transfinite, Cantor never returned to the problem of uniqueness of trigonometric series. Characterizing the sets of uniqueness turned out to be an extremely difficult problem, and research has been continuing on it for more than hundred years. Interestingly, the use of set theory and transfinite ordinals in the investigation in the problem of uniqueness has returned, through an area of set theory known as *Descriptive Set Theory.*

For more details on the connection between Cantor's creation of set theory and the problem of uniqueness of trigonometric series, we refer the reader to Dauben's book [9] and the article of Kechris [39], where further references can be found.

Chapter 17
Brouwer's Theorem and Sierpinski's Theorem

Abstract In this chapter we apply the theory of orders from Chap. 8, especially Cantor's theorem on countable dense orders, to prove two classical theorems: Brouwer's topological characterization of the Cantor set, and Sierpinski's topological characterization of the rationals.

17.1 Brouwer's Theorem

The Cantor set is an example of a perfect bounded nowhere dense subset of **R**.

Problem 1097. *Show that if $E \subseteq$ **R** is homeomorphic to the Cantor set then E is perfect bounded and nowhere dense.*

[Hint: The Bolzano–Weierstrass theorem and the Intermediate Value Theorem may help.]

Theorem 1098 (Brouwer). *Any two perfect bounded and nowhere dense subsets of* **R** *are homeomorphic to each other, and hence to the Cantor set.*

Since two closed subsets of **R** having the same order type are homeomorphic, Brouwer's theorem follows from the following stronger result.

Theorem 1099. *Let E be a perfect bounded and nowhere dense subset of* **R**. *Then there is an order-isomorphism of* **R** *onto* **R** *which maps E onto the Cantor set.*

Proof. Let $G = \mathbf{R} \setminus E$ so that G is open, and hence G is the union of a unique countable family of pairwise disjoint nonempty open intervals. Since E is bounded, there are two unbounded open intervals in the decomposition of G, one of the form $(-\infty, a)$ and another of the form (b, ∞) where $a = \inf E$ and $b = \sup E$. All other open intervals in the decomposition of G are bounded. Let C be the countable family of all bounded open intervals in the decomposition of G. Since the intervals in C are pairwise disjoint, the family C is naturally ordered, where for intervals $I, J \in$ C

A. Dasgupta, *Set Theory: With an Introduction to Real Point Sets*,
DOI 10.1007/978-1-4614-8854-5_17,

we have $I < J$ if and only if $x < y$ for all $x \in I$ and $y \in J$. Since E is nowhere dense, the ordering of C is dense order without endpoints.

Similarly if D is the family of bounded open intervals removed in the construction of the Cantor set, we find that using the ordering on the intervals, D forms a countable dense order without endpoints.

Hence, by Cantor's theorem on countable dense orders without endpoints, there is an order isomorphism ϕ from C onto D. For each $I \in$ C there is unique increasing linear function f_I mapping the interval I onto the interval $\phi(I)$. Also, let f_0 denote the unique translation map $x \mapsto x-a$ mapping the interval $(-\infty, a]$ onto the interval $(-\infty, 0]$ and let f_1 denote the unique translation map $x \mapsto x - b + 1$ mapping the interval $[b, \infty)$ onto the interval $[1, \infty)$. Combining all the mappings f_I ($I \in$ C), f_0 and f_1, we get an order preserving bijection f^* mapping the set G onto the complement of the Cantor set. Now note that since E is nowhere dense, for each $x \in E$ and any u, v with $u < x < v$, there exist $s, t \in G$ with $u < s < x < t < v$ so that f^* is defined at s and t. The same is true for the Cantor set and its complement. Moreover **R** is a complete order. Hence f^* extends uniquely to an order preserving bijection f mapping **R** onto **R**, which can be defined as

$$f(x) := \sup\{f^*(t) \mid t < x, t \in G\} = \inf\{f^*(t) \mid t > x, t \in G\}.$$

Clearly, f is then an order-isomorphism of **R** onto **R** which maps E onto the Cantor set. □

We had seen that every generalized Cantor set (i.e., any set generated by a Cantor system) is bounded, perfect, and nowhere dense. We now have the converse result.

Corollary 1100. *Let E be a bounded perfect nowhere dense set. Then E is a generalized Cantor set, that is, there is a Cantor system of intervals which generates E.*

Proof. By the theorem, we can fix a bijective order-isomorphism $f : \mathbf{R} \to \mathbf{R}$ which maps the Cantor set onto E. Let $\langle I_u \mid u \in \{0,1\}^* \rangle$ be the standard Cantor system which generates the Cantor set. We show that $\langle f[I_u] \mid u \in \{0,1\}^* \rangle$ is a Cantor system which generates E.

Since f is an order-isomorphism and I_u is a bounded proper closed interval, so $f[I_u]$ is a bounded proper closed interval, for any $u \in \{0,1\}^*$. Since f is a bijection and $I_{u^\frown 0} \cap I_{u^\frown 1} = \emptyset$, hence $f[I_{u^\frown 0}] \cap f[I_{u^\frown 1}] = \emptyset$. Again, since f is a bijection, if $b \in \{0,1\}^{\mathbf{N}}$ and x is the unique member of the singleton $\cap_n I_{b|n}$, so $f(x)$ is the unique member of the singleton $\cap_n f[I_{b|n}]$. Thus E is generated by $\langle f[I_u] \mid u \in \{0,1\}^* \rangle$. Finally, for any $b \in \{0,1\}^{\mathbf{N}}$, we have $\lim_n \operatorname{len}(f[I_{b|n}]) = 0$, since if the intersection of a nested sequence of intervals results in a singleton, then the lengths of the intervals in the sequence must approach zero. Hence $\langle f[I_u] \mid u \in \{0,1\}^* \rangle$ is a Cantor system. □

Corollary 1101. *A set is bounded perfect nowhere dense if and only if it is a generalized Cantor set generated by some Cantor system of intervals.*

Thus a perfect bounded nowhere dense not only is order-isomorphic and homeomorphic to the Cantor set, but also has the same structure in terms of definition via interval trees. So *any* perfect bounded nowhere dense subset of $\mathbf{R}$ (equivalently any generalized Cantor set) will be called *a Cantor set.*

Note the difference between the term "*a Cantor set*" (any perfect bounded nowhere dense subset of $\mathbf{R}$) and the term "*the Cantor set*" (the specific subset $\mathbf{K}$ of $[0, 1]$ obtained by repeatedly removing middle-third open intervals).

17.2 Homeomorphic Permutations of the Cantor Set

We will now introduce some especially nice homeomorphisms of the Cantor set onto itself.

Recall the natural bijection $\mathbf{F}$ between the set $\mathbf{2^N} := \{0, 1\}^{\mathbf{N}}$ of all infinite binary sequences and the Cantor set $\mathbf{K}$ given by the mapping

$$\langle b_1, b_2, \ldots, b_n, \ldots \rangle \longmapsto \mathbf{F}(\langle b_n \rangle) = \sum_{n=1}^{\infty} \frac{2b_n}{3^n}.$$

Using this bijection, we will *identify* $\mathbf{2^N}$ with the Cantor set. This means that elements of $\mathbf{2^N}$ will represent points of the Cantor set, subsets of $\mathbf{2^N}$ will represent subsets of the Cantor set, and so on.

Now consider the subset $\mathbf{K}_1 := [0, \frac{1}{3}] \cup [\frac{2}{3}, 1]$ of $[0, 1]$. The two closed intervals of $\mathbf{K}_1$ are translates of each other, and using these translations we can get a homeomorphism of F_1 onto itself which interchanges these two intervals. More precisely, this homeomorphism f_1 can be defined as:

$$f_1(x) = \begin{cases} x + \frac{2}{3} & \text{if } x \in [0, \frac{1}{3}], \\ x - \frac{2}{3} & \text{if } x \in [\frac{2}{3}, 1]. \end{cases}$$

Now, the Cantor set is a subset of $\mathbf{K}_1$, and note that f_1 maps the Cantor set onto itself. Restricting f_1 to the Cantor set, we get a map g_1 which is a homeomorphic permutation of the Cantor set.

In view of the identification of the Cantor set with $\mathbf{2^N}$, this homeomorphic permutation g_1 of the Cantor set admits a simpler definition in terms of elements of $\mathbf{2^N}$:

$$g_1(\langle b_1, b_2, b_3, \ldots, b_n, \ldots \rangle) = \langle 1 - b_1, b_2, b_3, \ldots, b_n, \ldots \rangle,$$

or more informally by saying that the permutation g_1 of $\mathbf{2^N}$ onto $\mathbf{2^N}$ transforms a binary sequence into another one by "flipping the first bit of the sequence."

Consider again the set $\mathbf{K}_2 := [0, \frac{1}{9}] \cup [\frac{2}{9}, \frac{1}{3}] \cup [\frac{2}{3}, \frac{7}{9}] \cup [\frac{8}{9}, 1]$ found in the second stage of the construction of the Cantor set. The intervals $[0, \frac{1}{9}]$ and $[\frac{2}{9}, \frac{1}{3}]$ are translates of each other and can be interchanged via these translations, and similarly the intervals $[\frac{2}{3}, \frac{7}{9}]$ and $[\frac{8}{9}, 1]$ can be interchanged via similar translations to yield a homeomorphic permutation f_2 of $\mathbf{K}_2$:

$$f_2(x) = \begin{cases} x + \frac{2}{9} & \text{if } x \in [0, \frac{1}{9}] \cup [\frac{2}{3}, \frac{7}{9}], \\ x - \frac{2}{9} & \text{if } x \in [\frac{2}{9}, \frac{1}{3}] \cup [\frac{8}{9}, 1]. \end{cases}$$

Once again we can restrict f_2 to the Cantor set to obtain a homeomorphic permutation g_2 of the Cantor set, and in view of the identification of the Cantor set with $\mathbf{2}^{\mathbf{N}}$, the homeomorphic permutation g_2 can be defined in terms of elements of $\mathbf{2}^{\mathbf{N}}$ as:

$$g_2(\langle b_1, b_2, b_3, \dots, b_n, \dots \rangle) = \langle b_1, 1 - b_2, b_3, \dots, b_n, \dots \rangle,$$

or more informally by saying that the permutation g_2 of $\mathbf{2}^{\mathbf{N}}$ onto $\mathbf{2}^{\mathbf{N}}$ transforms a binary sequence into another one by "flipping the second bit of the sequence."

In general, for each $n \in \mathbf{N}$, the operation of "flipping the n-th bit of an infinite binary sequence" gives a a homeomorphic permutation g_n of the Cantor set.

Even more generally, for each subset A of $\mathbf{N}$, we can define the map $g_A : \mathbf{2}^{\mathbf{N}} \to \mathbf{2}^{\mathbf{N}}$ by setting

$$g_A(\langle b_1, b_2, \dots, b_n, \dots \rangle) = \langle b'_1, b'_2, \dots, b'_n, \dots \rangle,$$

where

$$b'_n = \begin{cases} 1 - b_n & \text{if } n \in A, \\ b_n & \text{otherwise.} \end{cases}$$

Problem 1102. *Prove that for each $A \subseteq \mathbf{N}$ the mapping g_A defined above is a homeomorphic permutation of the Cantor set.*

[Hint: Using the ternary expansion representation for the elements of the Cantor set may help. Also note that $g_A^{-1} = g_A$, and so it suffices to show that g_A is continuous.]

Endpoints and Internal Points of the Cantor Set

Consider the closed intervals obtained in the formation of the Cantor set, namely

$$[0, 1], \quad [0, \tfrac{1}{3}], \quad [\tfrac{2}{3}, 1], \quad [0, \tfrac{1}{9}], \quad [\tfrac{2}{9}, \tfrac{1}{3}], \quad [\tfrac{2}{3}, \tfrac{7}{9}], \quad [\tfrac{8}{9}, 1], \quad \dots$$

The endpoints of these intervals

$$0,\ 1,\ \tfrac{1}{3},\ \tfrac{2}{3},\ \tfrac{1}{9},\ \tfrac{2}{9},\ \tfrac{7}{9},\ \tfrac{8}{9},\ \dots$$

will be called *the endpoints of the Cantor set.* All other points of the Cantor set will be called *the internal points of the Cantor set.* Note that the internal points of the Cantor set are precisely its two-sided limit points, while the endpoints of the Cantor set are its one-sided limit points. (More generally, for any dense-in-itself set A, we can define the endpoints of A to be the one-sided limit points of A, and the internal points of A to be the two-sided limit points of A.)

Problem 1103. *Show that*

1. $1/4$ *is an internal point of the Cantor set.*
2. *Under the identification of the Cantor set with* $\mathbf{2}^{\mathbf{N}}$, *an infinite binary sequence is an endpoint of the Cantor set if and only if it is eventually constant.*

The following technical result will be needed in the proof of Sierpinski's theorem:

Theorem 1104. *Let E be the set of endpoints of the Cantor set* $\mathbf{K}$. *Given any countable subset C of* $\mathbf{K}$, *there is a homeomorphic permutation f of* $\mathbf{K}$ *such that $f(x)$ is an internal point of* $\mathbf{K}$ *for every $x \in C$, that is, such that $f[C] \cap E = \emptyset$.*

Proof. We use the identification of the Cantor set with $\mathbf{2}^{\mathbf{N}}$, and work exclusively in $\mathbf{2}^{\mathbf{N}}$.

Fix a bijection $k\colon \mathbf{N} \times \mathbf{N} \to \mathbf{N}$.

Now let $C = \{x^{(1)}, x^{(2)}, \ldots, x^{(n)}, \ldots\}$ be a countable subset of $\mathbf{2}^{\mathbf{N}}$, and define $A \subseteq \mathbf{N}$ by:

$$A := \{k(m, 2n-1) \mid m, n \in \mathbf{N} \text{ and } x^{(m)}(k(m, 2n-1)) = 0\} \\ \cup \{k(m, 2n) \mid m, n \in \mathbf{N} \text{ and } x^{(m)}(k(m, 2n)) = 1\},$$

and put $f = g_A$ as above. In other words, $f\colon \mathbf{2}^{\mathbf{N}} \to \mathbf{2}^{\mathbf{N}}$ is defined by setting

$$f(\langle b_1, b_2, \ldots, b_n, \ldots \rangle) = \langle b'_1, b'_2, \ldots, b'_n, \ldots \rangle,$$

where

$$b'_n = \begin{cases} 1 - b_n & \text{if } n \in A, \\ b_n & \text{otherwise.} \end{cases}$$

Put $y^{(m)} = f(x^{(m)})$. Then for each $m \in \mathbf{N}$ we have $y^{(m)}(k(m, 2n-1)) = 1$ and $y^{(m)}(k(m, 2n)) = 0$ for all n. Since k is injective, it follows that $y^{(m)}(j) = 1$ for infinitely many j and $y^{(m)}(j) = 0$ for infinitely many j. Hence $y^{(m)} = f(x^{(m)})$ is an internal point of $\mathbf{2}^{\mathbf{N}}$ for each $m \in \mathbf{N}$. □

17.3 Sierpinski's Theorem

We begin by the following immediate consequence of Cantor's theorem on countable dense orders.

Proposition 1105. *Any countable dense subset of* $\mathbf{R}$ *is homeomorphic to* $\mathbf{Q}$.

Proof. This follows since if A is a countable dense subset of $\mathbf{R}$, then by Cantor's theorem we can get an order isomorphism f between A and $\mathbf{Q}$. But both in A and in $\mathbf{Q}$, every point is a two-sided limit point, and so both sets are continuously order-embedded in $\mathbf{R}$. Hence f is a homeomorphism. □

Now given any countable set $A \subseteq \mathbf{R}$, we can take $A \cup \mathbf{Q}$ to get a countable dense subset of $\mathbf{R}$ which contains A. Hence we have:

Corollary 1106. *Any countable subset of* $\mathbf{R}$ *is homeomorphic to a subset of* $\mathbf{Q}$.

Proposition 1107. *If A is a dense subset of the Cantor set consisting only of internal points of the Cantor set, then every point of A is a two-sided limit point of A (in* $\mathbf{R}$*) and hence A is continuously order-embedded in* $\mathbf{R}$.

Corollary 1108. *If C is a countable subset of the Cantor set consisting only of internal points and C is dense in the Cantor set, then C is a continuously order-embedded subset of* $\mathbf{R}$ *of order type η, and so C is homeomorphic to* $\mathbf{Q}$.

Theorem 1109. *If C is a countable subset of the Cantor set which is dense in it, then C is homeomorphic to* $\mathbf{Q}$.

Proof. By Theorem 1104 there is a homeomorphic permutation of the Cantor set mapping C to a subset D of the Cantor set consisting only of its internal points. Then D is a countable dense subset of the Cantor set consisting only of its internal points, and hence homeomorphic to $\mathbf{Q}$ by Corollary 1108. □

Problem 1110. *The set of endpoints of the Cantor set is a countable dense subset of it, and hence is homeomorphic to* $\mathbf{Q}$.

Theorem 1111 (Sierpinski). *Every countable dense-in-itself subset of* $\mathbf{R}$ *is homeomorphic to* $\mathbf{Q}$.

Proof. Let A be a countable dense-in-itself subset of $\mathbf{R}$. Then A is homeomorphic to a subset of $\mathbf{Q}$. Since the Cantor set has subsets homeomorphic to $\mathbf{Q}$, it follows that A is homeomorphic to a subset B of the Cantor set. B is then a dense-in-itself subset of the Cantor set, and hence its closure $\overline{B}$ is a perfect subset of the Cantor set, and hence $\overline{B}$ is perfect bounded nowhere dense set. By Brouwer's Theorem, $\overline{B}$ is homeomorphic to the Cantor set and hence B is homeomorphic to a countable dense subset C of the Cantor set. By Theorem 1109, C is homeomorphic to $\mathbf{Q}$. Therefore B, and hence A, is homeomorphic to $\mathbf{Q}$. □

Problem 1112. *Show that* $1 + \eta$, $1 + \eta + 1$, *and* $\eta + 2 + \eta$ *are all homeomorphic to* η.

Problem 1113. *Prove that 2η is homeomorphic to η.*

[Hint: The set of endpoints of the Cantor set other than 0 and 1 has order type 2η and is continuously order-embedded in **R**.]

17.4 Brouwer's and Sierpinski's Theorems in General Spaces

Although in this text we are restricting ourselves to subsets of **R** and will not define topological spaces or even metric spaces, let us point out that both Brouwer's Theorem and Sierpinski's Theorem can be stated in much more general settings.

The general statement of Brouwer's Theorem for metric spaces says: *Any totally disconnected perfect compact metric space is homeomorphic to the Cantor set.* Since any compact totally disconnected metric space is a zero-dimensional separable metric space and since any such space can be homeomorphically embedded in **R**, the general version of Brouwer's Theorem follows from our special version for subsets of **R**.

Sierpinski's Theorem for general metric spaces says: *Any countable metric space without isolated points is homeomorphic to* **Q**. Again, since any countable metric space is a zero-dimensional separable metric space and since any such space can be homeomorphically embedded in **R**, the general version of Sierpinski's Theorem follows from our special version for subsets of **R**.

Chapter 18
Borel and Analytic Sets

Abstract This chapter covers some of the basic theory of Borel and Analytic Sets in the context of the real line. We define analytic sets using the Suslin operation, and show that they have all the regularity properties (measurability, Baire property, perfect set property), and therefore satisfy the continuum hypothesis—the best result possible without additional axioms. Along the way we obtain the Lusin Separation Theorem, Suslin's theorem, the boundedness theorem, and an example of a non-Borel analytic set.

18.1 Sigma-Algebras and Borel Sets

Definition 1114. A nonempty collection S of subsets of $\mathbf{R}$ is called a *Sigma-Algebra* if

1. S is closed under taking complements: if $A \in S$ then $\mathbf{R} \setminus A \in S$; and
2. S is closed under countable unions: if $A_n \in S$ for all $n \in \mathbf{N}$ then $\cup_n A_n \in S$.

Trivially, the two-element family $\{\emptyset, \mathbf{R}\}$ and the power set $\mathbf{P}(\mathbf{R})$ are sigma-algebras. More importantly, we have: *The collection* $\mathbf{L}$ *of Lebesgue measurable sets is a sigma-algebra, and so is the collection* $\mathbf{Y}$ *of sets with Baire property.*

Problem 1115. *Show that the collection S below is a sigma algebra:*

$$S := \{A \in \mathbf{P}(\mathbf{R}) \mid \textit{Either } A \textit{ or } \mathbf{R} \setminus A \textit{ is countable}\},$$

and that S is the smallest sigma-algebra containing all singletons of $\mathbf{R}$.

Problem 1116. *Show that if S is a sigma-algebra then*

1. $\emptyset \in S$, $\mathbf{R} \in S$.
2. *S is closed under countable intersections: If* $A_n \in S$ *for all* $n \in \mathbf{N}$ *then* $\cap_n A_n \in S$.
3. *S is closed under finite unions, finite intersections, and set differences.*

A. Dasgupta, *Set Theory: With an Introduction to Real Point Sets*,

DOI 10.1007/978-1-4614-8854-5_18, © Springer Science+Business Media New York 2014

Problem 1117. *Show that the intersection of any nonempty family of sigma-algebras is a sigma-algebra. Deduce that given any family* C *of subsets of* $\mathbf{R}$*, there is a smallest sigma-algebra containing* C.

Definition 1118. Given a family C of subsets of $\mathbf{R}$ the smallest sigma-algebra containing C is called the *sigma-algebra generated by* C.

Problem 1119. *Show that*

1. *The sigma-algebra generated by the measure zero sets together with the open sets equals the collection* $\mathbf{L}$ *of all measurable sets.*
2. *The sigma-algebra generated by the meager sets together with the open sets equals the collection* $\mathbf{Y}$ *of all sets with Baire property.*

Definition 1120 (Borel Sets). $\mathbf{B}$ denotes the sigma-algebra generated by the open sets, and sets in $\mathbf{B}$ are called the *Borel sets.*

Being a sigma-algebra, $\mathbf{B}$ includes, along with open sets, all closed sets, F_σ sets, G_δ sets, and so on. Also, since $\mathbf{L}$ is a sigma-algebra containing all open sets, we have $\mathbf{B} \subseteq \mathbf{L}$, that is, *all Borel sets are Lebesgue measurable.* Similarly, $\mathbf{B} \subseteq \mathbf{Y}$, so *all Borel sets have Baire property.* Thus $\mathbf{B} \subseteq \mathbf{L} \cap \mathbf{Y}$.

Most effectively defined subsets of $\mathbf{R}$ normally encountered in analysis, including all examples we have seen so far, are Borel sets.

Problem 1121. *Let* $\langle A_n \mid n \in \mathbf{N}\rangle$ *be a sequence of Borel sets. Show that the set* $\{x \mid x \in A_n \text{ for all but finitely many } n\}$ *is a Borel set. Similarly, the set* $\{x \mid x \in A_n \text{ for infinitely many } n\}$ *is a Borel set.*

Problem 1122. *Let* $f_n\colon \mathbf{R} \to \mathbf{R}$ *be a continuous function for each* $n \in \mathbf{N}$*. Show that each of the following sets is Borel.*

1. $\{x \in \mathbf{R} \mid \text{the sequence } \langle f_n(x) \mid n \in \mathbf{N}\rangle \text{ is increasing but bounded}\}$.
2. $\{x \in \mathbf{R} \mid \lim_n f_n(x) = 0\}$.
3. $\{x \in \mathbf{R} \mid \lim_n f_n(x) \text{ exists}\}$.

Problem 1123. *If* $A, B \in \mathbf{B}$ *and* $f\colon A \to \mathbf{R}$ *is continuous, then* $f^{-1}[B] \in \mathbf{B}$.

For a collection C of sets, we let C_σ denote the collection of all countable unions, and C_δ the collection of all countable intersections, of sets in C. This is consistent with our notations F_σ and G_δ, where F is the collection of closed sets and G is the collection of open sets. Thus from F_σ and G_δ we can go to

$$\mathsf{F}_{\sigma\delta} := (\mathsf{F}_\sigma)_\delta \quad \text{and} \quad \mathsf{G}_{\delta\sigma} := (\mathsf{G}_\delta)_\sigma,$$

and so on through $\mathsf{F}_{\sigma\delta\sigma}$, $\mathsf{G}_{\delta\sigma\delta}$, $\mathsf{F}_{\sigma\delta\sigma\delta}$, $\mathsf{G}_{\delta\sigma\delta\sigma}$, etc. Clearly all these collection of sets consist only of Borel sets and we have

$$\mathsf{F} \subseteq \mathsf{F}_\sigma \subseteq \mathsf{F}_{\sigma\delta} \subseteq \mathsf{F}_{\sigma\delta\sigma} \subseteq \cdots \subseteq \mathbf{B} \quad \text{and} \quad \mathsf{G} \subseteq \mathsf{G}_\delta \subseteq \mathsf{G}_{\delta\sigma} \subseteq \mathsf{G}_{\delta\sigma\delta} \subseteq \cdots \subseteq \mathbf{B}.$$

The above collections (obtained by iterating the operations of countable union and intersection through finitely many steps) still do not exhaust the Borel sets, and the process can be continued into the transfinite through the ordinals. It can be shown that such iterations keep generating newer and newer Borel sets through all the countable ordinals until when it stops at iteration ω_1, giving precisely the collection of all Borel sets.

The iterative definition of Borel sets above make them a class of *effectively defined sets* (Sect. 5.5). Roughly speaking, among Borel sets, we can distinguish between *degrees of effectiveness* (or complexity). The open and closed sets are the simplest and most effective kinds of Borel sets. Sets which are F_σ or G_δ (but neither open nor closed) are at the next degree of effectiveness, and so on.

Proposition 1124. *If* C *is a collection of subsets of* $\mathbf{R}$ *containing the open sets and closed under both countable unions and countable intersections, then* C *contains all Borel sets.*

Proof. Since C contains all open sets and is closed under countable intersections, therefore C contains all G_δ sets and hence all closed sets.

Now let A be the collection all subsets of E of $\mathbf{R}$ such that both E and $\mathbf{R} \setminus E$ are in C. Since C contains both all the open sets and all the closed sets, it follows that A contains all open sets. It is readily verified that A is closed under complements and under countable unions. Hence A is a sigma-algebra containing the open sets and so $\mathbf{B} \subseteq \mathsf{A}$. Therefore $\mathbf{B} \subseteq \mathsf{C}$. □

Corollary 1125. *The collection of Borel sets is the smallest collection containing the open sets and closed under both countable unions and countable intersections.*

Problem 1126. *An infinite sigma-algebra* S *contains at least* $\mathfrak{c}$ *many sets.*

[Hint: Fix distinct $A_1, A_2, \dots \in \mathsf{S}$. For $E \subseteq \mathbf{N}$, put $B_E := \bigcup_{n \in E} A_n \setminus \bigcup_{n \notin E} A_n$. Then infinitely many of the (pairwise disjoint) sets $B_E \in \mathsf{S}$ are non-empty.]

Problem 1127. *For* $f : \mathbf{R} \to \mathbf{R}$, $\{a \in \mathbf{R} \mid \lim_{x \to a} f(x)$ *exists*$\}$ *is a Borel set.*

[Hint: By Problem 1037 the set of points of continuity of f is a G_δ set, and by Problem 985 the set of points of removable discontinuity of f is countable.]

Problem 1128. *Let* $f : \mathbf{R} \to \mathbf{R}$ *be continuous. Show that the set of points* $H := \{a \in \mathbf{R} \mid f'(a) = 0\}$ *at which the derivative of* f *vanishes is Borel.*

[Hint: $a \notin H \Leftrightarrow \exists p \in \mathbf{Q}^+ \forall \delta \in \mathbf{Q}^+ \exists r \in \mathbf{Q}(0 < |r - a| < \delta$ and $|\frac{f(r)-f(a)}{r-a}| \geq p)$, and quantifiers ranging over countable sets can be "converted into" countable unions and intersections.]

Problem 1129. *Let* $f : \mathbf{R} \to \mathbf{R}$ *be continuous. Then the set of points* $D := \{x \in \mathbf{R} \mid f'(x)$ *exists*$\}$ *at which the derivative of* f *exists is Borel.*

18.2 Analytic Sets

If $f: \mathbf{R} \to \mathbf{R}$ is a continuous function, then by the last problem (Problem 1129), the *domain* $\mathrm{dom}(f')$ of its derivative f' is a Borel set. However, the *range* $\mathrm{ran}(f')$ of f' in general *may not be a Borel set.* The great mathematician Lebesgue made a famous error thinking that such sets are Borel. Suslin, a young student of Lusin, caught Lebesgue's error and introduced a larger class of naturally and effectively defined sets which include sets such as $\mathrm{ran}(f')$ (where f is continuous). This is the class of *analytic sets,* and Suslin used a special operation, now called the *Suslin operation,* to define such sets. In this section, we will define the Suslin operation and analytic sets.

Review of Trees over N, Terminology and Notation

$\mathbf{N}^*$ is the set of all strings (finite sequences) from the set $\mathbf{N} = \{1, 2, 3, \ldots\}$. Then $\mathbf{N}^*$ is a *tree* under the relation "u is a prefix of v," in which every node branches into infinitely many immediate extensions (Sect. 11.5). A portion of the tree $\mathbf{N}^*$ is shown below. To simplify notation, we will often write a finite or infinite sequence $\langle n_1, n_2, n_3, \ldots \rangle$ simply as the string $n_1 n_2 n_3 \cdots$. For example, the string "231" denotes the finite sequence $\langle 2, 3, 1 \rangle$.

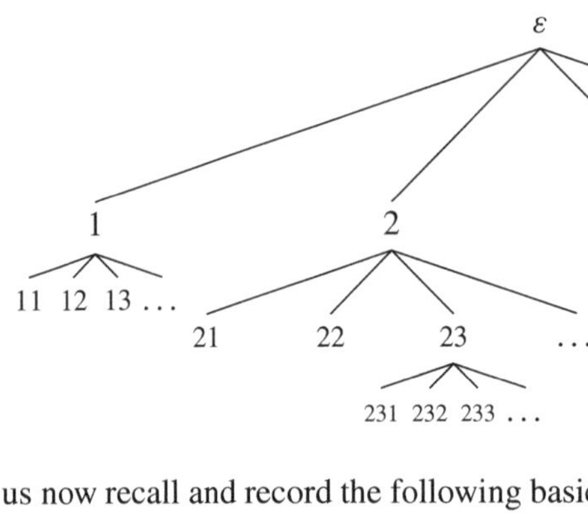

Let us now recall and record the following basic definitions.

If $u = u_1 u_2 \cdots u_m \in \mathbf{N}^*$ and $v = v_1 v_2 \cdots v_n \in \mathbf{N}^*$ are finite strings of natural numbers, and $x = x_1 x_2 \cdots x_n \cdots \in \mathbf{N}^{\mathbf{N}}$ is an infinite string of natural numbers, then:

1. The number m is called the *length of u,* denoted by $\mathrm{len}(u)$.
2. The *empty string* is denoted by ε, so that $\mathrm{len}(\varepsilon) = 0$.
3. u is an *initial segment* or *prefix* of v, or that v is an *extension* of u, if $m \leq n$ and $u_k = v_k$ for all $k \leq m$. If, in addition, $n = m + 1$, i.e., $\mathrm{len}(v) = \mathrm{len}(u) + 1$, then v is an *immediate extension* of u.

4. Similarly, u is an *initial segment* or *prefix* of x, or that x is an *extension* of u, if $u_k = x_k$ for all $k \leq m$.
5. If $u * v := u_1 u_2 \cdots u_m v_1 v_2 \cdots v_n$ is the *concatenation* of u and v.
6. For each $r \in \mathbf{N}$, we write $u^\frown r$ to denote the "immediate extension of u obtained by appending r," that is $u^\frown r := u_1 u_2 \cdots u_m r \ = u * \langle r \rangle$.
7. $x|m$ is the finite string $x_1 x_2 \cdots x_m$ obtained by truncating x to its first m values (the unique finite sequence of length m which is a prefix of x).

Definitions of trees, infinite branches, well-founded trees:

1. A subset $T \subseteq \mathbf{N}^*$ is a *tree* if any prefix of any string in T is in T.
2. An *infinite branch* B is an infinite tree $\subseteq \mathbf{N}^*$ which is linearly ordered:

$$B = \{\varepsilon,\ u_1,\ u_1u_2,\ u_1u_2u_3,\ \ldots\} \subseteq \mathbf{N}^*.$$

As in Sect. 13.3, infinite branches are identified with elements of $\mathbf{N}^\mathbf{N}$, with $x \in \mathbf{N}^\mathbf{N}$ determining the infinite branch $\{x|n \mid n = 0, 1, 2, \ldots\}$.

3. A tree $T \subseteq \mathbf{N}^*$ is *ill-founded* if T contains an infinite branch. Otherwise T is *well-founded*.
4. If $T \subseteq \mathbf{N}^*$ and $u \in \mathbf{N}^*$, then $T^{(u)}$, the *truncation of T at u*, is defined as $T^{(u)} := \{v \in \mathbf{N}^* \mid u * v \in T\}$. Note that if T is a tree then so is $T^{(u)}$.

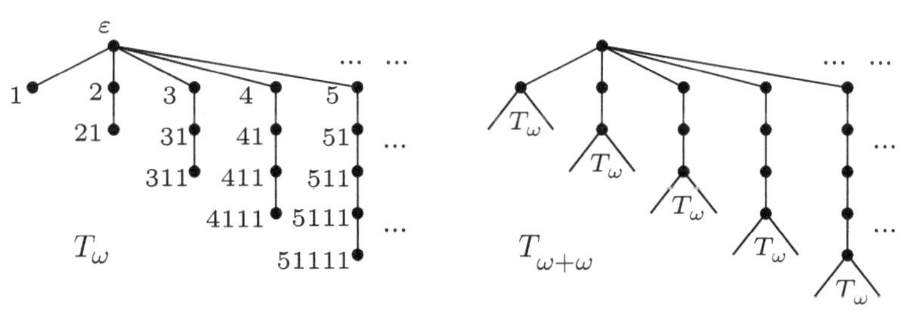

T_ω: A well founded tree of rank ω $T_{\omega+\omega}$: A well founded tree of rank $\omega+\omega$

Facts on well-founded trees (recall from Sect. 11.5):

1. A tree $T \subseteq \mathbf{N}^*$ is well-founded $\Leftrightarrow$ the string extension relation $\supset$ is well-founded on T, in which case rank(T) = rank of $\langle T \smallsetminus \{\varepsilon\}, \supset \rangle$. If $T \neq \emptyset$, then rank(T) = $\rho_T(\varepsilon)$, where ρ_T is the canonical rank function on $\langle T, \supset \rangle$.
2. If T is a well-founded tree then so is $T^{(u)}$, with rank($T^{(u)}$) $\leq$ rank(T). If $T^{(\langle n \rangle)}$ is well-founded for all $n \in \mathbf{N}$, then so is T.
3. The rank of a well-founded tree T satisfies the following recursion:

$$\text{rank}(T) = \sup\left\{\text{rank}(T^{(\langle n \rangle)}) + 1 \mid \langle n \rangle \in T,\ n \in \mathbf{N}\right\} \quad (\sup \emptyset := 0).$$

4. For $0 < \alpha < \omega_1$, $\text{rank}(T) = \alpha \Leftrightarrow \text{rank}(T^{(\langle n \rangle)}) < \alpha$ for all $n \in \mathbf{N}$ and for all $\xi < \alpha$ there is $u \in \mathbf{N}^* \setminus \{\varepsilon\}$ with $\text{rank}(T^{(u)}) = \xi$ (Problem 854).
5. Let T_1 and T_2 be trees and $f : T_1 \to T_2$ be strictly increasing, i.e., $u \subsetneq v \Rightarrow f(u) \subsetneq f(v)$. If T_2 is well-founded then so is T_1 with $\text{rank}(T_1) \leq \text{rank}(T_2)$ (Problem 819).
6. For each $\alpha < \omega_1$ there is a well-founded tree $T \subseteq \mathbf{N}^*$ with $\text{rank}(T) = \alpha$ (Problem 857).

Problem 1130. *Recall the Kleene–Brouwer ordering on* $\mathbf{N}^*$: $u <_{\text{KB}} v \Leftrightarrow$ *either* u *lexicographically precedes* v *or* u *is a proper extension of* v. *Put* $r(\varepsilon) := 0$ *and* $r(\langle n_1, n_2, \ldots, n_k \rangle) := \frac{1}{2^{n_1}} + \frac{1}{2^{n_1+n_2}} + \cdots + \frac{1}{2^{n_1+n_2+\cdots+n_k}}$. *Then (1)* r *is a bijection between* $\mathbf{N}^*$ *and the dyadic rationals in* $[0, 1)$. *(2)* $u <_{\text{KB}} v \Leftrightarrow r(u) > r(v)$. *(3)* $\langle \mathbf{N}^*, <_{\text{KB}} \rangle$ *is a linear order of type* $\eta + 1$. *(4) If* $T \subseteq \mathbf{N}^*$ *is a tree, then* T *is well-founded* $\Leftrightarrow$ T *is well-ordered under* $<_{\text{KB}}$.

The Suslin Operation

Recall how a Cantor set is generated from a family of closed intervals (a Cantor system) indexed by the full binary tree: Each infinite branch through the binary tree determines a nested sequence of closed intervals whose intersection—the "branch intersection"—is a singleton, and then the Cantor set is obtained by taking the union of all such branch intersections.

We now generalize the formation of the Cantor set to the case where arbitrary sets are used in place of the special closed intervals of a Cantor system, and where these sets are indexed by the infinitely branching tree $\mathbf{N}^*$ (instead of the finitely branching binary tree).

Let $\langle E_u \mid u \in \mathbf{N}^* \rangle$ be a family of sets indexed by $\mathbf{N}^*$. Then given any infinite branch $x = \langle x_n \rangle_{n \in \mathbf{N}} \in \mathbf{N}^{\mathbf{N}}$, we can form the "branch intersection"

$$\bigcap_{n=1}^{\infty} E_{x|n} = E_{x_1} \cap E_{x_1 x_2} \cap E_{x_1 x_2 x_3} \cap \cdots \cap E_{x_1 x_2 \ldots x_n} \cap \cdots$$

The union of all such branch intersections (as x ranges over all possible infinite branches through $\mathbf{N}^*$) will be called the result of the *Suslin Operation* applied to the family $\langle E_u \mid u \in \mathbf{N}^* \rangle$. More precisely, we have:

Definition 1131 (The Suslin Operation). If $\langle E_u \mid u \in \mathbf{N}^* \rangle$ is a family of sets indexed by nodes in the tree $\mathbf{N}^*$, then the result of the *Suslin operation* applied to $\langle E_u \mid u \in \mathbf{N}^* \rangle$, denoted by

$$\mathsf{A}(\langle E_u \mid u \in \mathbf{N}^* \rangle) = \mathsf{A}_u E_u,$$

is defined to be the set

$$\mathrm{A}_u E_u := \bigcup_{y \in \mathbf{N}^{\mathbf{N}}} \bigcap_{n=1}^{\infty} E_{y|n}$$
$$= \{x \in \mathbf{R} \mid (\exists y \in \mathbf{N}^{\mathbf{N}})(\forall n \in \mathbf{N})(x \in E_{y|n})\}.$$

Thus $x \in \mathrm{A}_u E_u$ if and only if there is a branch $y_1 y_2 \cdots y_n \cdots \in \mathbf{N}^{\mathbf{N}}$ such that

$$x \in \bigcap_{n=1}^{\infty} E_{y_1 y_2 \cdots y_n}.$$

The figure below shows a family $\langle E_u \mid u \in \mathbf{N}^* \rangle$ of sets indexed by $\mathbf{N}^*$ and the sets corresponding to the branch $3221 \cdots = \{\varepsilon, 3, 32, 322, 3221, \ldots\}$:

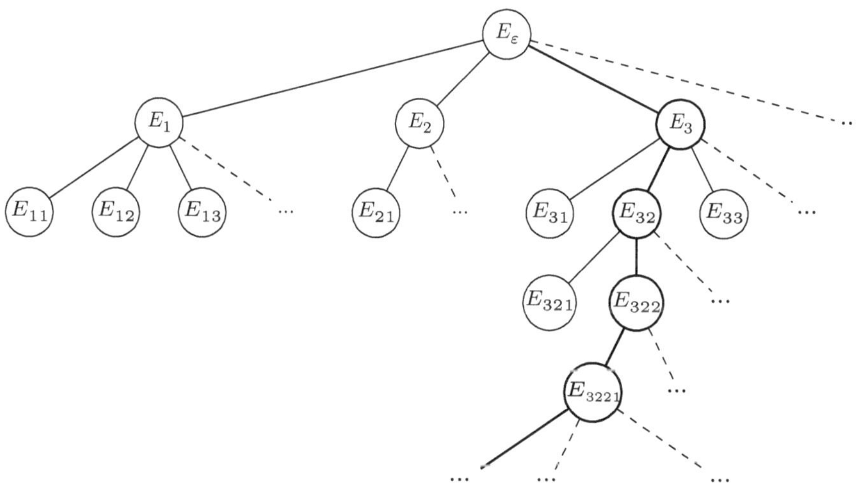

Note that the result $\mathrm{A}_u E_u$ of the Suslin operation performed on the family $\langle E_u \mid u \in \mathbf{N}^* \rangle$ does not depend on the set E_ε. The following problems show that the Suslin operation is "more powerful" than the operations of countable union and countable intersection of sets.

Problem 1132. *Show that if* $C = \cup_{n=1}^{\infty} C_n$, *then* $C = \mathrm{A}_u E_u$ *for some family* $\langle E_u \mid u \in \mathbf{N}^* \rangle$ *where each* $E_u = C_n$ *for some* n.

[Hint: Put $E_{u_1 u_2 \cdots u_n} = C_{u_1}$.]

Problem 1133. *Show that if* $C = \cap_{n=1}^{\infty} C_n$, *then* $C = \mathrm{A}_u E_u$ *for some family* $\langle E_u \mid u \in \mathbf{N}^* \rangle$ *where each* $E_u = C_n$ *for some* n.

[Hint: Put $E_u = C_{\mathrm{len}(u)}$.]

Definition 1134 (Suslin Systems). A *Suslin system* is a family of closed intervals $\langle F_u \mid u \in \mathbf{N}^* \rangle$ indexed by $\mathbf{N}^*$, satisfying the following conditions:

1. For each $u \in \mathbf{N}^*$, F_u is a closed interval in $\mathbf{R}$, possibly unbounded or empty.
2. $F_u \supseteq F_v$ whenever u is prefix of v.
3. For any infinite sequence of natural numbers $x \in \mathbf{N}^{\mathbf{N}}$, $\text{len}(F_{x|n}) \to 0$ as $n \to \infty$.

Note that we regard $\mathbf{R}$ and $\varnothing$ as closed intervals, with $\text{len}(\varnothing) = 0$.

An *analytic set* is now defined to be one which can be obtained as the result of the Suslin operation applied to a Suslin system.

Definition 1135 (Analytic Sets). A subset A of $\mathbf{R}$ is an *analytic set* if

$$A = \mathsf{A}_u F_u$$

for some Suslin system $\langle F_u \mid u \in \mathbf{N}^* \rangle$.

Clearly, the notion of a Suslin system is a generalization of that of a Cantor system, but it differs from a Cantor system in two important ways: (a) The sets F_u ($u \in \mathbf{N}^*$) in a Suslin system are indexed by the *infinitely-branching* tree $\mathbf{N}^*$ (instead of the finitely branching binary tree $\{0,1\}^*$), and (b) the sets F_u are arbitrary closed intervals (possibly unbounded or empty), and $F_{u^\frown m}$ and $F_{u^\frown n}$, $m \neq n$, are no longer required to be disjoint. (Problem 1145 below shows that both these requirements are necessary if the collection of analytic sets is going to be sufficiently comprehensive.)

Problem 1136. *Let $\langle F_u \mid u \in \mathbf{N}^* \rangle$ be a Suslin system. Show that*

1. *The set $\{u \in \mathbf{N}^* \mid F_u \neq \varnothing\}$ is tree over $\mathbf{N}$ (subtree of $\mathbf{N}^*$).*
2. *For each $x \in \mathbf{R}$, the set $\{u \in \mathbf{N}^* \mid x \in F_u\}$ is also a tree over $\mathbf{N}$.*
3. *$x \in \mathsf{A}_u F_u \Leftrightarrow$ the tree $\{u \in \mathbf{N}^* \mid x \in F_u\}$ is not well-founded.*

Problem 1137. *Every closed interval, possibly unbounded, is an analytic set.*

Problem 1138. *Any countable set is analytic.*

Problem 1139. *All closed sets and all open sets are analytic.*

Theorem 1140. *The collection of analytic sets is closed under the Suslin operation: If E_u is analytic for each $u \in \mathbf{N}^*$, then $\mathsf{A}_u E_u$ is analytic.*

Proof. Fix a bijection $\pi: \mathbf{N}^2 \to \mathbf{N}$ satisfying

$$m < m' \Rightarrow \pi(m,n) < \pi(m',n) \quad \text{and} \quad n < n' \Rightarrow \pi(m,n) < \pi(m,n').$$

Fix a function $g: \mathbf{N} \to \mathbf{N}$ such that for all n, $g(n) \leq n$ and there exist infinitely many m with $g(m) = n$. (The sequence $1, 1, 2, 1, 2, 3, 1, 2, 3, 4, \ldots$ is such a function.)

Let $h(n) := \big|\{k \mid k \leq n \text{ and } g(k) = g(n)\}\big|$.

Define $k \ll n \Leftrightarrow k \leq n$, $\pi(1, g(k)) \leq n$, and $\pi(g(k)+1, h(k)) \leq n$.

Given $w = w_1 w_2 \cdots w_n \in \mathbf{N}^*$, define

$$\alpha(w,k) = \begin{cases} w_{\pi(1,1)} w_{\pi(1,2)} \cdots w_{\pi(1,g(k))} & \text{if } k \ll n, \\ \varepsilon & \text{otherwise,} \end{cases}$$

and

$$\beta(w,k) = \begin{cases} w_{\pi(g(k)+1,1)} w_{\pi(g(k)+1,2)} \cdots w_{\pi(g(k)+1,h(k))} & \text{if } k \ll n, \\ \varepsilon & \text{otherwise.} \end{cases}$$

Now let E_u be analytic for each $u \in \mathbf{N}^*$. We need to show that the set $\mathrm{A}_u E_u$ is analytic. For each $u \in \mathbf{N}^*$ there is a Suslin system $\langle F_v^u \mid v \in \mathbf{N}^* \rangle$ such that

$$E_u = \mathrm{A}_v F_v^u.$$

Define a Suslin system $\langle D_w \mid w \in \mathbf{N}^* \rangle$ as:

$$D_w = \bigcap_{k \ll \mathrm{len}(w)} F_{\beta(w,k)}^{\alpha(w,k)}.$$

It is routine to verify that this is indeed a Suslin system (details left for the reader.) We claim that

$$\mathrm{A}_u E_u = \mathrm{A}_w D_w.$$

Suppose first that $a \in \mathrm{A}_u E_u$. Then there is $x = x_1 x_2 \cdots x_n \cdots \in \mathbf{N}^{\mathbf{N}}$ such that $a \in E_{x_1 x_2 \cdots x_n}$ for all n. Since $E_{x|n} = \mathrm{A}_v F_v^{x|n}$, so for each n there is an infinite sequence $y^{(n)} = y_{n,1} y_{n,2} \cdots y_{n,k} \cdots$ of natural numbers such that

$$a \in \bigcap_{k=1}^{\infty} F_{y^{(n)}|k}^{x|n} = \bigcap_{k=1}^{\infty} F_{y_{n,1} y_{n,2} \cdots y_{n,k}}^{x_1 x_2 \cdots x_n}.$$

Now consider the infinite matrix

$$\begin{matrix} x_1 & x_2 & \cdots & x_n & \cdots \\ y_{1,1} & y_{1,2} & \cdots & y_{1,n} & \cdots \\ y_{2,1} & y_{2,2} & \cdots & y_{2,n} & \cdots \\ \cdots & \cdots & \cdots & \cdots & \cdots, \end{matrix}$$

and combine it into a single infinite sequence $z_1 z_2 \cdots z_n \cdots$ using the pairing function π as follows:

$$z_{\pi(1,n)} := x_n \quad \text{and} \quad z_{\pi(m+1,n)} := y_{m,n}.$$

Then it is readily verified that $a \in D_{z|n}$ for any n, and so $a \in \mathrm{A}_w D_w$.

Conversely, suppose that $a \in \mathrm{A}_w D_w$. Fix an infinite sequence $z_1 z_2 \cdots z_n \cdots$ such that $a \in D_{z|n}$ for all n. Define infinite sequences $x = x_1 x_2 \cdots x_n \cdots$ and $y^{(n)} = y_{n,1} y_{n,2} \cdots$ (for each n) by setting

$$x_n := z_{\pi(1,n)} . y_{n,m} := z_{\pi(n+1,m)}.$$

We claim that $a \in E_{x|n}$ for all n. Fix any n. Since $E_{x|n} = \mathrm{A}_v F_v^{x|n}$, so it suffices to show that for all m,

$$a \in F^{x|n}_{y_{n,1} y_{n,2} \cdots y_{n,m}}$$

Enumerate $\{k \mid g(k) = n\}$ as $\{k_1 < k_2 < \cdots\}$, so that $g(k_m) = n$ and $h(k_1) = 1$, $h(k_2) = 2$, etc. For each m, fix any n_m such that $k_m \ll n_m$. Hence $\alpha(z|n_m, k_m) = x_1 x_2 \cdots x_n = x|n$ for all m, and $\beta(z|n_m, k_m) = y_{n,1} y_{n,2} \cdots y_{n,m}$. Since $a \in D_{z|n_m}$ and $k_m \ll n_m$, it follows that

$$a \in F^{\alpha(z|n_m,k_m)}_{\beta(z|n_m,k_m)} = F^{x|n}_{y_{n,1} y_{n,2} \cdots y_{n,m}}.$$

Thus $a \in \mathrm{A}_u E_u$. □

Corollary 1141. *The collection of analytic sets is closed under countable unions and countable intersections.*

A set is called *coanalytic* if its complement is analytic.

Corollary 1142. *Every Borel set is both analytic and coanalytic.*

Proof. Since every closed interval is analytic and every open interval is a countable union of closed intervals, therefore every open interval is analytic. Since every open set is a countable union of open intervals, therefore every open set is analytic. Thus the collection of all analytic sets contains all open sets and is closed under countable unions and countable intersections. Hence by Proposition 1124 the collection of analytic sets contains all Borel sets.

Now the complement of any Borel set is still Borel, and hence analytic. Therefore every Borel set is coanalytic as well. □

In the next section, we will prove the converse of the above result.

Problem 1143. *There are exactly* $\mathfrak{c} = 2^{\aleph_0}$ *analytic sets. Hence each of the following families of sets has cardinality* $\mathfrak{c} = 2^{\aleph_0}$*: The coanalytic sets, the Borel sets, the* F_σ *sets, the* G_δ *sets, the closed sets, and the open sets.*

Problem 1144. *There is a set which is both meager and of measure zero but neither analytic nor coanalytic. Conclude that there measurable sets and sets with Baire property which are not Borel, i.e.,* $\mathbf{B} \subsetneq \mathbf{L}$ *and* $\mathbf{B} \subsetneq \mathbf{Y}$.

Problem 1145. *Let us say that a Suslin system* $\langle F_u \mid u \in \mathbf{N}^* \rangle$ *is* finitely branching *if the tree* $\{u \in \mathbf{N}^* \mid F_u \neq \emptyset\}$ *is finitely branching, and that it is* disjointed *if*

$F_{u^\frown m} \cap F_{u^\frown m} = \emptyset$ whenever $m \neq n$. Show that if $\langle F_u \mid u \in \mathbf{N}^* \rangle$ is either a finitely branching or a disjointed Suslin system, then

$$\mathsf{A}_u F_u = \bigcap_{n=1}^{\infty} \bigcup_{\operatorname{len}(u)=n} F_u,$$

and hence the set $\mathsf{A}_u F_u$ must be an $\mathsf{F}_{\sigma\delta}$ set.

[Hint: For the finitely branching case, use König's Infinity Lemma.]

Problem 1146. *Show that if A is analytic and $f : \mathbf{R} \to \mathbf{R}$ is continuous, then $f^{-1}[A]$ and $f[A]$ are analytic sets.*

[Hint: Problem 983 may help.]

Remark. There are many characterizations of analytic sets that we will not cover. For example, it can be shown that *A is analytic if and only if $A = f[B]$ for some continuous f and Borel B* (in fact B can be taken to be a G_δ). Earlier we mentioned that $\operatorname{ran}(f')$ may not be Borel for a continuous f. A result of Poprougenko says that A is analytic $\Leftrightarrow A = \operatorname{ran}(f')$ for some continuous f. See [38, 45, 46, 55, 64] for many other characterizations.

18.3 The Lusin Separation Theorem

Definition 1147. To each Suslin system $\langle F_u \mid u \in \mathbf{N}^* \rangle$, we associate a family $\langle F_u^{(*)} \mid u \in \mathbf{N}^* \rangle$ of sets indexed by $\mathbf{N}^*$ by setting, for each $v \in \mathbf{N}^*$:

$$F_v^{(*)} := \{x \in \mathbf{R} \mid \text{There is } y \in \mathbf{N}^{\mathbf{N}} \text{ extending } v \text{ such that } x \in \cap_n F_{y|n}\}.$$

Thus $x \in F_{u_1 u_2 \cdots u_m}^{(*)}$ if and only if there is $y_1 y_2 \cdots y_n \cdots \in \mathbf{N}^{\mathbf{N}}$ such that

$$u_k = y_k \text{ for all } k \leq m \quad \text{and} \quad x \in \bigcap_{n=0}^{\infty} F_{y_1 y_2 \cdots y_n}.$$

Note that we have $F_\varepsilon^{(*)} = \mathsf{A}_u F_u$, and writing $v * u$ for the concatenation of v and u, we can express $F_v^{(*)}$ as $F_v^{(*)} = \mathsf{A}_u F_{v*u}$.

Problem 1148. *If $\langle F_u \mid u \in \mathbf{N}^* \rangle$ is a Suslin system and $v \in \mathbf{N}^*$, then:*

1. *$F_v^{(*)}$ is analytic.*
2. $F_v^{(*)} = \bigcup_{n=1}^{\infty} F_{v^\frown n}^{(*)}$.

Definition 1149. We say that C *separates* A *from* B if $A \subseteq C$ and $C \cap B = \emptyset$. The sets A and B are called *Borel separable* if there exists a Borel set C separating A from B.

Proposition 1150. *If* $A = \cup_m A_m$, $B = \cup_n B_n$, *and* A_m *and* B_n *are Borel separable for all* $m, n \in \mathbf{N}$, *then* A *and* B *are Borel separable.*

Proof. For each m, n fix a Borel set $C_{m,n}$ separating A_m from B_n. Put

$$D_m := \bigcap_n C_{m,n} \quad \text{(for each } m\text{)}, \quad \text{and} \quad E := \bigcup_m D_m.$$

Then the set E is Borel. Now note that for each m we have $A_m \subseteq D_m$ and $D_m \cap B = \emptyset$. Hence $A \subseteq E$ and $E \cap B = \emptyset$, so E separates A from B. □

The following result is called the *Lusin Separation Theorem.*

Theorem 1151 (Lusin). *If* A *and* B *are analytic sets which are not Borel separable, then* $A \cap B \neq \emptyset$. *Hence disjoint analytic sets are Borel separable.*

Proof. Suppose that $A = \mathsf{A}_u E_u$ and $B = \mathsf{A}_v F_v$ are analytic sets which are not Borel separable, where $\langle E_u \mid u \in \mathbf{N}^* \rangle$ and $\langle F_v \mid v \in \mathbf{N}^* \rangle$ are Suslin systems. Since

$$A = E_\varepsilon^{(*)} = \bigcup_m E_{\langle m \rangle}^{(*)} \quad \text{and} \quad B = F_\varepsilon^{(*)} = \bigcup_n F_{\langle n \rangle}^{(*)},$$

by the last proposition there must exist m_1 and n_1 such that $E_{\langle m_1 \rangle}^{(*)}$ and $F_{\langle n_1 \rangle}^{(*)}$ are not Borel separable. Again, since

$$E_{\langle m_1 \rangle}^{(*)} = \bigcup_m E_{\langle m_1, m \rangle}^{(*)} \quad \text{and} \quad F_{\langle n_1 \rangle}^{(*)} = \bigcup_n F_{\langle n_1, n \rangle}^{(*)},$$

there are m_2 and n_2 such that $E_{\langle m_1, m_2 \rangle}^{(*)}$ and $F_{\langle n_1, n_2 \rangle}^{(*)}$ are not Borel separable.

Continuing the process, we get two infinite sequences $\langle m_1, m_2, \ldots, m_k, \ldots \rangle$ and $\langle n_1, n_2, \ldots, n_k, \ldots \rangle$ such that for every k,

$$E_{\langle m_1, m_2, \ldots, m_k \rangle}^{(*)} \text{ and } F_{\langle n_1, n_2, \ldots, n_k \rangle}^{(*)} \text{ are not Borel separable.}$$

But if the above two sets are not Borel separable, then for every k, the closed intervals $I_k := E_{\langle m_1, m_2, \ldots, m_k \rangle}$ and $J_k := F_{\langle n_1, n_2, \ldots, n_k \rangle}$ cannot be disjoint, since if we had $I_k \cap J_k = \emptyset$ then the Borel set I_k would separate $E_{\langle m_1, m_2, \ldots, m_k \rangle}^{(*)}$ from $F_{\langle n_1, n_2, \ldots, n_k \rangle}^{(*)}$. Thus I_k and J_k are nested sequences of closed intervals with $I_k \cap J_k \neq \emptyset$ and $\operatorname{len}(I_k) \to 0$ and $\operatorname{len}(J_k) \to 0$ as $n \to \infty$, and so

$$\bigcap_k I_k = \{p\} \quad \text{and} \quad \bigcap_k J_k = \{q\}$$

with $p \in A$ and $q \in B$. Now if we had $p \neq q$, we could choose k sufficiently large so that $\mathrm{len}(I_k), \mathrm{len}(J_k) < |p-q|/2$, implying $I_k \cap J_k = \emptyset$, which is a contradiction. Hence $p = q$ and so $A \cap B \neq \emptyset$. □

Corollary 1152 (Suslin's Theorem). *A set is Borel if and only if it is both analytic and coanalytic.*

In Sect. 18.6 we will explicitly define a non-Borel analytic set.

18.4 Measurability and Baire Property of Analytic Sets

Definition 1153. An *Ulam matrix* is a family of sets $\langle E^n_\alpha \mid n \in \mathbf{N},\ \alpha < \omega_1 \rangle$ with $\aleph_0$ rows and $\aleph_1$ columns as shown below such that:

1. The sets in each row are pairwise disjoint: $E^n_\alpha \cap E^n_\beta = \emptyset$ for $\alpha \neq \beta$.
2. The union of the sets in each column contains any set to its right in the first row: $\bigcup_{n\in\mathrm{N}} E^n_\alpha \supseteq E^1_\beta$ for any $\beta > \alpha$.

An Ulam Matrix:

$E^n_\alpha,\ n \in \mathbf{N},\ \alpha < \omega_1$

$$
\begin{array}{c|cccccc}
 & \multicolumn{6}{c}{\xrightarrow{\quad\alpha\quad}} \\
 & E^1_0 & E^1_1 & \cdots & E^1_\alpha & \cdots\cdots \\
 & E^2_0 & E^2_1 & \cdots & E^2_\alpha & \cdots\cdots \\
n & \vdots & \vdots & & \vdots & \\
 & E^n_0 & E^n_1 & \cdots & E^n_\alpha & \cdots\cdots \\
\downarrow & \vdots & \vdots & & \vdots &
\end{array}
$$

Note. The ordering of the rows really does not matter, so the rows could be indexed by any countable set instead of $\mathbf{N}$ so long as one fixed row is designated as "the first row."

To prove that analytic sets are Lebesgue measurable and have the Baire property, we will use the following result, which is also of considerable interest in itself as it provides coanalytic sets with "nice ordinal ranks."

Theorem 1154 (Ulam Matrix Decomposition of Coanalytic Sets). *For any coanalytic set C there is an Ulam matrix of Borel sets whose first row has union C.*

More specifically, there is a family of Borel sets $\langle C^u_\alpha \mid u \in \mathbf{N}^*, \alpha < \omega_1 \rangle$ satisfying the following, where we use the abbreviation $C_\alpha := C^\varepsilon_\alpha$:

1. *For each $u \in \mathbf{N}^*$ and all ordinals $\alpha, \beta < \omega_1$, $C^u_\alpha \cap C^u_\beta = \emptyset$ if $\alpha \neq \beta$.*
2. *If $\beta > \alpha$ and $x \in C_\beta$ then $x \in C^u_\alpha$ for some $u \in \mathbf{N}^*$.*
3. *$C = \bigcup_{\alpha<\omega_1} C_\alpha$, where by (1) the sets $C_\alpha, \alpha < \omega_1$, are pairwise disjoint.*

Proof. Let $\langle F_u \mid u \in \mathbf{N}^* \rangle$ be a Suslin system with $\mathbf{R} \setminus C = \mathcal{A}_u F_u$. For each $x \in \mathbf{R}$ and $u \in \mathbf{N}^*$, let

$$T_x := \{v \in \mathbf{N}^* \mid x \in F_v\}, \qquad \text{and} \qquad T_x^{(u)} := \{v \in \mathbf{N}^* \mid x \in F_{u*v}\},$$

From the definition of a Suslin system, T_x and $T_x^{(u)}$ are trees over $\mathbf{N}$ with $T_x = T_x^{(\varepsilon)}$. Also, we have $x \in \mathbf{R} \smallsetminus C \Leftrightarrow x \in \mathcal{A}_u F_u \Leftrightarrow$ there is an infinite branch through $T_x \Leftrightarrow$ T_x is not well-founded. Thus we have:

$$\text{For all } x \in \mathbf{R}, \quad x \in C \Leftrightarrow T_x \text{ is well-founded.}$$

By Problem 854 (Fact 2 on page 325), it follows that:

$$\text{If } x \in C, \text{ then } T_x^{(u)} \text{ is well-founded and } \operatorname{rank}(T_x^{(u)}) \leq \operatorname{rank}(T_x).$$

Now define the set C_α^u, for each $u \in \mathbf{N}^*$ and ordinal $\alpha < \omega_1$, by the condition:

$$x \in C_\alpha^u \Leftrightarrow T_x^{(u)} \text{ is well-founded of rank } \alpha \quad (x \in \mathbf{R}),$$

and put, as in the statement of the theorem, $C_\alpha := C_\alpha^\varepsilon$.

Condition (1) of the theorem is now immediate.

Condition (2) follows from Problem 854 (Fact 4, page 325).

Now, note that $x \in C_\alpha \Leftrightarrow T_x$ is well-founded of rank α, so $x \in \bigcup_{\alpha<\omega_1} C_\alpha \Leftrightarrow T_x$ is well-founded, hence condition (3) of the theorem follows.

It remains to show that C_α^u is Borel. We prove this by transfinite induction on α for the statement "C_α^u is Borel for all $u \in \mathbf{N}^*$." Recall that the only trees of rank 0 are the empty tree and the singleton tree $\{\varepsilon\}$ consisting of the root node ε alone, i.e., T has rank $0 \Leftrightarrow T \subseteq \{\varepsilon\}$.

For $\alpha = 0$, note that $x \in C_0^u \Leftrightarrow T_x^{(u)}$ has rank $0 \Leftrightarrow T_x^{(u)} \subseteq \{\varepsilon\} \Leftrightarrow x \notin F_u$ or $x \in F_u \smallsetminus \bigcup_n F_{u\frown n}$, so $C_0^u = (\mathbf{R} \smallsetminus F_u) \cup \left(F_u \smallsetminus \bigcup_n F_{u\frown n}\right)$, which is Borel.

For $\alpha > 0$, assume that for every $\xi < \alpha$, C_ξ^u is Borel for all $u \in \mathbf{N}^*$ (induction hypothesis). Then, by Problem 854 again (Fact 4, page 325):

$$\begin{aligned} x \in C_\alpha^u \Leftrightarrow T_x^{(u)} &\text{ has rank } \alpha \\ &\Leftrightarrow \forall n \in \mathbf{N}, \operatorname{rank}(T_x^{(u\frown n)}) < \alpha, \text{ and } \forall \xi < \alpha\, \exists v \in \mathbf{N}^* \operatorname{rank}(T_x^{(u*v)}) = \xi \\ &\Leftrightarrow \forall n \in \mathbf{N}\, \exists \xi < \alpha \operatorname{rank}(T_x^{(u\frown n)}) = \xi, \text{ and} \\ &\qquad\qquad \forall \xi < \alpha\, \exists v \in \mathbf{N}^* \operatorname{rank}(T_x^{(u*v)}) = \xi. \end{aligned}$$

Writing the above in terms of set unions and intersections:

$$C_\alpha^u = \Big[\bigcap_{n\in\mathbf{N}} \bigcup_{\xi<\alpha} C_\xi^{u\frown n}\Big] \cap \Big[\bigcap_{\xi<\alpha} \bigcup_{v\in\mathbf{N}^*} C_\xi^{u*v}\Big],$$

which by induction hypothesis is Borel since all the unions and intersections involved are countable unions and intersections. □

Corollary 1155. *Every coanalytic set is the union $\aleph_1$-many Borel sets.*

Definition 1156. A sigma algebra S containing a σ-ideal Z is said to be *CCC modulo* Z if for any family $\langle A_i \mid i \in I \rangle$ of pairwise disjoint sets in S, we have $A_i \in$ Z for all but countably many $i \in I$.

Natural examples of sigma algebras which are CCC (modulo a σ-ideal) are:

1. The sigma algebra **L** of Lebesgue measurable sets is CCC modulo the σ-ideal of measure zero sets (Corollary 1029.6).
2. The sigma algebra **Y** of sets with Baire property is CCC modulo the σ-ideal of meager sets (Corollary 1062).

Theorem 1157. *Let* S *be a sigma algebra containing all Borel sets and* Z *be a σ-ideal contained in* S *such that* S *is CCC modulo* Z*. Then every coanalytic set (and so every analytic set) is in* S.

Proof. Let C be a coanalytic set. By Theorem 1154, fix an "Ulam matrix" of Borel sets $\langle C^u_\alpha \mid u \in \mathbf{N}^*, \alpha < \omega_1 \rangle$ such that, using the abbreviation $C_\alpha := C^\varepsilon_\alpha$,

1. For each $u \in \mathbf{N}^*$ and all ordinals $\alpha, \beta < \omega_1$, $C^u_\alpha \cap C^u_\beta = \emptyset$ if $\alpha \neq \beta$.
2. If $\beta > \alpha$ and $x \in C_\beta$ then $x \in C^u_\alpha$ for some $u \in \mathbf{N}^*$.
3. $C = \bigcup_{\alpha<\omega_1} C_\alpha$.

Since each C^u_α is Borel, so $C^u_\alpha \in$ S. For each $u \in \mathbf{N}^*$, the family $\langle C^u_\alpha \mid \alpha < \omega_1 \rangle$ is pairwise disjoint by condition (1), and since S is CCC modulo Z, so there exists $\alpha_u < \omega_1$ such that $C^u_\beta \in$ Z for all $\beta \geq \alpha_u$. Since there are only countably many $u \in \mathbf{N}^*$, we can fix some $\bar{\alpha} < \omega_1$ with $\bar{\alpha} > \alpha_u$ for all $u \in \mathbf{N}^*$. Then $C^u_{\bar{\alpha}} \in$ Z *for all* $u \in \mathbf{N}^*$, and so the countable union $\bigcup_{u\in\mathbf{N}^*} C^u_{\bar{\alpha}}$ is in Z. By (2), $\bigcup_{u\in\mathbf{N}^*} C^u_{\bar{\alpha}} \supseteq \bigcup_{\beta>\bar{\alpha}} C_\beta$, so $\bigcup_{\beta>\bar{\alpha}} C_\beta$ is in Z $\subseteq$ S. Also, $\bigcup_{\beta\leq\bar{\alpha}} C_\beta$ is Borel and so is in S. Hence $C = \left(\bigcup_{\beta\leq\bar{\alpha}} C_\beta\right) \cup \left(\bigcup_{\beta>\bar{\alpha}} C_\beta\right)$ is in S. □

Corollary 1158. *All analytic sets and all coanalytic sets are Lebesgue measurable and have Baire property.*

Corollary 1159. *No Vitali or Bernstein set is analytic or coanalytic.*

Remark. Ulam matrices are useful in showing that certain uncountable unions of measure zero sets can still have measure zero. Ulam used them to show that no nontrivial measure can be defined on a set whose cardinality is a successor cardinal (like $\aleph_1$, $\aleph_2$, etc). Ulam's proof is given in Theorem 1196, and it may be instructive to compare it with the above proof.

18.5 The Perfect Set Property for Analytic Sets

Theorem 1160. *Every uncountable analytic set contains a perfect set and hence has cardinality* $\mathfrak{c}$.

Proof. The proof of the theorem will be a variant of the proof of the corresponding theorem for dense-in-itself G_δ sets, but note that we cannot directly copy that proof since a dense-in-itself analytic set may be countable. Let

$$A = \mathsf{A}_u F_u$$

be an uncountable analytic set, where $\langle F_u \mid u \in \mathbf{N}^* \rangle$ is a Suslin system. The heart of the proof of the theorem is in the following lemma.

Lemma 1161. *For each $u \in \mathbf{N}^*$ and $\delta > 0$, if $F_u^{(*)}$ is uncountable then there exist v and w in $\mathbf{N}^*$ extending u such that F_v and F_w are disjoint nonempty closed intervals of length $< \delta$ with both $F_v^{(*)}$ and $F_w^{(*)}$ uncountable.*

Proof (of Lemma). Since $F_u^{(*)}$ is uncountable, and by Theorem 970 all but countably points of $F_u^{(*)}$ are condensation points, we can pick two distinct condensation points $p < q$ in $F_u^{(*)}$. Let $r = \min\left(\frac{1}{4}(q-p), \delta\right)$, and put

$$L := F_u^{(*)} \cap (p - r, p + r) \quad \text{and} \quad U := F_u^{(*)} \cap (q - r, q + r)$$

so that L and U are disjoint uncountable subsets of $F_u^{(*)}$. Finally, put

$$S := \{v \mid v \text{ extends } u,\ \operatorname{len}(F(v)) < r,\ F(v) \cap L \neq \emptyset\}, \qquad \text{and}$$
$$T := \{w \mid w \text{ extends } u,\ \operatorname{len}(F(w)) < r,\ F(w) \cap U \neq \emptyset\}.$$

We claim that

$$L \subseteq \bigcup_{v \in S} F_v^{(*)}.$$

To see this, let $x \in L$ and pick $z \in \mathbf{N}^{\mathbf{N}}$ extending u such that $x \in \bigcap_m F(z|m)$. Fix a large enough m to make $\operatorname{len}(F(z|m)) < r$ and $m > \operatorname{len}(u)$, and put $v = z|m$. Then $x \in F(v) \cap L$, $\operatorname{len}(F(v)) < r$, and v extends u, so $v \in S$, and hence $x \in F_v^{(*)}$. Thus the claim is established.

Since L is uncountable, it follows that $F_v^{(*)}$ is uncountable for some $v \in S$.

Similarly, $F_w^{(*)}$ is uncountable for some $w \in T$.

For such v and w, $F(v)$ and $F(r)$ are closed intervals of length $< r$. But since $\inf U - \sup L \geq 2r$ and $F(v) \cap L$ and $F(w) \cap U$ are both nonempty, it follows that $F(v)$ and $F(w)$ must be disjoint. □

Note that the v and w of the lemma must be proper extensions of u.

To continue with the proof of the theorem, we repeatedly apply the lemma to build a Cantor system by associating with each binary string $u \in \{0,1\}^*$ a string $t(u) \in \mathbf{N}^*$ such that for all $u, u' \in \{0,1\}^*$:

1. $F^{(*)}_{t(u)}$ is uncountable and $\text{len}(F_{t(u)}) < \frac{1}{\text{len}(u)+1}$.
2. If u' properly extends u then $t(u')$ properly extends $t(u)$.
3. $F_{t(u^\frown 0)} \cap F_{t(u^\frown 1)} = \emptyset$.

To do this, note that since $A = F^{(*)}_{\varepsilon}$ is uncountable we can use the lemma to choose v_0 with $\text{len}(F_{v_0}) < 1$ and $F^{(*)}_{v_0}$ uncountable, and define $t(\varepsilon) := v_0$. Then, having defined $t(u)$ for $u \in \{0,1\}^*$ with $F^{(*)}_{t(u)}$ uncountable, we can use the lemma to choose v and w properly extending $t(u)$ such that F_v and F_w are disjoint intervals of length $< 1/(\text{len}(u)+2)$ with both $F^{(*)}_v$ and $F^{(*)}_w$ uncountable, and define $t(u^\frown 0) := v$ and $t(u^\frown 1) := w$.

This makes the family $\langle F_{t(u)} \mid u \in \{0,1\}^* \rangle$ a Cantor system. Hence if $\varphi(z)$ denotes the unique member of $\bigcap_n F_{t(z|n)}$, then $\varphi\colon \{0,1\}^{\mathbf{N}} \to \mathbf{R}$ is an injective mapping whose range $\text{ran}(\varphi)$ is a generalized Cantor set. Also if $z \in \{0,1\}^{\mathbf{N}}$, then $t(z|1), t(z|2), t(z|3), \ldots$ form an infinite sequence of members of $\mathbf{N}^*$ where each term properly extends all the preceding ones, and so they define a unique $y \in \mathbf{N}^{\mathbf{N}}$ such that $y|n$ is a prefix of $t(z|n)$ for each n. Hence $\varphi(z) \in \bigcap_n F_{t(z|n)} \subseteq \bigcap_n F_{y|n} \subseteq A$. Thus $\text{ran}(\varphi) \subseteq A$, and therefore A contains the generalized Cantor set $\text{ran}(\varphi)$ (which is perfect and of cardinality $\mathfrak{c}$). □

This theorem was the final achievement in the classical program of showing that a class of effectively defined sets has the perfect set property and therefore the Continuum Hypothesis holds when restricted to that class of sets. There were early attempts to incrementally extend such restricted forms of CH to larger and larger collections of sets of reals—by showing that the collection in question has the perfect set property. Of course, Bernstein showed that there are sets which do not have the perfect set property, but such sets are not effectively defined, and so one could still hope for larger collections of *effectively defined* sets to possess the perfect set property.

The earliest major result along this line was the Cantor–Bendixson theorem (Corollary 973): The class of closed sets has the perfect set property. Alexandrov and Hausdorff extended the result to the class of Borel sets.

The last theorem says that the collection of analytic sets (which includes the Borel sets) has the perfect set property. This is essentially the best result that can be proved using the usual axioms of set theory, since without additional set-theoretic assumptions it cannot be proved that the coanalytic sets have the perfect set property. The classical program of extending the perfect set property for effectively defined sets thus came to a stall, and mathematicians such as Lusin realized that a limit has been reached.

Regularity Properties of Analytic Sets

The perfect set property is a *regularity property* of a set. Combining Corollary 1158 and Theorem 1160, we get the following classical result.

Theorem 1162 (Regularity Properties of Analytic Sets). *Every analytic set is measurable, has the Baire property, and the perfect set property.*

18.6 A Non-Borel Analytic Set

Coding Subsets of N* by Elements of the Cantor Set

We first need a special effective one-to-one enumeration of the set $\mathbf{N}^*$ of all finite strings of natural numbers.

Proposition 1163. *There is an enumeration of* $\mathbf{N}^*$ *without repetitions*

$$\mathbf{N}^* = \{\mathbf{u}_1, \mathbf{u}_2, \ldots, \mathbf{u}_n, \mathbf{u}_{n+1}, \ldots\},$$

such that if $\mathbf{u}_m$ *is a proper initial segment of* $\mathbf{u}_n$ *then* $m < n$.

Proof. Let p_n denote the n-th prime, so that $p_1 = 2$, $p_2 = 3$, etc.

First take $\mathbf{u}_1$ to be the empty sequence, i.e., put $\mathbf{u}_1 := \varepsilon$.

Next, for each $n > 1$ let k be the largest integer such that $p_k \mid n$. Then n can be written as

$$n = p_1^{n_1 - 1} p_2^{n_2 - 1} \cdots p_{k-1}^{n_{k-1}-1} p_k^{n_k}$$

for a unique sequence of k positive integers $n_1, n_2, \ldots, n_k \in \mathbf{N}$. Now define $\mathbf{u}_n := n_1 n_2 \cdots n_k$.

It is now readily verified that the strings $\mathbf{u}_1, \mathbf{u}_2, \ldots$ form a one-to-one enumeration of $\mathbf{N}^*$ and satisfy the condition of the proposition. □

We now fix, once and for all, an enumeration $\mathbf{u}_1, \mathbf{u}_2, \ldots, \mathbf{u}_n, \ldots$ of $\mathbf{N}^*$ as in the above proposition:

Definition 1164. $\mathbf{u}_1, \mathbf{u}_2, \ldots, \mathbf{u}_n, \ldots$ will denote the specific sequence of strings of Proposition 1163 which enumerate $\mathbf{N}^*$ without repetitions and satisfy the condition: If $\mathbf{u}_m$ is a proper initial segment of $\mathbf{u}_n$, then $m < n$.

As usual (review Sect. 6.6), we identify each point x in the Cantor set $\mathbf{K} \subseteq \mathbf{R}$ with the (unique) infinite binary sequences $\langle x_n \rangle$ in $\{0, 1\}^{\mathbf{N}}$ such that $x = \sum_{n=1}^{\infty} \frac{2x_n}{3^n}$. Conversely, any $\langle x_n \rangle \in \{0, 1\}^{\mathbf{N}}$ is mapped to $x = \sum_{n=1}^{\infty} \frac{2x_n}{3^n}$. For $x \in \mathbf{K}$ and $\langle x_n \rangle \in \{0, 1\}^{\mathbf{N}}$, we will use the notation "$x \sim \langle x_n \rangle$" to express this identification, i.e., as an abbreviation for "$x = \sum_{n=1}^{\infty} \frac{2x_n}{3^n}$".

Definition 1165. For $\langle x_n \rangle \in \{0, 1\}^{\mathbf{N}}$ and $x \in \mathbf{K}$, we write

$$x \sim \langle x_n \rangle \Leftrightarrow x = \sum_{n=1}^{\infty} \frac{2x_n}{3^n} \qquad (x \in \mathbf{K},\ \langle x_n \rangle \in \{0,1\}^{\mathbf{N}}).$$

Definition 1166 (Coding subsets of $\mathbf{N}^*$ by points of the Cantor set). For each x in the Cantor set $\mathbf{K}$ with $x \sim \langle x_n \rangle \in \{0,1\}^{\mathbf{N}}$, define the subset $\mathrm{U}(x) \subseteq \mathbf{N}^*$ by:

$$\mathrm{U}(x) := \{\mathbf{u}_n \mid x_n = 1\}.$$

We say that $\mathrm{U}(x)$ *is the subset of* $\mathbf{N}^*$ *coded by* x. More generally, if x is a member of the Cantor set $\mathbf{K}$ and $A \subseteq \mathbf{N}^*$ then we say that x *codes* A (or x *is a code for* A) if $A = \mathrm{U}(x)$.

So for $x \in \mathbf{K}$ and $A \subseteq \mathbf{N}^*$, we have: x codes $A \Leftrightarrow \mathrm{U}(x) = A$, and thus we have a natural bijection between the Cantor set $\mathbf{K}$ and $\mathrm{P}(\mathbf{N}^*)$, the power set of $\mathbf{N}^*$, via the one-to-one correspondence:

$$x \longleftrightarrow \mathrm{U}(x) \qquad (x \in \mathbf{K},\quad \mathrm{U}(x) \subseteq \mathbf{N}^*).$$

Now consider the set of $x \in \mathbf{K}$ for which $\mathrm{U}(x) \subseteq \mathbf{N}^*$ is a tree over $\mathbf{N}$:

Problem 1167. *Show that the set $\{x \in \mathbf{K} \mid \mathrm{U}(x)$ is a tree$\}$ is closed, i.e., the set of those members of the Cantor set which code trees is a closed set.*

Problem 1168. *Show that the following subsets of the Cantor set are Borel:*

1. *The set of codes for trees in which every node has a proper extension.*
2. *The set of codes for trees in which every node has at least two immediate extensions.*
3. *The set of codes for well-founded trees having rank < 2.*
4. *The set of codes for well-founded trees of finite rank.*
5. *The set of codes for well-founded trees of rank $< \alpha$, where $\alpha < \omega_1$.*

[Hint: Reviewing the proof of Theorem 1154 may help.]

Definition 1169. $\mathbf{WF}$ denotes the set of codes for well-founded trees, $\mathbf{IF}$ denotes the set of codes for ill-founded (non-well-founded) trees, and $\mathbf{WF}_\alpha$ denotes the set of codes of well-founded trees having rank $< \alpha$.

In other words, if $\mathbf{K}$ denotes the Cantor set then

$$\mathbf{WF} = \{x \in \mathbf{K} \mid \mathrm{U}(x) \text{ is a well-founded tree}\},$$

$$\mathbf{IF} = \{x \in \mathbf{K} \mid \mathrm{U}(x) \text{ is an ill-founded tree}\}, \text{ and}$$

$$\mathbf{WF}_\alpha = \{x \in \mathbf{WF} \mid \mathrm{rank}(\mathrm{U}(x)) < \alpha\}.$$

Recall that every countable well-founded tree has countable rank, so that

$$\mathbf{WF} = \bigcup_{\alpha<\omega_1} \mathbf{WF}_\alpha .$$

Moreover, since for each countable ordinal α there is a countable tree of rank $> \alpha$, it follows that $\mathbf{WF} \smallsetminus \mathbf{WF}_\alpha$ is nonempty for all $\alpha < \omega_1$. Also, we saw in the previous problem that $\mathbf{WF}_\alpha$ is a Borel set for each countable ordinal α.

Definition 1170. We say that a Suslin system $\langle S_u \mid u \in \mathbf{N}^* \rangle$ *dominates* $\mathbf{WF}$ if for all $x \in \mathbf{WF}$, the set $\{u \in \mathbf{N}^* \mid x \in S_u\}$ is a well-founded tree of rank $\geq \mathrm{rank}(\mathbf{U}(x))$. In other words, $\langle S_u \mid u \in \mathbf{N}^* \rangle$ dominates $\mathbf{WF}$ if for any $x \in \mathbf{WF}_{\alpha+1} \smallsetminus \mathbf{WF}_\alpha$, $\{u \in \mathbf{N}^* \mid x \in S_u\}$ is a well-founded tree of rank $\geq \alpha$.

Note that if a Suslin system $\langle S_u \mid u \in \mathbf{N}^* \rangle$ dominates $\mathbf{WF}$, then the analytic set $\mathsf{A}_u S_u$ generated by it must be a *subset of* $\mathbf{IF}$ (where $\mathbf{IF}$ = the set of codes for ill-founded trees). We now show that $\mathsf{A}_u S_u$ can actually be *equal to* $\mathbf{IF}$ for a suitably chosen Suslin system $\langle S_u \mid u \in \mathbf{N}^* \rangle$ dominating $\mathbf{WF}$, i.e., we can have $\mathbf{IF} = \mathsf{A}_u S_u$ for some $\langle S_u \mid u \in \mathbf{N}^* \rangle$ which dominates $\mathbf{WF}$.

Proposition 1171. *The set* $\mathbf{IF}$ *of codes for ill-founded trees is an analytic set. Moreover, there is a Suslin system* $\langle S_u \mid u \in \mathbf{N}^* \rangle$ *which dominates* $\mathbf{WF}$ *and generates* $\mathbf{IF}$ *(so that* $\mathbf{IF} = \mathsf{A}_u S_u$*).*

Proof. To simplify notation, a finite sequence $\langle n_1, n_2, \ldots, n_k \rangle$ will be denoted simply by the string $n_1 n_2 \cdots n_k$. Note the mapping

$$\langle i, j \rangle \;\mapsto\; 2j - i$$

is a natural effective bijection from $\{0,1\} \times \mathbf{N}$ onto $\mathbf{N}$, and its inverse is the mapping

$$n \;\mapsto\; \langle \bar{n}, \hat{n} \rangle ,$$

where we are writing, for any $n \in \mathbf{N}$:

$$\hat{n} := \lfloor (n+1)/2 \rfloor, \qquad \text{and} \qquad \bar{n} := 2\hat{n} - n .$$

The above bijection also induces other bijections. For example, the mapping

$$n_1 n_2 \cdots n_k \;\mapsto\; \langle \bar{n}_1 \bar{n}_2 \cdots \bar{n}_k,\; \hat{n}_1 \hat{n}_2 \cdots \hat{n}_k \rangle$$

is an effective bijection from $\mathbf{N}^*$ onto $\{\langle u, v \rangle \in \{0,1\}^* \times \mathbf{N}^* \mid \mathrm{len}(u) = \mathrm{len}(v)\}$. Similarly, it also induces a natural effective bijection between $\mathbf{N}^{\mathbf{N}}$ and $\{0,1\}^{\mathbf{N}} \times \mathbf{N}^{\mathbf{N}}$. In this last bijection, an element of $\langle y_n \rangle \in \mathbf{N}^{\mathbf{N}}$ can be thought to be coding the pair of sequences $\langle \langle \bar{y}_n \rangle, \langle \hat{y}_n \rangle \rangle$, where the "left sequence" $\langle \bar{y}_n \rangle \in \{0,1\}^{\mathbf{N}}$ is an infinite binary sequence which can be used to code a tree, while the "right sequence" $\langle \hat{y}_n \rangle \in \mathbf{N}^{\mathbf{N}}$ can be used code an infinite branch through that tree.

Let $\langle I[u] \mid u \in \{0,1\}^* \rangle$ be the classical Cantor system that was defined in Sect. 6.6 as:

$$I[\varepsilon] := [0,1], \quad \text{and for all } u \in \{0,1\}^*: \quad \begin{cases} I[u^\frown 0] = \text{left-third of } I[u], \\ I[u^\frown 1] = \text{right-third of } I[u]. \end{cases}$$

Thus for a real x in the Cantor set $\mathbf{K}$ with $x \sim \langle x_n \rangle$, we have $x \in I[n_1 n_2 \cdots n_k]$ if and only if $x_j = n_j$ for $j = 1, 2, \ldots, k$.

Now define the required Suslin system $S = \langle S_u \mid u \in \mathbf{N}^* \rangle$ as follows:

$$S_{n_1 n_2 \cdots n_k} = \begin{cases} \emptyset & \text{if } \exists i, j \leq k \; (\mathbf{u}_i \subseteq \mathbf{u}_j \wedge \bar{n}_i = 0 \wedge \bar{n}_j = 1), \\ \emptyset & \text{if } \exists i, j \leq k \; (\mathbf{u}_i = \hat{n}_1 \hat{n}_2 \cdots \hat{n}_j \wedge \bar{n}_i = 0), \\ I[\bar{n}_1 \bar{n}_2 \cdots \bar{n}_k] & \text{otherwise}, \end{cases}$$

where $\mathbf{u}_n$, $n = 1, 2, 3, \ldots$, is the enumeration of $\mathbf{N}^*$ as in Definition 1164.

Note that if $\langle n_k \rangle_{k \in \mathbf{N}} \in \mathbf{N}^{\mathbf{N}}$ and $x \in \bigcap_k S_{n_1 n_2 \cdots n_k}$, then x is a member of the Cantor set with $x \sim \langle \bar{n}_k \rangle_{k \in \mathbf{N}}$, so x codes a tree (by the first clause of the definition of S), and $\hat{n}_1 \hat{n}_2 \cdots \hat{n}_k \cdots$ is an infinite branch through the tree $\mathbf{U}(x)$ coded by x (by the second clause). Conversely, if $x \sim \langle x_k \rangle_{k \in \mathbf{N}}$ is in $\mathbf{IF}$ then the set $\mathbf{U}(x)$ coded by x is a tree containing some infinite branch, say $m_1 m_2 \cdots m_k \cdots$, which implies $x \in \bigcap_k S_{n_1 n_2 \cdots n_k}$ where $n_k := 2m_k - x_k$. It follows that $\mathbf{IF} = \mathsf{A}_u S_u$.

Next, suppose that $x \sim \langle x_n \rangle$ codes a well-founded tree, i.e., $x \in \mathbf{WF}$. For each $u = m_1 m_2 \cdots m_k \in \mathbf{U}(x)$, define $f(u) := n_1 n_2 \cdots n_k$ where

$$n_j = 2m_j - x_j, \quad j = 1, 2, \ldots, k.$$

For each $u \in \mathbf{U}(x)$, we have $x \in S_{f(u)}$, and so f is a function from $\mathbf{U}(x)$ to $\{v \mid x \in S_v\}$, i.e., $f\colon \mathbf{U}(x) \to \{v \mid x \in S_v\}$. Also f is strictly increasing (i.e., if u' is a proper extension of u then $f(u')$ is a proper extension of $f(u)$), and so by Problem 819 the rank of $\mathbf{U}(x)$ is at most the rank of $\{v \mid x \in S_v\}$. □

Corollary 1172. **WF** *is a coanalytic set.*

The Boundedness Theorem

Theorem 1173. *If $B \subseteq \mathbf{WF}$ is analytic then $B \subseteq \mathbf{WF}_\alpha$ for some $\alpha < \omega_1$.*

Proof. Let $B \subseteq \mathbf{WF}$ be analytic, and let $\langle B_u \mid u \in \mathbf{N}^* \rangle$ be a Suslin system with $B = \mathsf{A}_u B_u$. Let $\langle S_u \mid u \in \mathbf{N}^* \rangle$ be a Suslin system as in above proposition: That is, $\langle S_u \mid u \in \mathbf{N}^* \rangle$ generates $\mathbf{IF}$ and dominates $\mathbf{WF}$, so that for each $x \in \mathbf{WF}$, the rank of $\mathbf{U}(x)$ is at most that of the well-founded tree $\{u \mid x \in S_u\}$. Define a tree $S \otimes B$ on $(\mathbf{N} \times \mathbf{N})^*$ by the condition that

$$\langle m_1, n_1\rangle\,\langle m_2, n_2\rangle \cdots \langle m_k, n_k\rangle \in S \otimes B \Leftrightarrow S_{m_1 m_2 \cdots m_k} \cap B_{n_1 n_2 \cdots n_k} \neq \emptyset.$$

Then $S \otimes B$ is well-founded since $\mathbf{IF} \cap B = \emptyset$. Let α be the rank of $S \otimes B$. Now for each $x \in B$, we can fix an element $\langle n_k\rangle \in \mathbf{N}^{\mathbf{N}}$ such that $x \in B_{n_1 n_2 \cdots n_k}$ for all $k \in \mathbf{N}$, and then define $f_x: \{u \mid x \in S_u\} \to S \otimes B$ by setting

$$f_x(m_1 m_2 \cdots m_k) = \langle m_1, n_1\rangle\,\langle m_2, n_2\rangle \cdots \langle m_k, n_k\rangle\,.$$

Then f_x is strictly increasing, so by Problem 819, $\operatorname{rank}(\{u \mid x \in S_u\}) \leq \alpha$. But then $\operatorname{rank}(\mathbf{U}(x)) \leq \operatorname{rank}(\{u \mid x \in S_u\}) \leq \alpha$, so $x \in \mathbf{WF}_\alpha$. □

If $\alpha < \omega_1$ then $\mathbf{WF}_\alpha$ must be a proper subset of **WF**, since by Problem 857 there are well-founded trees $T \subseteq \mathbf{N}^*$ with $\operatorname{rank}(T) > \alpha$. Hence no analytic subset of **WF** can equal **WF**, and we have the following immediate corollary.

Corollary 1174. ***WF*** *is not analytic, and hence not Borel. Consequently,* ***IF*** *is an analytic set which is not Borel.*

Notice the effective nature of the proof that **WF** is not analytic: Each Suslin system $\langle B_u \mid u \in \mathbf{N}^*\rangle$ effectively determines the tree $S \otimes B$ of the proof above, which we regard as a subset of $\mathbf{N}^*$ (by identifying $(\mathbf{N} \times \mathbf{N})^*$ with $\mathbf{N}^*$ via any fixed bijection between $\mathbf{N} \times \mathbf{N}$ and $\mathbf{N}$). Moreover, for each tree $T \subseteq \mathbf{N}^*$ let $T^+ := \{\varepsilon\} \cup \{\langle 1, n_1, n_2, \ldots, n_k\rangle \mid \langle n_1, n_2, \ldots, n_k\rangle \in T\}$, so that T^+ is well-founded whenever T is well-founded, with $\operatorname{rank}(T^+) = \operatorname{rank}(T) + 1$. Finally, let $h(\langle B_u \mid u \in \mathbf{N}^*\rangle)$ be the code for the tree $(S \otimes B)^+$. We thus have an effective function h which assigns to each Suslin system $\langle B_u \mid u \in \mathbf{N}^*\rangle$ the real number $h(\langle B_u \mid u \in \mathbf{N}^*\rangle)$ satisfying the following property:

For every Suslin system $\langle B_u \mid u \in \mathbf{N}^*\rangle$,

$$\text{if } \mathsf{A}_u B_u \subseteq \mathbf{WF} \text{ then } h(\langle B_u \mid u \in \mathbf{N}^*\rangle) \in \mathbf{WF} \smallsetminus \mathsf{A}_u B_u.$$

Note also that the set **WF**, being a subset of the Cantor set, has measure zero and is nowhere dense, and therefore is Lebesgue measurable and has Baire property. Thus **WF** *is an explicitly defined example of a Lebesgue measurable set which is not a Borel set.*

Corollary 1174 is a modern version of a result of Lusin. It was originally stated in terms of continued fractions, but can be reformulated as follows.

Theorem 1175 (Lusin). *Let L be the set of all $x \sim \langle x_n\rangle$ in the Cantor set for which there are positive integers $n_1 < n_2 < \cdots < n_k < \cdots$ such that for all $k \in \mathbf{N}$, $x_{n_k} = 1$ and n_k divides n_{k+1}. Then L is analytic but not Borel.*

Problem 1176. *Prove Theorem 1175.*

[Hint: First find an injection $f: \mathbf{N}^* \to \mathbf{N}$ such that u is an initial prefix of $v \Leftrightarrow f(u)$ divides $f(v)$. Problems 996, 1102, and 1123 can then help.]

Problem 1177 (Borel Codes). *Fix an enumeration $\langle I_n \mid n \in \mathbf{N}\rangle$ of all open intervals with rational endpoints. For each well-founded tree $T \subseteq \mathbf{N}^*$, define the set $B(T) \subseteq \mathbf{R}$ by recursion on* $\operatorname{rank}(T)$ *as follows:*

$$B(T) := \begin{cases} \bigcup\{I_n \mid \langle n\rangle \in T,\ n \in \mathbf{N}\} & \text{if } \operatorname{rank}(T) \le 1, \\ \mathbf{R} \smallsetminus \bigcap\{B(T^{(\langle n\rangle)}) \mid n \in \mathbf{N}\} & \text{if } \operatorname{rank}(T) > 1. \end{cases}$$

Then E is Borel if and only if $E = B(T)$ for some well-founded tree $T \subseteq \mathbf{N}^*$.

Problem 1178 ($\mathbf{\Sigma}^0_\alpha$ and $\mathbf{\Pi}^0_\alpha$). *Let $B(T)$ be as above. For each $\alpha < \omega_1$, put:*

$$\mathbf{\Sigma}^0_\alpha := \{B(T) \mid \operatorname{rank}(T) \le \alpha\}, \quad \text{and} \quad \mathbf{\Pi}^0_\alpha := \{\mathbf{R} \smallsetminus E \mid E \in \mathbf{\Sigma}^0_\alpha\}.$$

Then $\mathbf{\Sigma}^0_1$ is the class of open sets, $\mathbf{\Sigma}^0_2$ the F_σ sets, $\mathbf{\Sigma}^0_3$ the $\mathsf{G}_{\delta\sigma}$ sets, and so on. Similarly, $\mathbf{\Pi}^0_1$ = closed, $\mathbf{\Pi}^0_2 = \mathsf{G}_\delta$, $\mathbf{\Pi}^0_3 = \mathsf{F}_{\sigma\delta}$, etc. Furthermore, we have: $\mathsf{B} = \{B(T) \mid T \subseteq \mathbf{N}^*$ is a well-founded tree$\} = \bigcup_{\alpha<\omega_1} \mathbf{\Sigma}^0_\alpha = \bigcup_{\alpha<\omega_1} \mathbf{\Pi}^0_\alpha$.

One also defines $\mathbf{\Sigma}^1_1 :=$ the class of analytic sets, $\mathbf{\Pi}^1_1 :=$ the coanalytic sets, and $\mathbf{\Delta}^1_1 := \mathbf{\Sigma}^1_1 \cap \mathbf{\Pi}^1_1$ = sets which are both analytic and coanalytic. In this notation, Suslin's theorem is expressed by the equation $\mathbf{\Delta}^1_1 = \mathsf{B}$.

Borel and Analytic Sets in More General Spaces*

We have limited ourselves to $\mathbf{R}$, but the concepts of Borel and analytic sets can be readily extended to the higher dimensional spaces $\mathbf{R}^n$. One can then show that *$A \subseteq \mathbf{R}$ is analytic if and only if A is the projection of a Borel set $B \subseteq \mathbf{R}^2$*, i.e., $A = \{x \in \mathbf{R} \mid \langle x, y\rangle \in B$ for some $y\}$ for some Borel B.

In the higher dimensional spaces, one can obtain *universal sets*. For example, there is a *universal open* $G \subseteq \mathbf{R}^2$ such that every open subset $U \subseteq \mathbf{R}$ is a section of G, i.e., $U = \{x \mid \langle x, y\rangle \in G\}$ for some y. Such universal sets are available for every level $\mathbf{\Sigma}^0_\alpha$, $\alpha < \omega_1$, of the Borel hierarchy. A simple Cantor diagonalization then shows that the Borel hierarchy "keeps producing new sets": There are sets in each level which do not belong to any lower level.

Similarly, there are universal analytic sets which easily produce non-Borel analytic sets. Our proof to produce such sets was much harder, but had the benefit of obtaining a highly effective form of the boundedness theorem.

Separable complete metric spaces (*Polish spaces*) provide an even more general and natural setting for studying Borel and analytic sets. Some examples are the Baire space $\mathbf{N}^{\mathbf{N}}$, the Cantor space $\{0, 1\}^{\mathbf{N}}$, and the space $C[0, 1]$ of continuous functions on $[0, 1]$ under uniform convergence. For example, in $C[0, 1]$, the set of everywhere differentiable functions is coanalytic but not Borel (Mazurkiewicz), and the set of functions which satisfy Rolle's theorem is non-Borel analytic (Woodin). More examples occur in various areas such as analysis and topology. See [38, 45, 46, 55, 64].

Chapter 19
Postscript III: Measurability and Projective Sets

Abstract In this postscript, we describe two important classical problems of real analysis that could not be settled using the usual axioms of set theory: (1) The Measure Problem on extending Lebesgue measure to all of $\mathbf{P}(\mathbf{R})$, and (2) Lusin's Problem on properties of PCA sets and projective sets. Ulam's analysis of Problem 1 (Measure Problem) led to large cardinals known as *measurable cardinals,* which, surprisingly enough, was shown by Solovay to have remarkable implications for Problem 2 (Lusin's Problem) as well. The independence results mentioned here illustrate the prophetic nature of Lusin's conviction that the problems of PCA and projective sets are unsolvable. This also sets up the background for Postscript IV which will describe how larger cardinals and determinacy essentially "solve" (!) Lusin's Problem.

19.1 The Measure Problem and Measurable Cardinals

Banach and other Polish mathematicians investigated the question *whether there is an extension of Lebesgue measure defined on all sets of reals.* It is clear that the question remains equivalent if we replace $\mathbf{R}$ by $[0, 1]$, so the problem can be stated as follows.

The Measure Extension Problem (Lebesgue). *Does there exist a countably additive set function* $\mu: \mathbf{P}([0, 1] \to [0, 1]$ *defined on all of* $\mathbf{P}([0, 1])$ *which extends Lebesgue measure?*

Banach and Kuratowski showed that if the Continuum Hypothesis (CH) holds, then the above measure extension problem has a negative answer.

Problem 1179. *Let* $\mu: \mathbf{P}(\mathbf{R}) \to [0, \infty]$ *be set function which is countably additive and such that for any bounded interval* I*, the measure* $\mu(I)$ *equals the length of the interval* I*. Then show that* μ *extends Lebesgue measure.*

A. Dasgupta, *Set Theory: With an Introduction to Real Point Sets*,
DOI 10.1007/978-1-4614-8854-5_19,

We saw that Vitali sets and Bernstein sets cannot be Lebesgue measurable. The crucial property of measure used in the Vitali proof is *translation invariance,* and that in the Bernstein proof is *outer regularity.* Indeed:

Theorem 1180 (Vitali and Bernstein). *Let μ be any countably additive nonnegative set function defined on a sigma algebra $\mathsf{S} \subseteq \mathbf{P}(\mathbf{R})$ containing all intervals and with $\mu(\mathbf{R}) > 0$. Then the following hold.*

1. *If μ is translation invariant and bounded on the intervals ($\mu([a,b]) < \infty$ for all $a < b$), then no Vitali set is in S.*
2. *If $\mu(\{x\}) = 0$ for all x and μ is outer regular (for all $E \in \mathsf{S}$ and $\epsilon > 0$ there is open $G \supseteq E$ with $\mu(G \setminus E) < \epsilon$), then any $E \in \mathsf{S}$ with $\mu(E) > 0$ contains an uncountable closed set, and so no Bernstein set is in S.*

Proof. The proof of (1) is exactly same as the original proof for Lebesgue measure: Let V be a Vitali set, so that $(V + r) \cap (V + s) = \emptyset$ for all rational $r \neq s$, and $\bigcup_{r \in \mathbf{Q}}(V + r) = \mathbf{R}$. If V were μ-measurable (i.e., $V \in \mathsf{S}$) then we have $\mu(V) > 0$, so there are $a < b$ with $\mu(V \cap [a,b]) > 0$. Put $W := V \cap [a,b]$. Then the $\langle W + r \mid r \in \mathbf{Q} \cap [0,1] \rangle$ is a family of pairwise disjoint μ-measurable sets all having constant measure $\mu(W) > 0$ and all contained in $[a, b+1]$, which is impossible since $\mu([a, b+1]) < \infty$.

For (2), let E be μ-measurable (i.e., $E \in \mathsf{S}$) with $\mu(E) > 0$. Fix $a < b$ such that $\mu([a,b] \cap E) > 0$, and put $A := [a,b] \cap E$, $B := [a,b] \setminus E$ so that $\mu(A) + \mu(B) = \mu([a,b])$. Since $\mu(A) > 0$ and μ is outer regular, there is open $G \supseteq B$ with $\mu(G \setminus A) < \mu(A)$. Put $F := [a,b] \setminus G$. Then F is closed with $F \subseteq A$. Now $[a,b] \subseteq F \cup (G \setminus B) \cup B$, so $\mu([a,b]) \leq \mu(F) + \mu(G \setminus B) + \mu(B) < \mu(F) + \mu(A) + \mu(B) = \mu(F) + \mu([a,b])$, hence $\mu(F) > 0$. Since $\mu(\{x\}) = 0$ for all x, so F must be uncountable. □

By the Vitali–Bernstein results above, if there is an extension of Lebesgue measure defined on all sets of reals, then such an extension can neither be translation invariant nor be outer regular. But if we drop these two requirements then the measure extension problem remains valid.

Banach generalized the problem further for measures defined on arbitrary sets (instead of $[0,1]$) which satisfy the *continuity condition* $\mu(\{p\}) = 0$ for all p and have a *normalized value* for the measure of the whole set:

The Measure Problem (Banach). *Is there a nonempty set X and a countably additive set function $\mu \colon \mathbf{P}(X) \to [0,1]$ defined on all of $\mathbf{P}(X)$ such that $\mu(X) = 1$ and $\mu(\{p\}) = 0$ for all $p \in X$?*

We will refer to this problem of Banach as the (general) *measure problem.*

The Banach–Kuratowski result was vastly improved by Ulam who did a full analysis of the measure problem. Among other things, Ulam showed that if the size of the continuum is less than the first weakly inaccessible cardinal then there is no extension of Lebesgue measure define on all sets of reals. Ulam's work had significant implications for future research in set theory, and we will now describe his work in detail.

Let us begin with some official definitions.

Definition 1181. By a *total measure on a set* X we mean a set function $\mu: \mathbf{P}(X) \to [0, \infty]$ defined on all subsets of X which is countably additive: If $\langle E_n \mid n \in \mathbf{N} \rangle$ is a pairwise disjoint family of subsets of X then $\mu\left(\bigcup_n E_n\right) = \sum_n \mu(E_n)$. We also say that

1. μ is *nontrivial* if $\mu(X) > 0$ and $\mu(\emptyset) = 0$.
2. μ is *finite* if $\mu(X) < \infty$.
3. μ is *continuous* if $\mu(\{p\}) = 0$ for all $p \in X$.
4. μ is a *probability measure* if $\mu(X) = 1$.

We will focus our attention to nontrivial finite continuous total measures. Note that by normalizing if necessary, the existence of such a measure is equivalent to the existence of a *continuous total probability measure*—which is exactly what the measure problem is asking.

Problem 1182. *Let μ be a finite total measure. If $n \in \mathbf{N}$ then any family of pairwise disjoint sets each of measure $\geq \frac{1}{n}$ is finite. Any family of pairwise disjoint sets of positive measure is countable.*

Definition 1183 (Atomless and Two-Valued Measures). Let μ be a total measure on X. $A \subseteq X$ is an *atom* for μ if $\mu(A) > 0$ and for all $E \subseteq A$ either $\mu(E) = 0$ or $\mu(A \setminus E) = 0$. The measure μ is *atomless* if there is no atom for μ, and μ is called a *two-valued* measure if X itself is an atom for μ.

An atomless measure is continuous and a two-valued measure is nontrivial.

Problem 1184. *Let μ be a finite atomless total measure on X. Show that for any E with $\mu(E) > 0$ and any $\epsilon > 0$ there is $S \subseteq E$ such that $0 < \mu(S) < \epsilon$.*

[Hint: If $\mu(E) > 0$, then there is $E' \subseteq E$ with $0 < \mu(E') \leq \frac{1}{2}\mu(E)$. Repeat.]

Definition 1185 (Separating Families). If $E_i \subseteq X$ for all $i \in I$ then $\langle E_i \mid i \in I \rangle$ is called a *separating family of subsets of* X if for all $p \neq q$ in X, there is $i \in I$ such that $x \in E_i$ and $y \notin E_i$, or $y \in E_i$ and $x \notin E_i$.

For example the family of intervals $\langle (-\infty, r] \mid r \in \mathbf{Q} \rangle$ is a countable separating family for $\mathbf{R}$. This can be generalized as follows.

Problem 1186. *For any cardinal ξ, any set of cardinality at most 2^ξ has a separating family of size at most ξ.*

[Hint: A set X of size at most 2^ξ can be regarded as a subset of $\{0,1\}^A$ for some A with $|A| = \xi$. Then $\langle E_a \mid a \in A \rangle$ is a separating family for $\{0,1\}^A$ where $E_a := \{f \in \{0,1\}^A \mid f(a) = 1\}$ $(a \in A)$.]

Definition 1187 (κ-complete measures). Let μ be a total measure on X and κ be a cardinal. μ is *κ-complete* if for any family $\langle E_i \mid i \in I \rangle$ of subsets of X with $|I| < \kappa$ and $\mu(E_i) = 0$ for all $i \in I$ we have $\mu\left(\bigcup_{i \in I} E_i\right) = 0$.

Note that by definition (countable additivity), every measure is $\aleph_1$-complete.

Proposition 1188. *Let μ be a continuous κ-complete total measure on a set X. If X has a separating family of size less than κ, then μ is atomless.*

Proof. Suppose $A \subseteq X$ is an atom for μ, and fix a separating family $\langle E_i \mid i \in I \rangle$ with $|I| < \kappa$. Since A is an atom, for each $i \in I$ we have either $\mu(A \cap E_i) = 0$ or $\mu(A \setminus E_i) = 0$. Define:

$$A_i := \begin{cases} A \cap E_i & \text{if } \mu(A \cap E_i) = 0 \\ A \setminus E_i & \text{otherwise.} \end{cases}$$

Since $\mu(A_i) = 0$ for all i and μ is κ-complete, $\mu\left(\bigcup_{i \in I} A_i\right) = 0$. The set $A \setminus \left(\bigcup_{i \in I} A_i\right)$ has at most one element, so it has μ-measure zero by continuity of μ. But then $\mu(A) = 0$, contradicting the fact that A is an atom. □

Since **R** has a countable separating family, this immediately gives:

Corollary 1189. *Any continuous total measure on a set of size at most $2^{\aleph_0}$ is atomless.*

Thus if a continuous *two-valued* total measure exists, it can only be defined on a set of cardinality larger than that of the continuum. We will see below that the cardinality of such a set must actually be greater than or equal to a strongly inaccessible cardinal!

A useful result about atomless measures is the following.

Proposition 1190. *Let μ be an atomless total measure on X. Then there is a family $\langle B_u \mid u \in \{0,1\}^* \rangle$ of subsets of X indexed by nodes u of the binary tree $\{0,1\}^*$ satisfying: $B_{\langle\rangle} = X$, $B_u = B_{u^\frown 0} \cup B_{u^\frown 1}$, $B_{u^\frown 0} \cap B_{u^\frown 1} = \emptyset$, and $\mu(B_{u^\frown 0}) = \mu(B_{u^\frown 1}) = \frac{1}{2}\mu(B_u)$.*

Proof. The result easily follows from the following lemma.

Lemma. For any $E \subseteq X$ there is $S \subseteq E$ such that $\mu(S) = \frac{1}{2}\mu(E)$.

Proof (Lemma). Call a family C of subsets of E to be *adequate* if each set in C has positive measure, distinct sets in C are disjoint, and whenever $E_1, E_2, \ldots, E_n \in$ C, we have $\mu\left(\bigcup_{k=1}^{n} E_k\right) \leq \frac{1}{2}\mu(E)$. By Problem 1182, each adequate family is countable. Now consider the collection of all adequate families partially ordered by inclusion, and apply Zorn's lemma to get a maximal adequate family C. Put $S := \bigcup$ C. Then $\mu(S) \leq \frac{1}{2}\mu(E)$ (this is easily seen by enumerating C without repetition and applying countable additivity of μ). We claim $\mu(S) = \frac{1}{2}\mu(E)$. Otherwise we could use Problem 1184 to choose $A \subseteq (E \setminus S)$ with $0 < \mu(A) < \frac{1}{2}\mu(E) - \mu(S)$. Then C $\cup \{A\}$ would be an adequate family properly extending C, contradicting the maximality of C and finishing the proof of the lemma. □

Now construct the family $\langle B_u \mid u \in \{0,1\}^* \rangle$ as follows: Let $B_{\langle\rangle} := X$, and having defined B_u for $u \in \{0,1\}^*$ use the lemma to choose $S \subseteq B_u$ with $\mu(S) = \frac{1}{2}\mu(B_u)$, and then put $B_{u^\frown 0} := S$ and $B_{u^\frown 1} := B_u \setminus S$. □

We now have a "converse" to Corollary 1189.

Corollary 1191. *If there is a κ-complete atomless total probability measure μ on a set X of cardinality κ, then $\kappa \leqslant 2^{\aleph_0}$.*

Proof. Fix $\langle B_u \mid u \in \{0,1\}^* \rangle$ as in Proposition 1190. For each $a \in \{0,1\}^{\mathbf{N}}$, let $X_a := \bigcap_{n \in \mathbf{N}} B_{a|n}$, so that $\mu(X_a) = 0$, and $X = \bigcup \{X_a \mid a \in \{0,1\}^{\mathbf{N}}\}$. If we had $2^{\aleph_0} < \kappa$, then κ-completeness would give $\mu(X) = 0$, a contradiction. □

Corollary 1192. *Let μ be a κ-complete continuous total probability measure on a set of cardinality κ. Then μ is atomless if and only if $\kappa \leqslant 2^{\aleph_0}$.*

The following theorem shows that the original measure extension problem for Lebesgue measure has little to do with the real numbers themselves, but is really a problem of abstract set theory that depends only on the cardinal number of the underlying set.

Theorem 1193 (Banach–Ulam). *The following are equivalent:*

1. *There is a total measure $\mu \colon \mathbf{P}([0,1]) \to [0,1]$ extending Lebesgue measure.*
2. *There is a total measure $\mu \colon \mathbf{P}(\mathbf{R}) \to [0,\infty]$ extending Lebesgue measure.*
3. *There is a continuous total probability measure on a set of size $\leqslant 2^{\aleph_0}$.*
4. *There is an atomless total probability measure on some set.*

Proof. $1 \Rightarrow 2$: Assume that $\mu \colon \mathbf{P}([0,1]) \to [0,1]$ is a total measure which extends Lebesgue measure on $[0,1]$. Then $\overline{\mu} \colon \mathbf{P}(\mathbf{R}) \to [0,\infty]$ is a total measure extending Lebesgue measure on $\mathbf{R}$, where, for each $E \subseteq \mathbf{R}$, we define:

$$\overline{\mu}(E) := \sum_{n=-\infty}^{\infty} \mu([0,1] \cap (E+n)).$$

$2 \Rightarrow 3$: This is immediate: Restrict μ to the unit interval.

$3 \Rightarrow 4$: Immediate by Corollary 1189.

$4 \Rightarrow 1$: Let $\mu \colon \mathbf{P}(X) \to [0,1]$ be an atomless total probability measure on X, and fix $\langle B_u \mid u \in \{0,1\}^* \rangle$ as in Proposition 1190. Each $x \in X$ determines a unique infinite branch $\{u \in \{0,1\}^* \mid x \in B_u\}$ through the binary tree $\{0,1\}^*$ which can be identified with an element $b_x = \langle b_x(1), b_x(2), \dots \rangle \in \{0,1\}^{\mathbf{N}}$. Now define $h \colon X \to [0,1]$ and a total measure ν on $[0,1]$ by:

$$h(x) = \sum_{n=1}^{\infty} \frac{b_x(n)}{2^n}, \qquad \nu(E) := \mu\left(h^{-1}[E]\right).$$

ν is easily verified to be a total measure on $[0,1]$ such that $\nu\left(\left[\frac{k-1}{2^n}, \frac{k}{2^n}\right]\right) = \frac{1}{2^n}$ for $1 \leq k \leq 2^n$. Thus the total measure ν agrees with Lebesgue measure at all dyadic intervals, and so is an extension of Lebesgue measure on $[0,1]$. □

We now show that if the measure problem has a solution, then such a solution defined on a set of least possible cardinality κ must be κ-complete:

Proposition 1194. *Let μ be a continuous total probability measure on a set X with least possible cardinality κ (i.e., there is no continuous total probability measure on any set of cardinality $< \kappa$). Then μ is κ-complete.*

Proof. Otherwise, we can get a well-ordered set I with $|I| < \kappa$ and a family $\langle E_i \mid i \in I \rangle$ such that $\mu(E_i) = 0$ for all $i \in I$ yet $\mu(E) > 0$ where $E := \bigcup_{i \in I} E_i$. Define $f: E \to I$ by setting $f(x) :=$ the least $i \in I$ such that $x \in E_i$. Define a total probability measure ν on I by

$$\nu(A) := \frac{\mu(f^{-1}[A])}{\mu(E)} \qquad (A \subseteq I).$$

ν is continuous since for any $i \in I$ we have $f^{-1}[\{i\}] \subseteq E_i$, and so $\nu(\{i\}) = \mu(f^{-1}[\{i\}])/\mu(E) \leq \mu(E_i)/\mu(E) = 0$. So ν is a continuous total probability measure on the set I with $|I| < \kappa$, contradicting the minimality of κ. □

Thus, as far as the existence of a solution to the measure problem is concerned, *without loss of generality we can and will assume that a continuous total probability measure defined on set of cardinality κ is κ-complete.*

Note that if μ is continuous and κ-complete, then $\mu(S) = 0$ for any set S with $|S| < \kappa$. As an immediate corollary we have:

Corollary 1195. *Let μ be a continuous κ-complete total probability measure on a set X of cardinality κ. Then κ is a regular cardinal.*

Proof. If $X = \bigcup_{i \in I} X_i$ with $|I| < \kappa$ and $|X_i| < \kappa$ for all $i \in I$, continuity and κ-completeness would give $\mu(X_i) = 0$ and so $\mu(X) = 0$, a contradiction. □

Theorem 1196 (Ulam). *Let μ be a continuous κ-complete total probability measure on a set X of cardinality κ. Then κ is a limit cardinal. Consequently, κ must be weakly inaccessible.*

Proof. (Cf. Theorem 1157.) If possible, let $\kappa = \aleph_{\alpha+1}$ be a successor cardinal. Well-order X with order type $\omega_{\alpha+1}$, and fix a set Y with $|Y| = \aleph_\alpha$. For each $a \in X$, the set $\mathrm{Pred}_X(a) = \{x \in X \mid x < a\}$ has cardinality $\leqslant \aleph_\alpha$, so we can fix an *injection* $f_a: \mathrm{Pred}_X(a) \to Y$. For each $x \in X$ and $y \in Y$, put:

$$E_x^y := \{a \in X \mid x < a \text{ and } f_a(x) = y\} \qquad (\textit{Ulam matrix}).$$

Then for each $y \in Y$, the sets $\langle E_x^y \mid x \in X \rangle$ form a family of *pairwise disjoint* subsets of X, so only countably many of these sets can have positive measure, and hence (as X has uncountable cofinality) we can fix $x_y \in X$ such that $\mu(E_x^y) = 0$ for all $x > x_y$. Since $\omega_{\xi+1}$ is regular and $|\{x_y \mid y \in Y\}| \leqslant \aleph_\alpha$, there is $p \in X$ with $p > x_y$ for all $y \in Y$. Then $\mu(E_p^y) = 0$ for all $y \in Y$, so by κ-completeness of μ, $\mu\left(\bigcup_{y \in Y} E_p^y\right) = 0$. Hence the set $\{a \in X \mid a > p\}$ (being $\subseteq \bigcup_{y \in Y} E_p^y$) also has

μ-measure 0. But by κ-completeness of μ again, the set $\{a \in X \mid a \leq p\}$ also has μ-measure 0, which is a contradiction since X cannot be the union of two sets of μ-measure 0. □

Corollary 1197 (Ulam). *If there is a total measure extending Lebesgue measure, then there is a weakly inaccessible cardinal* $\leqslant 2^{\aleph_0}$.

Thus if Lebesgue measure has a total extension, then $2^{\aleph_0} > \aleph_1$, $2^{\aleph_0} > \aleph_2, \ldots,$ $2^{\aleph_0} > \aleph_{\omega_1}$, etc, and so the Continuum Hypothesis is severely violated.

Conversely, if no cardinal $\leqslant 2^{\aleph_0}$ is weakly inaccessible, e.g., if $2^{\aleph_0} \leqslant \aleph_{\omega_1}$, then Lebesgue measure cannot have a total extension. Observe how this dramatically improves the Banach–Kuratowski result, which derived the same conclusion assuming the hypothesis $2^{\aleph_0} = \aleph_1$.

Finally, we consider the case of *two-valued* measures. The following result implies that a two-valued continuous total measure must be defined on a set of cardinality greater than or equal to some *strongly* inaccessible cardinal:

Corollary 1198 (Tarski–Ulam). *If a* κ*-complete continuous total measure* μ *on a set* X *of cardinality* κ *has an atom, then* κ *is strongly inaccessible.*

Proof. By Corollary 1195 κ is regular, so it suffices to show that κ is a strong limit: If $\xi < \kappa \leqslant 2^{\xi}$, then by Problem 1186 X has a separating family of size at most $\xi < \kappa$, so by Proposition 1188, μ must be atomless, a contradiction. □

Combining this with Corollary 1192 culminates in Ulam's major result:

Corollary 1199 (Ulam). *Let* μ *be a continuous* κ*-complete total probability measure on some set of cardinality* κ*. Then we have:*

(A) $\kappa \leqslant 2^{\aleph_0} \Leftrightarrow$ *μ is atomless* $\Leftrightarrow$ *κ is not strongly inaccessible.*

(B) $\kappa > 2^{\aleph_0} \Leftrightarrow$ *μ has an atom* $\Leftrightarrow$ *κ is strongly inaccessible.*

Moreover, if any (and so all) of the conditions in (A) *holds, then Lebesgue measure has a total extension, and if any (and so all) of the conditions in* (B) *holds, then a two-valued continuous total measure exists.*

Note the *dichotomy* involved here: If κ is as above, then either $\kappa \leqslant 2^{\aleph_0}$ or $\kappa > 2^{\aleph_0}$ but not both; hence either all of the equivalent conditions in (A) hold, or else all of the equivalent conditions in (B) hold, but not both.

Definition 1200 (Real Valued Measurable Cardinals). A cardinal κ is called a *real valued measurable cardinal* if there is a continuous κ-complete total probability measure on some set of cardinality κ.

Thus a real valued measurable cardinal exists if and only if the measure problem has a positive solution. Every real valued measurable cardinal is weakly inaccessible (and so cannot be shown to exist using usual axioms).

Definition 1201 (Measurable Cardinals). A cardinal κ is called a *measurable cardinal* if there is a two-valued continuous κ-complete total measure on some set of cardinality κ.

It follows that a cardinal is measurable if and only if it is real valued measurable and strongly inaccessible. Measurable cardinals turn out to be "very large." For example, it can be shown that if κ is measurable then κ is not only weakly compact, but is also preceded by κ-many weakly compact cardinals.

Problem 1202. *Every measurable cardinal has the tree property.*

Ulam's results can now be stated in terms of measurable cardinals as follows.

Corollary 1203 (Ulam).

(A) A total extension of Lebesgue measure exists
  $\Leftrightarrow$ *there is a cardinal which is real valued measurable but not measurable*
  $\Leftrightarrow$ *there is a real valued measurable cardinal* $\leqslant 2^{\aleph_0}$.

(B) If Lebesgue measure has no total extension, then a nontrivial continuous total measure on some set exists $\Leftrightarrow$ *there is a measurable cardinal.*[1]

Ulam's definitive work above was the first example of a natural problem of classical mathematics whose solution is *equivalent to a large cardinal hypothesis.* Measurable cardinals gave birth to the field of large cardinals, and have considerable implications for several areas of mathematics. Theorem 1206 below is one such application. Postscript IV (Chapter 22) describes surprising connections between measurable cardinals and *infinite games* (Theorem 1316).

19.2 Projective Sets and Lusin's Problem

In Chap. 18, we studied Borel and analytic sets. These sets are quite effectively defined, and we saw that they enjoy many *regularity properties,* such as being Lebesgue measurable, having the Baire property, and the perfect set property. The subject area dealing with the study of such effectively defined sets is known as *descriptive set theory,* which originated in the work of Borel, Baire, Lebesgue, Lusin, Sierpinski, Suslin, Hausdorff, etc.

It can be shown that analytic sets are precisely the continuous images of Borel sets. Using this idea, Lusin defined larger classes of sets called *projective sets* as follows. The analytic sets were called A sets and the coanalytic sets CA sets. Continuous images of coanalytic sets were called PCA sets, and complements of PCA sets CPCA sets, and so on.

[1] Solovay proved that the relative consistency (with the usual axioms) of each of the alternatives of Corollary 1203 implies the relative consistency of the other.

In modern notation, introduced by Addison, the projective hierarchy is defined recursively as follows.

Definition 1204 (The Projective Hierarchy). E is a $\mathbf{\Sigma}^1_1$ set if E is an analytic set, and for each $n \in \mathbf{N}$, E is a $\mathbf{\Sigma}^1_{n+1}$ set if E is the continuous image of the complement of some $\mathbf{\Sigma}^1_n$ set.

A $\mathbf{\Pi}^1_n$ set is one whose complement is a $\mathbf{\Sigma}^1_n$ set, and a $\mathbf{\Delta}^1_n$ set is one which is both $\mathbf{\Sigma}^1_n$ and $\mathbf{\Pi}^1_n$.

A set is *projective* if it is $\mathbf{\Sigma}^1_n$ (or equivalently $\mathbf{\Pi}^1_n$ or $\mathbf{\Delta}^1_n$) for some n.

Thus a $\mathbf{\Sigma}^1_1$ set is simply an analytic set, a $\mathbf{\Pi}^1_1$ set is nothing but a coanalytic set, and $\mathbf{\Delta}^1_1$ sets are those which are both analytic and coanalytic. By Suslin's theorem, therefore, the $\mathbf{\Delta}^1_1$ sets coincide with the class $\mathbf{B}$ of Borel sets. Also, we have $\mathbf{\Sigma}^1_n \cup \mathbf{\Pi}^1_n \subseteq \mathbf{\Delta}^1_{n+1} \subseteq \mathbf{\Sigma}^1_{n+1} \cap \mathbf{\Pi}^1_{n+1}$, so we have a hierarchy of ever more comprehensive collection of sets. It can be shown that this hierarchy increases properly as n increases, and so we get the following picture:

$$\begin{array}{ccccccccccccc}
 & & \mathbf{\Sigma}^1_1 & & \mathbf{\Sigma}^1_2 & & & & & \mathbf{\Sigma}^1_n & & & \\
 & \subsetneq & & \subsetneq \;\; \subsetneq & & \subsetneq & & & \subsetneq & & \subsetneq & & \\
\mathbf{B} = \mathbf{\Delta}^1_1 & & & \mathbf{\Delta}^1_2 & & & \mathbf{\Delta}^1_3 & \cdots \;\; \cdots & \mathbf{\Delta}^1_n & & & \mathbf{\Delta}^1_{n+1} & \cdots \\
 & \subsetneq & & \subsetneq \;\; \subsetneq & & \subsetneq & & & \subsetneq & & \subsetneq & & \\
 & & \mathbf{\Pi}^1_1 & & \mathbf{\Pi}^1_2 & & & & & \mathbf{\Pi}^1_n & & &
\end{array}$$

Thus $\mathbf{\Sigma}^1_2$ sets are precisely the PCA sets and $\mathbf{\Pi}^1_2$ sets the CPCA sets, etc. The projective sets are thus example of effectively defined sets and the total number of projective sets is $2^{\aleph_0}$. (The reason they are called "projective" is that they can alternatively defined by taking projections, instead of continuous images, of sets in higher dimensional Euclidean spaces.)

Classical descriptive set theory obtained the following major results, some of which we proved in Chap. 18:

Theorem 1205 (Lusin–Sierpinski–Suslin).

1. *Every $\mathbf{\Sigma}^1_1$ (analytic) set has the perfect set property, is Lebesgue measurable, and have the Baire property.*
2. *Any two disjoint $\mathbf{\Sigma}^1_1$ sets can be separated by a $\mathbf{\Delta}^1_1$ set, and any two disjoint $\mathbf{\Pi}^1_2$ sets can be separated by a $\mathbf{\Delta}^1_2$ set.*
3. *Any $\mathbf{\Sigma}^1_2$ set (PCA set) can expressed as a union of $\aleph_1$ Borel sets.*

There were some other *structural properties* that were established about $\mathbf{\Pi}^1_1$ and $\mathbf{\Sigma}^1_2$ sets (called reduction and uniformization that we have not covered), but little else was known about the higher projective classes.

Lusin and other mathematicians tried to obtain similar properties for the next levels of the projective hierarchy. They asked the following questions, which we will collectively refer to as *Lusin's Problem*:

Lusin's Problem. *What are the regularity and structural properties for the higher level projective classes? In particular:*

1. *Does every uncountable* $\mathbf{\Pi}_1^1$ *(coanalytic) set contain a perfect subset?*
2. *Is every* $\mathbf{\Sigma}_2^1$ *set Lebesgue measurable?*
3. *Does every* $\mathbf{\Sigma}_2^1$ *set have Baire property?*

Despite major efforts by all the leading descriptive set theorists of that period, none of the above questions could be answered, and essentially nothing about the higher level projective classes could be said.

This caused Lusin to remark that "one does not know *and one will never know*" the answer to the above questions about projective sets, even though projective sets are effectively defined and form an infinitesimally small part of $\mathbf{P}(\mathbf{R})$. As we will see, Lusin's remark turned out to be prophetic.

Independence Results for Lusin's Problem

The first results that partially explained why Lusin's Problems could not be solved came from Gödel, who showed [22] that one cannot prove that PCA ($\mathbf{\Sigma}_2^1$) sets are Lebesgue measurable or that coanalytic sets have the perfect set property using the usual axioms of set theory (provided these axioms are consistent). Gödel's work will be briefly discussed in Postscript IV. Much later, Solovay and Martin showed, by extending Cohen's technique of forcing, that one cannot disprove that PCA sets Lebesgue measurable (or that they have Baire property). What they showed is that MA+not-CH is relatively consistent with the usual axioms of set theory. Since MA+not-CH combined with Sierpinski's result Theorem 1205(3) implies that every $\mathbf{\Sigma}_2^1$ set is Lebesgue measurable and has Baire property, the relatively consistency of the latter also follows. Moreover, Solovay also showed, assuming the consistency of existence of an inaccessible cardinal, that one can consistently assume that every projective set is Lebesgue measurable, has the Baire property, and has the perfect set property (see Theorem 1308).

The Gödel–Martin–Solovay independence results showed that all of Lusin's Problems are essentially *unsolvable* using the standard axioms of mathematics, thus fully confirming Lusin's prediction.

19.3 Measurable Cardinals and PCA ($\mathbf{\Sigma}_2^1$) Sets

Surprisingly, Solovay also showed that the existence of measurable cardinals does resolve Lusin's Problem, and in a desirable positive way:

Theorem 1206 (Solovay). *If there is a measurable cardinal, then all* $\mathbf{\Sigma}_2^1$ *sets have the perfect set property, are measurable, and have the Baire property.*

After Ulam's settlement of the measure problem, this was another remarkable example of a large cardinal axiom resolving an unsolvable problem of classical mathematics, and raised the hopes of discovering large cardinal axioms which may be added as new axioms of mathematics. At the same time, Silver showed that Lusin's Problems for projective classes of *levels higher than* $\mathbf{\Sigma}_2^1$ cannot be resolved using measurable cardinals alone.

Solovay, however, conjectured that *stronger* large cardinal axioms might resolve the problems for the projective sets, and the search for such large cardinals—or other possible axioms—became one of the greatest problems of modern descriptive set theory. Outstanding work by many people culminated in a truly remarkable resolution of the problem. But this is a topic for Postscript IV, where we will discuss connections between large cardinals and determinacy of infinite games.

Goldring [25] is a very accessible survey of the topics of this postscript.

Part IV
Paradoxes and Axioms

Introduction to Part IV

In Parts I–III of this book we developed cardinals, order, ordinals, and real point set theory, and also indicated how these relate to some classical areas of mathematics. We carried out that development in an informal and naive way as we would do for any standard area of mathematics such as geometry, exploring structural details and obtaining views and intuitions about the subject matter. In this sense, Parts I–III of the book were purely mathematical.

The naive theory of sets, however, can lead to contradictions unless suitable restrictions are placed on the simple principles forming the basis of the theory. This requires a careful scrutiny of the logical foundations of set theory. To stay focused on the mathematical aspects of our topics, we had so far avoided getting into this *metamathematical* problem.

In this part, we will give an overview of such logical and foundational matters, starting with some famous contradictions of naive set theory and two early responses to them (Chapter 20). Our coverage will necessarily be very elementary and introductory, and we will refer the reader to more comprehensive works for further details.

In Chap. 21, we briefly present Zermelo–Fraenkel set theory (ZF) and the von Neumann ordinals, providing only bare outlines for the formal development of some of the basic notions of set theory such as order and cardinals. However, the reader who has mastered the theories of numbers, cardinals, ordinals, and the real continuum developed in Parts I–III, will find the re-development of all these theories within the formal framework of ZF a relatively routine matter, and we encourage the reader to take up this project of replicating the results of Parts I–III formally in ZF.

Finally, the postscript to this part (Chapter 22) provides glimpses into some landmark results of set theory of the past 75 years.

Chapter 20
Paradoxes and Resolutions

Abstract Unless carefully restricted, the informal naive set theory that we have so far been using can produce certain contradictions, known as *set theoretic paradoxes*. These contradictions generally result from consideration of certain very large sets whose existence can be derived from the unrestricted comprehension principle. This chapter discusses three such classical paradoxes due to Burali-Forti, Cantor, and Russell, which showed the untenability of naive set theory and the need for more careful formalizations. The two earliest responses to the paradoxes, namely Russell's theory of types and Zermelo's axiomatization of set theory, are discussed.

20.1 Some Set Theoretic Paradoxes

The Burali-Forti Paradox

One of the oldest paradoxes of set theory is the Burali-Forti paradox. It shows that a contradiction can be derived from the assumption that there is a set containing *all* ordinals.

Proposition 1207 (Burali-Forti's Paradox). *The assumption that there is a set of all ordinals leads to a contradiction.*

Proof. If there were a set Ω consisting of *all* ordinals, it will be an initial set of ordinals, so we will have $\Omega = \{\beta \mid \beta < \alpha\}$ where α is the order type of Ω (recall that any set of ordinals is well-ordered). Since Ω contains all ordinals, we will have $\alpha \in \Omega = \{\beta \mid \beta < \alpha\}$, whence $\alpha < \alpha$, a contradiction. □

When viewed as a proof by contradiction, this result can be put in the following form: *The set of all ordinals does not exist.*

The above proof is so simple and clear that it already forces us to doubt unrestricted comprehension. By unrestricted comprehension there is a set of all

A. Dasgupta, *Set Theory: With an Introduction to Real Point Sets*,
DOI 10.1007/978-1-4614-8854-5_20,

ordinals, and by Burali-Forti's theorem, there is no set of all ordinals. Since the proof of Burali-Forti's theorem is highly rigorous while the unrestricted comprehension axiom of uses the vague notion of "arbitrary property applicable to any object whatsoever," this causes us to question the axiom.

The Cantor–Russell Paradoxes

Cantor's paradox result from the assumption that there is a set containing all sets: By Cantor's theorem there is no largest cardinal since $2^\kappa > \kappa$ for any cardinal κ, but on the other hand the cardinality of the set of all sets must be the largest cardinal.

Proposition 1208 (Cantor's Paradox). *The assumption that there is set containing all sets leads to a contradiction.*

Proof. If there were a set V containing all sets, we would have $\mathbf{P}(V) \subseteq V$, hence $|\,\mathbf{P}(V)| \leq |V|$, contradicting Cantor's theorem that $|\,\mathbf{P}(V)| > |V|$. □

When viewed as a proof by contradiction, Cantor's paradox can be put in the following form: *There is no set containing all sets,* or, as famously paraphrased by Halmos: "Nothing contains everything."

The following problem gives a result closely related to Cantor's paradox.

Problem 1209. *The assumption that there is a set containing all cardinal numbers leads to a contradiction.*

[Hint: Use Hartogs' theorem and the fact that for any ordinal α there is $\beta > \alpha$ with $\omega_\beta = \beta$.]

Cantor's paradox is based on Cantor's theorem, which we present again in the following "diagonalization" form: *For any function F with domain X, there is a member of* $\mathbf{P}(X)$ *which is not in* $\operatorname{ran}(F)$.

Theorem 1210 (Cantor Diagonalization). *If X is any set and F is any function with domain X, then there is a subset of X not in the range of F.*

Proof. Take the Cantor diagonal set for the function F, namely:

$$D := \{x \in X \mid x \notin F(x)\}.$$

If we had $D = F(a)$ for some $a \in X$, then we get $a \in D \Leftrightarrow a \notin F(a) \Leftrightarrow a \notin D$, which is a contradiction. □

The famous *Russell's paradox* is closely related to Cantor's theorem above. In fact, as Russell comments in [69, p. 58], and as we will see now, it is obtained as a special case when the function F is taken to be the identity function on the set of all sets.

Given any set X, the *Russell set R_X for X* is defined to be the Cantor diagonal set for the identity function on X:

$$R_X := \{x \in X \mid x \notin x\}.$$

Hence, by Cantor's theorem in the form stated above, for any set X, its Russell set R_X cannot be in the range of the identity function on X, i.e., R_X is not a member of X.[1]

Proposition 1211 (Property of the Russell Set). *For any set X, its Russell set $R_X := \{x \in X \mid x \notin x\}$ is not a member of X.*

Call a set x to be *normal* if $x \notin x$ (most sets encountered in practice are normal). Thus the Russell set R_X of X consists of all normal members of X. Russell's paradox is obtained by taking the set R of all normal sets; in other words, R is the Russell set of the set of all sets, i.e., $R = R_V$, where V is the set of all sets.

Proposition 1212 (Russell's Paradox). *The assumption that there is a set R consisting of all normal sets leads to a contradiction.*[2]

Proof. We have $R \in R \Leftrightarrow R$ is normal $\Leftrightarrow R \notin R$. □

Thus Russell's Paradox, obtained by applying Cantor's theorem to the identify function defined on the set of all sets, results in a contradiction of a strikingly simple form. A popular version of Russell's paradox talks of a certain barber in a certain town who shaves those and only those who do not shave themselves. We then get a contradiction since the assumption that the barber shaves himself leads to its negation and vice versa. Viewed as a proof by contradiction, this means that such a barber cannot exist.

Impact on the Frege–Russell Logicist Program

The goal of the original *logicist* program—pioneered by Frege during 1879–1903 and championed by Russell—was to develop mathematics purely from logic using the central notion of *the extension* of a property (or of a concept). Roughly speaking, the extension of a property P is what we would now call the set of all objects having the property P, embodied in the *axiom of extensionality*: If two properties P and Q satisfy the condition that x has property P if and only if x has property Q for all objects x, then P and Q have the same extension (i.e., they determine the same set). In addition to the axiom of extensionality, the logicist system used the *axiom of*

[1] This is equivalent to saying $X \cup \{R_X\}$ is a proper superset of X. Under the Axiom of Foundation, R_X equals X itself, and so the Axiom of Foundation implies that the set $X \cup \{X\}$, called the successor set of X, is a strictly larger set.

[2] The paradox was also discovered independently by Zermelo (unpublished).

unlimited comprehension, which says that every property P has an extension (i.e., the set of all objects satisfying P exists). The logical deductive system based on these two axioms—which we will call the naive Frege–Russell logicist system—was used to develop significant bodies of mathematics (such as arithmetic and the theory of cardinals), until Russell discovered his paradox in 1901 showing that the naive Frege–Russell logicist system was inconsistent and therefore must be modified in some way or other. Remarkably, Russell reported his paradox in a famous letter to Frege in 1902, precisely when the second and final volume of Frege's completed development of the system was in press for publication.

Russell's paradox is perhaps the simplest one among all set-theoretic paradoxes. It can be quickly derived from unrestricted comprehension using the set-membership relation alone, without any need for using more advanced defined notions such as ordinals, cardinals, or the power set. In addition to the abandonment of the original naive Frege–Russell logicist system, it led to the permanent prohibition of the use of the unrestricted comprehension principle, and thus to revisions of the methods for new set formation.

Resolution of the Paradoxes

The set theoretic paradoxes almost invariably resulted from consideration of very large collections such as the collection of all sets or the collection of all ordinal numbers. The axiom of extensionality was uncontroversial, but as pointed out above, unlimited comprehension was highly suspect, and it soon became clear that this axiom had to be modified.

It was also clear that the informal nature of the naive set theory of Cantor and Frege carried risk of generating contradictions, and if contradictions were to be avoided then a more careful formalization of the principles of set theory was needed. Several such formal approaches developed over the years, and we now discuss the two earliest responses to the paradoxes, the first by Russell himself, and the second by Zermelo, the other mathematician who had independently encountered Russell's paradox.

20.2 Russell's Theory of Types

The first proposed way to address the paradoxes was introduced by Russell himself in his 1903 book *The Principles of Mathematics* [66]. Russell called his solution *the Theory of Types*, and developed and extended it fully in 1908 [67] and 1910 [81]. The theory of types was later modified and considerably simplified by Ramsey, Carnap, Tarski, Gödel, Church, and Quine. We will give a very rough description of type theory in its simplest form.

The theory of types classifies objects into a hierarchy according to their *logical type*. Objects which are not sets, called *individuals*, or *atoms*, have type 0. Sets consisting of individuals alone are of type 1. Sets consisting of sets of type 1, i.e., sets of sets of individuals, are of type 2, and so on. In general, the set of objects of type $n+1$ is simply the "power set" of the set of objects of type n. The main principle of type theory is that *the expression "$x \in y$" cannot be meaningful unless we have type of $y = 1 +$ type of x*. Thus the expression "$x \in x$," required in formulating Russell's paradox, is simply meaningless. Moreover, a set of type $n + 1$ can only contain objects of type n, and so objects of distinct types cannot be mixed. It follows that the collection of all sets is also meaningless. This results in a resolution of Russell's paradox and other paradoxes involving the set of all sets such as Cantor's paradox. One immediate oddity of this theory, however, is the "duplication of the empty set" across various types: There is an empty set of type 1, an empty set of type 2, etc. This may appear to be a violation of the axiom of extensionality, but note that in type theory there is a separate axiom of extensionality for each type!

Principia Mathematica

Russell and Whitehead's *Principia Mathematica* (or PM), published in three volumes [81], was a revival of the original naive Frege–Russell logicist program of developing mathematics deductively from a few "logical" axioms. Using a more careful revision (based now on type theory) of the axioms of the naive system, it made a heroic attempt to demonstrate that mathematics is an extension of logic. More specifically, it deductively developed portions of mathematics starting from axioms which they claimed to be principles of logic itself.

Principia Mathematica served as a reference point for the fact that mathematics can be deductively developed from a few axioms, logical or not, and had a tremendous impact on the development of set theory and logic in the subsequent years. However, while it succeeded in demonstrating that portions of mathematics can be developed from a few axioms, criticisms were raised that it failed in its logicist program. Some of the axioms used did not appear to be logical. The axiom of infinity, which states that there is an infinite set, appears more quantitative than logical. Most significantly, PM needed a special axiom called *the axiom of reducibility*, which did not appear to have any logical or intuitive justification at all, and was rejected even by other supporters of the logicist program.

Because of the complexities involved and the resulting need for the axiom of reducibility which had little justification, the form of type theory in PM, known as *ramified type theory*, never gained acceptability. Later work by Ramsey, Carnap, Tarski, Gödel, Church, Quine,[3] etc., resulted in a more satisfactory *simple theory*

[3] Type theory also lurks behind Quine's *New Foundations* (discussed in Sect. 21.9), where type distinction plays a basic role in forming sets via comprehension.

of types, but type theory itself is not popular for the deductive development of mathematics. Far more successful has been Zermelo's axiomatic formulation of 1908, where all kinds of sets, including those having distinct hierarchical *ranks,* can be freely mixed.

Although not popular in axiomatic developments of mathematics, type theory, nevertheless, has had far reaching implications for other areas such as logic, philosophy, and computability. For example, with its close relation with Church's *lambda-calculus*, type theory today finds significant applications in modern computer science.

20.3 Zermelo's Axiomatization

In 1908, Zermelo [85] introduced a completely different set of axioms in which the objects of the theory form a *single domain* consisting of all sets (and possibly also individuals, or atoms, which are not sets). Unlike type theory, there is no longer an a priori classification of objects into individuals, sets, sets of sets, etc. Instead, all these objects (having "mixed types") are put together in the single domain and are regarded to be of the same sort to start with. Along with this *domain of all sets* the only other primitive notion used in Zermelo's system is the relation of *set membership,* where x has this relation to y if and only if $x \in y$.

Zermelo's system was the first formulation of *axiomatic set theory* in the modern sense of the term. With its single domain of objects containing sets of different types freely mixed together, it has a considerably simpler setup than Russell's type theory.

A most important feature of Zermelo's system is the removal of the unrestricted comprehension principle. Instead of forming sets defined by arbitrary properties, Zermelo's system limits comprehension by only allowing formation of *subsets* of a set which is already known to exist. This restricted axiom of comprehension says that given a set A and a property P, we can form the subset $\{x \mid x \in A \text{ and } P(x)\}$ of A. This subset is said to be *separated* out of A via the property P. Thus Zermelo's axiom of *restricted comprehension* is also known as the *axiom of separation* (*Aussonderungsaxiom*) or as the *axiom of subsets.* This limited principle of comprehension prevents the entire domain of all sets from being a set itself, avoiding contradictions involving "the set of all sets."

Of course, with such a limited form of comprehension, additional axioms are needed to form new sets. One such axiom allows to form the power set of any given set. Even from assuming only the existence of the empty set, one can iterate the formation of power sets to get more and more sets, giving us a rich supply of finite sets. Thus, another important feature of Zermelo's system is that it forms new sets by building them from ground up in stages, rather than by sweeping uses of comprehension. There are other axioms for forming new sets, such as the axiom of union, which states that we can form the union of the members of any given set (of sets). Since we will discuss these and other axioms in more detail in Chap. 21, we finish this short discussion of Zermelo's axiomatization with some comments on subsequent modifications of Zermelo's system.

Zermelo's system was later improved by Skolem (who made the vague notion of "definite property" precise by replacing it with the formal syntactic notion of "first-order formula"), and was made more comprehensive by Fraenkel (who introduced the axiom of replacement necessary for the development of the theory of the transfinite). The enlarged system which is now standard is sometimes called the Zermelo–Skolem–Fraenkel system, but we will follow current usage and call it the *Zermelo–Fraenkel system,* abbreviated as ZF set theory. This notation assumes that the axiom of choice is not included in ZF, and ZF augmented with the axiom of choice is called ZFC, or *Zermelo–Fraenkel set theory with Choice.*

ZFC has turned out to be the most popular formulation of axiomatic set theory, and has become the standard axiomatization of set theory today. It is a very powerful (perhaps too powerful) system certainly capable of providing a framework for all of mathematics: All mathematical statements can be expressed in its language and all theorems of classical mathematics can be proved in ZFC.

Van Heijenoort [78, p. 199] illustrates the striking difference between the two early responses to the paradoxes by contrasting the pragmatic foundation for mathematics provided by Zermelo's axiomatization with the wide logico-philosophical ramifications of Russell's type theory.

Chapter 21
Zermelo–Fraenkel System and von Neumann Ordinals

Abstract In this chapter we present the *Zermelo–Fraenkel axiom system* which is an enhancement of Zermelo's 1908 system by Fraenkel, Skolem and others, as well as the *von Neumann ordinals,* which assigns, in a remarkably simple, constructive, and canonical fashion, a unique representative well-order to each ordinal number.

21.1 The Formal Language of ZF

For a precise formulation of the ZF axioms we first need to formalize its language, which we call the *language of ZF set theory.*

Expressions in the language of ZF set theory will be certain strings of a specific group of symbols.

First we will need symbols for *variables,* for which we will use letters such as $a, b, c, \ldots, u, v, w, x, y, z$, etc. Next we need *logical connectives,* namely $\neg$ (not), $\vee$ (or), $\wedge$ (and), $\rightarrow$ (implies), and $\leftrightarrow$ (if and only if). The other type of logical symbol we will need are the *quantifiers,* namely $\forall$ (for all) and $\exists$ (there exists). For grouping expressions, we will also use the two special symbols "(" and ")" (parentheses).

The above types of symbols are called *logical symbols,* and widely used in mathematical contexts.

The most important symbol—the only nonlogical primitive symbol in the language of ZF set theory—will be the symbol "$\in$," representing set membership.

Using "$\in$," we can form *atomic formulas* such as $x \in y$, $u \in u$, etc, where the letters are used for variables ranging over sets. By combining atomic formulas using connectives and using quantifiers over them, we can form general formulas of arbitrary complexity such as:

$$u \notin u, \quad \forall y(y \notin x), \quad \forall z(z \in x \rightarrow z \in y), \quad \text{etc.}$$

A. Dasgupta, *Set Theory: With an Introduction to Real Point Sets,*
DOI 10.1007/978-1-4614-8854-5_21,

Formally, a *ZF formula* is any expression (i.e., a string consisting of the above types of symbols) that can be built starting from the atomic formulas using a finite number of applications of the following formation rules:

1. Any expression of the form "$u \in v$" is a ZF-formula, where u and v are variable letters (any atomic formula is a ZF-formula)
2. If α, β are ZF formulas and v is a variable letter, then each of the following is a ZF formula:

$$\neg(\alpha),\ (\alpha) \wedge (\beta),\ (\alpha) \vee (\beta),\ (\alpha) \rightarrow (\beta),\ (\alpha) \leftrightarrow (\beta),\ \forall v(\alpha),\ \exists v(\alpha).$$

The occurrence of a variable within a formula becomes *bound* if it is quantified by some quantifier (more precisely if it falls under the scope of a quantifier). The occurrence of a variable which is not quantified by some quantifier of the formula is called *free*. For example, in the formula below (which says "x is a subset of y"):

$$\forall z(z \in x \rightarrow z \in y),$$

all occurrences of the variable z are bound, while the variables x and y are free. In the formula

$$\exists y(y \in x) \wedge \forall x(x \notin x),$$

the first occurrence of x is free, but all other occurrences are bound. Thus a variable may be both free and bound in a formula.

This language is used to formulate ZF set theory as a *first order theory,* in which all variables range over—and so all quantification are limited to—a *single domain of objects* consisting of pure sets. Thus in ZF, '$\forall x(\ldots)$' is interpreted as "for all sets $x, \ldots,$" '$\exists x(\ldots)$' as "for some set $x, \ldots,$" and '$x \in y$' as "set x is a member of set y."

With this overview of the language of ZF set theory, we now turn to the ZF axioms.

21.2 The First Six ZF Axioms

ZF 1 (Axiom of Extensionality). $\forall x \forall y(\forall z(z \in x \leftrightarrow z \in y) \rightarrow x = y)$.

ZF 2 (Axiom of Empty Set). *There exists a set which has no members:* $\exists x \forall y(y \notin x)$.

By extensionality, there is a unique set which has no members, and this set (the empty set) is denoted as usual by Ø.

At this point, we can define in ZF the subset relation as:

$$x \subseteq y \leftrightarrow \forall z(z \in x \rightarrow z \in y),$$

and basic properties of the subset relation can be established in ZF: For all a, b, c we have $\emptyset \subseteq a$, $a \subseteq a$, and $a \subseteq b$ and $b \subseteq c \Rightarrow a \subseteq c$.

One crucial axiom of ZF is the *axiom of separation* or the *axiom of restricted comprehension* (Zermelo's *Aussonderungsaxiom*). It is sometimes called the *axiom of subsets* since the axiom states that, *given a set a* and a "definite property" P, we can form the *subset of a* defined by

$$\{x \mid x \in a \text{ and } P(x)\} \qquad \textbf{(Restricted Comprehension)}.$$

This is thus an *axiom scheme,* that, is an infinite list of axioms, one for each definite property P.

Contrast this with the old *naive unrestricted comprehension* which allows forming a set

$$\{x \mid P(x)\} \qquad \textbf{(Unrestricted Comprehension)}$$

for any property P, without requiring that the elements of the set being formed be restricted within some set a already known to exist. This allowed the formation of sets like "the set of all sets" by taking $P(x)$ to be '$x = x$', or "Russell's set of all sets not containing themselves" by taking $P(x)$ to be '$x \notin x$', which lead to paradoxes. Separation (restricted comprehension) is designed to avoid such large problematic sets, see Theorem 1213 below.

The term "definite property" was used in Zermelo's original paper of 1908, but it was not precisely defined. One of Skolem's many important contributions to axiomatic set theory was to make the vague notion of "property" precise: In each instance of the separation scheme, the expression "$P(x)$" is taken to be any formal ZF formula of set theory in which x occurs free.

We will say that P *is a ZF property* if P is a ZF formula with a specified free variable, and we use the notation $P = P(x)$ to indicated that the specified free variable is x. If P is a ZF property, then we will also use the notation $P(a)$ to denote the formula which results from P when the specified free variable is replaced by a.

More generally, other variables may occur free as well in the defining property P, and these variables are then called *parameters,* as indicated in the notation $P = P(x, y_1, y_2, \ldots, y_n)$. For example,

$$\{x \mid x \in a \text{ and } P(x, y_1, y_2, \ldots, y_n)\}$$

is the subset of a defined by the property $P = P(x, y_1, y_2, \ldots, y_n)$ with the parameters $y_1, y_2, \ldots, y_n$. Thus we have the following precise formulation of the separation axiom scheme due to Skolem.

ZF 3 (Axiom Scheme of Separation). *If $\varphi(x, y_1, y_2, \ldots, y_n)$ is a ZF formula in which the free variables are among $x, y_1, y_2, \ldots, y_n$, then the following is an instance of the separation scheme:*

$$\forall y_1 \forall y_2 \cdots \forall y_n \forall a \exists b \forall x (x \in b \leftrightarrow x \in a \wedge \varphi(x, y_1, y_2, \ldots, y_n)).$$

With just these three axioms, we can now turn Russell's paradox into a theorem of ZF which says that there is no set containing all sets:

Theorem 1213 (ZF). $\neg \exists x \forall y (y \in x)$.

The proof is left as an exercise.

The only set whose existence can be proved with these three axioms is the empty set, and so we clearly need more axioms for building new sets. Two more axioms are:

ZF 4 (Axiom of Power Set). $\forall x \exists y \forall z (z \in y \leftrightarrow z \subseteq x)$: *For any set X, there is a set y consisting precisely of the subsets of x.*

ZF 5 (Axiom of Union). $\forall x \exists y \forall z (z \in y \leftrightarrow \exists w (w \in x \wedge z \in w))$: *For any set x, there is a set y consisting precisely of the members of members of x.*

By extensionality, the set y in the power set axiom is uniquely determined by x, and so we can define the usual notion of power set. We let, as usual, $\mathbf{P}(x)$ denote the power set of x. Similarly the set y in the union axiom is uniquely determined by x, and will be denoted using the usual notation $\cup x$.

We will freely use the notations $\mathbf{P}(x)$ and $\cup x$, but note that these *defined notions* are not part of the ZF language and can be formally eliminated. Thus '$y = \mathbf{P}(x)$' is expressed in the language of ZF set theory as '$\forall z (z \in y \leftrightarrow \forall u (u \in z \to u \in x))$', and '$y = \cup x$' as '$\forall z (z \in y \leftrightarrow \exists u (z \in u \wedge u \in x))$'.

Another defined notion is that of *the singleton $\{x\}$ of x,* which is defined formally as $y = \{x\} \leftrightarrow \forall z (z \in y \leftrightarrow z = x)$.

Problem 1214. *Formulate and prove in ZF the assertion that for any a, the singleton set $\{a\}$ exists and is unique.*

[Hint: Note that $\{a\} \subseteq \mathbf{P}(a)$, and then use separation.]

Starting from the empty set, we can iterate the power set operation to get larger and larger sets, as in

$$\emptyset \subseteq \mathbf{P}(\emptyset) \subseteq \mathbf{P}(\mathbf{P}(\emptyset)) \subseteq \cdots$$

Let us put:

$$V_0 := \emptyset, \quad \text{and} \quad V_{n+1} := \mathbf{P}(V_n).$$

Note that if V_n has m elements then V_{n+1} has 2^m elements.

Problem 1215. *List all elements of* V_4.

From the axioms introduced so far, one can derive that each set V_n (and in fact every member of V_n) exists, but the infinite collection $\bigcup_n V_n$ itself cannot be shown to exist without using the axioms of Replacement and Infinity to be introduced later.

Another useful axiom is the *Axiom of Unordered Pairs* also called *the Axiom of Pairing,* which says that for any x, y we can form the unordered pair $\{x, y\}$. This amounts to $\forall x \forall y \exists z(z = \{x, y\})$, which in turn can be expressed as a ZF formula by replacing $z = \{x, y\}$ with '$\forall w(w \in z \leftrightarrow w = x \vee w = y)$':

ZF 6 (Axiom of Unordered Pairs). $\forall x \forall y \exists z \forall w(w \in z \leftrightarrow w = x \vee w = y)$.

Problem 1216. *Using the ZF axioms introduced so far, show that for any* a, b, c *the unordered triple* $\{a, b, c\}$ *exists. That is, it is a theorem of ZF that*

$$\forall a \forall b \forall c \exists z \forall w(w \in z \leftrightarrow w = a \vee w = b \vee w = c).$$

We now have enough axioms to develop most of the important notions of set theory, although infinite sets cannot yet be shown to exist. Using Pairing and Union, we can define $a \cup b$ simply as $\bigcup\{a, b\}$, while $a \cap b$ and $a \smallsetminus b$ can be defined more simply using Separation (once again, these defined notions can be formally eliminated).

Definition 1217. We define in ZF:

$$a \cup b := \bigcup\{a,b\}, \quad a \cap b := \{x \in a \mid x \in b\}, \quad a \smallsetminus b := \{x \in a \mid x \notin b\},$$

Repeatedly using Pairing shows that for each a, b the set $\{\{a\}, \{a, b\}\}$ exists and is unique, which gives the definition of ordered pair due to Kuratowski:

Definition 1218 (Ordered Pair). $\langle u, v\rangle := \{\{u\}, \{u, v\}\}$.

Problem 1219. *Prove that this is an acceptable definition of ordered pair, i.e., show in ZF that* $\langle a, b\rangle = \langle c, d\rangle \rightarrow (a = c \wedge b = d)$.

As in the case of the singleton and the unordered pair, the defined notion of ordered pair can be eliminated in the sense that "$x = \langle u, v\rangle$" can be replaced by a pure ZF formula.

Note that $\langle a, b\rangle \in \mathbf{P}(\mathbf{P}(\{a, b\}))$, so we can use Separation to define the Cartesian product of two sets.

Definition 1220. $a \times b := \{x \in \mathbf{P}(\mathbf{P}(a \cup b)) \mid \exists u, v(u \in a \wedge v \in b \wedge x = \langle u, v\rangle)\}$.

Now we can define relations, functions, domains, ranges, equinumerosity between sets, equivalence relations and partitions, order relations, order-isomorphisms, well-orders, etc., using the ZF axioms we have so far.

Definition 1221. The following notions can be formally defined in ZF:

1. R is a relation $\leftrightarrow \exists a \exists b(R \subseteq a \times b)$
2. $R^{-1} := \{\langle y, x\rangle \in \mathbf{P}(\mathbf{P}(\bigcup \bigcup R)) \mid \langle x, y\rangle \in R\}$.

3. $R \circ R := \{\langle x, z\rangle \in \mathbf{P}(\mathbf{P}(\bigcup\bigcup R)) \mid \exists y(\langle x, y\rangle \in R \wedge \langle y, z\rangle \in R)\}$.
4. $\operatorname{dom}(R) := \{x \in \bigcup\bigcup R \mid \exists y(\langle x, y\rangle \in R)\}$.
5. $\operatorname{ran}(R) := \operatorname{dom}(R^{-1})$.
6. R is symmetric $\leftrightarrow R^{-1} \subseteq R$.
7. R is asymmetric $\leftrightarrow R^{-1} \cap R = \emptyset$.
8. R is transitive $\leftrightarrow R \circ R \subseteq R$.
9. $\Delta_A := \{z \in A \times A \mid \exists x \in A(z = \langle x, x\rangle)\}$.
10. R is a relation on $A \leftrightarrow R \subseteq A \times A$.
11. R is reflexive on $A \leftrightarrow R$ is a relation on $A \wedge \Delta_A \subseteq R$.
12. R is irreflexive on $A \leftrightarrow R$ is a relation on $A \wedge \Delta_A \cap R = \emptyset$.
13. R is connected on $A \leftrightarrow A \times A \subseteq R \cup R^{-1} \cup \Delta_A$.
14. E is an equivalence relation on $A \leftrightarrow E$ is symmetric, transitive, and reflexive on A.
15. R orders $A \leftrightarrow R$ is asymmetric, transitive, and connected on A.
16. R well-orders $A \leftrightarrow R$ orders A and $\forall B \subseteq A(B \neq \emptyset \rightarrow \exists b \in B(\neg\exists c \in B(\langle c, b\rangle \in R)))$.
17. f is a function $\leftrightarrow f$ is a relation $\wedge \forall x, y, z(\langle x, y\rangle \in f \wedge \langle x, z\rangle \in f \rightarrow y = z)$.
18. f is one-one $\leftrightarrow$ both f and f^{-1} are functions.
19. $A \approx B \leftrightarrow \exists f(f$ is one-one $\wedge \operatorname{dom}(f) = A \wedge \operatorname{ran}(f) = B)$.
20. $\langle A, R\rangle \cong \langle B, S\rangle \leftrightarrow R \subseteq A\times A \wedge S \subseteq B\times B \wedge \exists f(f$ is one-one $\wedge \operatorname{dom}(f) = A \wedge \operatorname{ran}(f) = B \wedge \forall \langle u, v\rangle, \langle x, y\rangle \in f(\langle u, x\rangle \in R \leftrightarrow \langle v, y\rangle \in S))$.

Problem 1222. *Using the axioms of ZF introduced so far, prove in ZF the principle of transfinite induction for well-orders formalized as follows: If R well-orders A, $B \subseteq A$, and*

$$\forall x \in A((\forall y \in A(\langle y, x\rangle \in R \rightarrow y \in B)) \rightarrow x \in B),$$

then $B = A$.

Problem 1223. *Formulate and prove in ZF the comparability theorem for well-orders that was proved informally in Theorem 636: For any two well-orders one must be isomorphic to an initial segment of the other. Use only the axioms of ZF that have already been introduced,*

More basic results as above about sets and orders that were proved informally in the initial parts of this book can be replicated formally in ZF using the axioms we have so far. We leave it to the reader to pursue this project: Find and prove (in ZF) as many basic results as possible from the axioms of ZF we have introduced so far, while defining and developing (formally in ZF) appropriate notions necessary for those results.

Define a well-order X to be an *infinite well-order* if either X or some initial segment of X is a nonempty order without a greatest element. Recall that a set is *Dedekind infinite* (or *reflexive*) if there is a one-to-one mapping of the set into one of its proper subsets. Here are the formal versions of these notions in ZF.

Definition 1224 (ZF). $\langle A, R\rangle$ is an *infinite well-order* if and only if

$$R \text{ well-orders } A \ \wedge\ A \neq \emptyset \ \wedge\ \big([\forall x \in A\, \exists y \in A(\langle x, y\rangle \in R)]$$
$$\vee\ [\exists z \in A(\forall x \in A(\langle x.z\rangle \in R \rightarrow \exists y \in A(\langle x, y\rangle \in R \wedge \langle y, z\rangle \in R)))]\big).$$

A is *Dedekind infinite* (or *reflexive*) if and only if

$$\exists f(f \text{ is a bijection} \wedge \operatorname{dom}(f) = A \ \wedge\ \operatorname{ran}(f) \subseteq A \ \wedge\ A \smallsetminus \operatorname{ran}(f) \neq \emptyset).$$

Problem 1225. *From the ZF axioms introduced so far, show that there is an infinite well-order if and only if there is a Dedekind infinite (reflexive) set.*

The development of ZF so far should already illustrate the ZF-paradigm "everything is a set": In ZF, the only type of objects that exist are sets. Ordered pairs, functions, and well-orders have all been shown to be sets in ZF, and all theorems of ZF are, in the end, results about sets and membership.

However, not everything that were done informally in Parts I–III can be formally developed in ZF yet. Numbers, in particular cardinals and ordinals, cannot be formed yet in full generality, and existence of infinite sets cannot be established. For these we will need the axioms of replacement and infinity which will be introduced in the next sections. But once those axioms are available, the formal development of ZF becomes capable of realizing the ZF-paradigm "everything is a set" to its fullest extent: Not only all types of numbers, such as natural, rational, real, and complex numbers, but also all objects of higher mathematics such as algebraic structures and spaces, can be constructed as sets. In fact, all mathematical statements can be expressed in the language of ZF set theory, and all theorems of classical mathematics can be proved in ZFC (ZFC = ZF augmented with the Axiom of Choice).

Classes, Relationals, and Functionals

We have seen that in ZF there is no set containing all sets. However, it is convenient to informally refer to such large collections as *classes*. For example, we will let V denote the collection of all sets,

$$V := \{x \mid x = x\},$$

so that V is a class which is not a set. Another example is the class of all ordered pairs, conveniently denoted by $V \times V$, which cannot be a set in ZF either (why?). A class such as V does not formally exist in ZF, and the term "class" really refers to a ZF property (i.e., a ZF formula with a specified free variable and possibly with additional parameters). For example, with x as the defining free variable, the class V above stands for the ZF property $V = V(x)$ where $V(x)$ is the ZF formula '$x = x$', and the class $V \times V$ can be replaced by the ZF formula '$\exists y \exists z(x = \langle y, z\rangle)$'. Of course some ZF-formulas do define sets, such as the formula '$x \neq x$' which defines the empty set Ø and the formula '$\forall z(z \notin x)$' which defines the set $\{\emptyset\}$. So some classes are sets. In fact, every set is a class since for any set a, the formula '$x \in a$' defines a. A class which is not a set, such as V or $V \times V$, is called a *proper class.*

As another example, the membership relation '$\in$', which is a *subclass* of $V \times V$, is a relation which is not a set:

$$\in = \{\langle x, y\rangle \mid x \in y\} \subseteq V \times V.$$

A relation which may or may not be set, i.e., a class consisting of ordered pairs alone, will be called a *relational.* Informally, a relational is simply any subclass of $V \times V$. Formally, a relational refers to a ZF formula of the form $\varphi(x, y)$ with two specified free variables. Another familiar relational which is not a set is equinumerosity, $\approx$, given by

$$\approx := \{\langle x, y\rangle \mid |x| = |y|\},$$

and it corresponds to the ZF formula "$\exists f(f$ is a bijection from x onto $y)$." The domain and range of a relational are classes which may not in general be sets. For example, we have $\operatorname{dom}(\in) = V$ and $\operatorname{ran}(\in) = V \smallsetminus \{\emptyset\}$.

A relational F will be called a *functional* if $xFy \wedge xFy' \Rightarrow y = y'$. For a functional F, we will use the functional notation $F(x)$ to denote the unique y such that xFy. Formally, a functional expression '$y = F(x)$' is really a ZF formula of the form $\varphi(x, y)$ for which it can be proved in ZF that

$$\forall x \forall y \forall y' (\varphi(x, y) \wedge \varphi(x, y') \rightarrow y = y') \qquad (\text{“}\phi \text{ is many-one”}).$$

In fact, we have already been using functionals and functional notation in expressions like $\{x\}$, $\mathbf{P}(x)$, and $\cup x$, which assign a unique set to every set x. These functionals all have domain V, but some functionals are not defined for all sets. For example, the functional π_2 which assigns to every ordered pair its second coordinate ($\pi_2(\langle a, b\rangle) = b$ for all a, b) has domain $\operatorname{dom}(\pi_2) = V \times V$.

21.3 The Replacement Axiom

The *Axiom of Replacement* (due to Fraenkel) says that if A is a set and F is any functional whose domain includes A, then the image $F[A]$ is also a set. In a slightly variant but equivalent form it says that for any set A and functional F we can form the set

$$\{F(x) \mid x \in A \cap \operatorname{dom}(F)\}.$$

Formally, it is an axiom scheme (like separation):

ZF 7 (Axiom Scheme of Replacement). *For every ZF formula* $\varphi = \varphi(x, y)$, *we have the axiom:*

$$\forall x \forall y \forall z (\varphi(x, y) \wedge \varphi(x, z) \rightarrow y = z) \rightarrow \forall a \exists b \forall u \forall v (u \in a \wedge \varphi(u, v) \rightarrow v \in b).$$

We had already implicitly used the Axiom of Replacement in some constructions such as the Hartogs' ordinal. Assume for the moment that we have defined ordinals and that every well-order X has a unique ordinal $\mathrm{Ord}(X)$. (This will be done in detail in the next section.) For any set A, we want to form the "Hartogs' set" $H(A)$ of ordinals as

$$H(A) := \{\alpha \mid \text{There is an injection from } W(\alpha) \text{ into } A\},$$

but in the above form it is not clear that this collection of ordinals will form a set. To address this issue, consider the set W_A of all well-orders defined on subsets of A. Then since $W_A \subseteq \mathbf{P}(\mathbf{P}(\mathbf{P}(A)))$, so W_A can be shown to be a set by the power set and separation axioms. Since W_A is a set, we can use Replacement to derive that

$$H(A) = \{\mathrm{Ord}(R) \mid R \in W_A\}$$

is a set too.

However, in most of ordinary mathematics, replacement is scarcely used.

21.4 The von Neumann Ordinals

In 1923, von Neumann [80] gave a remarkably simple, natural, absolute definition for ordinal numbers:

> The aim of this work is to present Cantor's notion of ordinal numbers in a clear and concrete form. ...
>
> We actually take the proposition "Every ordinal is the type of the set of all ordinals preceding it" as the basis of our considerations. But we avoid the vague notion of "type" by expressing it as follows: "Every ordinal is the set of all ordinals preceding it." ... Accordingly (O is the empty set, $(a, b, c, \ldots)$ is the set with the elements $a, b, c, \ldots$):
>
> $$0 = \mathrm{O},$$
> $$1 = (\mathrm{O}),$$
> $$2 = (\mathrm{O}, (\mathrm{O})),$$
> $$3 = (\mathrm{O}, (\mathrm{O}), (\mathrm{O}, (\mathrm{O}))),$$
> $$\ldots \quad \ldots \quad \ldots$$
> $$\omega = (\mathrm{O}, (\mathrm{O}), (\mathrm{O}, (\mathrm{O})), (\mathrm{O}, (\mathrm{O}), (\mathrm{O}, (\mathrm{O}))), \ldots)$$
> $$\ldots \quad \ldots \quad \ldots$$

[Author's translation of a part of [80, p. 199], done with permission from Acta Sci. Math. (Szeged).]

We now present von Neumann's results.

Definition 1226 (Von Neumann Well-Orders). A well-order X is said to be a *von Neumann well-order* if for every $x \in X$, we have $x = \{y \in X \mid y < x\}$ (that is x is equal to the set $\mathrm{Pred}(x)$ consisting of its predecessors).

Clearly the examples listed by von Neumann above, namely

$$\emptyset,\ \{\emptyset\},\ \{\emptyset,\{\emptyset\}\},\ \{\emptyset,\{\emptyset\},\{\emptyset,\{\emptyset\}\}\},\ \ldots$$

are all von Neumann well-orders if ordered by the membership relation "$\in$," and the process can be iterated through the transfinite. Our immediate goal is to show that these and only these are the von Neumann well-orders, with exactly one von Neumann well-order for each ordinal (order type of a well-order). This is called the existence and uniqueness result for the von Neumann well-orders.

An immediate consequence of the definition of a von Neumann well-order is:

Proposition 1227. *Let X be a well-order. If X is a von Neumann well-order then the ordering relation on X coincides with the membership relation "$\in$" restricted to X, that is, for all $x, y \in X$ we have $x < y \leftrightarrow x \in y$.*

Problem 1228. *Show that the converse of the above proposition fails.*

It is also immediate that every proper initial segment of a von Neumann well-order X is a member of X and vice versa:

Proposition 1229. *For a von Neumann well-order X, the proper initial segments of X coincide with the members of X; that is, Y is a proper initial segment of X if and only if $Y \in X$. Thus, the set of proper initial segments of X equals X itself.*

Since a set cannot be equal to any of its proper subsets, we get:

Corollary 1230. *If X is a von Neumann well-order then $X \notin X$.*

Corollary 1231. *Every member of a von Neumann well-order is a von Neumann well-order (under the membership relation).*

Corollary 1232. *If X is a von Neumann well-order, then the ordering relation the relation on X coincides with relation of proper set containment for the members of X, that is, for all $x, y \in X$ we have $x < y \leftrightarrow x \subsetneq y$.*

Problem 1233. *Show that the converse of the above result fails.*

We now proceed to prove the existence and uniqueness results for von Neumann well-orders.

Uniqueness

First, we have the following uniqueness theorem, which says that isomorphic von Neumann well-orders are identical:

Theorem 1234. *If X and Y are isomorphic von Neumann well-orders, then $X = Y$.*

Proof. Let f be an order isomorphism from X onto Y. We show by transfinite induction that $f(x) = x$ for all x, thereby establishing the result. Suppose that $x \in X$ and that $f(u) = u$ for all $u < x$. Then since f is an order isomorphism, we have

$$\begin{aligned} f(x) = \{v \mid v \in Y \wedge v < f(x)\} &= \{f(u) \mid u \in X \wedge f(u) < f(x)\} \\ &= \{f(u) \mid u \in X \wedge u < x\} \\ &= \{u \mid u \in X \wedge u < x\} = x. \end{aligned}$$ □

In other words, the uniqueness theorem says that every well-order is isomorphic to *at most one* von Neumann well-order.

Uniqueness has some immediate consequences.

Corollary 1235. *If a von Neumann well-order X is isomorphic to a proper initial segment of a von Neumann well-order Y, then $X \in Y$.*

Hence, using the comparability theorem for well-orders, we get:

Corollary 1236 (Comparability). *For any two von Neumann well-orders one must be an initial segment of the other. For any two distinct von Neumann well-orders one must be a member of the other. Thus if X and Y are von Neumann well-orders, then exactly one of $X \in Y$, $X = Y$, and $Y \in X$ holds.*

Corollary 1237. *If X and Y are von Neumann well-orders, then the order-type of X is less than the order type of Y if and only if $X \in Y$.*

Corollary 1238. *Every von Neumann well-order X consists of all von Neumann well-orders having order type less than that of X.*

Corollary 1239. *If X is a set of von Neumann well-orders and if for every member u of X any von Neumann well-order having order type less than that of u is also in X, then X is itself a von Neumann well-order.*

Definition 1240 (Successor of a Set). For any set X, we define *the successor of* X, denoted by X^+, as $X^+ := X \cup \{X\}$.

Proposition 1241. *If X is a von Neumann well-order of order type α, then X^+ is a von Neumann well-order of order type $\alpha + 1$.*

Proof. Since $X \notin X$, so with $\in$ as the ordering relation $X \cup \{X\}$ becomes a strict linear order with greatest element X in which the set of $\in$-predecessors of X is X itself. □

Corollary 1242. *The union of any set C of von Neumann well-orders is a von Neumann well-order whose order type is the supremum of the order types of members of C.*

Proof. C is a chain by comparability. Hence $\bigcup C$ is linearly ordered by $\in$. Regarding $\bigcup C$ to be a linear order with $\in$ as the ordering relation, we see that each member of C is an initial segment of $\bigcup C$, and also each proper initial segment of $\bigcup C$ is contained in a member of C. Hence $\bigcup C$ is well-ordered by $\in$, and its order type is the supremum of the order types of members of C. □

Existence

We now prove the existence theorem, whose proof uses Replacement in an essential way.[1]

Theorem 1243. *For each well-order X there is a (unique) von Neumman well-order which is isomorphic to X.*

Proof. The proof is by transfinite induction on well-orders. Suppose that X is well-order such that every proper initial segment of X is isomorphic to some von Neumann well-order. We show that then X itself is isomorphic to some von Neumann well-order.

For every $x \in X$ there is a unique von Neumann well-order Y_x isomorphic to $\mathrm{Pred}_X(x)$ (by induction hypothesis and by uniqueness). By Replacement, we can form the set

$$C := \{Y_x \mid x \in X\}.$$

Note that for any $x, y \in X$ we have $x < y$ if and only if order type of Y_x is less than the order type of Y_y, which holds if and only if $Y_x \in Y_y$. Hence the mapping $x \mapsto Y_x$ is an order-isomorphism from X onto $\langle C, \in\rangle$. Therefore C is well-ordered by $\in$ with every member of C equal to the set of its predecessors, and hence C must be a von Neumann well-order isomorphic to X. □

New Definition of Ordinal Number

By the existence and uniqueness theorems, for every well-order X there is a unique von Neumann well-order isomorphic to it, and two well-orders are isomorphic if

---

[1]Earlier than von Neumann, others (Zermelo, Mirimanoff, etc) had partially developed similar ideas, but the results were limited as the general existence theorem could not be proved due to the lack of the Replacement axiom.

and only their associated von Neumann well-orders are identical. We thus have a complete invariant for the equivalence relation of order-isomorphism between well-orders, which provides a remarkably simple, elegant, and concrete definition for ordinal numbers:

Definition 1244 (Von Neumann Ordinals). For each well-order X, we define *the von Neumann ordinal of* X, or simply *the ordinal of* X, denoted by $\mathrm{Ord}(X)$, to be the unique von Neumann well-order isomorphic to X.

α is called an *ordinal* if α is the ordinal of some well-order, i.e. if α is a von Neumann well-order. The class of all ordinals will be denoted by On.

Von Neumann's definition of ordinal numbers has become the standard in axiomatic set theory, and forms the backbone of the universe of sets and the theory of the transfinite. From now on we will follow the above definition and use the term "ordinal" to mean a von Neumann well-order.

Since any ordinal is equal to the set of smaller ordinals, the older set $W(\alpha) = \{\beta \mid \beta < \alpha\}$ becomes identical to α itself:

$$W(\alpha) = \{\beta \mid \beta < \alpha\} = \{\beta \mid \beta \in \alpha\} = \alpha.$$

We can therefore dispense with the notation "$W(\alpha)$," replacing it with the simpler α.

By the Burali-Forti paradox, the class On is not a set and does not exist formally in ZF, and the expression "$x \in \mathrm{On}$" really stands for the ZF formula "x is an ordinal." However, if a subclass A of On is a *proper* initial segment of On then A is a set, since A then equals the least ordinal not in A.

Definition 1245 (Zero, Successor, and Limit Ordinals). We define the smallest ordinal *zero* as $0 := \emptyset$. An ordinal α is a *successor ordinal* if $\alpha = \beta^+$ for some ordinal β. An ordinal which is neither zero nor a successor is a *limit ordinal.*

Definition 1246. We say that x is an *initial set of ordinals* if every member of x is an ordinal and for all $y \in x$, if $z < y$ then $z \in x$.

Problem 1247. *x is an ordinal if and only if x is an initial set of ordinals.*

Theorem 1248 (The Principle of Transfinite Recursion). *For each functional G with domain V, there exists a unique functional F with domain* On *such that for every $\alpha \in$* On,

$$F(\alpha) = G(\langle F(\beta) \mid \beta < \alpha \rangle) \qquad (i.e., F(\alpha) = G(F \restriction \alpha)).$$

Problem 1249. *Prove Theorem 1248 in ZF.*

[Hint: The proof is the same as that of Theorem 622, but note the following. Formally, one cannot quantify over classes in ZF, so phrases like "for each functional G" and "there exists a functional F" are not officially allowed. So Theorem 1248 is really a *theorem scheme*: For each class G (really a formula), another class (formula)

F can be formed (in terms of G) for which the corresponding *instance* of the theorem can be proved in ZF.

However, there is little danger in informally treating relationals and functionals as ordinary sets as long as one is careful about them. (Functionals defined on proper initial segments of On are really sets, so they can be quantified over at will.) Thus it is alright to pattern the proof as follows:

> Given $G: V \to V$, define F to be the class of those ordered pairs $\langle x, y\rangle$ such that there is an ordinal α and a function h with $\mathrm{dom}(h) = \alpha$, $x \in \mathrm{dom}(h)$, $h(x) = y$, and which satisfies $h(\beta) = G(h \restriction \beta)$ for all $\beta < \alpha. \ldots$

The rest of the proof, which consists of showing that F is a functional satisfying the theorem, is identical to that of Theorem 622.]

Transitive Sets

Definition 1250. A set x is called *transitive* if for all y, z, $z \in y \wedge y \in x \to z \in x$.

Problem 1251. *Show that a set x is transitive if and only if every element of x is a subset of x if and only if $\bigcup x \subseteq x$.*

Problem 1252. *Show that the empty set is transitive and that for any set x, if x is transitive then so are $x^+ = x \cup \{x\}$, $\mathbf{P}(x)$, and $\bigcup x$.*

Problem 1253. *Show that every ordinal is transitive.*

The following problem characterizes von Neumann ordinals as transitive sets well-ordered by the membership relation $\in$, and thus in some treatments it is taken as the *definition* for von Neumann ordinals.

Problem 1254 (Characterization of Von Neumann Ordinals). *Show that x is an ordinal if and only if x is transitive and x is well-ordered by the membership relation $\in$.*

21.5 Finite Ordinals and the Axiom of Infinity

Finite Ordinals

The smallest von Neumann well-order is $0 = \emptyset$, and repeatedly applying the successor operation we get:

$$0 := \emptyset$$

$$1 := 0 \cup \{0\} = \{\emptyset\}$$

$$2 := 1 \cup \{1\} = \{0, 1\} = \{\emptyset, \{\emptyset\}\},$$

$$3 := 2 \cup \{2\} = \{0, 1, 2\} = \{\emptyset, \{\emptyset\}, \{\emptyset, \{\emptyset\}\}\}, \text{ etc.}$$

The above are the finite von Neumann well-orders. Formally, we have

Definition 1255 (Finite and Infinite Ordinals). n is a *finite ordinal* if either $n = 0$ or n is a successor ordinal and every nonzero element of n is a successor ordinal.

An *infinite ordinal* is an ordinal which is not finite.

Problem 1256. *0 is a finite ordinal. If x is a finite ordinal then so is x^+.*

Problem 1257. *If n is a finite ordinal and $m \in n$ then m is a finite ordinal. Thus any initial segment of a finite ordinal is a finite ordinal.*

Problem 1258. *If n is a nonzero finite ordinal then $n = m^+$ for some finite ordinal $m \in n$.*

The existence of infinite ordinals cannot be proved yet, since the collection of all finite ordinals is not known to be a set—at this point it is only a class. But we still have:

Problem 1259. *Any infinite ordinal contains all finite ordinals as members.*

The class of finite ordinals together with the successor operation can now be verified to satisfy the Dedekind–Peano axioms, including the principle of finite induction.

Theorem 1260 (The Principle of Finite Induction). *Let P be a ZF property for which we have $P(0)$ and $\forall n(P(n) \rightarrow P(n^+))$. Then we have $\forall n(n$ is a finite ordinal $\rightarrow P(n))$.*

Proof. Assume the hypothesis, and to derive a contradiction, suppose that n is a finite ordinal for which we have $\neg P(n)$. Then $n \neq 0$ by hypothesis, and so $n = m^+$ for some finite ordinal $m \in n$. Now we must have $\neg P(m)$ since otherwise by hypothesis we would have $P(n)$. Hence the set $\{k \in n \mid \neg P(k)\}$ is nonempty and must have a least element q. Since $P(0)$ is true, we have $q \neq 0$, and so, since q is a nonzero finite ordinal, $q = r^+$ for some finite ordinal $r \in q$. We now must have $P(r)$ by minimality of q, which implies $P(q)$ is true, a contradiction. □

Therefore, we can define operations on the finite ordinals such as addition and multiplication, and the entire theory of "Peano Arithmetic" can be developed based on the finite ordinals.

We kept our development of finitude in ZF independent of the one given in Sect. 5.3, but they are really equivalent in the following sense.

Problem 1261. *Call a set x* inductive *if for any $y \in \mathbf{P}(\mathbf{P}(x))$ the two conditions $\emptyset \in y$ and $\forall z \in y\, \forall u \in x(z \cup \{u\} \in y)$ imply $x \in y$. Show that a set x is inductive if and only if $x \approx n$ for some (unique) finite ordinal n.*

The Axiom of Infinity

We now want to get ω, the supremum of all the finite ordinals, which, as an ordinal, must equal to the set of all smaller ordinals. In other words, ω must consist of all finite ordinals, as in:

$$\omega := \{0, 1, 2, \dots\}.$$

But how can we justify that this exists as a set and is not a proper class? More precisely, how can it be proved that there is a set consisting precisely of the finite ordinals?

To discuss this question, let "$x = \omega$" stand for "x is the set of all finite ordinals," or more precisely for the ZF formula '$\forall y(y \in x \leftrightarrow y$ is a finite ordinal)', and let "ω exists" stand for the ZF formula '$\exists x(x = \omega)$'.

Note that to show that ω exists, it suffices to show that there is some set b containing all finite ordinals, since then by Separation we can get $\omega = \{y \in b \mid y \text{ is a finite ordinal}\}$. By the principle of finite induction, any set b satisfying $0 \in b \wedge \forall n(n \in b \to n^+ \in b)$ will contain all finite ordinals. Moreover, since every infinite ordinal contains all finite ordinals, the existence of ω can be seen to be equivalent to the existence of an infinite ordinal. These equivalences do not need the replacement axiom, and so we have:

Proposition 1262. *From the ZF axioms introduced so far one can derive, without using the replacement axiom, that the following conditions are equivalent to each other:*

1. *ω exists.*
2. *There is a set b such that $0 \in b \wedge \forall n(n \in b \to n^+ \in b)$.*
3. *There is an infinite ordinal.*

It turns out that none of the statements above (in particular the existence of ω) can be derived from the axioms introduced so far. Zermelo, in his 1908 paper [85] introduced the *Axiom of Infinity* precisely for this purpose.[2]

ZF 8 (Axiom of Infinity). *ω exists, or equivalently, there is an infinite ordinal, or equivalently $\exists b(0 \in b \wedge \forall n(n \in b \to n^+ \in b))$.*

We will now show, using Replacement, that the Axiom of Infinity is equivalent to the existence of an infinite set, where by an infinite set we mean a Dedekind infinite (reflexive) set.

First, since the mapping $n \mapsto n^+$ maps ω into a proper subset of ω, so the Axiom of Infinity implies (even without Replacement) that there is a Dedekind infinite (reflexive) set.

[2]Zermelo used the operation $x \mapsto \{x\}$ instead of the successor operation, and his version of the axiom stated that there is a set b such that $\emptyset \in b$, and $\{x\} \in b$ for every $x \in b$.

To get the converse, we will use Replacement. If we carefully examine the proof of the existence theorem, we see that the existence of an ordinal α was obtained from the existence of *some* well-order of type α, through the use of the replacement axiom. Thus the existence of ω will follow from the existence of a well-order of type ω, and the existence of an infinite ordinal will follow from the existence of an infinite well-order. Now, earlier we had seen that the existence of a Dedekind infinite (reflexive) set is equivalent to the existence of an infinite well-order, hence under replacement all our conditions become equivalent.

Theorem 1263. *From the ZF axioms introduced prior to Infinity (thus including Replacement), one can derive that the following are equivalent:*

1. *The Axiom of Infinity, that is ω exists, or equivalently that there is an infinite ordinal.*
2. *There is an infinite well-order.*
3. *There is a Dedekind infinite (reflexive) set.*

Problem 1264. *Prove, based on the ZF axioms introduced so far, that the ordinal $\omega + \omega$ exists.*

The Replacement axiom, together with the Axiom of Infinity, guarantees access to the transfinite. All the infinite cardinals and ordinals that we had studied earlier can be shown to exist using Replacement.

21.6 Cardinal Numbers and the Transfinite

In addition to providing a canonical representative for every well-order, the von Neumann ordinals can give a complete invariant for the equivalence relation of equinumerosity so long as the Axiom of Choice is assumed.

Definition 1265 (Equinumerosity, Initial Ordinals). We write $a \approx b$ to denote a is equinumerous (bijective) with b. An ordinal α is said to be an *initial ordinal* if there is no $\beta < \alpha$ such that $\beta \approx \alpha$.

All finite ordinals are initial ordinals. The first few transfinite initial ordinals are $\omega, \omega_1, \omega_2, \dots$.

By AC, every set can be well-ordered, and so must be equinumerous to some ordinal, and therefore also to some initial ordinal. Also, in Cantor's original conception of the transfinite (which implicitly assumed AC so that every infinite cardinal was an aleph), cardinals correspond naturally and uniquely with the initial ordinals. We thus obtain the classic *Cantor–Von Neumann definition of cardinal numbers.*

Definition 1266 (Cantor–Von Neumann Cardinals (AC)). For any set x, define $|x|$, *the cardinality of* x, to be the least ordinal α such that $\alpha \approx x$.

We say that κ is a *cardinal* if $\kappa = |x|$ for some set x.

Thus in the Cantor–Von Neumann definition, *every cardinal is an ordinal.*

Problem 1267. *α is an initial ordinal if and only if α is a cardinal.*

The Cantor–Von Neumann definition is readily seen to satisfy the condition

$$|x| = |y| \leftrightarrow x \approx y,$$

and so we have an adequate definition of cardinal numbers under AC. In particular, we have $\aleph_0 = \omega$, $\aleph_1 = \omega_1$, etc., but the ordinal $|\mathbf{R}|$ cannot be effectively determined because of the independence of the continuum hypothesis even under AC.

The Cumulative Hierarchy

We had earlier defined the sets V_n for finite ordinals n, where

$$V_0 = \emptyset \quad \text{and} \quad V_{n^+} = \mathbf{P}(V_n).$$

Now, by Infinity, ω exists, and so by Replacement and Union, we can define

$$V_\omega := \bigcup_{n<\omega} V_n.$$

Since each V_n is finite for $n \in \omega$, so V_ω is countably infinite. We can next define

$$\begin{aligned} V_{\omega+1} &:= \mathbf{P}(V_\omega), \\ V_{\omega+2} &:= \mathbf{P}(V_{\omega+1}) \\ &\vdots \end{aligned}$$

so that $V_{\omega+1}$ has cardinality $\mathfrak{c} = 2^{\aleph_0}$, $V_{\omega+2}$ has cardinality $2^{\mathfrak{c}}$, and so on.

Definition 1268 (The Cumulative Hierarchy). Define the *cumulative hierarchy of sets* $V_\alpha, \alpha \in \text{On}$, by transfinite recursion on α as:

$$\begin{aligned} V_0 &:= \emptyset, \\ V_{\alpha+1} &:= \mathbf{P}(V_\alpha), \quad \text{and} \\ V_\alpha &:= \bigcup_{\beta<\alpha} V_\beta \text{ if } \alpha \text{ is a limit ordinal.} \end{aligned}$$

The sets V_α increase with the ordinal α, that is,

$$V_0 \subseteq V_1 \subseteq \cdots \subseteq V_\alpha \subseteq V_{\alpha+1} \subseteq \cdots.$$

In fact, we have:

Problem 1269. *For any ordinal* α, $V_\alpha \subsetneq V_{\alpha+1}$, *and*

$$V_\alpha = \bigcup_{\beta<\alpha} V_{\beta+1}.$$

Problem 1270. $x \in V_\alpha$ *if and only if* $x \subseteq V_\beta$ *for some* $\beta < \alpha$.

Problem 1271. *If* α *is an ordinal, then* $\alpha \in V_{\alpha+1} \setminus V_\alpha$.

Problem 1272. *Show that* V_α *is transitive for each ordinal* α.

Set Theory Without Replacement

Definition 1273 (Zermelo Set Theory, Z). The axiom system consisting of all the ZF axioms mentioned so far, including Infinity, but *without Replacement,* is known as *Zermelo Set Theory* and is denoted by Z.

The axiom system Z, introduced by Zermelo[3] in 1908 [85], was the first formulation of axiomatic set theory in the modern sense. As we will indicate below, Z has sufficient power for the development of almost all of the ordinary mathematics.

Problem 1274. *Prove the following in Z. Given any sets* A *and* B, *the set* $A \times B$ *(cartesian product) and the set* B^A *of all functions from* A *to* B *both exist. Also, given any set* A, *the set* $A^{<\omega}$ *of all finite sequences from* A *exists. (A* finite sequence from *A is a function whose domain is a finite ordinal and whose range is contained in A.)*

[Hint: By Union, Power Set, and Separation, note that $A \times B \subseteq \mathbf{P}(\mathbf{P}(A \cup B))$, $B^A \subseteq \mathbf{P}(A \times B)$, and $A^{<\omega} \subseteq \mathbf{P}(\omega \times A)$.]

Developing Mathematics in Z

By the Axiom of Infinity, the von Neumann ordinal ω exists in Z. So we can iterate the power set operation and get the existence of each of the following sets in Z: $\mathbf{P}(\omega)$, $\mathbf{P}(\mathbf{P}(\omega))$, $\ldots$, $\mathbf{P}^n(\omega)$, $\ldots$, where we define $\mathbf{P}^0(\omega) = \omega$ and $\mathbf{P}^{n+1}(\omega) := \mathbf{P}(\mathbf{P}^n(\omega))$. Also the sets $\omega\times\omega$, $\mathbf{P}(\omega\times\omega)$, ω^ω, all exist as well (with $\omega\times\omega \subseteq \mathbf{P}^2(\omega)$, $\omega^\omega \subseteq \mathbf{P}^3(\omega)$, etc.). This allows the construction of the positive rationals (ratios) as

[3]As mentioned earlier, Zermelo's original system used the mapping $x \mapsto \{x\}$ instead of the successor operation in its formulation of the Axiom of Infinity, and the infinite set whose existence was asserted was $\{\emptyset, \{\emptyset\}, \{\{\emptyset\}\}, \ldots\}$. In this respect our system Z is strictly distinct from Zermelo's original; see Kunen [44, p.125], Exercise II.4.21.

ordered pairs from ω, so the set of rational numbers $\mathbf{Q}$ is in $\mathbf{P}^n(\omega)$ for some finite n. Therefore the set $\mathbf{R}$ of real numbers (defined as Dedekind cuts of rationals numbers so that $\mathbf{R} \subseteq \mathbf{P}(\mathbf{Q})$) is also in $\mathbf{P}^n(\omega)$ for some finite n. So $\mathbf{P}(\mathbf{R} \times \mathbf{R})$, i.e., the set of all relations on $\mathbf{R}$ is a set in Z. Hence also the set of all real functions is a set whose existence can be established in Z. Complex numbers and functions can evidently be constructed too. In this way, all common mathematical objects can be seen to be present in $\mathbf{P}^n(\omega)$ for some finite n, and almost all of ordinary mathematics can be carried out in Z.

Limitations on the Ordinals and the Transfinite in Z

Since the ordinal ω exists in Z, so do the ordinals $\omega + 1, \omega + 2, \ldots, \omega + n, \ldots$, by repeatedly applying the successor operation. But, due to lack of Replacement, we cannot prove in Z that the ordinal $\omega + \omega$ exists. In Z, the sum of two ordinals may fail to exist! On the other hand, the "sum of well-orders" exists in the following sense.

Problem 1275. *Prove in Z that if a and b are well-orders then there is a well-order c which can be partitioned into an initial segment u and a final segment v such that a is order isomorphic to u and b is isomorphic to v.*

Thus in Z one can prove that there exist well-orders of type $\omega + \omega$, although the von Neumann ordinal $\omega + \omega$ does not exist and in fact the only ordinals which can be shown to exist in Z are the ones less than $\omega + \omega$. This means that the existence theorem (that every well-order is isomorphic to some von Neumann ordinal) fails in Z: Although the *definition* of von Neumann ordinals does not need Replacement, the *proof* the existence theorem needs Replacement in an essential way.

Note that uncountable well-orders can be shown to exist in Z.

Problem 1276. *Construct a well-order of type ω_1 in Z.*

Thus even though most countable ordinals (those $\geq \omega + \omega$) are not available in Z, we can fix a well-order of type ω_1 and take its proper initial segments to serve as representatives for the countable ordinals in Z. Unless we insist on the "absolutist" approach of von Neumann ordinals, all countable ordinals exist in Z in this "structuralist" sense used in ordinary mathematics.

More generally, the Hartogs construction can be done in Z:

Problem 1277. *For any set A, there is a well-order X such that every proper initial segment of X, but not X itself, is equinumerous to some subset of A.*

By induction, well-orders of type ω_n (and so sets of cardinality $\aleph_n$) exist for all $n \in \omega$. Thus, in spite of the lack of von Neumann ordinals $\geq \omega + \omega$, representative well-orders for every ordinal $\alpha < \omega_\omega$ are available in Z. For cardinals, put $\beth_n := |\mathbf{P}^n(\omega)|$, and note that for any cardinal $\kappa \leqslant \beth_n$ $(n \in \omega)$, sets of cardinality κ exist in Z.

However, this is where the development of the transfinite stops in Z. No well-order of type ω_ω or set of cardinality $\aleph_\omega$ can be shown to exist in Z. In fact, it can be shown that $V_{\omega+\omega}$ works as a "set-theoretic universe" or "model" for Z.

Other Limitations in Mathematics Without Replacement

Another problem due to lack of Replacement appears when defining functions on ω (or on $\mathbf{N}$) by recursion, i.e., where one defines a term $F(n)$ for $n \in \omega$. If all the values $F(n)$ of the function F belong to a set B which already exists in Z, then F becomes a subset of $\omega \times B$, and so by separation F exists in Z as an ordinary function (i.e., a set whose existence can be established in Z). If, however, the values of F are not restricted to be within such a predetermined set, then in general F will only be a functional, not necessarily a function; and its range will be a class, not necessarily a set. For example, the functional $n \mapsto \omega + n$ $(n \in \omega)$ cannot be proved to be a function in Z. (More generally, this problem affects definition by transfinite recursion.)

Such constructions are common in ordinary mathematics (e.g., forming algebraic closures of fields, infinite direct products of groups, etc.) where one defines (recursively) sets $A_0, A_1, A_2, \ldots$, and in the end needs to combine them somehow, e.g., form the union $\cup_n A_n$. In general, this cannot be done in Z, unless the sets A_n are all subsets of a fixed set already known to exist in Z. For example, if we let $A_n = \omega + n$, then, as mentioned before, the union $\cup_n A_n = \omega + \omega$ is a class in Z which cannot be shown to exist as a set in Z.

However, in most ordinary cases, this problem can be addressed as follows: It is usually possible to find suitable isomorphic copies of the sets A_n within a predetermined set known to exist in Z, and work with these copies instead. In the structuralist approach of ordinary mathematics, such isomorphic replacements for the sets A_n are acceptable. For example, in Z one can find well-ordered subsets X_n of $\mathbf{Q}$ where each X_n is an initial segment of X_{n+1} and X_n is order isomorphic to $\omega + n$, so that their union $\cup_n X_n$ has type $\omega + \omega$.

There is one scenario in which this approach does not work: When the sets A_n grow bigger in cardinality without bounds, there may be no set in Z which has room to fit them (or their copies) all in. This is the case, e.g., if we put $A_n = \mathbf{P}^n(\omega)$, since for any set E which contains copies of A_n we must have $|E| > \beth_n$ for all n, and such an E cannot be guaranteed to exist in Z. But such situations are rare[4] in ordinary mathematics. For most of ordinary mathematics, one does not need the Axiom of Replacement and Zermelo's system Z turns out to be quite adequate.

[4]There are a few theorems of mathematics, such as the determinacy of Borel games (proved by Martin), where such situations are indeed encountered and the use of the Axiom of Replacement becomes necessary.

21.7 Regular Sets and Ranks

Definition 1278. For a relation R on X, we say that $\langle X, R\rangle$ has the *von Neumann property* if the set of R-predecessors in X of any element of $x \in X$ coincides with x, that is if for any $x \in X$, we have $(\forall y)(y \in x \Leftrightarrow y \in X \wedge yRx)$.

Problem 1279. *Let R be a relation on X. Then $\langle X, R\rangle$ has the von Neumann property if and only if X is a transitive set and R is the membership relation restricted to X.*

Thus the membership relation restricted to a set X has the von Neumann property if and only if X is transitive.

Extensional Well-Founded Relations

A relation R on a set X is called *extensional* if distinct elements of X have distinct sets of R-predecessor, that is if for all $a, b \in X$, $(\forall x)(xRa \Leftrightarrow xRb) \to a = b)$. (Thus the axiom of extensionality says that the set membership relation $\in$ defined on the class of all sets is extensional.) For well-founded extensional relations, a generalization of the existence theorem for well-orders (representation by von Neumann ordinals) holds. This result, due to Mostowski, states: *If R is a relation on X which is well-founded and extensional on X, then there exists a unique transitive set M such that $\langle X, R\rangle$ is isomorphic to $\langle M, \in_M\rangle$, where $\in_M$ is the set membership relation restricted to M.*

Problem 1280. *Prove Mostowski's Theorem as stated above.*

[Hint: Recall the uniqueness-existence proofs for von Neumann well-orders.]

Transitive Closures

Given any set x, its *transitive closure* is formed by collecting together the members of x along with the members of members of x, the members of members of members of x, and so on. In the language of Chap. 10 Sect. 11.4, the transitive closure of x is the "$\in$-ancestry" of x. In other words, y is in the transitive closure of x if there is a positive integer n and $y_0, y_1, \ldots, y_n$ such that $y_0 = y$, $y_n = x$, and $y_k \in y_{k+1}$ for $k = 0, 1, \ldots, n-1$. This can be formalized in ZF as follows.

Definition 1281 (Transitive Closure).

$$\begin{aligned} y \in \mathrm{tc}(x) \leftrightarrow \exists f \exists n (n \in \omega \wedge n \neq 0 \wedge f \text{ is a function} \wedge \mathrm{dom}(f) = n^{+} \\ \wedge f(0) = y \wedge f(n) = x \wedge \forall k (k < n \to f(k) \in f(k+1)). \end{aligned}$$

We then have

Proposition 1282. *One can deduce the following results using the ZF axioms introduced so far*

1. $\text{tc}(x)$ *is transitive and is the smallest transitive set containing x as a subset.*
2. $\text{tc}(\text{tc}(x)) = \text{tc}(x)$.
3. $\text{tc}(x) \cup \{x\} = \text{tc}(x^+)$ *is transitive and is the least transitive set containing x as a member.*
4. $\text{tc}(x) = x \cup (\bigcup x) \cup (\bigcup \bigcup x) \cup \cdots$.
5. $\text{tc}(x) = \bigcup_{y \in x} \text{tc}(y^+)$.

Regular Sets

Recall that we saw that under AC, a relation R is well-founded if and only if there is no descending infinite R-chain, that is there is no infinite sequence $\langle x_n \rangle$ such that $x_{n+1} R x_n$ for all n. In particular, if R is well-founded then R is irreflexive and there are no R-cycles such as xRy and yRx, or xRy, yRz, and zRx.

In the context of sets, we will say that x *contains a descending $\in$-chain* if there is an infinite sequence $\langle x_n \rangle$ such that

$$\cdots \in x_{n+1} \in x_n \in \cdots \in x_3 \in x_2 \in x_1 \in x.$$

If x contained a descending chain as above, then all the elements x_n are in the transitive closure of x, and so $\text{tc}(x)$ would not be well-founded under the membership relation $\in$.

Definition 1283 (Regular Sets). x is *regular* if $\text{tc}(x)$ is well-founded under the membership relation $\in$.

A regular set is often called a *well-founded set.*

Under AC, a set is regular if and only if it does not contain a descending $\subset$ chain, but even without AC a regular set cannot contain a descending $\in$-chain.

Problem 1284. *Every ordinal is regular.*

In addition to the ordinals, all sets we have encountered so far were regular.[5] Regularity of a set prevents it from satisfying unusual conditions such as self-membership. In particular, if x is regular then the singleton $\{x\}$ will be distinct from x. Similarly, we cannot have "$\in$-cycles" such as $x \in y \wedge y \in x$, or $x \in y \wedge y \in z \wedge z \in x$, if any of the sets in the cycle is regular.

[5]It can be shown that the existence of non-regular sets cannot be demonstrated on the basis of the ZF axioms introduced so far, unless those axioms are contradictory.

Since $\mathrm{tc}(\mathrm{tc}(x)) = \mathrm{tc}(x)$, we see that x is regular if and only if $\mathrm{tc}(x)$ is regular. Moreover, we have:

Problem 1285. *For any set x,*

1. *x is regular if and only if every member $y \in x$ is regular.*
2. *If x is regular then any subset of x is regular.*
3. *If x is regular then $\mathbf{P}(x)$ is regular.*

Problem 1286. *If $x \in V_\alpha$ for some ordinal α then x is regular.*

Ranks

Recall that every well-founded relation R on a set A has a unique rank. If x is regular, so that $\mathrm{tc}(x)$ is well-founded under $\in$, we define *the rank of x* to be the rank of the well-founded structure $\langle \mathrm{tc}(x), \in \rangle$ (we can think of $\mathrm{tc}(x)$ providing the structure by which x is "built up" from $\emptyset$).

Definition 1287 (Rank of Sets). For any regular x, the *rank of x*, denoted by $\mathrm{rnk}(x)$, is the rank of the well-founded structure $\langle \mathrm{tc}(x), \in \rangle$.

From the results of Chap. 10 Sect. 11.4, we have the following result.

Problem 1288. *If x be regular, then*

$$\mathrm{rnk}(x) = \sup\{\mathrm{rnk}(y) + 1 \mid y \in x\}.$$

Corollary 1289. *If x is regular and $y \in x$ then* $\mathrm{rnk}(y) < \mathrm{rnk}(x)$.

The following theorem establishes the link between ranks and the levels of the cumulative hierarchy.

Theorem 1290. *$x \in V_\alpha$ if and only if x is regular and* $\mathrm{rnk}(x) < \alpha$.

Proof. We use transfinite induction on α. Suppose that the result holds for all $\beta < \alpha$.

First, let $x \in V_\alpha$. Then $x \subseteq V_\beta$ for some $\beta < \alpha$, so $y \in V_\beta$ for all $y \in x$, so by induction hypothesis we get: y is regular and $\mathrm{rnk}(y) < \beta$ for all $y \in x$. Since every member of x is regular, so x is regular, and since $\mathrm{rnk}(y) < \beta$ for all $y \in x$, we get

$$\mathrm{rnk}(x) = \sup_{y \in x} \mathrm{rnk}(y) + 1 \leq \beta < \alpha.$$

Conversely, suppose that x is regular and $\mathrm{rnk}(x) = \beta < \alpha$. Hence for every $y \in x$, y is regular and $\mathrm{rnk}(y) < \mathrm{rnk}(x) = \beta$, so by induction hypothesis $y \in V_\beta$ for all $y \in x$, which implies $x \subseteq V_\beta$, and so $x \in \mathbf{P}(V_\beta) = V_{\beta+1} \subseteq V_\alpha$. □

By the following, the regular sets are exactly the ones that can be obtained through the progression of the cumulative hierarchy.

Corollary 1291. *x is regular with* $\operatorname{rnk}(x) = \alpha$ *if and only if* $x \in V_{\alpha+1} \smallsetminus V_\alpha$. *Hence for regular x,* $\operatorname{rnk}(x)$ *is the least ordinal μ such that $x \in V_{\mu+1}$. Moreover, x is regular if and only if $x \in V_\alpha$ for some ordinal α.*

21.8 Foundation and the Set Theoretic Universe V

As we have mentioned before, non-regular sets (sets with descending $\in$-chain) not only are unnecessary for the development of mathematics, but such sets also are not naturally encountered since the existence of non-regular sets cannot be derived from the ZF axioms we have mentioned so far. The last axiom of ZF is used to explicitly rule out such extraneous sets.

ZF 9 (Axiom of Foundation). *Every set is regular, or equivalently every set belongs to V_α for some ordinal α.*

Problem 1292. *The Axiom of Foundation is equivalent to the statement that every nonempty set x has a member which is disjoint from x.*

The following problems need the Axiom of Foundation.

Problem 1293. *x is an ordinal if and only if x is transitive and linearly ordered by the membership relation $\in$.*

Problem 1294. *A set x is said to* hereditarily *have a property P if x and all members of* $\operatorname{tc}(x)$ *have property P. Show that x is hereditarily transitive if and only if x is an ordinal and x is hereditarily finite if and only if $x \in V_\omega$.*

Recall that $V := \{x \mid x = x\}$ denotes the class of all sets, also called *the set theoretic universe* (we had seen that V cannot be a set). Then the Axiom of Foundation can be stated in the following suggestive form:

$$V = \bigcup_{\alpha \subset \mathrm{On}} V_\alpha .$$

Since the sets V_α increase with α, the above equation shows that the universe V of all sets is arranged in a hierarchy of sets of ever increasing ranks. Foundation thus enables us to prove that a property is true of *all sets* using the convenient method of transfinite induction on the rank of a set.

Note that the complement of any set is a class which will contain all sets of sufficiently large rank. Thus in ZF set theory each set is a miniscule infinitesimal part of the universe, and the universe V of all sets is *truly large* compared to any particular set.

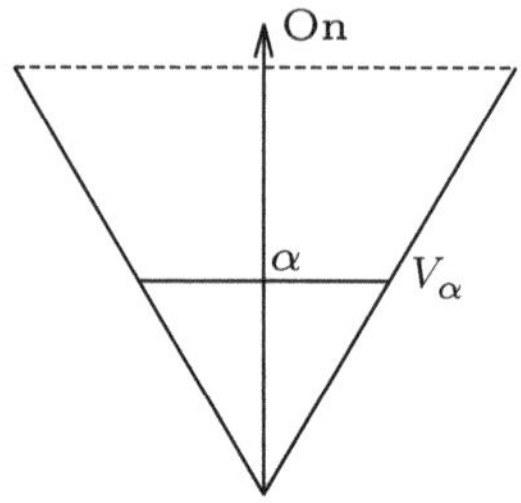

The Universe V

Problem 1295 (ZFC). *For each cardinal κ, put $H(\kappa) := \{x \mid |\mathrm{tc}(x)| < \kappa\}$.*

1. *$H(\aleph_0) = V_\omega$ (which is countable).*
2. *$H(\aleph_\alpha) \subseteq V_{\omega_\alpha}$, so $H(\kappa)$ is a set.*
3. *The set of hereditarily countable sets has cardinality $\mathfrak{c}$: $|H(\aleph_1)| = 2^{\aleph_0}$.*
4. *There is a natural surjection from the set W of well-founded trees over ω onto $H(\aleph_1)$ preserving rank (between trees and sets).*

Defining Cardinals Without Choice

Although V is not a set in ZF, V_α is a set for each ordinal α. This is useful in effectively extracting a nonempty subset from any large collection which may not be a set. In particular, we have the method of Scott, in which the Axiom of Foundation is used to define the notion of cardinal number for arbitrary sets *without using the Axiom of Choice*. In the Frege–Russell definition, the cardinal number $|a|$ of a set a is the equivalence class of a under the equinumerosity relation. Thus $|a|$, consisting of all sets equinumerous to a, is a problematic large collection which is not a set.

Problem 1296 (ZF). *For any nonempty set a, there is no set which contains all sets equinumerous with a.*

Scott's method allows us to effectively select a nonempty subcollection of the equivalence class which is small enough to be a set.

Definition 1297 (The Frege–Russell–Scott Cardinal). For any set x, let $|x|^-$ be the collection of all sets of *least rank* which are equinumerous to x:

$$|x|^- := \{y \mid y \approx x \wedge \forall z(z \approx x \to \mathrm{rnk}(y) \le \mathrm{rnk}(z))\}.$$

Problem 1298. *For any x, $|x|^- \subseteq V_{\mathrm{rnk}(x)+1}$ and so $|x|^-$ is a set.*

Moreover, the functional $x \mapsto |x|^-$ is a complete invariant for equinumerosity ($|x|^- = |y|^- \leftrightarrow x \approx y$ for all sets x, y), and therefore is a satisfactory definition of "cardinal number."[6]

More generally, for any relational R, one can define the *Frege–Russell–Scott invariant for* R as follows.

Definition 1299 (The Frege–Russell–Scott Invariant). If R is a relational and x is any set, let $[x]_R^-$ be the collection of all sets y of *least rank* for which yRx holds:

$$[x]_R^- := \{y \mid yRx \wedge \forall z(zRx \to \text{rnk}(y) \leq \text{rnk}(z))\}.$$

Problem 1300. *Let $\equiv$ be an equivalence relational on a class C. Then $[x]_\equiv^-$ is a set for all x, and the Frege–Russell–Scott invariant mapping*

$$x \mapsto [x]_\equiv^- \qquad (x \in C)$$

acts as a complete invariant for the equivalence relational $\equiv$, that is, for all $x, y \in C$ we have $x \equiv y$ if and only if $[x]_\equiv^- = [y]_\equiv^-$.

The above Frege–Russell–Scott method works for any equivalence relational whatsoever, but is of particular relevance when some of the equivalence classes are too large to be sets. For example, we can conveniently define the *order type* OrdTyp(X) *of an arbitrary order* X using the Frege–Russell–Scott invariant by taking OrdTyp$(X) := [X]_\cong^-$, where $\cong$ denotes similarity (isomorphism) of orders.

21.9 Other Formalizations of Set Theory

Other than the Zermelo–Fraenkel system, there have been several other major axiomatizations of set theory which use the formal framework of first order logic. We briefly discuss a few of them.

Von Neumann–Bernays Set Theory

This popular axiomatization of set theory was initiated by von Neumann in 1925 [78, pp. 393–413], and later developed extensively by Bernays (see [3]). We will abbreviate the name "Von Neumann–Bernays Set Theory" as VNB. Further work on

[6]One can even use a "hybrid method" to define cardinal numbers without Choice, where $|x|$ is defined as the least ordinal equinumerous to x if x can be well-ordered, and as the Frege–Russell–Scott cardinal of x otherwise [48]. Under Choice, this definition reduces to the Cantor-von Neumann definition.

VNB was done by Robinson and Gödel, and VNB is often called the *Von Neumann–Bernays–Gödel Set Theory* (abbreviated as NBG).

In ZF, we had used the term "class" to informally talk about large collections which were not sets, but classes did not exist as formal objects in ZF. A key feature of VNB is that it formally allows talking about large collections such as the collection V of all sets—which are now allowed as objects that formally exist. Since such collections cannot be sets, VNB uses the formal term *class* for general collections, and the system is developed as a theory of classes (subcollections are called *subclasses,* e.g.). In VNB, every object is a class, and so all sets are classes, but there are classes which are too large to be sets, such as the class V of all sets and the class On of all ordinals, which are called *proper classes.* A proper class cannot be a member of any class. A *set* is defined as a class which can be member of some class. Thus VNB divides classes into two distinct and exclusive sorts, sets and proper classes.

The notion of class in VNB is a formal way of representing Zermelo's vague notion of "definite property," and is similar to Skolem's use of the formal notion of a first-order formula in place of a "property." VNB has a "class comprehension scheme" for forming new classes, but the quantifiers in the formulas used in the formation of classes cannot range over arbitrary classes—they are limited to range over sets. Using classes, the axiom of replacement can be stated simply as follows: *If F is a class which is a function and A is a set then the image $F[A]$ is a set.* Note that in ZF, replacement was an axiom scheme—an infinite list of individual axioms (instances of the scheme), but in VNB it is a single axiom. This brings us to another important aspect of VNB: It is an axiom system which turns out to be *finitely axiomatizable,* that is, it is possible to find a finite list of individual axioms (not schemes) which will axiomatize VNB. ZF cannot be axiomatized with a finite set of axioms (unless ZF turns out to be inconsistent).

In spite of its appearance to be a more extensive theory than ZF —allowing a larger collection of objects that it can formally talk about—VNB turns out to be essentially equivalent to ZF in the following sense: Any statement mentioning only sets (no classes) that can be proved in VNB can be proved in ZF, and vice versa. In other words, VNB cannot prove any new fact about sets that ZF cannot already prove (which is technically stated by saying that VNB is a *conservative extension* of ZF). In particular, this implies that ZF and VNB are *equiconsistent* theories: if one of them is consistent, then so must be the other.

More on VNB can be found in the references, such as Bernays [3], which has a detailed presentation of VNB.

Another system due to Morse and Kelley, known as *Morse–Kelley* Set Theory or MK, was first introduced in 1955 as an appendix to Kelley's book on topology [40]. MK is closely related to VNB, but it allows quantification over arbitrary classes in its class comprehension scheme, resulting in a system which is strictly stronger than ZF and VNB (assuming ZF is consistent).

Quine's New Foundations

From the systems of set theory discussed so far (Type-Theory/PM, ZF, VNB, MK), one may think that in order to avoid contradictions, a formal system should never allow the collection of all sets to be a set itself—an idea expressed by Russell as the *vicious circle principle* and famously paraphrased by Halmos as "nothing contains everything." Yet, in 1937, Quine [61] introduced an axiomatization of set theory called *New Foundations,* or NF, which has a set containing all sets (and thus containing itself)! In this and several other respects, NF is a formal system strikingly different from ZF and VNB.

NF allows the set of all sets (or the class of all classes): The entire universe $V := \{x \mid x = x\}$ is itself a set, and we have $V \in V$ in NF. Every set A has a global complement $V \smallsetminus A := \{x \mid x \notin A\}$, and so with V as the universal set, the collection $\mathbf{P}(V)$ of all subsets of V becomes a Boolean algebra of sets.

Like Frege's original system NF has only two axioms: Extensionality and Comprehension. The language of NF is identical to that of ZF—any NF formula is a ZF-formula and vice versa. Also, The axiom of extensionality says, as usual, $x = y \Leftrightarrow \forall z(z \in x \Leftrightarrow z \in y)$.

Instead of building sets from bottom up via mechanisms such as the power set operation, NF forms sets only via applications of the comprehension scheme, just as in Frege's original (inconsistent) system. For example, for any set A, its power set is formed as $\mathbf{P}(A) := \{x \mid \forall y(y \in x \Rightarrow y \in A)\}$, which exists by NF-comprehension—no special power set axiom is needed.

However, to avoid contradictions, NF puts a restriction on the syntax of the formulas that can be used in its comprehension scheme: In order to form the set $\{x \mid \phi(x)\}$ via comprehension, the formula $\phi = \phi(x)$ must be "stratified." An NF-formula ϕ is said to be *stratified* if there is a mapping f from the instances of variable letters occurring in ϕ to $\mathbf{N}$ such that for any substring of ϕ having the form "$x \in y$" we have $f(y) = f(x) + 1$ and for any substring of ϕ having the form "$x = y$" we have $f(x) = f(y)$. This avoids forming sets such as Russell's set $\{x \mid \neg(x \in x)\}$, since the formula "$x \in x$" is not stratified.

In NF, cardinal and ordinal numbers are defined using the original Frege–Russell global invariant: The cardinal number of a set A is the set of all sets equinumerous to A, and the order-type number of an order X is the set of all orders isomorphic to X. Unlike ZF, there is no problem in NF of these notions being "too large to be sets."

On the other hand, NF sharply deviates from classical "Cantorian" set theory in several ways. For example, Cantor's Theorem $|X| < |\mathbf{P}(X)|$ cannot be proved in NF in full generality, and NF refutes the axiom of choice. Some of NF's oddities are addressed in a newer theory due to Jensen known as NFU, which allows non-set atoms (individuals).

For more on NF and NFU, see Quine [61], Fraenkel, Bar-Hillel, and Levy [21], Forster [19], and Holmes [31].

21.10 Further Reading

Part IV was a rather brief and sketchy introduction to axiomatic set theory, so the reader is encouraged to consult the excellent works given below.

Alternative treatments of some or all of the topics covered in the first parts of our book, following a naive informal approach (and with little or no coverage of axiomatic systems at all), can be found in the following older classic works: Hausdorff [29], Kamke [36], Sierpinski [73], Russell [68], Fraenkel [20], and Kuratowski and Mostowski [46].

For more coverage on the formal systems of ZF or VNB, see Stoll [76], Suppes [77], Bernays [3], Devlin [13], Levy [48], Hrbacek and Jech [32], Fraenkel, Bar-Hillel, and Levy [21], Rotman and Kneebone [65], Halmos [27], Enderton [15], Vaught [79], Moschovakis [54], Kunen [43], Just and Weese [35], Bourbaki [5], Hajnal and Hamburger [26], Hamilton [28], Schimmerling [70], and Goldrei [24].

The original work of Cantor and Dedekind can be found in Cantor [6] and Dedekind [12]. An excellent collection of primary sources on the historical development of axiomatic set theory and logic is van Heijenoort's *From Frege to Gödel: A Source Book in Mathematical Logic, 1879–1931* [78], where one can find English translations of Zermelo's and von Neumann's original papers introducing their axiomatic set theories, as well as subsequent enhancements of their systems by Skolem, Fraenkel, and Bernays. Zermelo's 1908 paper [85] still serves as an excellent introduction to his system.

For Part III, the introductory topics on the basic topology of $\mathbf{R}$ can be found in any standard real analysis text. An excellent review of the analogies between Lebesgue measure and Baire category is Oxtoby [58]. The topic of Borel and Analytic sets belongs to the area of Descriptive Set Theory, for which two standard modern references are Kechris [38] and Moschovakis [55]. Some of these topics are also covered in the older texts of Hausdorff [29], Sierpinski [74], Kuratowski [45], and Kuratowski and Mostowski [46]. See also Rogers [64].

More advanced treatments of set theory covering topics such as Gödel's constructible universe $\mathbf{L}$ and Cohen's 1963 technique of forcing for obtaining independence proofs (which revolutionized modern set theory and has been in constant use since then) require some background in mathematical logic. A basic early text on this topic is Cohen [8], while two highly standard references with expositions of constructibility and forcing are Kunen [42] and Jech [34]. Bell [2] focuses on Boolean-valued models. In 2011, a new rewritten version [44] of Kunen's 1980 book has been published.

For the topic of large cardinals, Kanamori [37] is a current standard reference (more forthcoming volumes expected), but the encyclopaedic Jech [34] and the older Drake [14] are also helpful.

A volume containing many interesting articles by set theorists is Link [49]. A recent handbook containing highly advanced technical surveys of current research in set theory is [18].

Chapter 22
Postscript IV: Landmarks of Modern Set Theory

Abstract This part contains brief informal discussions (with proofs and most details omitted) of some of the landmark results of set theory of the past 75 years. Topics discussed are constructibility, forcing and independence results, large cardinal axioms, infinite games and determinacy, projective determinacy, and the status of the Continuum Hypothesis.

Note: *Many of the topics discussed below are metamathematical in nature and so their precise and rigorous definitions depend on mathematical logic, which is beyond the scope of this text. Therefore the descriptions below are necessarily sketchy and incomplete. Most of the details can be found in Jech [34], Kunen [42, 44], and Kanamori [37].*

22.1 Gödel's Axiom of Constructibility

All efforts to settle the Continuum Hypothesis by late nineteenth and early twentieth century mathematicians failed. Then, in the late 1930s, Gödel made a major breakthrough by introducing the notion of *constructible sets* and the *axiom of constructibility*. We now briefly describe Gödel's results [22].

Relativization

Let C be a given class (which can be a set or a proper class). Then for each ZF-formula $\phi = \phi(x_1, x_2, \ldots, x_n)$, the *relativization of* ϕ *to* C, denoted by $\phi^C = \phi^C(x_1, x_2, \ldots, x_n)$, is obtained by restricting all the quantifiers in ϕ to range over C. For example if σ is '$\forall x \exists y (x \in y)$' then σ^C is '$\forall x \in C\, \exists y \in C\, (x \in y)$'. This can be formally defined by recursion on formulas by taking $\phi^C = \phi$ if ϕ is an atomic formula, $(\phi \wedge \psi)^C = \phi^C \wedge \psi^C$, $(\neg\phi)^C = \neg(\phi^C)$, and $(\exists v\, \phi)^C = \exists v (v \in C \wedge \phi^C)$.

A. Dasgupta, *Set Theory: With an Introduction to Real Point Sets*,
DOI 10.1007/978-1-4614-8854-5_22, © Springer Science+Business Media New York 2014

Definition 1301. Let A be a set. We say that a subset $E \subseteq A$ is *definable from parameters in* A if there exist a ZF-formula $\phi(x, x_1, x_2, \ldots, x_n)$ and elements $a_1, a_2, \ldots, a_n \in A$ such that

$$E = \{x \in A \mid \phi^A(x, a_1, a_2, \ldots, a_n)\}.$$

Note: This notion can actually be defined formally in ZF (Tarski–Gödel).

Definition 1302. Def(A) denotes the collection of all subsets of A which are definable from parameters in A.

It is clear that Def$(A) \subseteq \mathbf{P}(A)$, and $\emptyset$, A itself, and all finite subsets of A are members of Def(A). Thus if A is finite then Def$(A) = \mathbf{P}(A)$. On the other hand, if A is infinite then $|\operatorname{Def}(A)| = |A|$ but $|\mathbf{P}(A)| > |A|$ so Def(A) is a relatively small part of $\mathbf{P}(A)$ when A is infinite. In addition, if A is transitive, then $A \subseteq \operatorname{Def}(A) \subseteq \mathbf{P}(A)$, and so Def$(A)$ is itself transitive.

We now define the hierarchy of constructible sets by transfinite recursion.

Definition 1303 (The Constructible Hierarchy). Define

$$L_0 := \emptyset, \qquad L_{\alpha+1} := \operatorname{Def}(L_\alpha), \qquad L_\alpha := \bigcup_{\beta<\alpha} L_\beta \quad \text{if } \alpha \text{ is limit.}$$

A set A is said to be *constructible* if $A \in L_\alpha$ for some ordinal α. The class of all constructible sets is denoted by L.

The *axiom of constructibility* says that every set is constructible. Since the class of all sets is denoted by V and the class of constructible sets by L, so the axiom of constructibility can be stated as "V=L."

Note that each L_α is transitive, and $L_\alpha \subseteq V_\alpha$ for all α. Also, since Def$(A) = \mathbf{P}(A)$ whenever A is a finite set, so we have $L_\alpha = V_\alpha$ for $\alpha < \omega$. It follows that $L_\omega = V_\omega$.

However, $L_{\omega+1}$ is much smaller than $V_{\omega+1}$, since $L_{\omega+1} = \operatorname{Def}(L_\omega)$ is countable while $|V_{\omega+1}| = |\mathbf{P}(V_\omega)| = 2^{\aleph_0}$. Next, $L_{\omega+2}$ is still countable but $|V_{\omega+2}| = 2^{2^{\aleph_0}}$. In fact, L_α stays countable for all $\alpha < \omega_1$, while the sets V_α grow enormously in size! On the other hand, for every ordinal α we have $\alpha \in L_{\alpha+1} \setminus L_\alpha$ and $\alpha \in V_{\alpha+1} \setminus V_\alpha$, so L_α and V_α have the same rank or "height," i.e., L_α is as "tall" as V_α. This gives the following picture.

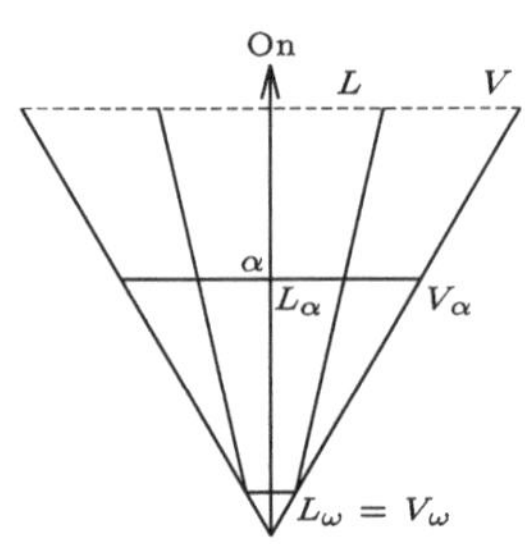

Gödel proved two major facts about constructibility:

1. Both the Axiom of Choice (AC) and the Generalized Continuum Hypothesis (GCH) can be derived from ZF augmented with the axiom of constructibility, i.e.,

$$\text{ZF} \vdash \ V{=}L \rightarrow \text{AC} + \text{GCH},$$

   where we write "ZF $\vdash \sigma$" to mean that σ can be formally derived in ZF.
2. The axiom of constructibility is relatively consistent with ZF, i.e., if ZF is consistent then so is ZF + $V{=}L$.

As an immediate consequence, we have the following:

Theorem 1304 (Gödel). *If* ZF *is consistent then so is* ZF + AC + GCH.
In particular, the Continuum Hypothesis cannot be disproved from ZFC, *unless* ZF *itself is inconsistent.*

We now sketch how Gödel's results can be derived.

First note that if a transitive set A can be well-ordered, then we can also well-order Def(A) effectively from the well-order on A. This is because the ZF-formulas, countable in number, can be effectively enumerated and the set A^* of finite sequences of parameters from A can also be effectively well-ordered (from the well-order on A). This way, all the sets L_α can be well-ordered in a uniform and effective fashion such that if $\alpha < \beta$ then L_α is an initial segment of L_β, which gives an effective global well-order on all of L. Hence if $V{=}L$, then the class of all sets gets equipped with a global well-order and the axiom of choice immediately follows.

To get an idea on how $V{=}L$ implies the Continuum Hypothesis, note that we have $|L_{\omega_1}| = \aleph_1$ effectively using the nice well-order of L described above. Gödel also proved that $\mathbf{P}(\omega) \cap L \subseteq L_{\omega_1}$ (we will not prove this fact here). Hence, if $V{=}L$, then $\mathbf{P}(\omega) \subseteq L_{\omega_1}$, so $|\mathbf{P}(\omega)| \leqslant \aleph_1$, and CH follows.

GCH is derived from $V{=}L$ in an exactly similar fashion.

Finally, to prove the second part (relative consistency of $V{=}L$ with ZF), Gödel first established that for every ZF-sentence σ:

$$\text{If ZF} \vdash \sigma \ \text{ then ZF} \vdash \sigma^L.$$

This is expressed by saying that "L is a model of ZF." Gödel then showed:

$$\text{ZF} \vdash (V{=}L)^L,$$

which is expressed by saying "L is a model of $V{=}L$." From these two facts, it easily follows that if ZF is consistent then so is ZF+$V{=}L$, or equivalently that $V{\neq}L$ cannot be derived from ZF unless ZF itself is inconsistent. To see this, suppose that ZF $\vdash \neg(V{=}L)$. Then, since L is a model of ZF, we would get ZF $\vdash \neg(V{=}L)^L$. But we saw above that ZF $\vdash (V{=}L)^L$, so ZF is inconsistent.

For the details that our proof-sketch has left out, see [34,42], or [44].

The axiom of constructibility ($V{=}L$) is a very strong axiom which settles many unsolved problems of set theory. For example, $V{=}L$ implies $\diamondsuit$, and so the Suslin Hypothesis is false under $V{=}L$.

The axiom of constructibility also decides Lusin's important unsolved questions about regularity properties of the projective sets, but the answers are rather negative (for a class of effectively defined sets). Essentially, the highly effective well-ordering present under $V{=}L$ can be used to produce "pathological" Bernstein sets in the low levels of the projective hierarchy, giving:

Theorem 1305 (Gödel). *If $V{=}L$, then there exist PCA ($\mathbf{\Sigma}^1_2$) sets which are not measurable and do not have the Baire property, and there exist uncountable coanalytic ($\mathbf{\Pi}^1_1$) sets which do not have any perfect subset.*

22.2 Cohen's Method of Forcing

After Gödel proved his relative consistency result, the problem of independence of CH remained open[1] until 1963, when Paul Cohen showed that ZFC cannot prove CH either (assuming ZF is consistent). The Gödel–Cohen results are known as *the independence of the Continuum Hypothesis.* Cohen's proof introduced a new method called *forcing*, which immediately flourished as an extremely powerful and versatile technique for obtaining general independence results. Since then, the method of forcing has been extended in many ways, and a vast body of independence results have been obtained. Forcing remains the most fruitful tool for building models of set theory.

Forcing is best understood in the context of models of set theory, where it is viewed as a method for extending a given model. This is beyond the scope of this text, and we will only give a bare bones sketch of forcing in purely syntactic terms with most of the details left out. Modern expositions of forcing are [2,34,42], and [44]. Cohen's original text is [8].

A *forcing poset* $\langle \mathbb{P}, \preceq, \mathbb{1} \rangle$ is a partial order with a largest element $\mathbb{1}$. Given such a poset, define the (ranked) class $V^{\mathbb{P}}$ of $\mathbb{P}$-*names* as $\bigcup_{\alpha \in \mathrm{On}} V^{\mathbb{P}}_\alpha$, where

$$V^{\mathbb{P}}_0 := \emptyset, \qquad V^{\mathbb{P}}_{\alpha+1} := \mathbf{P}(V^{\mathbb{P}}_\alpha \times \mathbb{P}), \qquad V^{\mathbb{P}}_\alpha := \bigcup_{\xi<\alpha} V^{\mathbb{P}}_\xi \quad \text{for limit } \alpha.$$

A $\mathbb{P}$-*sentence* is a ZF-formula in which all free variables have been replaced by $\mathbb{P}$-names. In particular, each ZF-sentence is a $\mathbb{P}$-sentence. One can then define, for each $\mathbb{P}$-sentence σ, the *forcing relation* (read "p forces σ"):

[1] Gödel's method of showing relative consistency, known as the method of *inner models,* cannot be used to show the relative consistency of the negation of CH (or of the negation of any statement provable from $V{=}L$). The reason is that the only inner model of L containing the ordinals is L itself.

$$p \Vdash_{\mathbb{P}} \sigma \qquad (p \in \mathbb{P}),$$

first for atomic $\mathbb{P}$-sentences by a suitable recursion on the $\mathbb{P}$-names μ, ν as:

- $p \Vdash_{\mathbb{P}} \mu = \nu \Leftrightarrow \forall \rho \in \mathrm{dom}(\mu) \cup \mathrm{dom}(\nu)\, \forall q \preceq p\, (q \Vdash_{\mathbb{P}} \rho \in \mu \leftrightarrow q \Vdash_{\mathbb{P}} \rho \in \nu)$,
- $p \Vdash_{\mathbb{P}} \mu \in \nu \Leftrightarrow \forall q \preceq p\, \exists r \preceq q\, \exists \langle \rho, s \rangle \in \nu\, (r \preceq s \wedge r \Vdash_{\mathbb{P}} \mu = \rho)$,

and then for more complex $\mathbb{P}$-sentences by:

- $p \Vdash_{\mathbb{P}} \sigma \wedge \tau \Leftrightarrow p \Vdash_{\mathbb{P}} \sigma \wedge p \Vdash_{\mathbb{P}} \tau$,
- $p \Vdash_{\mathbb{P}} \neg\sigma \Leftrightarrow \neg\exists q \preceq p\, (q \Vdash_{\mathbb{P}} \sigma)$,
- $p \Vdash_{\mathbb{P}} \exists x \phi(x) \Leftrightarrow \forall q \preceq p\, \exists r \preceq q\, \exists \nu \in V^{\mathbb{P}} (r \Vdash_{\mathbb{P}} \phi(\nu))$.

Let us note that this really is a *definition scheme*—an infinite list of definitions, one for each $\mathbb{P}$-sentence σ.

We can see that $p \Vdash_{\mathbb{P}} \neg\sigma \Rightarrow \neg\,(p \Vdash_{\mathbb{P}} \sigma)$, and $q \preceq p$ and $p \Vdash_{\mathbb{P}} \sigma \Rightarrow q \Vdash_{\mathbb{P}} \sigma$, but it takes a lot of work to establish the following theorem-scheme of ZFC:

$$(*) \qquad \text{If}\quad \mathrm{ZFC} \vdash \sigma, \quad \text{then} \quad \mathrm{ZFC} \vdash \mathbb{1} \Vdash_{\mathbb{P}} \sigma.$$

The $\mathbb{P}$-names are thought of as "labels" for a "virtual extension" of the universe V. There are a lot of duplications in the $\mathbb{P}$-names, but one can identify duplicate $\mathbb{P}$-names using the equivalence $\mu \sim_{\mathbb{P}} \nu \Leftrightarrow \mathbb{1} \Vdash_{\mathbb{P}} \mu = \nu$ (for $\mu, \nu \in V^{\mathbb{P}}$). For each $x \in V$, one can define a canonical $\mathbb{P}$-name $\check{x} \in V^{\mathbb{P}}$ for x by the recursion $\check{x} := \{\langle \check{y}, \mathbb{1} \rangle \mid y \in x\}$. Then it can be verified that the mapping $x \mapsto [\check{x}]_{\sim_{\mathbb{P}}}$ "embeds" the universe V into the "virtual extension" $V^{\mathbb{P}}/{\sim_{\mathbb{P}}}$. Here it is convenient to think of the set $\{p \in \mathbb{P} \mid p \Vdash_{\mathbb{P}} \sigma\}$ as the "generalized truth value" for the $\mathbb{P}$-sentence σ, so that by $(*)$ above, the generalized truth value of σ equals $\mathbb{P}$ ("true") if ZFC $\vdash \sigma$, and equals $\emptyset$ ("false") if ZFC $\vdash \neg\sigma$.

Let $\mathrm{Fn}(I, J)$ denote the poset consisting of all functions with domain a finite subset of I and range contained in J, ordered by reverse inclusion so that $f \preceq g \Leftrightarrow f \supseteq g$ and $\emptyset$ is the greatest element. Let $\mathbb{P} := \mathrm{Fn}(\omega \times \omega_2, \{0, 1\})$. Cohen showed that, for this poset $\mathbb{P}$,

$$\mathbb{1} \Vdash_{\mathbb{P}} 2^{\aleph_0} \geqslant \omega_2, \qquad \text{i.e.,} \qquad \mathbb{1} \Vdash_{\mathbb{P}} \neg\,\mathrm{CH}.$$

In other words, the generalized truth value of CH is "false" ($\emptyset$) under the poset $\langle \mathrm{Fn}(\omega \times \omega_2, 2), \supseteq, \emptyset \rangle$. It follows that CH is not a theorem of ZFC, since otherwise we would have both $\mathbb{1} \Vdash_{\mathbb{P}}$ CH and $\neg\,(\mathbb{1} \Vdash_{\mathbb{P}}$ CH) (since $\mathbb{1} \Vdash_{\mathbb{P}} \neg\,$CH gives $\neg\,(\mathbb{1} \Vdash_{\mathbb{P}}$ CH)), implying that ZFC is inconsistent. This gives:

Theorem 1306 (Cohen). *If* ZFC *is consistent then so is* ZFC + $\neg$CH.

In particular, the Continuum Hypothesis cannot be proved from ZFC, *unless* ZFC *itself (and so* ZF *as well) is inconsistent.*

The combinatorial properties of a forcing poset $\langle \mathbb{P}, \preceq, \mathbb{1} \rangle$ are crucial in determining which statements will be forced, and a great variety of independence results have been obtained using various kinds of forcing posets. Forcing can also be used to show the independence of AC, and the relative consistency of "$2^{\aleph_0} = \aleph_\alpha$" for any α so long as $\aleph_\alpha$ has uncountable cofinality.

The forcing method has been extended considerably, using which, e.g., the relative consistency of MA+not-CH was obtained. This shows that Suslin's Hypothesis (SH) is relatively consistent with ZFC. Combined with the fact that SH is false under V=L, we see that SH is independent of the ZFC axioms.

In Postscript III, we mentioned that MA+not-CH implies that all $\mathbf{\Sigma}^1_2$ (PCA) sets are measurable and have the Baire property. Combined with Gödel's Theorem 1305, it follows that the measurability (and Baire property) of $\mathbf{\Sigma}^1_2$ sets is *independent* of ZFC. This shows that Lusin's conviction that these problems are unsolvable was correct, and that Lusin and the mathematicians of his time had reached the limits of what could be proved about projective sets using the usual axioms of set theory. See also Theorem 1308 below.

We conclude our brief discussion of forcing by stating a landmark result obtained by Solovay using the method of forcing. Let I denote the assertion that there is an inaccessible cardinal.

Theorem 1307 (Solovay). *If* ZFC+I *is consistent, then so is* ZF+DC *together with all of the following assertions:*

1. *All subsets of* **R** *have the perfect set property.*
2. *All subsets of* **R** *are Lebesgue measurable.*
3. *All subsets of* **R** *have the Baire property.*

Solovay also proved the following consistency result about the projective sets:

Theorem 1308 (Solovay). *If* ZFC+I *is consistent, then so is* ZFC+GCH *together with all of the following assertions:*

1. *All projective sets have the perfect set property.*
2. *All projective sets are Lebesgue measurable.*
3. *All projective sets have the Baire property.*

Theorems 1307 and 1308 raised the question whether the assumption of the existence of an inaccessible cardinal in the hypothesis of the theorems was really necessary. It was already known that conclusion (1) does need the assumption of an inaccessible (since the perfect set property for coanalytic sets implies the relative consistency of inaccessible cardinals with ZFC, see Theorem 1310 below). Surprisingly, Shelah later proved that in both theorems (2) needs the inaccessible assumption, but (3) does not!

22.3 Gödel's Program and New Axioms

Extending Cohen's results, Solovay showed that any assertion of the form $2^{\aleph_0} = \aleph_\alpha$ is consistent relative to ZFC, so long as α is a successor ordinal or has uncountable cofinality. Thus any one of these statements can be taken as an additional axiom to get a formal extension of set theory. This suggests the consideration of two possible approaches to the Continuum Problem.

The first view, sometimes called *pluralism* or *formalism,* is that there is no pre-existing intrinsic reason to prefer any of these assertions over another. Cohen himself expressed support for this view. Pluralism is applicable not only to the Continuum Problem, but also to any of the many problems known to be independent of the ZFC axioms. Formalists may regard the study of the multitude of possible axiomatic set theories as new human constructions or *inventions* that had never existed before.

The other view, supported by Gödel himself, is that the ideal universe of sets exists in a reality which is independent of axioms. Gödel believed that "in this reality, Cantor's conjecture must be either true or false, and its undecidability from the axioms as known today can only mean that these axioms do not contain a complete description of this reality" [23]. This view is thus sometimes called *platonism* or *realism.* Gödel suggested a search for the *discovery* of new natural axioms of set theory[2] which will be powerful enough to determine the "correct truth value" of problems currently known to be independent of ZFC. This is known as *Gödel's Program.*[3]

22.4 Large Cardinal Axioms

One possible candidate for a new axiom could be the axiom of constructibility ($V=L$). We saw that $V=L$ is powerful enough to settle many of the major undecidable problems of set theory such as CH and SH, and, as Jensen points out, is a form of Occam's razor since *it denies the existence of any set other than the constructible ones.* It gives a very "narrow" universe of sets.

Very different from the axiom of constructibility are *large cardinal axioms* or *axioms of strong infinity.* Existence of a large cardinal implies the consistency of ZF. For example, let I denote the assertion that there is an inaccessible cardinal. Then it can be shown that

$$\text{ZFC+I} \rightarrow \text{Con(ZFC)},$$

where "Con(ZFC)" stands for "ZFC is consistent." By a result known as Gödel's second incompleteness theorem on unprovability of consistency, existence of such cardinals (or even the relative consistency of their existence) cannot be proved in ZFC.

---

[2]Similar to the search for discovering true principles in physics.

[3]Set theorists differ widely on these matters, and pluralists and believers of Gödel's program represent only two of many possible viewpoints. Feferman has expressed that the Continuum Hypothesis is not even a definite mathematical problem. See [16] for a panoramic debate, [50] for some background, and [51] for more references. See also the EFI project web site http://logic.harvard.edu/efi.php.

The smallest large cardinals are the inaccessible cardinals, but we have also met two other types in the earlier postscripts, namely the weakly compact cardinals encountered in infinitary combinatorics, and the measurable cardinals that arose in Ulam's work on extensions of Lebesgue measure. One can use Gödel's second incompleteness theorem again to distinguish between "strengths" of large cardinal axioms. For example, let M denote "there is a measurable cardinal" and W denote "there is a weakly compact cardinal." It can be shown that ZFC+M $\to$ Con(ZFC+W) and ZFC+W $\to$ Con(ZFC+I). Hence (the existence of) a measurable cardinal is *strictly stronger in consistency strength* than (the existence of) a weakly compact cardinal, which in turn is strictly stronger than (the existence of) an inaccessible.

Now let PP denote "every projective set has the perfect set property." Then from Solovay's results it follows that Con(ZFC+I) is equivalent to Con(ZFC+PP), and so the perfect set property for projective sets is *equiconsistent,* relative to ZFC, with (the existence of) an inaccessible.

Most set theorists, starting from the inventor Gödel himself, find the axiom of constructibility to be highly unacceptable as an axiom. Gödel's results showed that the axiom of constructibility does answer Lusin's question about regularity properties of $\mathbf{\Sigma}^1_2$ (PCA) sets, but in a "negative" way: If $V{=}L$ then there are $\mathbf{\Sigma}^1_2$ sets which are not Lebesgue measurable, there are uncountable $\mathbf{\Pi}^1_1$ (coanalytic) sets without perfect subsets, etc. More generally, most set theorists find the restriction on set existence placed by the axiom of constructibility as too severe to be acceptable.

On the other hand, large cardinal axioms in general have been far more attractive to set theorists. They often resolve problems of ordinary mathematics in more "pleasant" ways. For example, recall Solovay's result:

Theorem 1309 (Solovay). *If there is a measurable cardinal, then all $\mathbf{\Sigma}^1_2$ sets have the perfect set property, are measurable, and have the Baire property.*

Thus constructibility and large cardinals seem to be naturally opposed to each other:[4] In the low levels of the projective hierarchy, the former implies some pathological phenomena, while the latter is intimately connected with regularity properties. In fact, we have the following partial reversal:

Theorem 1310 (Solovay). *If all uncountable $\mathbf{\Pi}^1_1$ (coanalytic) sets have perfect subsets, then at most countably many real numbers are constructible and the existence of inaccessible cardinals is relatively consistent with ZFC.*

This indicates that large cardinals beyond ZFC are *necessary* for establishing the perfect set property for the higher projective classes, further vindicating Lusin's conviction that the regularity properties enjoyed by the analytic sets would be impossible to extend to the higher projective classes (using the usual axioms of set

[4] An earlier result of Scott had shown that the axiom of constructibility contradicts the existence of measurable cardinals. Gaifman, Rowbottom and Silver dramatically improved Scott's result to show that if a measurable cardinal exists then in a certain sense the vast majority of sets must be non-constructible.

theory). It led Solovay to conjecture that stronger large cardinal axioms will imply regularity properties for all projective sets—a conjecture that was spectacularly confirmed through later works of set theorists such as Martin, Steel, and Woodin.

22.5 Infinite Games and Determinacy

Closely related to large cardinal axioms in this regard are the *axioms* or *principles of determinacy.* Determinacy provides the key to understanding why large cardinals imply regularity properties for projective sets.

Given $A \subseteq \mathbf{N}^{\mathbf{N}}$, consider the *game* $G(A)$ played by two players I and II alternately choosing natural numbers $x_1, x_2, x_3, \ldots$ as follows, with Player I going first:

I	x_1		x_3		x_5		x_7		$\ldots$
II		x_2		x_4		x_6		x_8	$\ldots$

The resulting sequence $x = \langle x_1, x_2, x_3, \ldots \rangle \in \mathbf{N}^{\mathbf{N}}$ is called a *play* or *run* of the game, and we declare this play x to be a *win for Player I* if $x \in A$; otherwise we say that the play x is a win for Player II.

A *strategy for Player I* is a function $\sigma\colon \{u \in \mathbf{N}^* \mid \operatorname{len}(u) \text{ is even}\} \to \mathbf{N}$, and given a play $x = \langle x_1, x_2, \ldots \rangle \in \mathbf{N}^{\mathbf{N}}$ we say that *I plays according to* σ or *follows* σ if for all even n, $x_{n+1} = \sigma(\langle x_1, x_2, \ldots, x_n \rangle)$. We say that σ is a *winning strategy for I* if every play following σ is a win for Player I, i.e., I always wins by playing according to σ, no matter what II plays.

The corresponding notions for Player II (*strategy* τ *for Player II* etc) are similarly defined.

A game $G(A)$ (or the set A) is said to be *determined* if either I or II has a winning strategy, i.e., if one of the players can force a win no matter how the opponent plays.

It can be shown that games with only finitely long plays are always determined; but for an arbitrary (infinite) game it is not at all clear that it will necessarily be determined.

We will now make two identifications:

- We will identify $\mathbf{N}^{\mathbf{N}}$ with the real interval $(0, 1]$ using the bijective mapping $\mathbf{H}\colon \mathbf{N}^{\mathbf{N}} \to (0, 1]$ (Problem 421) given by:

$$\mathbf{H}(\langle n_1, n_2, n_3, \ldots \rangle) := \frac{1}{2^{n_1}} + \frac{1}{2^{n_1+n_2}} + \frac{1}{2^{n_1+n_2+n_3}} + \cdots$$

- The reals $\mathbf{R}$ can be identified with the open interval $(0, 1)$ via some very effective homeomorphism such as $x \mapsto \frac{1}{2} + \frac{x}{2(|x|+1)}$.

We can therefore talk about games $G(E)$ where E is a subset of $(0, 1]$ or of $\mathbf{R}$ (by "transferring" the set E to a subset of $\mathbf{N}^{\mathbf{N}}$ via the above identifications).

Problem 1311. *If A is countable then II has a winning strategy in $G(A)$.*

The *axiom of determinacy* (AD), first introduced by Mycielski and Steinhaus in 1962, says that every set of reals is determined. Properties of the axiom of determinacy were studied by Mycielski [56]. It is a very powerful axiom which implies regularity properties for all sets: Under AD, all sets of reals have the perfect set property, are measurable, and have the Baire property. Thus, AD is incompatible with the Axiom of Choice, but as an alternative to AC it is an extremely interesting axiom with some surprising implications. For example, Solovay proved that under AD, $\aleph_1$ is a measurable cardinal!

Problem 1312. *Let $A \subseteq (0,1]$. If I has a winning strategy in $G(A)$, then A has a perfect subset. If II has a winning strategy in $G(A)$, then the complement of A has a perfect subset.*

[Hint: If I has a winning strategy σ in $G(A)$, let $P \subseteq (0,1]$ be the set of reals corresponding to all plays according to σ in which II always plays 1 or 2:

$$P := \{\mathbf{H}(x) \mid \text{For all } n,\, x_{2n-1} = \sigma(\langle x_1, x_2, \dots, x_{2n-2}\rangle) \text{ and } x_{2n} \in \{1,2\}\}.$$

Then P is a perfect subset of A.]

Corollary 1313. *If B is a Bernstein set then $G(B)$ is not determined.*

Recall, however, that the construction of a Bernstein set is highly non-effective and requires heavy use of the full axiom of choice. We therefore consider games $G(A)$ with A restricted to some natural class of effectively defined sets—such as open, Borel, analytic, projective, etc—and ask if such games are necessarily determined. A very basic and early result in such restricted *definable determinacy principles* is the Gale–Stewart Theorem:

Theorem 1314 (Gale–Stewart 1953). *Every open game is determined. Every closed game is determined.*

Roughly speaking, the more effectively a set is defined, the easier it is to establish that it is determined. Thus, it is somewhat harder to prove that F_σ and G_δ games are determined, and still harder to prove that $\mathsf{F}_{\sigma\delta}$ and $\mathsf{G}_{\delta\sigma}$ games are determined. Work of Harvey Friedman indicated the reason behind such increasing levels of difficulty: To establish determinacy for each additional level of the Borel hierarchy one needs the existence of an additional level of the cumulative hierarchy of sets V_α for $\alpha > \omega$. Gale and Stewart had asked if all Borel games are determined, and by Friedman's result establishing Borel determinacy would require an uncountable number of iterations of the power set operation all the way through V_{ω_1}. This means Borel determinacy is a result that cannot be established in Zermelo set theory Z with choice (i.e., ZFC minus the replacement axiom). Martin established the celebrated result:

Theorem 1315 (Martin 1975). *Every Borel game is determined.*

Borel determinacy is the first major mathematical result provable in ZFC that requires the full strength of ZFC via essential use of the replacement axiom. It is the strongest determinacy principle for a natural class of definable sets that can be proved in ZFC. Determinacy of analytic ($\mathbf{\Sigma}^1_1$) games, as we will see now, requires stronger (large cardinal) assumptions.

Even before Borel determinacy was proved, work of Martin, Kechris, and Solovay had established fundamental connections between determinacy and large cardinals. We already mentioned Solovay's result that full AD implies that $\aleph_1$ is a measurable cardinal. The following two theorems show that determinacy of analytic games interpolates in between the hypothesis and conclusion of Solovay's earlier theorem Theorem 1309.

Theorem 1316 (Martin). *If a measurable cardinal exists, then all analytic games are determined.*

Theorem 1317. *If all analytic games are determined, then all $\mathbf{\Sigma}^1_2$ sets are measurable, have the perfect set property, and the Baire property.*

By Theorem 1310, the perfect set property for $\mathbf{\Sigma}^1_2$ sets implies relative consistency of inaccessible cardinals; hence, by Theorem 1317, analytic determinacy implies the consistency of inaccessibles as well, and so cannot be proved in ZFC. We have thus encountered a definable determinacy principle for a naturally arising class of effectively defined sets which is inextricably linked to large cardinals.

Actually, analytic determinacy implies relative consistency of much larger cardinals, in fact, larger than weakly compact cardinals. On the other hand, the hypothesis of measurable cardinals in Theorem 1316 is too strong, and analytic determinacy can be derived from smaller cardinals (see [57] for such a proof). By an exact characterization due to Martin and Harrington (in terms of so called *sharps* or *Silver indiscernibles*) analytic determinacy has consistency strength lying strictly between weakly compact and measurable cardinals.(Going further, determinacy of $\mathbf{\Sigma}^1_2$ sets implies regularity properties for $\mathbf{\Sigma}^1_3$ sets and has much stronger consistency strength, entailing the relative consistency of many measurable cardinals.)

These results indicate that determinacy principles provide the key for obtaining regularity properties for projective classes, and they themselves represent a form of large cardinal axioms. In fact, it turns out, beautifully, that determinacy principles establish a correlation between the projective hierarchy and large cardinal axioms such that determinacy for larger projective classes corresponds to stronger large cardinal axioms.

22.6 Projective Determinacy

Using determinacy principles, Theorem 1317 can be generalized through the entire projective hierarchy:

Theorem 1318 (Kechris–Martin, after Mazur, Banach, Mycielski, Swierczkowski, Davis). *If all $\mathbf{\Sigma}^1_n$ games are determined then all $\mathbf{\Sigma}^1_{n+1}$ sets are measurable, have the perfect set property, and have the Baire property.*

The assumption that all projective games are determined is known as *Projective Determinacy* (PD). Thus under PD, the regularity properties of analytic sets extend to all the higher projective classes—a result that Lusin believed (correctly) would be impossible to obtain using the ordinary axioms of mathematics.

Corollary 1319. *If all projective games are determined, then all projective sets are measurable, have the perfect set property, and have the Baire property.*

Another line of development which uses projective determinacy concerns *structural properties* of the projective classes. We had proved the Lusin separation theorem (Theorem 1151) for the class of analytic sets, or $\mathbf{\Sigma}^1_1$. The strongest classical separation theorem was for the class $\mathbf{\Pi}^1_2$. In 1967, Blackwell found a proof of the $\mathbf{\Sigma}^1_1$ separation theorem using determinacy of closed games. Assuming determinacy of projective sets, Addison, Martin, and Moschovakis generalized Blackwell's result through the entire projective hierarchy, and the separation property was found precisely in the classes $\mathbf{\Sigma}^1_{2n-1}$ and $\mathbf{\Pi}^1_{2n}$ ($n = 1, 2, \ldots$).[5]

Thus PD gives a complete structure theory for the projective classes, i.e., the entire theory of projective classes takes a remarkably canonical and coherent form under PD, with all questions about regularity and structural properties settled in an intuitively desirable and natural fashion. We can speculate that, perhaps, this is the best that Lusin could have hoped for.

The optimal large cardinal notion that implies determinacy for the projective classes is that of a *Woodin cardinal.* We will not define Woodin cardinals here (see [37] or [34] for a definition), but state the following seminal result:[6]

Theorem 1320 (Martin–Steel 1985). *If there are n Woodin cardinals and a measurable cardinal above them all, then all $\mathbf{\Sigma}^1_{n+1}$ games are determined.*

If there are infinitely many Woodin cardinals, then all projective games are determined.[7]

In the other direction, we have the following results.

Theorem 1321. *$\mathbf{\Sigma}^1_{n+1}$-determinacy implies the relative consistency of the existence n Woodin cardinals. Therefore, projective determinacy implies, for each $n \in \mathbf{N}$, the consistency of the existence of n Woodin cardinals.*

---

[5]Other stronger structural properties that we have not defined (such as reduction, pre-well ordering, uniformization, and scale) hold in the dual (opposite) classes.

[6]Deep research by several set theorists including Martin, Steel, Kechris, Foreman, Magidor, Shelah, and Woodin, culminated in the final ideas and results.

[7]Woodin showed that with a marginally stronger hypothesis (existence of a measurable cardinal above infinitely many Woodin cardinals) the determinacy of a much larger class of sets (than the projective sets) called $L(\mathbf{R})$ can be established.

Theorem 1322 (Woodin). *The full Axiom of Determinacy (AD) is consistent with ZF (without Choice) if and only if the existence of infinitely many Woodin cardinals is consistent with ZFC.*

These results further confirm our earlier statement that determinacy is a form of large cardinal axiom, via an almost perfect "correlation of strength" through the projective classes.

The remarkable results above (and many others that were not mentioned) indicate why most set theorists find that the axiom of projective determinacy (as opposed to $V=L$) gives the "true and correct" picture for the projective sets, and therefore can be regarded as a truly natural strong axiom vindicating Gödel's program—as far as the theory of projective sets is concerned.

The situation for CH is far more complex.

22.7 Does the Continuum Hypothesis Have a Truth Value?

As mentioned earlier, the Continuum Problem, which was first on Hilbert's famous list, is widely regarded as the greatest problem of set theory. It has remained unsettled after more than a hundred years of attack. Moreover, unlike Lebesgue and Banach's Measure Problem or Lusin's Problem involving the projective sets, the Continuum Problem cannot be resolved using the usual type of large cardinal hypotheses.[8]

Of course, pluralists (formalists) may not think that CH can ever be decided, and some of them may think that the Gödel–Cohen independence results have settled the matter for ever. For many pluralists, CH does not have an absolute or intrinsic truth value. For some, it may not even be a well-defined mathematical problem.

Supporters of Gödel's program, on the other hand, keep searching for strong natural axioms which might decide CH. Most of the known axioms which decide CH (such as $V=L$ and Martin's Maximum) are not considered sufficiently natural to be acceptable. Thus the Continuum Problem, unlike the theory of projective sets, remains open from the perspective of Gödel's program. However, some highly sophisticated recent work of Woodin and others has made the problem more tantalizing than ever by arguing that natural axioms settling the Continuum Problem may be around the corner. This has been a topic of much discussion (and debate) among set theorists, who differ widely in their mathematical and philosophical approaches to CH. For a general survey of this large subject, see Koellner's article [87] on CH in the online *Stanford Encyclopedia of Philosophy*.

The recent *EFI project* (Exploring the Frontiers of Incompleteness) of Koellner brought together major thinkers in a workshop on this foundational debate.[9]

---

[8]This was shown by Cohen, Levy, and Solovay.

[9]The web site http://logic.harvard.edu/efi.php has more information and resources. The project is funded by a grant from the John Templeton Foundation.

22.8 Further References

Introductory accounts of constructibility can be found in [14, 34, 42, 55], while Gödel's original presentation is [22].

Standard references for learning forcing are [2, 34, 42, 44], but Cohen's original text [8] is still in print.

For large cardinals, the definitive reference is [37] (see also [34]), but the older [14] is helpful as well.

The theory of determinacy of infinite games is covered in [34, 37, 55]. Inviting introductions to this area can be found in [53, 57].

In addition to Koellner [87] mentioned above, discussions on some of the recent approaches to the Continuum Problem are in [1, 17, 52], and in Woodin's own expository articles [82, 83].

A handbook containing highly advanced up to date surveys of current research in set theory is [18].

Appendix A
Proofs of Uncountability of the Reals

In this appendix, we summarize and review the proofs of uncountability of the reals given in the main text, and indicate how the methods of these proofs generalize and connect to other areas of mathematics. (This appendix is not an exhaustive list of such proofs.)

There were essentially three distinct proofs of uncountability of the reals given in the text. All proofs depend, in the end, on some form of order completeness of **R**, but they take very different forms and generalize in different ways to give other significant results in mathematics.

A.1 Order-Theoretic Proofs

Section 8.5 presented a proof of uncountability of the reals which follows immediately from Cantor's powerful theorem characterizing the order type η (which says any countable dense order without endpoints has order type η). That theorem also implies $\eta + \eta = \eta$, and so any countable dense order must have Dedekind gaps. Hence any dense linear order without Dedekind gaps, such as **R**, must be uncountable.

This proof is so short because it exploits a very powerful result of order theory. It is related to Cantor's first proof of uncountability of **R**, which directly shows that a countable dense order cannot be complete:

Proof (Cantor's first proof of uncountability of **R***).* To get a contradiction, suppose that the set of real numbers can be enumerated as $p_1, p_2, \ldots$ (without repetition). Recursively define two sequences of reals $\langle a_n \rangle$ and $\langle b_n \rangle$ with

$$a_1 < a_2 < \cdots < a_n < \cdots \qquad \cdots < b_n < \cdots < b_2 < b_1,$$

in the following manner. Let $a_1 = p_1$, and $b_1 = p_m$ where m is the least index such that $a_1 < p_m$. Having defined $a_1, a_2, \ldots, a_n$ and $b_1, b_2, \ldots, b_n$ with $a_n < b_n$, define

A. Dasgupta, *Set Theory: With an Introduction to Real Point Sets*,
DOI 10.1007/978-1-4614-8854-5, © Springer Science+Business Media New York 2014

$a_{n+1} = p_j$ where j is the least index such that $a_n < p_j < b_n$ and $b_{n+1} = p_k$ where k is the least index such that $a_{n+1} < p_k < b_n$. Then we have $a_n < a_{n+1} < b_{n+1} < b_n$, and the recursive definition is complete. In particular, for each n we have $a_n = p_{j_n}$ and $b_n = p_{k_n}$ for some indices j_n and k_n. Now, by completeness of **R**, there must be a real number p such that $a_n < p < b_n$ for all n, and so $p = p_i$ for some i. Since the indices j_n are all distinct, we can fix n with $j_{n+1} > i$. Note that $a_n < p_i < b_n$ and by definition of $a_{n+1} = p_{j_{n+1}}$, we see that j_{n+1} equals the least index j such that p_j lies between a_n and b_n, and so $j_{n+1} \leq i$, a contradiction. □

This was Cantor's first published proof of the uncountability of **R**. *Given any enumeration of a countable dense order, it effectively produces a gap in it.*

Both the proof of Sect. 8.5 based on Cantor's theorem characterizing the order type η and Cantor's first proof given above appeal to order completeness, but note that full completeness is not necessary. For both proofs, it suffices to assume that there are no $(\omega, {}^*\omega)$ gap in the ordering.

Proposition 1323. *A dense order without $(\omega, {}^*\omega)$ gaps has cardinality $> \aleph_0$.*

In this form, the proof generalizes to η_1 orders without $(\omega_1, {}^*\omega_1)$ gaps:

Proposition 1324. *Any η_1 order without $(\omega_1, {}^*\omega_1)$ gaps has cardinality $> \aleph_1$.*

Proof. Recall that any two η_1 orders of cardinality $\aleph_1$ must be isomorphic to each other. If there were an η_1 order X of cardinality $\aleph_1$ without $(\omega_1, {}^*\omega_1)$ gaps, then any suborder Y of X obtained by removing a single point of X would also be an η_1 order of cardinality $\aleph_1$ and so must be isomorphic to X. But Y has a $(\omega_1, {}^*\omega_1)$ gap, and so X has such a gap, a contradiction. □

Another related generalization is this: *Any dense-in-itself complete order contains an isomorphic copy of the real line and so has cardinality $\geqslant \mathfrak{c}$.*

Connected spaces and their uncountability. As mentioned in the text, the notion of connectedness in topology is a direct generalization of Dedekind's definition of linear continuum: An order is a continuum if and only if in any Dedekind partition of the order at least one of the sets contains a point which is a limit point of the other. A metric or topological space is connected if and only if for any partition of the space into two nonempty sets, at least one set contains a limit point of the other. Under certain regularity conditions, the uncountability of linear continuums carries over to connected spaces. To see this, note that the Intermediate Value Theorem generalizes: *The range of any continuous function from a connected space to an order must be a linear continuum.* Since the distance function on a metric space is continuous, any connected metric space with at least two points is uncountable.[1]

[1]By a basic topological result known as Urysohn's Lemma, this generalizes to any T_4 (normal Hausdorff) topological space, and in fact to any T_3 (regular Hausdorff) space: *Any connected T_3 space with at least two points must be uncountable.* All these generalizations are thus related to the order-based proof of uncountability of **R**.

A.2 Proof Using Cantor's Diagonal Method

Cantor discovered his "diagonal method" for proving uncountability several years after he obtained his order-based proof given above where he first discovered that **R** is uncountable. Unlike the order-theoretic proofs, the diagonal method is applicable in much more general situations where no order may be present.

In a sense, *diagonalization* means that given an infinite list of conditions, we construct a "counterexample" real number which refutes all those conditions. The nested intervals theorem gives a direct version of this form of diagonalization: Given a sequence of reals $\langle x_1, x_2, \ldots \rangle$, one builds nested closed intervals of shrinking length $I_1 \supseteq I_2 \supseteq \cdots$ such that $x_1 \notin I_1$ ("I_1 avoids x_1"), $x_2 \notin I_2$, and so on. The unique real x in their intersection then differs from all the given reals $x_1, x_2, \ldots$. Here the n-th given condition is "$x = x_n$," and the above method of diagonalization via nested intervals produces the real x which satisfies $x \neq x_n$ for all n. Therefore, we call x *the diagonal counterexample* for the given sequence $\langle x_1, x_2, \ldots \rangle$ of reals.

In this proof, we could, for definiteness, use the specific scheme for building nested closed intervals where the initial interval I_0 is the unit interval $I_0 = [0, 1]$, and each I_n is either the left-third or the right-third subinterval of I_{n-1} (whichever avoids the real x_n first). The diagonal counterexample will then always be a member of the Cantor set, and conversely, any member of the Cantor set can be seen to be a diagonal counterexample for a suitably given sequence of reals $\langle x_1, x_2, \ldots \rangle$. It follows that with this scheme of building nested intervals, *the Cantor set is the set of all possible diagonal counterexamples to various given sequences of real numbers.*

With a little modification, the above proof of uncountability of **R** yields the Baire Category Theorem, where the n-th condition to be met is to be inside an arbitrary given dense open set G_n (instead of the special dense open set of the form $\{x \mid x \neq x_n\}$). The Baire category theorem holds in complete metric spaces as well as in locally compact Hausdorff spaces, and thus any such space without isolated points must be uncountable (and in fact of cardinality at least $\mathfrak{c}$). This illustrates how Cantor's diagonal method leads to a powerful general theorem of very wide applicability.

In a more literal form of diagonalization we regard a family $\langle E_i \mid i \in E \rangle$ of subsets of a set E indexed by E itself as the following relation on E:

$$\{\langle i, j \rangle \in E \times E \mid j \in E_i\},$$

(or, using the identification via characteristic functions, as a binary array $\langle a_{i,j} \mid i, j \in E \rangle$ where each $a_{i,j}$ is 0 or 1). We then form the diagonal set $D := \{i \in E \mid i \in E_i\}$, and finally take its complement to get the "anti-diagonal" set $A := E \setminus D = \{i \in E \mid i \notin E_i\}$, which must differ from all the sets E_i. In other words, it shows that $\mathbf{P}(E)$ cannot be listed as a family of sets indexed by E. This is Cantor's theorem that $|E| < |\mathbf{P}(E)|$, another far reaching generalization (of the uncountability of **R**) which ensures existence of sets of arbitrarily large infinite cardinality.

This last version is a more abstract form of diagonalization which is usually referred to as *the Cantor diagonal method.*

The Cantor set establishes a close connection between these two forms of the diagonal method: It is constructed by a "binary tree of nested intervals" in which infinite branches (of nested intervals) through the tree correspond, on the one hand, to the points of the Cantor set, and, on the other hand, to infinite binary sequences, i.e., to members of $\{0, 1\}^{\mathbf{N}}$ or to subsets of **N**.

One thus obtains a variant of the diagonal proof of uncountability of **R** by identifying the Cantor set with **P**(**N**) (or with $\{0, 1\}^{\mathbf{N}}$) and then appealing to the abstract Cantor diagonal theorem that $|\mathbf{P}(\mathbf{N})| > |\mathbf{N}|$.

The more abstract version of the Cantor diagonal method has quite wide ramifications. It not only gives (via Cantor's theorem that $|\mathbf{P}(X)| > |X|$) sets of larger and larger infinite cardinalities by iterating the power set operation, but also is a method used in the proofs of many important theorems of logic and computability, such as Gödel's incompleteness theorem, the unsolvability of the Halting problem, and Tarski's undefinability theorem.

A.3 Proof Using Borel's Theorem on Interval Lengths

In Corollary 1018 it was shown that the interval $[a, b]$ is uncountable using properties of lengths of intervals. The length of a bounded interval in **R** is defined by

$$\text{len}([a,b]) = \text{len}((a,b]) = \text{len}([a,b)) = \text{len}((a,b)) = b - a \qquad (a \leq b).$$

The length function thus defined on the intervals has several natural properties (which are essential in obtaining the Lebesgue measure on **R**). For example, the lengths of intervals are easily seen to satisfy the condition of *finite additivity,* which says that if an interval I is partitioned into finitely many pairwise disjoint intervals $I_1, I_2, \ldots, I_n$, then

$$\text{len}(I) = \text{len}(I_1) + \text{len}(I_2) + \cdots \text{len}(I_n).$$

However, the key fact about lengths of intervals used in the uncountability proof mentioned above was *Borel's theorem,* which says that the interval $[a, b]$, which has length $b - a$, cannot be covered by countably many intervals of smaller total length. This important condition is known as *countable subadditivity* of length, which was established (in Borel's theorem) using the powerful Heine–Borel theorem. Since any countable set of reals can be covered by countably many intervals having arbitrarily small total length, countable subadditivity immediately implies that a proper interval must be uncountable.

The proof also readily generalizes to more abstract setups as follows. Let X be a fixed set. A nonempty collection S of subsets of X is called a *semiring* on X if for any $A, B \in \mathrm{S}$ the intersection $A \cap B$ is in S and the difference $A \setminus B$ can be expressed as the union of finitely many pairwise disjoint sets from S. By a *set-function* on a semiring S we mean a function μ defined on S which takes nonnegative extended real values (i.e., we allow $\mu(A)$ to be $+\infty$). A set-function μ on a semiring S on X is said to be *continuous* if for every $p \in X$ and every $\epsilon > 0$ there is a set $E \in \mathrm{S}$ with $p \in E$ and $\mu(E) < \epsilon$, and μ is said to be *countably subadditive* on S if whenever $E \in \mathrm{S}$ is covered by countably many sets $E_1, E_2, \cdots \in \mathrm{S}$, we have $\mu(E) \leq \sum_{n=1}^{\infty} \mu(E_n)$. Essentially the same proof that a countable set has measure zero now immediately gives:

Proposition 1325. *Suppose that μ is a nonnegative continuous set function on a semiring S of subsets of a fixed set X. If μ is countably subadditive on S, then E is uncountable for any $E \in \mathrm{S}$ for which $\mu(E) \neq 0$.*

Countable subadditivity is necessary here. For example, let X be the set $\mathbf{Q}$ of rational numbers. By a *rational half-open interval* we mean a set of the form $[a, b) \cap \mathbf{Q}$ with $a, b \in \mathbf{Q}$. The set of half-open rational intervals forms a semiring on $\mathbf{Q}$ on which the length function (defined as before) is continuous and finitely additive. But countable subadditivity fails and every rational interval is countable.

We conclude by noting that under finite additivity, the condition of countable subadditivity (as in Borel's theorem) actually entails a much stronger and important result known as the *measure extension theorem*, whose proof can be found in any standard textbook of measure theory. By a *measure* we mean a nonnegative extended real valued set-function ν defined on a sigma-algebra which vanishes on the empty set ($\nu(\emptyset) = 0$) and which satisfies the condition that if $\langle A_n \rangle$ is a pairwise disjoint sequence of sets from the sigma-algebra then $\nu(\cup_{n=1}^{\infty} A_n) = \sum_{n=1}^{\infty} \nu(A_n)$ (countable additivity).

Theorem 1326 (The Measure Extension Theorem). *Let μ be a finitely additive nonnegative extended real valued set-function on a semiring S of subsets of a fixed set X. Assume that $X = \cup_n A_n$ for some sets $A_n \in \mathrm{S}$ with $\mu(A_n) < \infty$ for all n. If μ is countably subadditive on S, then there is a unique measure defined on the sigma-algebra generated by S which extends μ.*

Taking S to be the semiring of all real intervals of the form $[a, b)$ and μ to be the length function on such intervals, we get the following immediate corollary of the theorem: *There is a unique measure defined on the Borel subsets of* **R** *for which the measure of any interval is its length.* This measure is known as *the Lebesgue measure,* and it also uniquely extends as a measure to the collection of all Lebesgue measurable sets (the sigma-algebra generated by the Borel sets together with the measure zero sets).

Appendix B
Existence of Lebesgue Measure

This appendix gives a proof of the existence of Lebesgue measure. That is, we prove Theorem 1028 whose statement is as below. Recall that $E \in \mathbf{L}$, or E is *measurable,* if for all $\epsilon > 0$ there exist closed F and open G with $F \subseteq E \subseteq G$ and intervals $I_1, I_2, \ldots$ covering $G \setminus F$ with $\sum_n \text{len}(I_n) < \epsilon$.

Theorem (Lebesgue). There is $m \colon \mathbf{L} \to [0, \infty]$ such that

1. m is *countably additive*: If $A_1, A_2, \ldots$ are pairwise disjoint measurable sets, then $m(\bigcup_n A_n) = \sum_n m(A_n)$.
2. $m(I) = \text{len}(I)$ for any interval I (thus $m(\emptyset) = 0$).

To prove the theorem, we first define the *outer measure* $m^*(E)$ of any set $E \subseteq \mathbf{R}$ (not necessarily measurable), and then restrict m^* to $\mathbf{L}$ to get m.

Definition 1327 (Outer Measure). For any $E \subseteq \mathbf{R}$, we define:

$$m^*(E) := \inf\left\{ \textstyle\sum_{n=1}^{\infty} \text{len}(I_n) \mid \langle I_n \rangle \text{ is a sequence of intervals covering } E \right\}.$$

m is m^* restricted to $\mathbf{L}$, so if $E \in \mathbf{L}$, then $m^*(E)$ is denoted by $m(E)$.

Recall Borel's theorem (Theorem 1011) which says $\text{len}(I) \leq m^*(I)$ for any interval I. The following facts are now immediate.

Problem 1328 (Monotonicity). *If $A \subseteq B$ then $m^*(A) \leq m^*(B)$.*

Proposition 1329. *For any interval I, $m^*(I) = \text{len}(I)$.*

Proof. $m^*(I) \leq \text{len}(I)$ is trivial and Borel's theorem says $m^*(I) \geq \text{len}(I)$. □

Proposition 1330 (Countable Subadditivity of Outer Measure). *For any sequence $E_1, E_2, \ldots$ of sets, $m^*(\bigcup_n E_n) \leq \sum_n m^*(E_n)$.*

Proof. Given $\epsilon > 0$, choose, for each n, a sequence of intervals $\langle I_{n,k} \mid k \in \mathbf{N} \rangle$ covering E_n and with $\sum_k \text{len}(I_{n,k}) \leq m^*(E_n) + \frac{\epsilon}{2^n}$. Combining all these sequences

A. Dasgupta, *Set Theory: With an Introduction to Real Point Sets*,
DOI 10.1007/978-1-4614-8854-5,

of intervals into a single sequence, we get a covering of $\bigcup_n E_n$ with total length $\leq \sum_n (m^*(E_n) + \frac{\epsilon}{2^n}) = \sum_n m^*(E_n) + \epsilon$. □

So we can (and will) prove equalities of the form $m^*(\bigcup_n E_n) = \sum_n m^*(E_n)$ by only showing $m^*(\bigcup_n E_n) \geq \sum_n m^*(E_n)$ (by countable subadditivity).

Corollary 1331. *If A is measurable and $\epsilon > 0$ then there are closed F and open G such that $F \subseteq A \subseteq G$, $m^*(A) \geq m^*(G) - \epsilon$, and $m^*(F) \geq m^*(A) - \epsilon$.*

Proof. Let $\epsilon > 0$. Fix closed F and open G such that $F \subseteq A \subseteq G$ and $m^*(G \setminus F) < \epsilon$. Then by countable subadditivity and monotonicity, $m^*(G) \leq m^*(A) + m^*(G \setminus A) \leq m^*(A) + m^*(G \setminus F) \leq m^*(A) + \epsilon$, so $m^*(A) \geq m^*(G) - \epsilon$. Similarly $m^*(F) \geq m^*(A) - \epsilon$. □

Proposition 1332. *Let G be an open set expressed as a disjoint union of open intervals $\cup_n J_n = G$. Then $m^*(G) = \sum_n \text{len}(J_n)$.*

Proof. Easily $m^*(G) \leq \sum_n \text{len}(J_n)$ (since the J_n's cover G).

For the other direction, let $\langle I_n \rangle$ be *any* sequence of open intervals covering G. Then for each n, $\langle I_n \cap J_m \mid m \in \mathbf{N} \rangle$ is a sequence of pairwise disjoint intervals all contained in I_n and so $\text{len}(I_n) \geq \sum_m \text{len}(I_n \cap J_m)$. Hence

$$\sum_n \text{len}(I_n) \geq \sum_n \sum_m \text{len}(I_n \cap J_m) = \sum_m \sum_n \text{len}(I_n \cap J_m) \geq \sum_m \text{len}(J_m),$$

where the last inequality follows by Borel's theorem since for each m, the intervals $\langle I_n \cap J_m \mid n \in \mathbf{N} \rangle$ cover J_m. □

Corollary 1333. *If G_1 and G_2 are disjoint open sets then $m^*(G_1 \cup G_2) = m^*(G_1) + m^*(G_2)$.*

Proposition 1334. *If F_1 and F_2 are disjoint closed sets then $m^*(F_1 \cup F_2) = m^*(F_1) + m^*(F_2)$.*

Proof. Let $\epsilon > 0$. Fix open G with $F_1 \cup F_2 \subseteq G$ and $m^*(F_1 \cup F_2) \geq m^*(G) - \epsilon$. Fix disjoint open G_1 and G_2 containing F_1 and F_2 respectively (Problem 938). Then we have $m^*(F_1 \cup F_2) \geq m^*(G) - \epsilon \geq m^*((G \cap G_1) \cup (G \cap G_2)) - \epsilon = m^*(G \cap G_1) + m^*(G \cap G_2) - \epsilon \geq m^*(F_1) + m^*(F_2) - \epsilon$. □

Proposition 1335 (Finite Additivity). *Let A and B be disjoint measurable sets. Then $m^*(A \cup B) = m^*(A) + m^*(B)$.*

Proof. Let $\epsilon > 0$. Fix closed sets $F_A \subseteq A$ and $F_B \subseteq B$ with $m^*(F_A) \geq m^*(A) - \frac{\epsilon}{2}$ and $m^*(F_B) \geq m^*(B) - \frac{\epsilon}{2}$. Then $m^*(A \cup B) \geq m^*(F_A \cup F_B) = m^*(F_A) + m^*(F_B) \geq m^*(A) + m^*(B)$. □

Proposition 1336 (Countable Additivity of Lebesgue Measure). *If A_1, A_2, … are disjoint measurable sets then $m^*(\bigcup_n A_n) = \sum_n m^*(A_n)$.*

Proof. $\sum_n m^*(A_n) = \sup_n \sum_{k=1}^{n} m^*(A_k) = \sup_n m^*(\bigcup_{k=1}^n A_k) \leq m^*(\bigcup_n A_n)$. □

The main theorem now follows from and Propositions 1329 and 1336.

Appendix C
List of ZF Axioms

ZF 1 (Extensionality). $\forall x \forall y(\forall z(z \in x \leftrightarrow z \in y) \rightarrow x = y)$.

ZF 2 (Empty Set). $\exists x \forall y(y \notin x)$.

ZF 3 (Separation Scheme). *If* $\varphi(x, t_1, t_2, \ldots, t_n)$ *is a ZF formula in which the free variables are among* $x, t_1, t_2, \ldots, t_n$, *then the following is an axiom:*

$$\forall t_1 \forall t_2 \cdots \forall t_n \forall a \exists b \forall x(x \in b \leftrightarrow x \in a \wedge \varphi(x, t_1, t_2, \ldots, t_n)).$$

ZF 4 (Power Set). $\forall x \exists y \forall z(z \in y \leftrightarrow \forall w(w \in z \rightarrow w \in x))$.

ZF 5 (Union). $\forall x \exists y \forall z(z \in y \leftrightarrow \exists w(w \in x \wedge z \in w))$.

ZF 6 (Unordered Pairs). $\forall x \forall y \exists z \forall w(w \in z \leftrightarrow w = x \vee w = y)$.

ZF 7 (Replacement Scheme). *If* $\varphi(x, y, t_1, t_2, \ldots, t_n)$ *is a ZF formula with free variables among the ones shown, then we have the axiom:*

$$\begin{aligned}\forall t_1 \forall t_2 \cdots \forall t_n \big(\forall x \forall y \forall z(\varphi(x, y, t_1, \ldots, t_n) \wedge \varphi(x, z, t_1, \ldots, t_n) \rightarrow y = z)\\ \rightarrow \forall a \exists b \forall u \forall v(u \in a \wedge \varphi(u, v, t_1, \ldots, t_n) \rightarrow v \in b)\big).\end{aligned}$$

ZF 8 (Infinity). $\exists b\big(\exists y(y \in b \wedge \forall z(z \notin y))\wedge$
$\forall x(x \in b \rightarrow \exists y(y \in b \wedge \forall z(z \in y \leftrightarrow z \in x \vee z = x)))\big)$.

ZF 9 (Foundation). $\forall x(\exists y(y \in x) \rightarrow \exists y(y \in x \wedge \neg \exists z(z \in y \wedge z \in x)))$.

ZFC is obtained by adding to ZF the *Axiom of Choice,* which says:

$$\begin{aligned}\forall x((\forall y(y \in x \rightarrow \exists z(z \in y))\wedge\\ \forall u \forall v(u \in x \wedge v \in x \wedge u \neq v \rightarrow \neg \exists y(y \in u \wedge y \in v)))\\ \rightarrow \exists w \forall y(y \in x \rightarrow \exists! z(z \in y \wedge z \in w))).\end{aligned}$$

A. Dasgupta, *Set Theory: With an Introduction to Real Point Sets*,
DOI 10.1007/978-1-4614-8854-5, © Springer Science+Business Media New York 2014

References

1. J. Bagaria. Natural axioms of set theory and the continuum problem. In *Proceedings of the 12th International Congress of Logic, Methodology, and Philosophy of Science*, pages 43–64. King's College London Publications, 2005.
2. J. L. Bell. *Boolean-Valued Models and Independence Proofs in Set Theory*, volume 12 of *Oxford Logic Guides*. Clarendon Press, Oxford, 2nd edition, 1985.
3. P. Bernays. *Axiomatic Set Theory*. North Holland, 1958.
4. G. Birkhoff and S. Mac Lane. *A Survey of Modern Algebra*. Macmillan, 4th edition, 1977.
5. N. Bourbaki. *Theory of Sets*. Springer, 2004.
6. G. Cantor. *Contributions to the Founding of the Theory of Transfinite Numbers*. Dover, 1955.
7. R. Carnap. *An Introduction to the Philosophy of Science*. Dover, 1995.
8. P. J. Cohen. *Set Theory and the Continuum Hypothesis*. WA Benjamin, 1966.
9. J. W. Dauben. *Georg Cantor: His Mathematics and Philosophy of the Infinite*. Princeton University Press, 1990.
10. R. Dedekind. Continuity and irrational numbers (1872). In *Essays on the Theory of Numbers* [12]. English translation of "Stetigkeit und irrationale Zahlen", Vieweg, 1872.
11. R. Dedekind. The nature and meaning of numbers (1888). In *Essays on the Theory of Numbers* [12]. English translation of "Was sind und was sollen die Zahlen?", Vieweg, 1888.
12. R. Dedekind. *Essays on the Theory of Numbers*. Public Domain (originally published by Open Court), 1901.
13. K. Devlin. *The Joy of Sets: Fundamentals of Contemporary Set Theory*. Springer, 2nd edition, 1993.
14. F. R. Drake. *Set theory: An Introduction to Large Cardinals*. North Holland, 1974.
15. H. B. Enderton. *Elements of Set Theory*. Academic Press, 1977.
16. S. Feferman, H. M. Friedman, P. Maddy, and J. R. Steel. Does mathematics need new axioms? *The Bulletin of Symbolic Logic*, 6(4):401–446, 2000.
17. M. Foreman. Has the Continuum Hypothesis been settled? Talk presented in Logic Colloquium 2003 (Helsinki).
18. M. Foreman and A. Kanamori. *Handbook of Set Theory*. Springer, 2010.
19. T. E. Forster. *Set Theory with a Universal Set*. Oxford University Press, 2nd edition, 1995.
20. A. A. Fraenkel. *Abstract Set Theory*. North Holland, 4th revised edition, 1976. Revised by A. Levy.
21. A. A. Fraenkel, Y. Bar-Hillel, and A. Levy. *Foundations of set theory*. North Holland, 2nd revised edition, 1973.
22. K. Gödel. *The Consistency of the Axiom of Choice and of the Generalized Continuum Hypothesis with the Axioms of Set Theory*. Number 3 in Annals of Mathematics Studies. Princeton, 1940. Seventh printing, 1966.

A. Dasgupta, *Set Theory: With an Introduction to Real Point Sets*,
DOI 10.1007/978-1-4614-8854-5,

23. K. Gödel. What is Cantor's continuum problem? In P. Benacerraf and H. Putnam, editors, *Philosophy of Mathematics: Selected Readings*, pages 470–485. Cambridge University Press, second edition, 1983.
24. D. C. Goldrei. *Classic Set Theory for Guided Independent Study*. CRC Press, 1996.
25. Noa Goldring. Measures: back and forth between point sets and large sets. *The Bulletin of Symbolic Logic*, 1(2):170–188, 1995.
26. A. Hajnal and P. Hamburger. *Set Theory*. Cambridge, 1999.
27. P. R. Halmos. *Naive Set Theory*. Springer, 1960.
28. A. G. Hamilton. *Numbers, Sets and Axioms*. Cambridge, 1983.
29. F. Hausdorff. *Set Theory*. Chelsea Publishing Company, 1957.
30. E. Hewitt and K. Stromberg. *Real and Abstract Analysis*. Springer, 1965.
31. R. Holmes. *Elementary Set Theory with a Universal Set*. Louvain-la-Neuve: Bruylant-Academia, 1998.
32. K. Hrbacek and T. Jech. *Introduction to Set Theory*. CRC Press, 3rd edition, 1999.
33. T. Jech. *The Axiom of Choice*. North Holland, 1973.
34. T. Jech. *Set theory: The Third Millennium Edition*. Springer, 2003.
35. W. Just and M. Weese. *Discovering Modern Set Theory, I and II*. American Mathematical Society, 1996, 1997.
36. E. Kamke. *Theory of Sets*. Dover, 1950.
37. A. Kanamori. *The Higher Infinite*. Springer, 2nd edition, 2003.
38. A. S. Kechris. *Classical Descriptive Set Theory*. Springer, 1995.
39. A. S. Kechris. Set theory and uniqueness for trigonometric series. *preprint*, 1997.
40. J. L. Kelley. *General Topology*. Van Nostrand, 1955.
41. K. Kunen. Combinatorics. In *Handbook of Mathematical Logic*, pages 371–401. North Holland, 1977.
42. K. Kunen. *Set Theory : An Introduction to Independence Proofs*, volume 102 of *Studies in Logic and the Foundations of Mathematics*. Elsevier, 1980.
43. K. Kunen. *The Foundations of Mathematics*. College Publications, 2009.
44. K. Kunen. *Set Theory*. College Publications, 2011.
45. K. Kuratowski. *Topology*, volume I. Academic Press, 1966.
46. K. Kuratowski and A. Mostowski. *Set Theory: With an Introduction to Descriptive Set Theory*. North Holland, 1976.
47. E. Landau. *Foundations of Analysis*. Chelsea Publishing Company, 1966.
48. A. Levy. *Basic Set Theory*. Dover, 2002.
49. G. Link, editor. *One Hundred Years of Russell's Paradox: Mathematics, Logic, Philosophy*, volume 6 of *de Gruyter Series in Logic and Its Applications*. Walter de Gruyter, 2004.
50. P. Maddy. Believing the axioms. I and II. *The Journal of Symbolic Logic*, 53(2, 3):481–511, 736–764, 1988.
51. P. Maddy. *Defending the axioms: On the philosophical foundations of set theory*. Oxford University Press Oxford, 2011.
52. M. Magidor. Some set theories are more equal. *preprint*, 201?
53. D. A. Martin and A. S. Kechris. Infinite games and effective descriptive set theory. In *Analytic Sets* [64], pages 403–470.
54. Y. N. Moschovakis. *Notes on Set Theory*. Springer, 2006.
55. Y. N. Moschovakis. *Descriptive Set Theory*, volume 155 of *Mathematical Surveys and Monographs*. American Mathematical Society, 2009.
56. J. Mycielski. On the axiom of determinateness. *Fund. Math*, 53:205–224, 1964.
57. J. Mycielski. Games with perfect information. In *Handbook of game theory with economic applications*, volume 1, pages 41–70. Elsevier, 1992.
58. J. C. Oxtoby. *Measure and Category: A Survey of the Analogies between Topological and Measure Spaces*. Springer, 2nd edition, 1980.
59. G. Peano. The principles of arithmetic, presented by a new method (1889). In *From Frege to Gödel* [78], pages 83–97. English translation of "Arithmetices principia, nova methodo exposita", Turin, 1889.

60. J. M. Plotkin, editor. *Hausdorff on Ordered Sets*. American Mathematical Society, Providence, 2005.
61. W. V. O. Quine. New foundations for mathematical logic. *The American Mathematical Monthly*, 44(2):70–80, 1937. Reprinted in [62], pages 80–101.
62. W. V. O. Quine. *From a Logical Point of View*. Harvard University Press, 1980.
63. E. H. Reck. Dedekind's structuralism: An interpretation and partial defense. *Synthese*, 137(3):369–419, 2003.
64. C. A. Rogers et al. *Analytic Sets*. Academic Press, 1980.
65. B. Rotman and G. T. Kneebone. *The Theory of Sets and Transfinite Numbers*. Oldbourne, 1966.
66. B. Russell. *The Principles of Mathematics*. WW Norton, 2nd edition, 1903.
67. B. Russell. Mathematical logic as based on the theory of types. *American Journal of Mathematics*, 30(3):222–262, 1908. Reprinted in [78], pages 150–182.
68. B. Russell. *Introduction to Mathematical Philosophy*. Public Domain (originally published by George Allen & Unwin and Macmillan), 2nd edition, 1920.
69. B. Russell. *My Philosophical Development*. Unwin Books, 1959.
70. E. Schimmerling. *A Course on Set Theory*. Cambridge, 2011.
71. S. Shapiro. *Philosophy of Mathematics: Structure and Ontology*. Oxford University Press, 1997.
72. S. Shapiro. *The Oxford Handbook of Philosophy of Mathematics and Logic*. Oxford University Press, 2005.
73. W. Sierpinski. *Cardinal and Ordinal Numbers*. PWN, Warsaw, 1958.
74. W. Sierpinski. *General Topology*. Dover, 2000.
75. S. G. Simpson. *Subsystems of Second Order Arithmetic*. Association for Symbolic Logic, 2nd edition, 2010.
76. R. R. Stoll. *Set Theory and Logic*. Dover, 1979.
77. P. Suppes. *Axiomatic Set Theory*. Dover, 1972.
78. J. Van Heijenoort. *From Frege to Gödel: A Source Book in Mathematical Logic, 1879–1931*. Harvard University Press, 1967.
79. R. L. Vaught. *Set Theory: An Introduction*. Birkhauser, 2001.
80. J. Von Neumann. Zur Einführung der transfiniten Zahlen. *Acta Scientiarum Mathematicarum (Szeged)*, 1(4-4):199–208, 1922–23. An English translation titled "On the introduction of transfinite numbers" appears in [78], pages 346–354.
81. A. N. Whitehead and B. Russell. *Principia Mathematica*, volume 1–3. The University Press, Cambridge, 1910–1913.
82. W. H. Woodin. The continuum hypothesis, Part I and Part II. *Notices of the American Mathematical Society*, 48(6, 7):567–576, 681–690, 2001.
83. W. H. Woodin. Set theory after Russell: The journey back to Eden. In Link [49], pages 29–47.
84. E. Zermelo. Proof that every set can be well-ordered (1904). In *From Frege to Gödel* [78], pages 139–141. English translation of "Beweis, daß jede Menge wohlgeordnet werden kann", *Mathematische Annalen,* 59(4):514–516, 1904.
85. E. Zermelo. Investigations in the foundations of set theory I (1908). In *From Frege to Gödel* [78], pages 199–215. English translation of "Untersuchungen über die Grundlagen der Mengenlehre I", *Mathematische Annalen,* 65(2):261–281, 1908.
86. E. Zermelo. A new proof of the possibility of a well-ordering (1908). In *From Frege to Gödel* [78], pages 183–198. English translation of "Neuer Beweis für die Möglichkeit einer Wohlordnung", *Mathematische Annalen,* 65(1):107–128, 1908.

Online Reference

87. P. Koellner. The Continuum Hypothesis. *The Stanford Encyclopedia of Philosophy* (Summer 2013 Edition), E. N. Zalta (ed.). http://plato.stanford.edu/archives/sum2013/entries/continuum-hypothesis/.

List of Symbols and Notations

A. Dasgupta, *Set Theory: With an Introduction to Real Point Sets*,
DOI 10.1007/978-1-4614-8854-5, © Springer Science+Business Media New York 2014

Index

A. Dasgupta, *Set Theory: With an Introduction to Real Point Sets*,
DOI 10.1007/978-1-4614-8854-5, © Springer Science+Business Media New York 2014

K

L

M

N

T

MIX
Papier aus verantwortungsvollen Quellen
Paper from responsible sources
FSC® C105338

If you have any concerns about our products,
you can contact us on
ProductSafety@springernature.com

In case Publisher is established outside the EU,
the EU authorized representative is:
Springer Nature Customer Service Center GmbH
Europaplatz 3, 69115 Heidelberg, Germany

Printed by Libri Plureos GmbH
in Hamburg, Germany